Biomedical Science: Computational and Mathematical Methods

Biomedical Science: Computational and Mathematical Methods

Editor: Joel Montgomery

FOSTER ACADEMICS

www.fosteracademics.com

www.fosteracademics.com

FOSTER ACADEMICS

Cataloging-in-Publication Data

Biomedical science : computational and mathematical methods / edited by Joel Montgomery.
p. cm.
Includes bibliographical references and index.
ISBN 978-1-63242-888-2
1. Medical sciences. 2. Biomedical engineering. 3. Medical sciences--Data processing.
4. Medical sciences--Mathematical models. I. Montgomery, Joel.
R129 .B56 2020
610--dc23

Foster Academics,
118-35 Queens Blvd., Suite 400,
Forest Hills, NY 11375, USA

ISBN 978-1-63242-888-2 (Hardback)

Contents

Preface

This book aims to highlight the current researches and provides a platform to further the scope of innovations in this area. This book is a product of the combined efforts of many researchers and scientists, after going through thorough studies and analysis from different parts of the world. The objective of this book is to provide the readers with the latest information of the field.

Biomedical sciences are the disciplines whose primary objective is the study of the human biology and diseases pathology. The scope of these fields extends beyond elementary studies to specialized domains of pharmacology, human nutrition and physiology. Various basic sciences are associated with biomedical sciences, such as biochemistry, genetics and molecular biology, anatomy, physiology, immunology, bioinformatics, mathematics and statistics. Biomedical engineering bridges the gap between medicine and engineering, by integrating the problem solving skills of engineering with medical biological sciences. Some significant biomedical engineering applications are diagnostic and therapeutic medical devices, common imaging equipment such as EKG/ECGs and MRIs, pharmaceutical drugs, regenerative tissue growth, etc. Studies in biomedical science are aided by advanced computational and mathematical models. An example of this is the field of bioinformatics which uses computer programming for biological studies, particularly in the field of genomics, for developing an understanding of the genetic basis of diseases, their adaptations, and differences between populations. This book covers in detail some existent theories and innovative concepts revolving around biomedical science. It brings forth some of the most innovative concepts and elucidates the computational and mathematical methods of use in biomedical science. Students, researchers, experts and all associated with this field will benefit alike from this book.

I would like to express my sincere thanks to the authors for their dedicated efforts in the completion of this book. I acknowledge the efforts of the publisher for providing constant support. Lastly, I would like to thank my family for their support in all academic endeavors.

Editor

1

Modelling of the Current Density Distributions during Cortical Electric Stimulation for Neuropathic Pain Treatment

S. Fiocchi, E. Chiaramello, P. Ravazzani, and M. Parazzini

CNR Consiglio Nazionale delle Ricerche, Istituto di Elettronica e di Ingegneria dell'Informazione e delle Telecomunicazioni IEIIT, Milan, Italy

Correspondence should be addressed to S. Fiocchi; serena.fiocchi@ieiit.cnr.it

Academic Editor: Reinoud Maex

In the last two decades, motor cortex stimulation has been recognized as a valuable alternative to pharmacological therapy for the treatment of neuropathic pain. Although this technique started to be used in clinical studies, the debate about the optimal settings that enhance its effectiveness without inducing tissue damage is still open. To this purpose, computational approaches applied to realistic human models aimed to assess the current density distribution within the cortex can be a powerful tool to provide a basic understanding of that technique and could help the design of clinical experimental protocols. This study aims to evaluate, by computational techniques, the current density distributions induced in the brain by a realistic electrode array for cortical stimulation. The simulation outcomes, summarized by specific metrics quantifying the efficacy of the stimulation (i.e., the effective volume and the effective depth of penetration) over two cortical targets, were evaluated by varying the interelectrode distance, the stimulus characteristics (amplitude and frequency), and the anatomical human model. The results suggest that all these parameters somehow affect the current density distributions and have to be therefore taken into account during the planning of effective electrical cortical stimulation strategies. In particular, our calculations show that (1) the most effective interelectrode distance equals 2 cm; (2) increasing voltage amplitudes increases the effective volume; (3) increasing frequencies allow enlarging the effective volume; and (4) the effective depth of penetration is strictly linked to both the anatomy of the subject and the electrode placement.

1. Introduction

Neuropathic pain (NP) is defined as pain caused by a lesion or a disease of the central (central neuropathic pain (CNP)) or peripheral (peripheral neuropathic pain (PNP)) somatosensory nervous system [1]. Despite the availability of different drugs, no more than 30–40% patients with chronic NP receive adequate pain reduction (≥50%) by currently available drug-based therapies [2, 3].

Among the nonpharmacological treatments, favourable effects of motor cortex stimulation (MCS) upon CNP have been reported [4–10]. This technique is based on the cortical stimulation with biphasic waves of electric current administered through two (the most used bipolar stimulation) or more electrodes. The stimulating electrode is placed over the motor cortex region where the contralateral painful body area is represented [11–13].

Since its first introduction in 1991 [14] for the treatment of thalamic pain, MCS has been applied worldwide on both animals and humans in the attempt to alleviate medically refractory CNP of different origins, such as the facial NP (including trigeminal neuralgia, trigeminal NP, trigeminal deafferentation pain, symptomatic trigeminal neuralgia, and postherpetic neuralgia), the phantom limb pain, the brachial plexus avulsion, the poststroke pain, the Wallenberg syndrome, the pain secondary to multiple sclerosis, and the posttraumatic brain injury pain (to this purpose see the review studies of both animal and human experiments [15–18]). All these studies have been elaborated based on empirical experience given the incomplete understanding of the pathophysiology of CNP and the difficulties to conduct double-blinded studies; therefore, the mechanism underlying the analgesic effect produced by MCS is still largely debated [11, 19, 20].

TABLE 1: Conductivities (S/m) of the head tissues at different stimulation frequencies [36, 37].

Tissue	Conductivity (S/m)			
	40 Hz	50 Hz	85 Hz	130 Hz
Air internal	0	0	0	0
Artery, blood vessels, vein	0.7	0.7	0.7	0.7
Mandible, marrow red, skull, teeth, vertebrae	0.0201	0.0201	0.0201	0.0201
Brain grey matter, hippocampus, hypothalamus, thalamus	0.0681	0.0753	0.0869	0.0915
Brain white matter, commissure	0.0506	0.0533	0.0573	0.0590
Cartilage, ear cartilage, intervertebral disks,	0.171	0.171	0.172	0.172
cerebellum	0.0881	0.0953	0.107	0.111
Cerebrospinal Fluid (CSF)	2	2	2	2
Connective tissue, tendon, ligament	0.263	0.270	0.295	0.322
Cornea,	0.421	0.421	0.422	0.422
Muscle	0.224	0.233	0.259	0.278
Ear skin, skin	0.0002	0.0002	0.0002	0.0002
Eye lens,	0.321	0.321	0.322	0.323
Eye sclera	0.503	0.503	0.503	0.503
Eye vitreous humor	1.5	1.5	1.5	1.5
Fat, subcutaneous adipose tissue (SAT)	0.0188	0.0196	0.0206	0.0210
Hypophysis, pineal body	0.521	0.521	0.522	0.523
Medulla oblongata, midbrain, pons	0.0594	0.0643	0.0721	0.0753
Mucosa	0.00042	0.00043	0.00045	0.00048
Nerve, spinal cord	0.0269	0.0274	0.0280	0.0281
Tongue	0.271	0.271	0.272	0.272

However, the most accredited theories addressing pain relief by MCS (for a review see [21]) attribute its efficacy to the neural modulation, given by the flow of current density in the motor cortex [22]. It is indeed believed that the electrical stimulation of the motor cortex inhibits, disrupts, or interferes with the allogenic signals coming from the thalamus and from other hyperactive areas in the brain networks that govern the nociception [14, 23].

In this context, the MCS delivered "dose," here intended as the current density quantification in the cortex and in the neural tissues, should be considered one of the key points for both the treatment optimization and a deeper understanding of the mechanisms lying behind its efficacy [24].

So far, studies addressing the characterization of the current density distributions due to MCS are scarce and limited to the analysis of different electrode types [25, 26] and/or placed on very simplified cortical models [12, 27–30].

In this study a precise quantification, using computational techniques and appropriately defined metrics, of the current density distributions in the brain of different-aged detailed anatomical human models was conducted. This was performed by reproducing the model of a realistic electrode array for MCS [31] used in clinic and placed on the motor cortex target area that corresponds to the somatic area of pain. Typically, for most of the clinical applications for the neuropathic pain treatment [18], these coincide with facial and upper limb cortical areas, placed in the lower part and in the middle part of the central gyrus, respectively. The electrode array was fed according to the typical frequency and voltage amplitude delivered in the medical practice [7, 9, 10, 32, 33] and the extent to which they affect specific current density-related parameters was quantified. The rigorous analysis of these parameters in realistic and detailed human head models could ultimately help to gain further insights into the evaluation of the clinical outcomes and the optimization of the treatment delivered through the MCS.

2. Materials and Methods

2.1. Human Models. Three realistic anatomical male models of the Virtual Population Family [34, 35] (Duke, 34 years old; Louis, 14 years old; and Glenn, 84 years old) were used in the study. They were obtained by the segmentation of high-resolution magnetic resonance (MR) images of healthy volunteers and then reconstructed based on a computer-aided design representation of the organ surfaces.

The use of these different anatomical models allowed assessing the characteristics of the current density distributions in three different brain morphologies. In each of them, more than 40 tissues can be distinguished at the head level (head tissues list in Table 1).

For all the human models, in the regions where the electrodes were positioned, it was possible to distinguish the grey and white brain matter and the CSF (Figure 1). The dielectric properties of each tissue were assigned based on the literature data at low frequency [36, 37]. Table 1 reports the

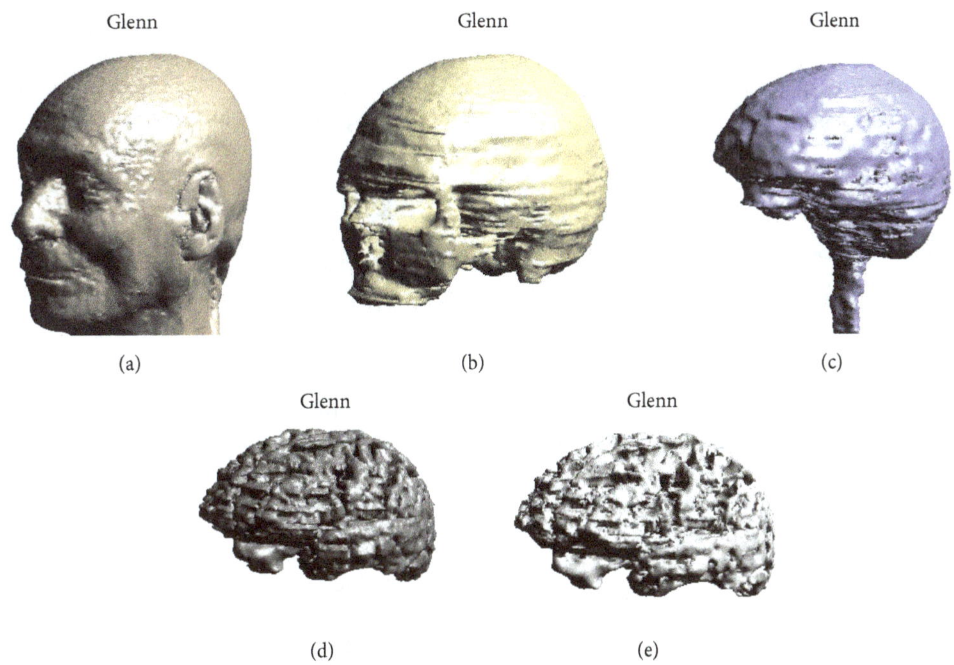

FIGURE 1: 3d rendering of Glenn head model. Tissues shown are (a) skin; (b) skull; (c) CSF; (d) brain grey matter; and (e) brain white matter.

conductivity values of each tissue segmented in the models for each frequency of the stimulation settings used in the study (see Section 2.4 below).

2.2. Electrode Modelling. Since the meninges are not segmented in our anatomical models, our computational model is simulating subdural cortical stimulation.

The geometrical and physical characteristics of the electrodes and their position over the cortex have been modelled based on the clinical literature which uses the MCS for the treatment of neuropathic pain (see, e.g., [7, 9, 10, 32, 33]). Briefly, it consists of a four-electrode array of Medtronic, named, Resume II [31]; each electrode has the following properties, as taken from the technical manual: 4 mm diameter and 1.9 mm height disk made of a platinum-iridium alloy (electrical conductivity: 5.278×10^6 S/m). The lower base of each electrode was positioned in contact with the cortex, whereas the side surface and the upper base are covered by a silicone layer (conductivity: $0.1 * 10^{-9}$ S/m) mimicking the silicone backing that is not modelled, following the approach already used in literature [25]. A 1 cm distance was kept between two neighbouring electrode centres, according to the manufacturer specifications.

The array was centred on the central sulcus alternatively in two different positions corresponding to the cortical somatotopic representation of the upper limb and of the face. More specifically, the electrodes were placed so that the first and the second (numbered 0 and 1 in Figure 2) lie on the motor cortex over the precentral gyrus, and the third and the fourth (numbered 2 and 3) lie on the somatosensory cortex over the postcentral gyrus and in the proximity of postcentral sulcus, respectively (Figure 2).

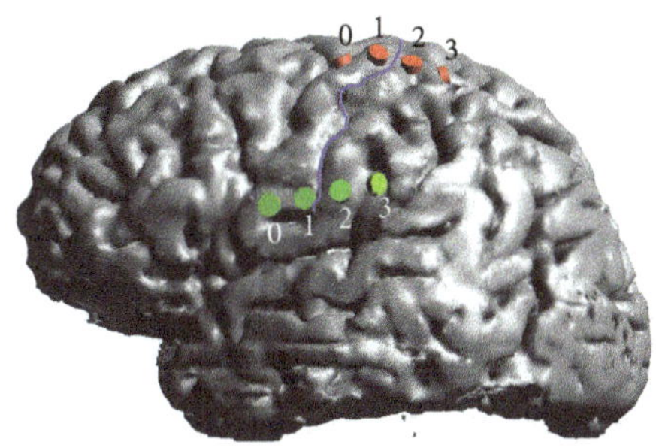

FIGURE 2: Electrode placement on Duke's cortex over the face target region (green) and over the upper limb target region (red). Blue line shows the central sulcus.

2.3. Current Density Numerical Simulations. In this study, a simulation-based approach to calculate the current density distributions within brain tissues was performed through the simulation platform SEMCAD X [38], which implements the Finite Element Method (FEM) in the low frequency range. In detail, it solves the Laplace equation to obtain the electric potential (φ) distribution:

$$\nabla \cdot (\sigma \nabla \varphi) = 0, \tag{1}$$

where σ (S/m) is the electrical conductivity of the human tissues. In the low frequency approximation adopted, ohmic currents dominate displacement currents and capacitive effects are disregarded. The electric field ($\mathbf{E}$) and then the current density ($\mathbf{J}$) distributions were obtained by means of the following relations:

$$\begin{aligned} \mathbf{E} &= -\nabla \varphi \\ \mathbf{J} &= \sigma \mathbf{E}. \end{aligned} \tag{2}$$

Implicitly, we assumed the neuronal excitability in the cortex is proportional to the **J** (and E) amplitude and that therefore the regions with a greater current density are more likely stimulated, while the regions with very low current will not directly feel the effects of stimulation [39].

For each simulation, the human head model was inserted in a surrounding bounding box filled with air. The tangential E field component, E_t, was set to be continuous ($E_{t1} = E_{t2}$, which is equivalent to $J_{t1}/\sigma_1 = J_{t2}/\sigma_2$) at the interface between two tissues. Current density was set to be parallel to the face at the interface between skin and air. The upper surface of each of the two active electrodes activated was set to a uniform electrical potential and the potential difference between the two electrodes was adjusted according to the settings listed in the following figure (Figure 3). The computation domain was discretized using a uniform rectilinear mesh with a grid discretization equal to 0.5 mm, which assures resolving the thinnest structures of the model. Relative tolerance for FEM convergence was set to 10^{-12}. To further optimize the quality of the grid and reduce the calculation time, the computational domain was truncated at the brainstem level. Both these choices (mesh step and dimension of the computational domain) were based on a sensitivity analysis showing that neither reducing the mesh step nor increasing the computational domain had a substantial effect on the field values. At the truncation section, we assigned the boundary condition of continuity of the current, whereas the other faces of the bounding box have been treated as insulated, that is, vanishing flux normal to the computational boundary.

2.4. Stimulation Scenarios and Settings. The analysis of the current density distributions was articulated in three distinct phases, corresponding to different scenarios and stimulation settings as summarized in Figure 3.

Phase 1. In Phase 1, we evaluated to which extent the inter-electrode distance affects the current density distributions on both the target regions. A bipolar electrode configuration was set: the cathode was kept fixed on the electrode positioned on the motor cortical representation of the pain region (i.e., electrode 0 in Figure 2), whereas the anode was changed in each simulation from electrodes 1 to 3 of the four-electrodes array (see Figure 2) so that the distance between the two active electrodes varied from 1 cm to 2 cm and 3 cm. This was done for both motor cortex targets, placing the electrode array alternatively on these two regions (i.e., once on the face and once on the upper limb cortical somatotopic representation). The potential difference between the active electrodes was kept fixed at 2 V and the stimulation frequency equal to 40 Hz (in the following these setting parameters will be referred to as "Reference"), as commonly used in chronic MCS therapy for neuropathic pain [5, 32] and the simulation was performed on the 34-year-old model ("Duke").

Phase 2. In Phase 2, we analysed the effect of both the amplitude and the frequency stimulus for a fixed interelectrode distance of 2 cm. In particular, the potential difference between cathode (Electrode 0 placed over the precentral gyrus in the motor cortex, Figure 2) and anode (Electrode 2 on postcentral gyrus in the somatosensory cortex, Figure 2) and the frequency of the stimulation signal have been varied in the range described by following literature studies:

(i) Setting A: taken from [32], who found a significant improvement in the clinical assessment for the evaluation of NP of various origins when the stimulation settings were as follows: amplitude, 2 V; frequency: 40 Hz; and pulsewidth, 60 μs.

(ii) Setting B: based on [10], who established the efficacy of electric stimulation on a central poststroke facial pain subject, with the following stimulation parameters: amplitude, 3.65 V; frequency, 50 Hz; and pulsewidth, 120 μs.

(iii) Setting C: [7] reported the successful application of bipolar MCS (amplitude: 4.5 V, frequency: 85 Hz, and pulsewidth: 210 μs) to patients suffering from thalamic neuropathic pain (TNP) and poststroke pain (PSP).

(iv) Setting D: [9] obtained a positive pain relief in 27 patients affected by chronic neuropathic pain, by increasing both amplitude and frequency and pulsewidth of the delivered pulse up to 5.3 V and 130 Hz and 210 μs, respectively.

Simulations were run by varying, for each stimulus amplitude (i.e., 2 V, 3.5 V, 4.5 V, and 5.3 V), all the four stimulus frequencies used in the above-mentioned clinical studies (40, 50, 85, and 130 Hz), thus allowing us to consider all the possible combinations of the two parameters. The stimulation signal was considered as a pure sinusoid, following an approach already used in the literature [25, 26]. The pulsewidth was then considered only to verify the stimulation stayed within the safety margins to avoid neuronal damage (see in Section 4 below). Also in this phase, the electrode array was placed over the face motor cortex area of the 34-year-old model (Duke).

Phase 3. The third phase was designed to investigate the age-dependent anatomical differences of the current density distributions generated in the cortex. Bipolar stimulation at the intensity equals 2 V and frequency to 40 Hz was delivered at the electrodes 0 and 2 (interelectrode distance of 2 cm) placed on Duke (34 years old), Louis (14 years old), Glenn (84 years old) on both cortical targets of the face and of the upper limb.

2.5. Data Analysis. The current density distributions were assessed in the cortex and in the white matter. The following parameters were then calculated, according to the study performed by Kim's group ([25–27, 40, 41]):

(i) Effective volume (EV_{50}): evaluated for both grey and white matter, it is the volume which has a current density greater than 50% of motor cortex threshold

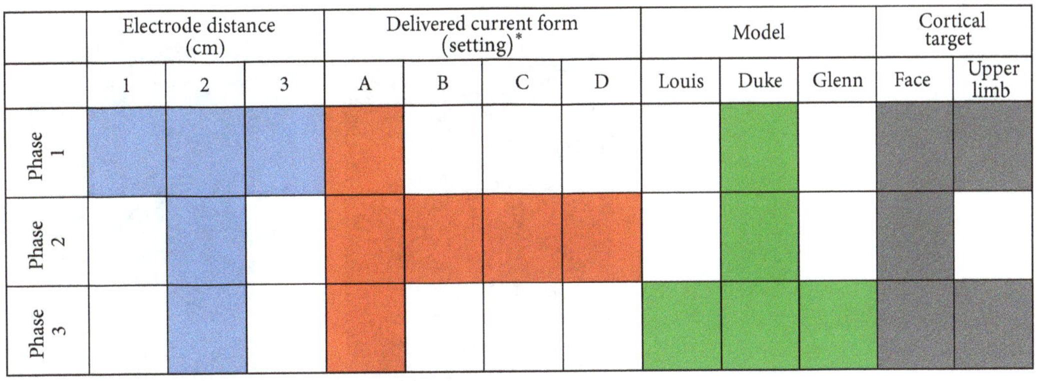

*Setting A [32]: amplitude 2 V, frequency 40 Hz, pulsewidth 60 μs
*Setting B [10]: amplitude 3.65 V, frequency 50 Hz, pulsewidth 120 μs
*Setting C [7]: amplitude 4.5 V, frequency 85 Hz, pulsewidth 210 μs
*Setting D [9]: amplitude 5.3 V, frequency 130 Hz, pulsewidth 210 μs

FIGURE 3: Schematic representation of the simulation scenarios and input signal settings used in the three phases.

(MCT) ($J_{MCT} = 2.5\ A/m^2$ as calculated by [42]). That threshold was chosen as the current density amplitude that can provoke analgesia without motor effects and lies in the range of the practical usage case [27]. This index therefore quantifies the tissue volume that undergoes neural modulation.

(ii) Effective depth of penetration (D_{JMCT}): it is the maximum depth [mm] from the cerebral cortex surface, reached by a current density higher than the 50% of J_{MCT}.

These indexes were analysed as surrogate of the stimulation effectiveness.

3. Results

Figure 4 shows some examples of the current density distributions over the axial slice at 4 cm below the projection of Cz over the cortex (where Cz is referred to according to the 10–20 EEG system) for some phases of the different stimulation settings. The colour maps represent the amplitude distribution of **J** and are all clipped above 50% of the J_{MCT} to favour the comparison between the amplitude distributions resulting from the settings used in the three different phases. The green arrows represent the direction of **J**, which is preferentially directed tangentially in the crowns of the gyri, whereas it is directed predominantly normally at the bottom of the sulci. In the following, the results will be presented by investigating the two indexes described above, and within each index analysis, by comparing the results of the three phases.

3.1. Effective Volume (EV_{50}). Figure 5 shows the effective volume (in cm^3) calculated varying the stimulation settings according to the three phases (see Section 2.4 above). The bars in the graphs are the sum of the cortex effective volume (dark colours) and the white matter effective volume (light colours). The total effective volume (EV_{50}) (i.e., the sum of cortex and white matter effective volume) for each simulation setting can be therefore read on the vertical axis. From that figure, one can notice that for a 2 cm interelectrode distance (Phase 1), the total effective volume is significantly higher (up to 30%) than the same quantity calculated for the other two distances and for both the cortical areas. Moreover, one could identify a comparable trend of the EV_{50} along the three distances for both the cortical regions. Similar considerations apply when we consider separately the effective volume trends on the cortex and white matter. Interestingly, the total effective volumes calculated on the facial cortical area are almost double the ones calculated on the upper limb area. This is mainly due to the contribution of the cortex effective volume, whereas the white matter effective volume is higher in the upper limb area. However, for both the electrode placements and for all the distances, the total effective volume stays below 4 cm^3. This result changes when increasing amplitude and frequency stimulus (Phase 2). In particular, increasing amplitude, the ratio between EV_{50} calculated with an applied potential difference of 3.65 V, 4.5 V, and 5.3 V with respect to EV_{50} calculated at 2 V equals 1.73, 2.10, and 2.46, respectively, independently of the stimulus frequency. Similarly, increasing frequency, the trend of the EV_{50} is the same for each applied potential difference. The ratio between EV_{50} calculated at 50, 85, and 130 Hz with respect to EV_{50} calculated at 40 Hz equals 1.1, 1.25, and 1.3, independently of the applied potential difference and that happens even for both cortex and white matter proportions.

On the contrary, the effective volume variation does not present a clear age-related trend across the three models (Phase 3). The levels, however, for both the targets, decrease in the younger (i.e., Louis) and older (i.e., Glenn) male model with respect to the adult male (i.e., Duke), and hence in this phase the total EV_{50} stays everywhere below 4 cm^3.

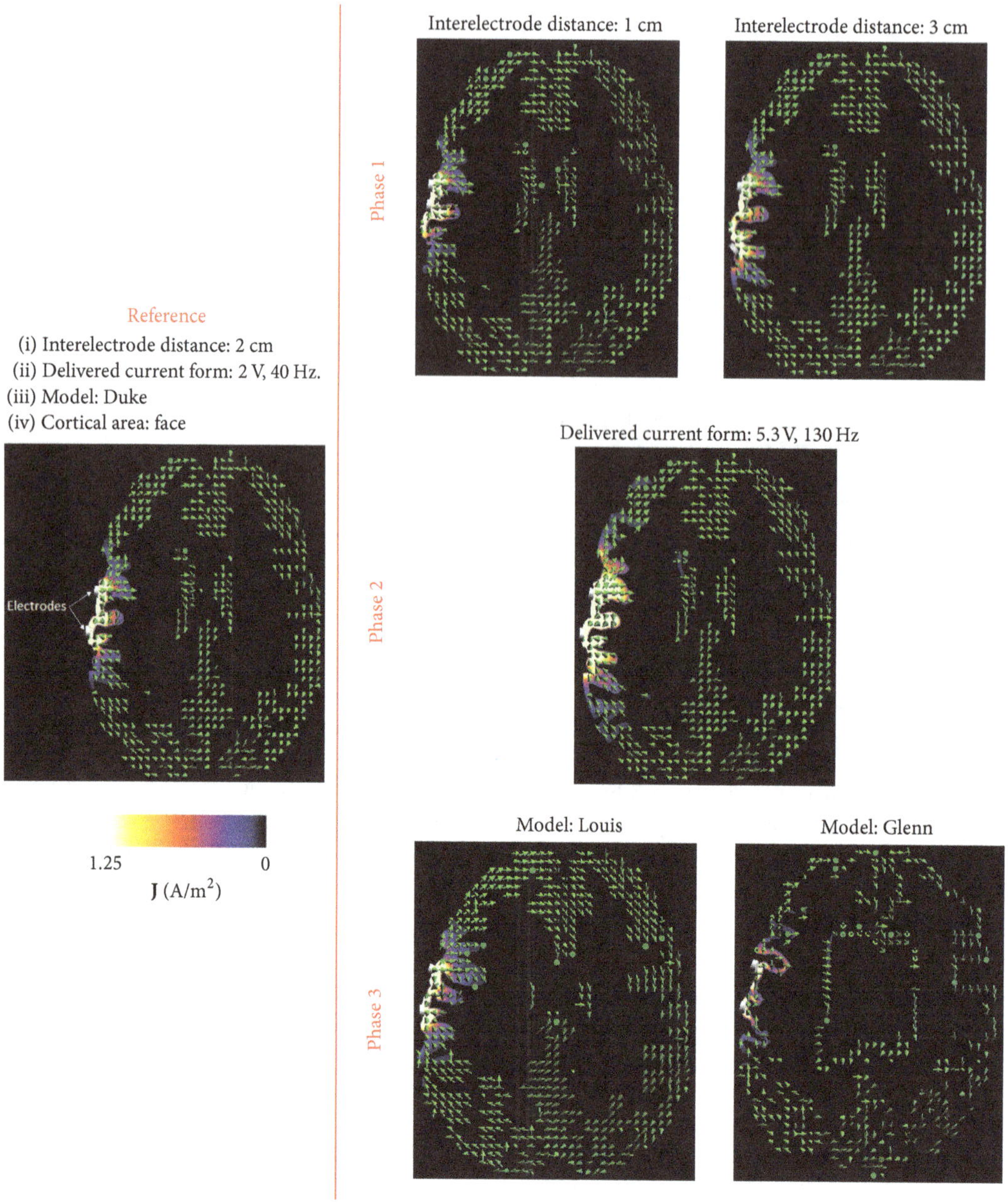

FIGURE 4: Examples of current density (J) distributions over the axial slices at 4 cm below the projection of Cz over the cortex. Colormaps are set between 0 and the 50% of MCT (i.e., 1.25 A/m^2); therefore, white areas correspond to the areas in which J is equal or higher to the scale upper limit. Green arrows represent the direction of the current density on the cortex. Panel on the left shows the J distribution produced in the cortex by "Reference" simulation.

3.2. Effective Depth of Penetration (D_{JMCT}). Figure 6 shows the penetration depth (D_{JMCT} in mm) calculated varying stimulation scenarios and settings of the three phases. This analysis was conducted considering the whole brain matter, that is, the cortex and the white matter taken together. By increasing the interelectrode distance (Phase 1), the penetration depth increases with some differences between the two cortical areas: in the face cortical area, the capability to penetrate the cortex ranges from 9.7 to 10.3 mm, whereas in the upper limb from 6.5 to 10.5 mm.

The increase in both amplitude and frequency (Phase 2) results in penetration depth increase, but no clear trend of those increases can be identified. Moreover, D_{JMCT} calculated on the old male model (Glenn) is reduced in both the cortical regions of about 15–20% with respect to the 34-years-old male model (Phase 3).

4. Discussion and Conclusions

4.1. Motor Threshold Definition. Although MCS for the neuropathic pain treatment is under investigation for more than two decades, the mechanism of action behind its effectiveness is still not clear. Moreover, no guidelines for the best set of stimulation parameters and electrode montages exist so

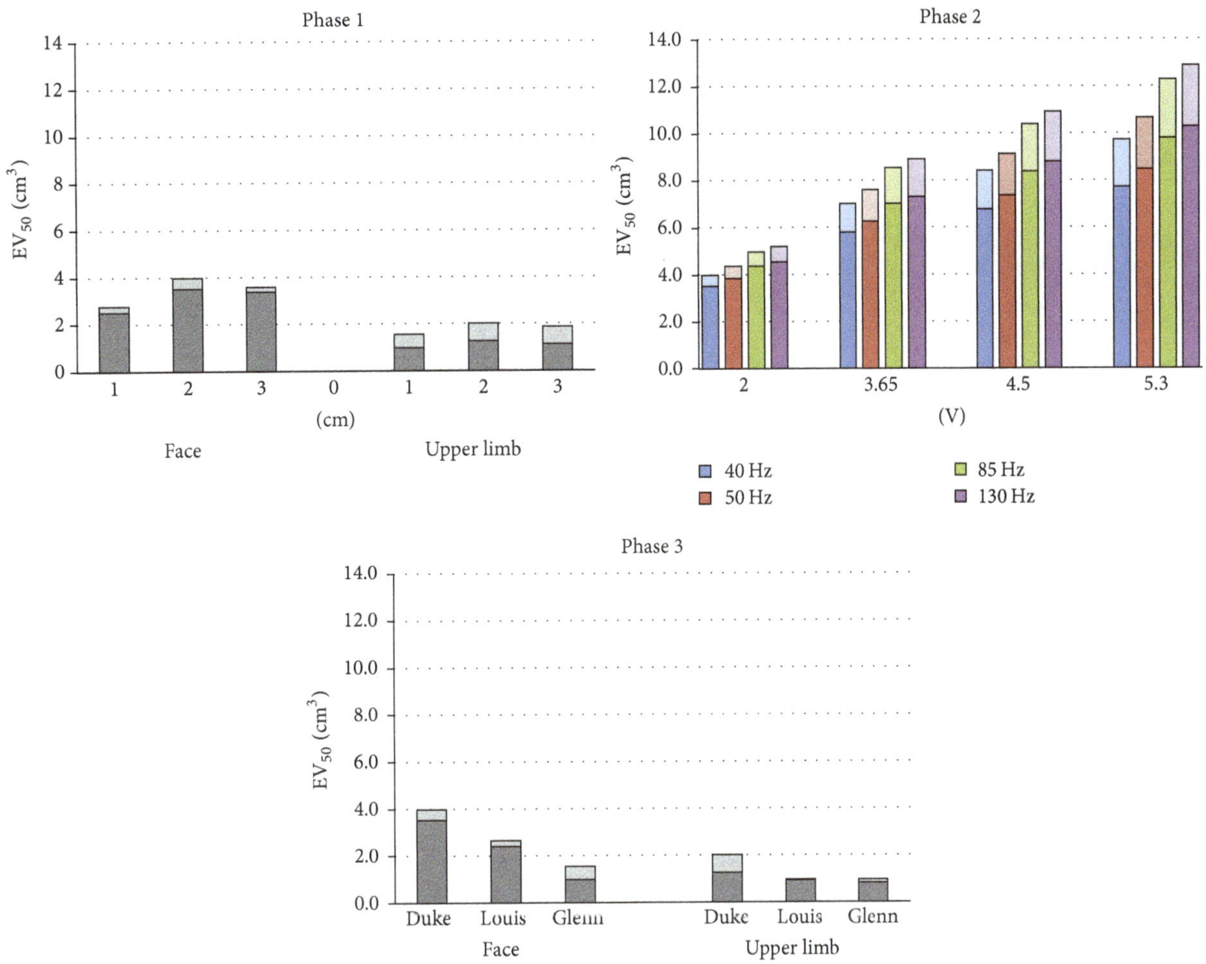

FIGURE 5: Effective volume calculated from the current density distributions by varying the interelectrode distance and positioning (Phase 1), delivered signal settings (Phase 2), and models and positioning (Phase 3). Each bar is the sum of the cortex effective volume (dark) and the white matter effective volume (light).

that all adjustments to them for improving MCS results still depend on iterative empirical testing [24, 43] and are highly individualized.

Lately, different authors [15, 24, 43] have proposed that individual stimulation parameters can be predicted as a percentage of motor threshold activation. A simple way to assess the reliability of this approach can be provided by numerical simulation, actually consistent with experimental data [26].

As discussed in the clinical neurobiology literature [44, 45] the mechanism underlying both the damaging and nondamaging effects of the stimulation are associated with the synchronous activity of a substantially large number of neurons. That supports that the modulation effects are driven by some "mass action" gathered from the simultaneous activation of a critical neuronal volume. Although in our simulations the maximum levels of current density exceed the MCT (see, i.e., white areas in Figure 4), the cortical volume with a current density higher than the MCT is very small and limited to a maximum volume of 1.5 cm^3 (across all the stimulation settings and scenarios examined) distributed mainly under the cathode and the anode. This volume, given that MCS clinical studies that had used the same stimulation parameters did not report side effects [7, 9, 10, 32, 33], is unlikely to induce a neural "mass action" which would lead to an undesired motor response. According to these considerations, in our study we used the 50% of the MCT as a modulation threshold and hence the level that limits the analgesic effect and the efficacy of the technique. However, the goodness of this threshold is an important key factor to be validated in further studies, also in view of the highly individualized actual device settings and the fact that here we modelled the subdural stimulation instead of the epidural stimulation from which stimulus parameters (mainly amplitude and frequency) are taken. One should also note that the MCS threshold was estimated by a transcranial magnetic stimulation (TMS) based study [42]: therefore, given the different J spatial derivative in TMS compared to MCS, it cannot be excluded that neuronal activation in subdural cortical stimulation could be also predicted by different metrics, such as its spatial derivative, as done in previous studies based on the activating function [12, 28, 30,

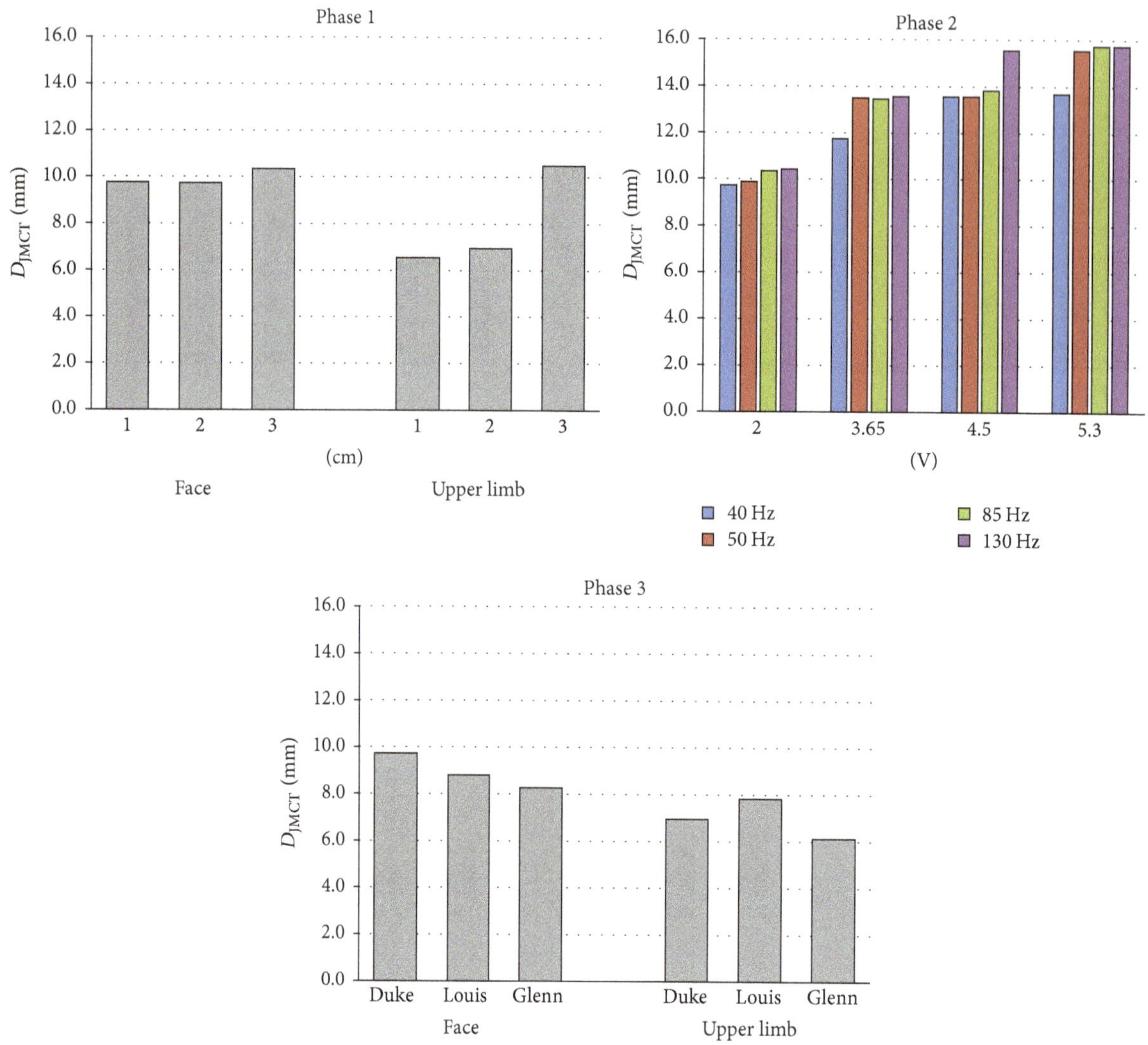

FIGURE 6: Effective depth of penetration calculated from the current density distributions by varying the interelectrode distance and positioning (Phase 1), the delivered signal settings (Phase 2), and the models and positioning (Phase 3).

46]. The authors of [42] numerically calculated the minimum current density peak in the cortex that elicits motor activation from which they estimated the rheobase current and the chronaxie time to determine the strength-duration curve. In the frequency range considered in this study, it slightly varies between 2.49 A/m^2 and 2.62 A/m^2. This range is in line with the levels for motor cortical activation proposed in the electric stimulation literature, and the 20–50% of that quantity (i.e., 0.5–1.5 A/m^2) corresponds to the widely accepted range for motor cortex modulation administered through direct current stimulation [30, 47–53].

4.2. Effective Volume Analysis. The first index of efficacy analysed, that is, the effective volume, allows quantifying the percentage of volume of the cerebral cortex and of the white matter that responds to the stimulus. In other words, higher volumes correspond to a broader area stimulated. From Figures 4 and 5, one can notice that, among the three interelectrode distances tested, in both the target areas and particularly for the facial cortical region, the 2 cm interelectrode distance can stimulate a broader volume. This is probably due to the anatomy of the cerebral cortex that, with its typical convolutions, strongly influences the current density flow, preventing a trivial prediction of its distribution. In an isotropic spherical model, an increasing interelectrode distance linearly raises the values of the volumetric parameters. Conversely, using a realistic anatomical model we show that it happens only as long as the second electrode does not fall in a sulcus of the motor cortex. The central sulcus, which separates the cathode and the anode, prevents indeed that the totality of the area of the cortex between the two electrodes feels a current density sufficient to generate motor cortex neural excitation.

Similarly, the anatomical differences between the different-aged models considerably change both the pattern of stimulation and the indexes examined, with a net contraction of the effective volume in the older men (Figures 5 and 6). That is probably due to the cortical atrophy that

TABLE 2: Schematic representation of the main findings of this study.

	EV_{50} (Figure 5)	D_{JMCT} (Figure 6)
Interelectrode distance (Phase 1)	Max @ 2 cm	↑ with distance
Stimulus amplitude (Phase 2)	↑↑ with amplitude	↑ with amplitude
Stimulus frequency (Phase 2)	↑ with frequency	↑ with frequency
Age (Phase 3)	No clear trend with age (less in the elderly)	No clear trend with age (but min in the elderly)

Neuronal response was activated in the volume where **J > 50% MCT** (→ modulation threshold); ↑ EV_{50} → ↑ spread of modulated tissue; ↑ D_{JMCT} → ↑ depth of modulated tissue.

usually affects the elderly [51, 52], producing an enlargement of the subarachnoid space (average CSF thickness at the electrodes levels is 0.21 mm, 0.94 mm, and 3.46 mm for Louis, Duke, and Glenn, resp.) and an increase of the CSF volume filling the empty space in the sulci (total CSF volume increases from 214 to 333 and to 619 cm^3 in Louis, Duke, and Glenn, resp.). Increasing CSF volume/thickness has a strong impact in epidural cortical stimulation by reducing the amount of current penetrating the cortex and hence reducing the stimulation efficacy, as reported in previous studies [12, 28, 29], but has little effect on the total current entering grey matter in subdural cortical stimulation [41]. However, it is important to consider the CSF volume-related shunting effect when translating the present study's results to epidural motor cortex stimulation.

The thinning of the cortex in Glenn (up to 1.5 mm cortical thickness decrease), visible also in Figure 4, enlarges the space between the convolutions (the sulci) and reduces the spread of the current density in the cerebral cortex in the direction perpendicular to the sulcus, thus increasing the focality of stimulation and hence reducing the effective volume. This is in line with previous modelling studies showing that anatomical differences would highly affect the stimulation efficacy in epidural cortex stimulation [12, 28] and further enforce the need to use anatomical and detailed human models in computational studies addressing current density distribution quantification.

In the case in which it is necessary to stimulate a larger area, our results suggest that the frequency or the amplitude of the stimulus should be increased (Figure 5, Phase 2). The increasing rate of the effective volume scales with the increasing rate of the stimulus amplitude and it agrees with previous calculations performed in two similar studies by Kim's group [27, 41], whereas it is much slower with respect to the increasing rate of the stimulus frequency. Limited effects of frequency could be given by the conductivity variation (Table 1), which however weakly contributes to the electric field distribution variability [54]. However, one should also take into account the fact that increasing frequencies could have other effects on neural activation such as the ones linked to the selective fibers recruitment or to the indirect/synaptic activation of neurons [55].

4.3. Effective Penetration Depth Analysis. The effective volume is strictly linked with the second index examined: the effective penetration depth (Figure 6). As with the previous index, it is greatest in the facial area (Phase 1 and Phase 3). It increases with increasing interelectrode distance, in particular in the upper limb cortical area, when the anode falls into the postcentral gyrus. However, one can notice its limited variability across the different interelectrode distances and across the models. Indeed, this index, which is a sulci-parallel propagation index, is less affected by the elderly cortical atrophy compared to the previous index.

As to the stimulus settings (Figure 6, Phase 2), the effective penetration depth increases very slowly with the frequency increase. While the increase in the frequency can more effectively enlarge the stimulation (Figure 5, Phase 2), it cannot deepen it. The only way to improve that distance of about the 50% within the settings range examined here is to increase the amplitude of the stimulus.

4.4. Relationship between Stimulation Efficacy Related Indexes and Simulation Scenarios/Settings. Table 2 summarizes the main relationships between the indexes and the stimulation scenario and settings analysed.

In terms of absolute values, the indexes of efficacy here evaluated are in good agreement with the values calculated by Kim and colleagues [41] with similar electrode geometry (4 mm versus 5 mm diameter), montage (interelectrode distance of 1 cm, placement over the upper limb area), and stimulation parameters (delivered voltage 2 V at 50 Hz). However, one should take into account the fact that the impact of using typical epidural cortical stimulus settings for a subdural cortical stimulation could not be negligible in terms of both current density values and spatial distribution. As a result of the first remark, the effective volume, the depth of penetration, and the cortical volume above MCT are most likely overestimated here compared to epidural cortical stimulation, thus representing a conservative estimate of side effects related to motor activation in case of epidural cortical stimulation. In addition, moving the electrode from above the dura to directly on the cortex strongly affects also the current density spatial derivative, which in turn impacts on activation of axons (i.e., activating function).

4.5. Stimulation Settings Selection and Safety Related Issues. In the clinical practice, the stimulation settings are chosen based on the available literature and are often modified during the treatment according to the individual response [43]. The need to use, for some patients, increasing frequency and amplitude is indeed probably driven both by the ineffectiveness of the stimulation when the starting settings are used and by the different patients' motor thresholds. As discussed above, the motor threshold was here kept fixed but the intersubjects physiological variability here examined has to be taken into account when we move from modelling to practice. In the same way, the upper limit of the stimulation settings should assure a sufficient risk margin to prevent the tissue damage. That is indeed the main constraining factor in the choice of stimulation settings for implantable devices used for the treatment of neurological disorders [56]. However, here we verified that the stimulation parameters used in our study stays within the recommended safety margins. Specifically, we evaluated the charge per phase (A·s) and the charge density per phase (measured on the surface of the electrode) ($A\cdot s/cm^2$) for each combination of amplitude and pulsewidth tested and we compared them with the couple limit values reported in the literature [56, 57]. It is believed indeed that both are factors that synergistically determine the stimulation threshold that induces neuronal damage [57]. The results of this analysis (not shown here in detail), indicated, however, that all the possible setting combinations here modelled produce a couple of charge per phase and charge density per phase that guarantees the compliance with the safety limits.

This discussion therefore confirms that our computational study can give important indications about the spatial distribution of the current density during MCS. The use of detailed anatomical models provides indeed a substantial advance in the computational results reliability and have the potential to be included in forward models [58], similar to that used to solve inverse problems in electroencephalography (EEG) analysis, for providing important suggestions about the planning of a more focused stimulation strategy.

Conflicts of Interest

The authors declare that there are no conflicts of interest regarding the publication of this paper.

Acknowledgments

The authors wish to thank Schmid & Partner Engineering AG (https://www.speag.com) for having provided the simulation software SEMCAD X/SIM4life. They are also grateful to Dr. Brunella Marzolini for her valuable help and advice.

References

[1] "International Association for the Study of Pain. IASP Taxonomy," http://www.iasp-pain.org/Taxonomy?navItemNumber=576#Neuropathicpain, 2016.

[2] N. B. Finnerup, S. H. Sindrup, and T. S. Jensen, "The evidence for pharmacological treatment of neuropathic pain," *PAIN*, vol. 150, no. 3, pp. 573–581, 2010.

[3] S. Tamburin, K. Borg, X. J. Caro et al., "Immunoglobulin G for the Treatment of Chronic Pain: Report of an Expert Workshop," *Pain Medicine*, vol. 15, no. 7, pp. 1072–1082, 2014.

[4] T. Tsubokawa, Y. Katayama, T. Yamamoto, T. Hirayama, and S. Koyama, "Chronic motor cortex stimulation in patients with thalamic pain," *Journal of Neurosurgery*, vol. 78, no. 3, pp. 393–401, 1993.

[5] J.-P. Nguyen, J.-P. Lefaucheur, P. Decq et al., "Chronic motor cortex stimulation in the treatment of central and neuropathic pain. Correlations between clinical, electrophysiological and anatomical data," *PAIN*, vol. 82, no. 3, pp. 245–251, 1999.

[6] D. Carroll, C. Joint, N. Maartens, D. Shlugman, J. Stein, and T. Z. Aziz, "Motor cortex stimulation for chronic neuropathic pain: A preliminary study of 10 cases," *PAIN*, vol. 84, no. 2-3, pp. 431–437, 2000.

[7] D. Rasche, M. Ruppolt, C. Stippich, A. Unterberg, and V. M. Tronnier, "Motor cortex stimulation for long-term relief of chronic neuropathic pain: a 10 year experience," *PAIN*, vol. 121, no. 1-2, pp. 43–52, 2006.

[8] D. Fontaine, C. Hamani, and A. Lozano, "Efficacy and safety of motor cortex stimulation for chronic neuropathic pain: Critical review of the literature," *Journal of Neurosurgery*, vol. 110, no. 2, pp. 251–256, 2009.

[9] W. J. Fagundes-Pereyra, M. J. Teixeira, N. Reyns et al., "Motor cortex electric stimulation for the treatment of neuropathic pain," *Arquivos de Neuro-Psiquiatria*, vol. 68, no. 6, pp. 923–929, 2010.

[10] A. Balossier, O. Etard, C. Descat, D. Vivien, and E. Emery, "Epidural electrical stimulation to improve chronic poststroke aphasia: A 5-year follow-up," *Brain Stimulation*, vol. 5, no. 3, pp. 364–368, 2012.

[11] J. A. Brown and N. M. Barbaro, "Motor cortex stimulation for central and neuropathic pain: Current status," *PAIN*, vol. 104, no. 3, pp. 431–435, 2003.

[12] L. Manola, B. H. Roelofsen, J. Holsheimer, E. Marani, and J. Geelen, "Modelling motor cortex stimulation for chronic pain control: Electrical potential field, activating functions and responses of simple nerve fibre models," *Medical & Biological Engineering & Computing*, vol. 43, no. 3, pp. 335–343, 2005.

[13] J. P. Nguyen, J. P. Lefaucher, C. Le Guerinel et al., "Motor cortex stimulation in the treatment of central and neuropathic pain," *Archives of Medical Research*, vol. 31, no. 3, pp. 263–265, 2000.

[14] T. Tsubokawa, Y. Katayama, T. Yamamoto, T. Hirayama, and S. Koyama, "Treatment of thalamic pain by chronic motor cortex stimulation," *Pacing and Clinical Electrophysiology*, vol. 14, no. 1, pp. 131–134, 1991.

[15] G. Monsalve, "Motor cortex stimulation for facial chronic neuropathic pain: A review of the literature," *Surgical Neurology International*, vol. 3, no. 5, pp. S290–S311, 2012.

[16] J. M. Henderson and S. P. Lad, "Motor cortex stimulation and neuropathic facial pain," *Neurosurg Focus*, vol. 21, no. 6, p. E6, 2006.

[17] J. M. Lucas, Y. Ji, and R. Masri, "Motor cortex stimulation reduces hyperalgesia in an animal model of central pain," *Pain*, vol. 152, no. 6, pp. 1398–1407, 2011.

[18] Y. Lazorthes, J. C. Sol, S. Fowo, F. E. Roux, and J. C. Verdié, "Motor cortex stimulation for neuropathic pain," in *Operative Neuromodulation. Acta Neurochirurgica Supplements*, D. E. Sakas and B. A. Simpson, Eds., vol. 97/2, Springer, Vienna, 2007.

[19] X. Drouot, J.-P. Nguyen, M. Peschanski, and J.-P. Lefaucheur, "The antalgic efficacy of chronic motor cortex stimulation is related to sensory changes in the painful zone," *Brain*, vol. 125, no. 7, pp. 1660–1664, 2002.

[20] J. P. Brasil-Neto, "Motor cortex stimulation for pain relief: Do corollary discharges play a role?" *Frontiers in Human Neuroscience*, vol. 10, no. 323, 2016.

[21] S. Zaghi, N. Heine, and F. Fregni, "Brain stimulation for the treatment of pain: A review of costs, clinical effects, and mechanisms of treatment for three different central neuromodulatory approaches," *Journal of Pain Management*, vol. 2, no. 3, pp. 339–352, 2009.

[22] J. M. Hoogendam, G. M. Ramakers, and V. Di Lazzaro, "Physiology of repetitive transcranial magnetic stimulation of the human brain," *Brain Stimulation*, vol. 3, no. 2, pp. 95–118, 2010.

[23] R. Melzack, "Pain and the neuromatrix in the brain.," *Journal of Dental Education*, vol. 65, no. 12, pp. 1378–1382, 2001.

[24] P. J. Slotty, S. Chang, and C. R. Honey, "Motor threshold: a possible guide to optimizing stimulation parameters for motor cortex stimulation," *Neuromodulation*, vol. 18, pp. 566–571, 2015.

[25] D. Kim et al., "Computational study on subdural cortical stimulation - the influence of the head geometry, anisotropic conductivity, and electrode configuration," *PLoS One*, vol. 17, Article ID e108028, p. 9, 2014.

[26] D. Kim, J. Jeong, S. Jeong, S. Kim, S. C. Jun, and E. Chung, "Validation of Computational Studies for Electrical Brain Stimulation With Phantom Head Experiments," *Brain Stimulation*, vol. 8, no. 5, pp. 914–925, 2015.

[27] H. Seo, D. Kim, and S. C. Jun, "Effect of anatomically realistic full-head model on activation of cortical neurons in subdural cortical stimulation – a computational study," *Scientific Reports*, vol. 6, Article ID 27353, pp. 1892–1895, 2016.

[28] A. Wongsarnpigoon and W. M. Grill, "Computational modeling of epidural cortical stimulation," *Journal of Neural Engineering*, vol. 5, no. 4, pp. 443–454, 2008.

[29] A. Wongsarnpigoon and W. M. Grill, "Computer-based model of epidural motor cortex stimulation: Effects of electrode position and geometry on activation of cortical neurons," *Clinical Neurophysiology*, vol. 123, no. 1, pp. 160–172, 2012.

[30] L. Manola, J. Holsheimer, P. Veltink, and J. R. Buitenweg, "Anodal vs cathodal stimulation of motor cortex: A modeling study," *Clinical Neurophysiology*, vol. 118, no. 2, pp. 464–474, 2007.

[31] ResumeTM, Medtronic, Inc., Minneapolis, Minnesota, USA – Technical Manual, https://www.fda.gov.tw/MLMS/ShowFile.aspx?LicId=06009114&Seq=001&Type=9.

[32] J. P. Lefaucheur et al., "Motor cortex stimulation for the treatment of refractory peripheral neuropathic pain," *Brain*, vol. 132, pp. 1463–1471, 2009.

[33] J. P. Nguyen et al., "Treatment of chronic neuropathic pain by motor cortex stimulation: results of a bicentric controlled crossover trial," *Brain stimulation*, vol. 1, pp. 89–96, 2008.

[34] A. Christ, W. Kainz, E. G. Hahn et al., "The Virtual Family—development of surface-based anatomical models of two adults and two children for dosimetric simulations," *Physics in Medicine and Biology*, vol. 55, no. 2, pp. N23–N38, 2010.

[35] M. C. Gosselin et al., "Development of a new generation of high-resolution anatomical models for medical device evaluation: the Virtual Population 3.0," *Physics in Medicine and Biology*, vol. 59, pp. 5287–5303, 2014.

[36] S. Gabriel, R. W. Lau, and C. Gabriel, "The dielectric properties of biological tissues: II. Measurements in the frequency range 10 Hz to 20 GHz," *Physics in Medicine and Biology*, vol. 41, no. 11, pp. 2251–2269, 1996.

[37] C. Gabriel, A. Peyman, and E. H. Grant, "Electrical conductivity of tissue at frequencies below 1 MHz," *Physics in Medicine and Biology*, vol. 54, no. 16, pp. 4863–4878, 2009.

[38] SEMCAD X v14.8 by SPEAG, Schmid & Partner Engineering, AG, Zurich, Switzerland, www.speag.com.

[39] M. Bikson and A. Datta, "Guidelines for precise and accurate computational models of tDCS," *Brain Stimulation*, vol. 5, no. 3, pp. 430-431, 2012.

[40] D. Kim, S. C. Jun, and H. I. Kim, "Computational study of subdural and epidural cortical stimulation of the motor cortex," in *Proceedings of the 2011 Annual International Conference of the IEEE Engineering in Medicine and Biology Society*, pp. 7226–7229, Boston, MA, USA, 2011.

[41] D. Kim, H. Seo, H-I. Kim, and S. C. Jun, "The computational study of subdural cortical stimulation: a quantitative analysis of voltage and current stimulation," in *Proceedings of the 34th Annual International Conference of the IEEE Engineering in Medicine and Biology Society (EMBC)*, pp. 867–870, San Diego, Calif, USA, 2012.

[42] T. Kowalski, J. Silny, and H. Buchner, "Current Density Threshold for the Stimulation of Neurons in the Motor Cortex Area," *Bioelectromagnetics*, vol. 23, no. 6, pp. 421–428, 2002.

[43] C. Nuti, R. Peyron, L. Garcia-Larrea et al., "Motor cortex stimulation for refractory neuropathic pain: four year outcome and predictors of efficacy," *Pain*, vol. 118, no. 1-2, pp. 43–52, 2005.

[44] W. F. Agnew, T. G. H. Yuen, D. B. McCreery, and L. A. Bullara, "Histopathologic evaluation of prolonged intracortical electrical stimulation," *Experimental Neurology*, vol. 92, no. 1, pp. 162–185, 1986.

[45] A. M. Lozano and M. Hallett, "Handbook of clinical neurology," *Brain Stimulation*, vol. 116, pp. 9-10, 2013.

[46] F. Rattay, "Analysis of Models for Extracellular Fiber Stimulation," *IEEE Transactions on Biomedical Engineering*, vol. 36, no. 7, pp. 676 682, 1989.

[47] J. G. R. Jefferys, "Nonsynaptic modulation of neuronal activity in the brain: Electric currents and extracellular ions," *Physiological Reviews*, vol. 75, no. 4, pp. 689–723, 1995.

[48] P. C. Miranda, A. Mekonnen, R. Salvador, and G. Ruffini, "The electric field in the cortex during transcranial current stimulation," *NeuroImage*, vol. 70, pp. 48–58, 2013.

[49] P. C. Miranda, M. Lomarev, and M. Hallett, "Modeling the current distribution during transcranial direct current stimulation," *Clinical Neurophysiology*, vol. 117, no. 7, pp. 1623–1629, 2006.

[50] A. R. Brunoni, M. A. Nitsche, and N. Bolognini, "Clinical research with transcranial direct current stimulation (tDCS): challenges and future directions," *Brain Stimulation*, vol. 5, no. 3, pp. 175–195, 2012.

[51] S. Fiocchi, "Assessment of the Electric Field induced by deep Transcranial Magnetic Stimulation in the elderly using H-coil," *ACES Journal*, vol. 31, pp. 636–643, 2016.

[52] Z. Nahas et al., "Safety and benefits of distance-adjusted prefrontal transcranial magnetic stimulation in depressed patients 55-75 years of age: a pilot study," *Depress Anxiety*, vol. 19, pp. 249–256, 2004.

[53] M. A. Nitsche, L. G. Cohen, E. M. Wassermann et al., "Transcranial direct current stimulation: State of the art 2008," *Brain Stimulation*, vol. 1, no. 3, pp. 206–223, 2008.

[54] S. Fiocchi, E. Chiaramello, M. Parazzini, P. Ravazzani, and R. M. Lafrenie, "Influence of tissue conductivity on foetal exposure to extremely low frequency magnetic fields at 50 Hz using stochastic dosimetry," *PLoS ONE*, vol. 13, no. 2, Article ID e0192131, 2018.

[55] C. McIntyre and W. Grill, "Extracellular stimulation of central neurons: Influence of stimulus waveform and frequency on neuronal output," *Journal of Neurophysiology*, vol. 88, no. 4, pp. 1592–1604, 2002.

[56] S. F. Cogan, K. A. Ludwig, C. G. Welle, and P. Takmakov, "Tissue damage thresholds during therapeutic electrical stimulation," *Journal of Neural Engineering*, vol. 13, no. 2, Article ID 021001, 2016.

[57] D. B. McCreery, W. F. Agnew, T. G. H. Yuen, and L. Bullara, "Charge density and charge per phase as cofactors in neural injury induced by electrical stimulation," *IEEE Transactions on Biomedical Engineering*, vol. 37, no. 10, pp. 996–1001, 1990.

[58] J. P. Dmochowski, A. Datta, M. Bikson, Y. Su, and L. C. Parra, "Optimized multi-electrode stimulation increases focality and intensity at target," *Journal of Neural Engineering*, vol. 8, no. 4, Article ID 046011, 2011.

2

Euclidean Distance Analysis Enables Nucleotide Skew Analysis in Viral Genomes

Formijn van Hemert,[1] Maarten Jebbink,[1] Andries van der Ark,[2] Frits Scholer,[3] and Ben Berkhout[1]

[1]*Laboratory of Experimental Virology, Medical Microbiology, Amsterdam UMC, University of Amsterdam, Amsterdam, Netherlands*
[2]*Research Institute of Child Development and Education, University of Amsterdam, Amsterdam, Netherlands*
[3]*Medical Microbiology, Amsterdam UMC, University of Amsterdam, Meibergdreef 9, 1105 AZ Amsterdam, Netherlands*

Correspondence should be addressed to Formijn van Hemert; f.j.vanhemert@amc.uva.nl

Academic Editor: Emil Alexov

Nucleotide skew analysis is a versatile method to study the nucleotide composition of RNA/DNA molecules, in particular to reveal characteristic sequence signatures. For instance, skew analysis of the nucleotide bias of several viral RNA genomes indicated that it is enriched in the unpaired, single-stranded genome regions, thus creating an even more striking virus-specific signature. The comparison of skew graphs for many virus isolates or families is difficult, time-consuming, and nonquantitative. Here, we present a procedure for a more simple identification of similarities and dissimilarities between nucleotide skew data of coronavirus, flavivirus, picornavirus, and HIV-1 RNA genomes. Window and step sizes were normalized to correct for differences in length of the viral genome. Cumulative skew data are converted into pairwise Euclidean distance matrices, which can be presented as neighbor-joining trees. We present skew value trees for the four virus families and show that closely related viruses are placed in small clusters. Importantly, the skew value trees are similar to the trees constructed by a "classical" model of evolutionary nucleotide substitution. Thus, we conclude that the simple calculation of Euclidean distances between nucleotide skew data allows an easy and quantitative comparison of characteristic sequence signatures of virus genomes. These results indicate that the Euclidean distance analysis of nucleotide skew data forms a nice addition to the virology toolbox.

1. Introduction

Nucleotide skew analysis [1] provides a powerful tool to visualize compositional aspects of a DNA/RNA sequence. For instance, the minimum and maximum of a G vs C skew can be used to predict the origin of replication and the location of the terminus, respectively, in prokaryotic genomes [2–4]. The GenSkew algorithm [1] calculates ratio values of the six nucleotide combinations (C vs G, G vs A, U vs G, U vs A, C vs A, and U vs C) in predefined windows and steps along a sequence. Ratio values are calculated according to (N1-N2)/(N1+N2), and hence, a positive value indicates that N1 wins over N2. Skew graphs are generally created by plotting the subsequent windows as numbers on the X-axis against the corresponding cumulative skew values on the Y-axis. In this way, we demonstrated for a representative collection of RNA viruses that the skew plots can be interpreted as "nucleotide compositional signatures" of the viral genomes and that these characteristic signatures are more prominently observed in the single-stranded regions than that in the base-paired, double-stranded regions of a viral RNA genome [5]. Likewise, we demonstrated that purine enrichment in the Zika virus RNA genome [6] is a general property of most but not all Flaviviridae and, surprisingly, prominently observed at the first position of the codons and not the silent 3rd codon position (unpublished results). It is, however, difficult, time-consuming, and nonquantitative to compare different skew graphs with respect to similarities and dissimilarities. We therefore developed a simple mathematical addition to GenSkew analysis that converts skew data into a pairwise Euclidean distance matrix, which can be formatted by means of clustering into a neighbour-joining tree, facilitating the identification of putative relationships, e.g., between viral sequences. The key result of this study is that this Euclidean algorithm offers an easy and quantitative interpretation of nucleotide skew data of virus genomes. The construction of Euclidean distance trees based on skewed nucleotide compositions does not require a prior alignment of the sequences. In contrast, "classical" phylogenetic trees are modelled after accurate nucleotide alignment of the sequences. As an additional result, we demonstrate that skew distance trees and phylogenetic trees are surprisingly similar but not identical.

2. Materials and Methods

Nucleotide sequences of the single-stranded (ss) RNA genomes of coronavirus, picornavirus, and HIV (reference strains and unclassified species) were downloaded from the ViralZone database (http://viralzone.expasy.org/) [7, 8]. Flaviviridae were selected and classified according to [9, 10] (Berkhout and van Hemert, unpublished results). GenBank IDs are provided in the figures. Nucleotide skew analysis was done by means of the GenSkew algorithm [1], of which Dr. T. Rattei (Technische Universität München) kindly provided a version that is not restricted by the length of a sequence. For normalization purposes, the overlapping window size was set 1% of the sequence length with a step size of 20% of the window size. The skew between two nucleotides N1 and N2 is defined by the ratio (N1 − N2)/(N1 + N2), and hence, a positive value of this ratio indicates that N1 proportionally exceeds N2. If the N1 versus N2 comparison results in a negative skew value, the same but positive skew value is true for these nucleotides in the reverse order (N2 versus N1). Algorithms converting skew data into a pairwise Euclidean distance matrix are provided as Additional File S1. The multiple sequence alignment of coronaviral, flaviviral, picornaviral, and HIV genomes was obtained by means of MAFFT [11]. Other alignments and Neighbor Joining (NJ) skew distance trees were built in MEGA v7 [12]. Phylogenetic histories were inferred by using the maximum-likelihood method based on the general time reversible model of nucleotide substitution in the viral genomes [12]. A discrete gamma distribution (5 categories) was used to model evolutionary rate differences among sites. The tree with the highest log likelihood is shown. Randomization of the rubella virus RNA genome (10 consecutive cycles to ensure complete nucleotide randomization) was performed by means of the BioWeb server (http://www.cellbiol.com/scripts/randomizer/dna_protein_sequence_randomizer.php). All calculations were performed in Excel.

3. Results

3.1. Nucleotide Skew Analysis of a Single Sequence. We used the rubella virus with its G- and C-rich RNA genome (JN635295) to illustrate the skew plot analysis (Figure 1). Skew profiles are shown of the nucleotide sequence before Figure 1(a) and after 10 consecutive cycles of randomization Figure 1(b). The profiles are nearly identical because skew profiles are determined solely by nucleotide composition and not by the nucleotide sequence. The rubella virus RNA genome size is 9761 nucleotides, and hence, the window size and step size are set to 98 (1% of sequence length) and 20 (20% of window size), respectively, generating 489 overlapping windows from the 5′- to the 3′-end (X-axis) with the corresponding skew values cumulatively plotted on the Y-axis. The skew lines start by definition at the origin of the plot. It should be noted that, in skew language, CG does not represent a CG base pair but a comparison of the C with the G nucleotide proportions. We adopted the notation C versus G (C vs G) in the text and the figures. The nucleotides C and G win over A (steep positive slope), and the U-nucleotide loses from C and G (steep negative slope) by approximately the same proportion. C is slightly more prominent than G, and the proportions of U and A are close to equality. Importantly, the skew profiles are straight lines, which indicates that the virus-specific nucleotide bias is quite constant along the sequence of the RNA genome. Therefore, skew values at the ultimate 3′ end of the genome (position 489) represent a reliable measure of the pairwise nucleotide compositional bias of the rubella virus genome. For instance, the skew endpoint values of C vs A (216.24) and U vs C (−213.37) are very similar but with the opposite sign. The same is true for G vs A (169.79) and U vs G (−168.02), and the skew endpoint values for C vs G and U vs A are 56.39 and 2.31, respectively. For the rubella virus genome, these values can be considered a characteristic signature of the nucleotide composition and presented as a vector in a six-dimensional Euclidean space (C vs G, G vs A, U vs G, U vs A, C vs A, U vs C) = (56.39, 169.79, −168.02, 2.31, 216.24, −213.37).

The skew lines (N1 − N2)/(N1 + N2) are shown of the rubella virus genome sequence (Figure 1(a)) and the same sequence after 10 consecutive cycles of nucleotide randomization (Figure 1(b)). Overlapping windows are plotted on the X-axis against the corresponding skew values cumulatively on the Y-axis.

3.2. Nucleotide Skew Analysis of Unrelated Sequences. The algorithm outlined above can be applied to sets of sequences provided that one carefully preserves the order of nucleotide comparisons as well as the normalization of window and step sizes to gain ±500 cumulative skew values. The resulting vectors can be formatted into a pairwise Euclidean distance matrix and visualized by means of cluster analysis as a neighbor-joining tree. It should be emphasized that the branch nodes of the skew value trees predict equality of the nucleotide skew values among the viral species involved. In contrast, branch nodes of evolutionary trees derived by means of nucleotide substitution models mark the expectation of sequence equality—the common ancestor—of the taxa involved. Hence, only partial similarity between the two tree building methods can be expected. Here, we apply

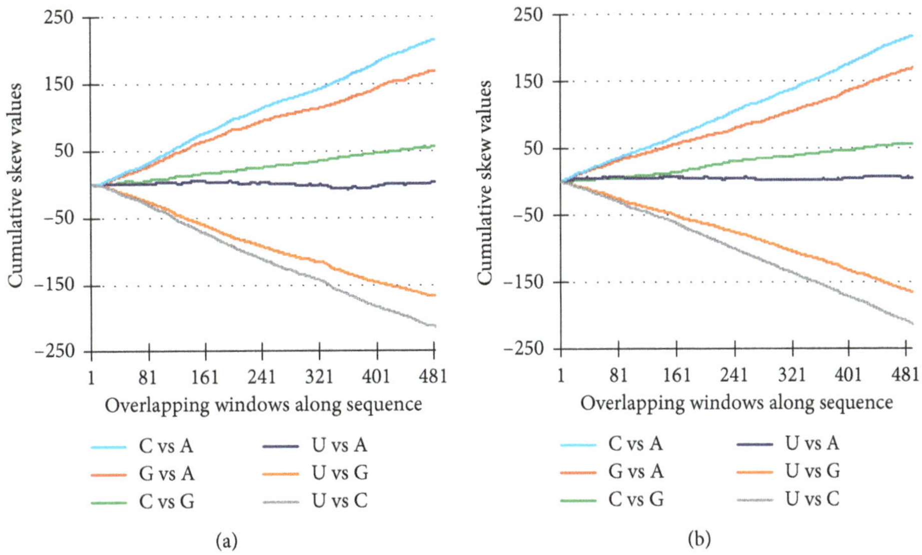

Figure 1: Nucleotide skew profiles of the rubella virus genome (JN635259). (a) Rubella virus. (b) 10x randomized.

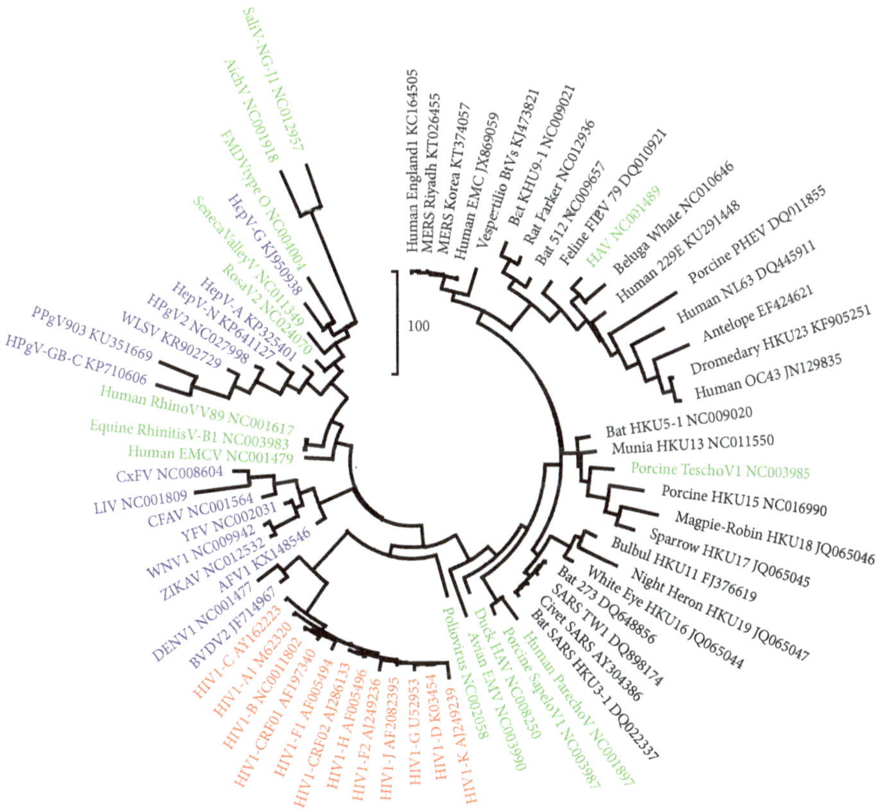

(a)

Figure 2: Continued.

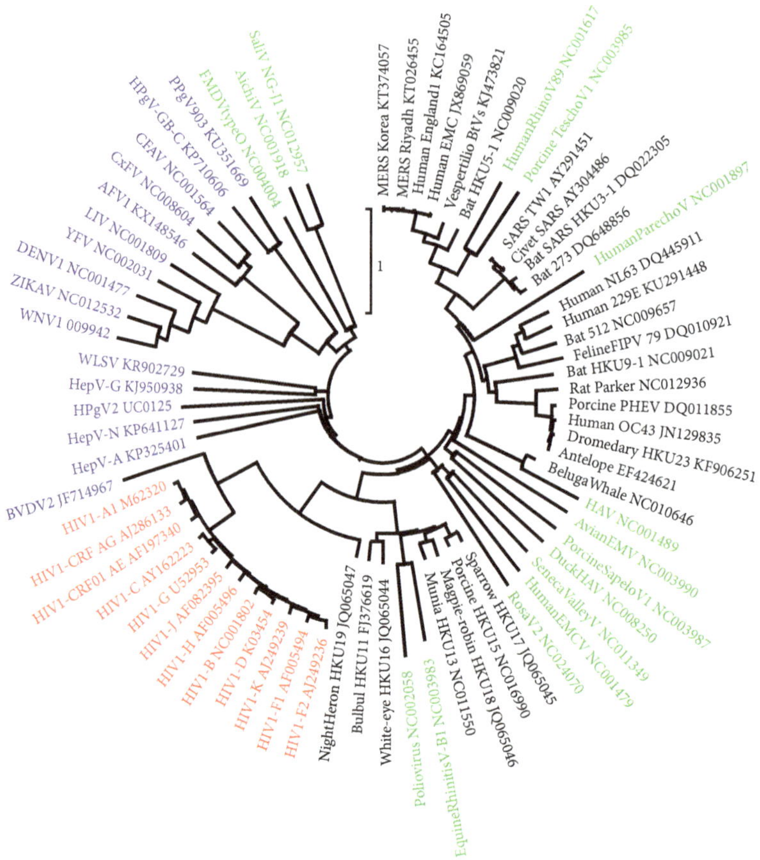

(b)

FIGURE 2: Trees of virus genomes based on nucleotide skew values (a) or nucleotide substitution (b).

the procedure to collections of arbitrarily chosen genomes of picornaviruses, flaviviruses, coronaviruses, and HIV-1 subtypes. First, we investigated whether the Euclidean skew algorithm is able to discern differences among the four families of viral genomes. Next, the four viral families are individually subjected to the Euclidean skew analysis.

Indeed, the neighbor-joining tree (Figure 2(a)) shows a consistent clustering of the HIV-1 subtypes (red), the coronaviruses (black), and the flaviviral species (blue). The picornaviruses (green) are significantly less resolved. Notably, HIV-1 genomes share A enrichment as the common feature of their nucleotide composition [13], coronavirus genomes typically tend to accumulate the U-nucleotide [14], and the Flaviviridae displays enhanced purine proportions in the first codon positions (unpublished results). The genomes of picornavirus species do not display such a prominent compositional feature. Apparently, the skew algorithm is able to discern viral families by giving credit to unique compositional features that are typical of a virus family. Viral genomes with a nucleotide composition that is more relaxed like picornavirus RNA are resolved poorly by the skew algorithm. A "classically" constructed phylogeny of the viruses shows a similar grouping (Figure 2(b)). HIV (red), coronavirus (black), and flavivirus species (blue) are clustered in separated clades, whereas representatives of picornavirus are found scattered among the flavi- and coronaviruses. With respect to salivirus, aichivirus, foot-and-mouth disease virus, and hepatitis A virus, both trees show similar topological positions, but in general, this is not the case. In addition, construction of "classical" phylogenetic trees relies a great deal on the accurate alignment of the sequences involved, which becomes less reliable as the evolutionary distance of the taxa increases. For that reason, we separately analyzed the four virus families.

Virus genomes are mentioned by short names and GenBank IDs and are colored black (coronaviruses), blue (flaviviruses), green (picornaviruses), and red (HIVs). The trees were constructed by means of neighbor-joining clustering (NJ, Figure 2(a)) and by maximum likelihood phylogeny (ML, Figure 2(b)). The scale bars are in skew units (Figure 2(a)) or estimate the number of nucleotide substitutions per site (Figure 2(b)).

3.3. Nucleotide Skew Analysis of Related Sequences

3.3.1. Coronavirus. Coronavirus RNA genomes show high U and low C proportions that are quite variable and in fact act like communicating vessels [14]. Indeed, a bipartition can be observed in the cluster tree based on Euclidean skew distances (Figure 3(a)). The upper group containing the MERS and SARS (in red) isolates is characterized by

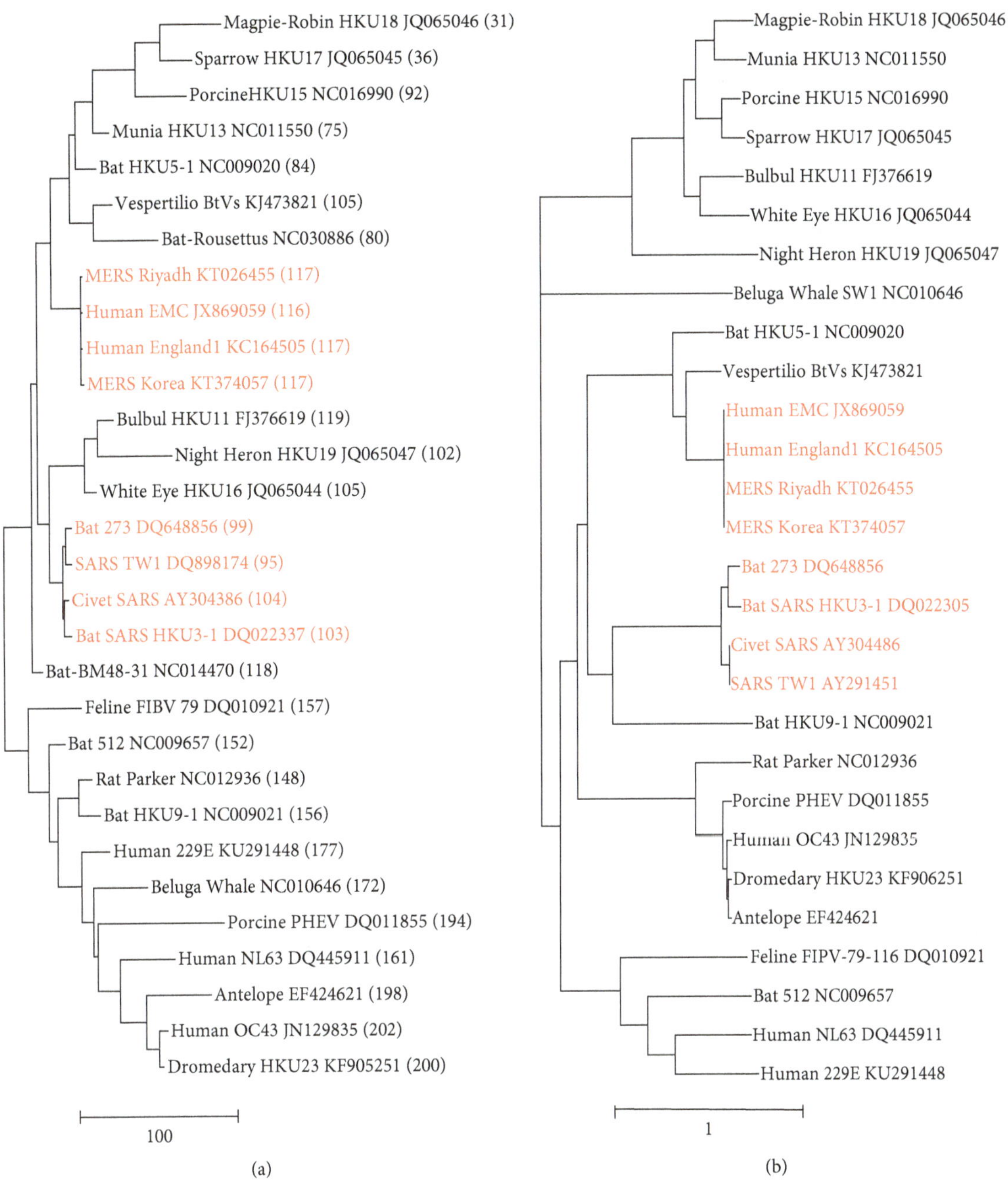

Figure 3: Trees of coronavirus genomes based on nucleotide skew values (a) or nucleotide substitution (b).

relatively low U vs C skew values varying from 120 to 31. The lower group with the human 229E, NL63, and OC43 coronaviruses displays U vs C skew values increasing from 140 to 201. Apparently, the opposite movement of the U and C proportions in coronaviral genomes is a distinctive feature, and hence, an important parameter in shaping coronaviral phylogenies is based on nucleotide substitution (Figure 3(b)). We included two unclassified virus isolates—marked in italic—to investigate whether the method can contribute to an initial virus classification. The unclassified *Bat-Rousettus* isolate is found near the *Vespertilio* coronavirus, and the unclassified *Bat-BM48-31* isolate is at the root position of in the upper clade.

Coronavirus genomes are mentioned by short names and GenBank IDs. Clustering of MERS and SARS coronaviruses is indicated by coloring in red. Numbers in brackets indicate U versus C skew values. Two clades differing by U vs C skew values can be discerned in the skew-based tree, the upper clade with U vs C proportions decreasing from 120 to 31 and the lower clade with U vs C proportions decreasing from 201 to 140. The two unclassified isolates are italicized. The trees were constructed by means of neighbor-joining clustering (NJ, Figure 3(a)) and by maximum likelihood phylogeny (ML, Figure 3(b)). The scale bars are in skew units Figure 3(a) or estimate the number of nucleotide substitutions per site Figure 3(b).

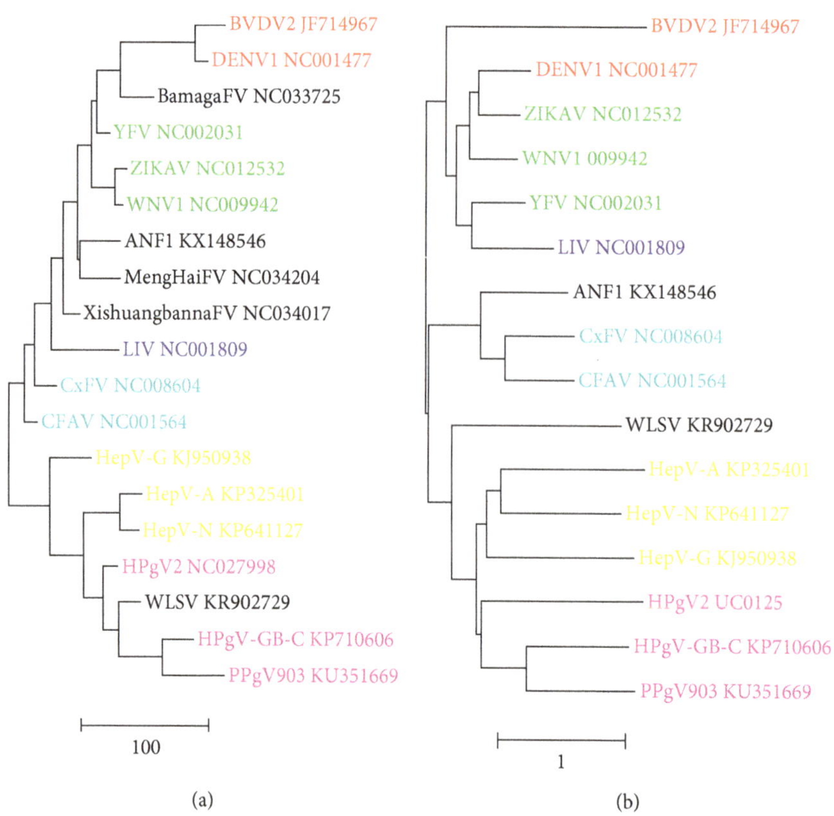

FIGURE 4: Trees of flavivirus genomes based on nucleotide skew values (a) or nucleotide substitution (b).

3.3.2. *Flavivirus.* The Flaviviridae represent an extensive group of ssRNA viruses with a diverse nucleotide composition. Yet, the pairwise distance tree based on skew values (Figure 4(a)) is in remarkable agreement with the topology obtained by means of evolutionary models (Figure 4(b)) [9]. Dengue virus (DENV), bovine viral diarrhea virus (BVDV), yellow fever virus (YFV), Zika virus (ZIKAV), West Nile virus (WNV) species, and the unclassified Bamaga flavivirus are clustered into the clade of mosquito-borne flaviviruses. The unclassified *Anopheles*, Menghai, and Xishuangbanna flaviviruses map in between the mosquito-borne viruses and the tick-borne representative (louping ill virus, LIV). The insect-specific culex flavivirus (CxFV) and cell fusing agent virus (CFAV) cluster together. Finally, the skew algorithm distinguishes the hepaciviruses from the pegiviruses (including Wenling shark virus, WLSV) consistent with more classical phylogeny [10].

Flavivirus genomes are mentioned by short names and GenBank IDs. Representatives of the flaviviruses are colored red (dengue virus and pestivirus), green (flavivirus), blue (tick-borne flavivirus), light-blue (insect-specific flavivirus), yellow (hepacivirus), and magenta (pegivirus). The five unclassified flavivirus genomes are italicized. The trees were constructed by means of neighbor-joining clustering (NJ, Figure 4(a)) and by maximum likelihood phylogeny (ML, Figure 4(b)). The scale bars are in skew units (Figure 4(a)) or estimate the number of nucleotide substitutions per site (Figure 4(b)).

3.3.3. *Picornavirus.* The skew cluster tree of picornaviruses (Figure 5(a)) also matches the tree topology based on nucleotide substitution (Figure 5(b)). As an exception, the rhinoviruses of humans and equines are close neighbors based on skew values that are separated more distantly in the standard evolutionary tree. Two of the three unclassified picornaviruses (Washington bat and bat crohivirus) map in the cluster of avian EMV and hepatitis A virus. The unclassified African bat icavirus takes a position in between rosavirus 2 and two rhinoviruses.

Picornavirus genomes are mentioned by short names and GenBank IDs. The three unclassified picornavirus genomes are italicized. The trees were constructed by means of neighbor-joining clustering (NJ, Figure 5(a)) and by maximum likelihood phylogeny (ML, Figure 5(b)). The scale bars are in skew units (Figure 5(a)) or estimate the number of nucleotide substitutions per site (Figure 5(b)).

3.3.4. *HIV.* The reference sequences of HIV-1 subtypes show much less compositional divergence than corona-, flavi-, and picornaviruses as indicated by the approximately 10x smaller size of the scale bar (Figure 6). In fact, the nucleotide skew profiles are nearly identical to the average skew value with small standard deviation (Table 1). The well-known A-accumulation in HIV-1 genomes [13] is prominently demonstrated by the negative values for G vs A, U vs A, and C vs A (Table 1: AVG: −101.30, −118.86, and −171.62,

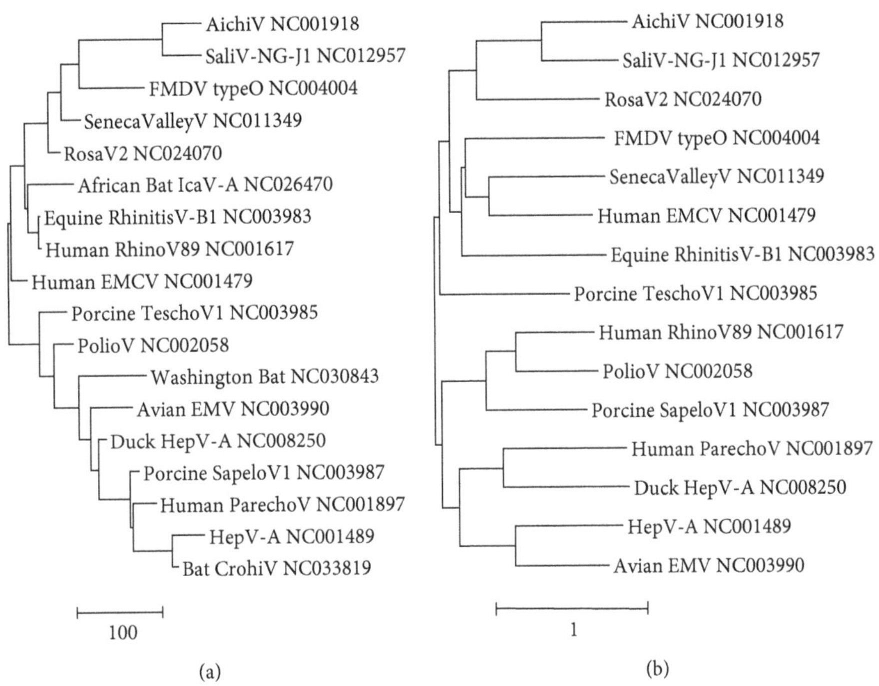

Figure 5: Trees of picornavirus genomes based on nucleotide skew values (a) or nucleotide substitution (b).

respectively). A-accumulation occurred mainly at the expense of C (C vs G: 77.15; U vs C: 59.48). Similar calculations for average skew values and standard deviations of coronavirus, flavivirus, and picornavirus genomes did not provide useful AVG and StDev data (Table 1).

HIV-1 subtype genomes are mentioned by subtype and GenBank IDs. The trees were constructed by means of neighbor-joining clustering (NJ, Figure 6(a)) and by maximum likelihood phylogeny (ML, Figure 6(b)). The scale bars are in skew units (Figure 6(a)) or estimate the number of nucleotide substitutions per site (Figure 6(b)).

Differences can be observed between the Euclidean skew tree of HIV-1 subtypes and the HIV-1 topology based on regular nucleotide substitution models (Figure 6). For instance, the evolutionary tree indicates a close relationship of F1 and F2 strains and between both HIV-1-CRF isolates, which is not observed in the skew-based tree. The dominant A-pressure in HIV may drive the compositional signatures of different HIV sequences towards similar skew values, by which the resolving power of the Euclidean skew algorithm becomes insufficient for a proper discrimination among HIV-1 subtypes. In addition, different sequences may have similar skew profiles. For example, randomization of a sequence does not alter the skew profile of the nucleotides (Figure 1). Finally, unclassified HIV-1 subtypes were not found in the ViralZone database.

4. Discussion

Pairwise Euclidean distance analysis facilitates an easy initial interpretation of the relationships among viral sequences based on skew values as a nucleotide compositional signature. The algorithm distinguishes related and unrelated sequences, particularly in case of compositional features that are characteristic for a certain virus family (e.g., A-accumulation in HIV, the U-C balance in coronavirus genomes, and the purine enrichment at the first codon position in flaviviral genomes). Within a family of viral genomes, the topology based on Euclidean skew distances is surprisingly similar (but not identical) to the topology based on regular evolutionary analysis (coronavirus, flavivirus, and picornavirus). This is remarkable because branch nodes in skew-based trees predict equal skew values of both sequences, and, in contrast, branch nodes in evolutionary trees mark nucleotide sequence equality of the two taxa. The resolving power of skew value analysis is diminished at high sequence similarity as shown for HIV-1 subtypes. This is due to the prominent A-pressure in all HIV genomes [15].

The key focus of this research was the Euclidean analysis of nucleotide skew data among virus genomes. In addition, we demonstrated similarity of skew data-based cluster trees with "classical" maximum-likelihood phylogenetic trees constructed via a general time reversible (GTR) model of nucleotide substitution during evolution. Confidence of ancestral nodes in "classical" trees is usually attained by means of bootstrap analysis, which requires repetitive resampling of sites in the original sequence alignment. Each bootstrap replicate generates a tree, and a consensus tree is constructed based on the collection of replicates. Each node in this consensus tree is decorated with the frequency number (%) that reflects the level of bootstrap support. It is evident that a skew-based pairwise distance matrix cannot be used for bootstrap analysis simply due to the absence of a sequence alignment. Also, resampling of sites in a sequence prior to skew analysis does not affect the skew data vector (Figure 1). Future investigation may reveal whether the

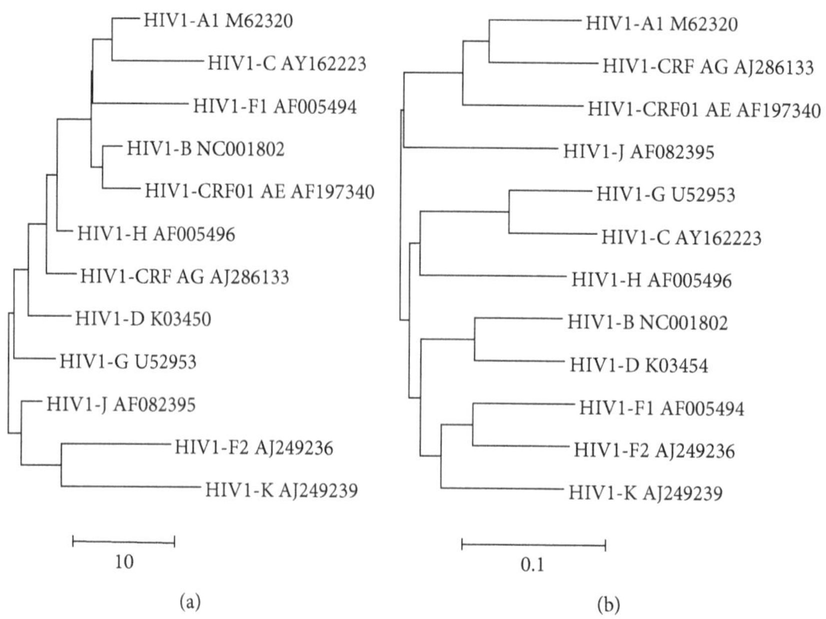

FIGURE 6: Trees of HIV-1 subtype genomes based on nucleotide skew values (a) or nucleotide substitution (b).

TABLE 1: Mean skew values of coronavirus, flavivirus, picornavirus, and HIV-1 subtype RNA genomes. HIV-1 subtype RNAs display much lower StDEV of skew values compared with those of the other viruses.

Skew values		C vs G	G vs A	U vs G	U vs A	C vs A	U vs C
Coronavirus	AVG	−29.09	−58.05	101.36	40.51	−86.34	124.36
	StDev	46.67	33.47	21.40	24.00	43.15	45.22
Flavivirus	AVG	−43.38	38.95	−44.53	−5.01	−3.66	−1.19
	StDev	29.92	53.33	32.36	59.98	71.05	25.50
Picornavirus	AVG	9.48	−32.05	37.12	2.81	−24.10	26.49
	StDeV	62.05	45.85	40.06	39.51	95.34	81.95
HIV	AVG	−77.15	−101.30	−17.95	−118.86	−171.62	59.48
	StDEV	3.82	7.00	3.46	6.20	7.32	5.26

building of skew-based trees may serve as a valuable addition to many sequence-independent methods of evolutionary tree reconstruction.

5. Conclusions

Euclidean skew analysis supports the study of nucleotide compositional features and facilitates the classification of unclassified viral sequences as nearest neighbors of classified ones. Compared with methods based on nucleotide substitution models, (Euclidean) skew analysis is very robust in that it tolerates sequence errors and shifts in reading frames.

Data Availability

The relevant databases ViralZone and GenBank used to support the findings of this study are included within the article (Materials and Methods). The GenBank IDs of the sequences are provided in the figures. The software used is mentioned by the references as provided by MAFFT and MEGA. Nucleotide randomization was performed by a BioWeb server application. The GenSkew algorithm was provided by Dr. T. Rattei. A web server version can be found at http://genskew.csb.univie.ac.at/.

Conflicts of Interest

The authors declare that they have no conflicts of interest.

Acknowledgments

Dr. Thomas Rattei (Department of Computational Systems Biology of the University of Vienna) kindly provided a GenSkew version without sequence length limitations.

References

[1] A. Grigoriev, "Analyzing genomes with cumulative skew diagrams," *Nucleic Acids Research*, vol. 26, no. 10, pp. 2286–2290, 1998.

[2] K. Arakawa, H. Suzuki, and M. Tomita, "Quantitative analysis of replication-related mutation and selection pressures in bacterial chromosomes and plasmids using generalised GC skew index," *BMC Genomics*, vol. 10, no. 1, p. 640, 2009.

[3] W. C. Li, Z. J. Zhong, P. P. Zhu et al., "Sequence analysis of origins of replication in the *Saccharomyces cerevisiae* genomes," *Front Microbiol*, vol. 5, p. 574, 2014.

[4] N. V. Sernova and M. S. Gelfand, "Identification of replication origins in prokaryotic genomes," *Briefings in Bioinformatics*, vol. 9, no. 5, pp. 376–391, 2008.

[5] F. van Hemert, A. C. van der Kuyl, and B. Berkhout, "Impact of the biased nucleotide composition of viral RNA genomes on RNA structure and codon usage," *Journal of General Virology*, vol. 97, no. 10, pp. 2608–2619, 2016.

[6] F. van Hemert and B. Berkhout, "Nucleotide composition of the Zika virus RNA genome and its codon usage," *Virology Journal*, vol. 13, no. 1, p. 95, 2016.

[7] C. Hulo, E. de Castro, P. Masson et al., "ViralZone: a knowledge resource to understand virus diversity," *Nucleic Acids Research*, vol. 39, no. 1, pp. D576–D582, 2011.

[8] P. Masson, C. Hulo, E. De Castro et al., "ViralZone: recent updates to the virus knowledge resource," *Nucleic Acids Research*, vol. 41, no. D1, pp. D579–D583, 2013.

[9] A. M. Schubert and C. Putonti, "Evolution of the sequence composition of Flaviviruses," *Infection, Genetics and Evolution*, vol. 10, no. 1, pp. 129–136, 2010.

[10] J. T. Stapleton, S. Foung, A. S. Muerhoff, J. Bukh, and P. Simmonds, "The GB viruses: a review and proposed classification of GBV-A, GBV-C (HGV), and GBV-D in genus Pegivirus within the family Flaviviridae," *Journal of General Virology*, vol. 92, no. 2, pp. 233–246, 2011.

[11] K. Katoh and D. M. Standley, "MAFFT: iterative refinement and additional methods," *Methods in Molecular Biology*, vol. 1079, pp. 131–146, 2014.

[12] S. Kumar, G. Stecher, and K. Tamura, "MEGA7: molecular evolutionary genetics analysis version 7.0 for bigger datasets," *Molecular Biology and Evolution*, vol. 33, no. 7, pp. 1870–1874, 2016.

[13] F. J. van Hemert, A. C. van der Kuyl, and B. Berkhout, "The A-nucleotide preference of HIV-1 in the context of its structured RNA genome," *RNA Biology*, vol. 10, no. 2, pp. 211–215, 2013.

[14] B. Berkhout and F. van Hemert, "On the biased nucleotide composition of the human coronavirus RNA genome," *Virus Research*, vol. 202, pp. 41–47, 2015.

[15] A. C. van der Kuyl and B. Berkhout, "The biased nucleotide composition of the HIV genome: a constant factor in a highly variable virus," *Retrovirology*, vol. 9, no. 1, p. 92, 2012.

3

Single-Frequency Ultrasound-Based Respiration Rate Estimation with Smartphones

Linfei Ge,[1] **Jin Zhang**,[1] **and Jing Wei**[2]

[1]*Department of Computer Science and Engineering, Southern University of Science and Technology, Shenzhen, Guangdong, China*
[2]*Department of Electrical and Electronic Engineering, Southern University of Science and Technology, Shenzhen, Guangdong, China*

Correspondence should be addressed to Jin Zhang; zhangj4@sustc.edu.cn

Academic Editor: Dingchang Zheng

Respiration monitoring is helpful in disease prevention and diagnosis. Traditional respiration monitoring requires users to wear devices on their bodies, which is inconvenient for them. In this paper, we aim to design a noncontact respiration rate detection system utilizing off-the-shelf smartphones. We utilize the single-frequency ultrasound as the media to detect the respiration activity. By analyzing the ultrasound signals received by the built-in microphone sensor in a smartphone, our system can derive the respiration rate of the user. The advantage of our method is that the transmitted signal is easy to generate and the signal analysis is simple, which has lower power consumption and thus is suitable for long-term monitoring in daily life. The experimental result shows that our system can achieve accurate respiration rate estimation under various scenarios.

1. Introduction

Respiration is one of the most important vital signs for human beings. Many chronic diseases can be detected through respiration monitoring, such as asthma and chronic obstructive pulmonary diseases [1, 2]. Respiration monitoring can also reflect the users' sleep and emotion status. Respiration rate detection is a key function in respiration monitoring [3]. Most existing respiration rate detection devices are wearable, for example, respiration belts [4] and oronasal airflow meters [5]. Users need to wear these devices on their chests or faces during respiration monitoring, which is extremely inconvenient. Therefore, these devices are mainly applicable for medical examination for disease diagnosis, but they are not suitable for long-term everyday monitoring.

To overcome the disadvantage of the wearable devices, recently, researchers started investigating noncontact respiration monitoring methods. Some works use wireless signals to detect the respiration rate. For example, Abdelnasser et al. leveraged the WiFi signals [6–10], Lazaro et al. used the UWB signals [11, 12], and Rahman et al. used a microwave radar to detect respiration [13–15]. These systems require deploying extra wireless transceivers to transmit and receive wireless signals, which makes the system expensive. Instead of specially deployed wireless transceivers, some researchers proposed using smartphones to detect vital signs, which is easier to access in daily life. Some works used built-in inertial sensors in smartphones to monitor vital signs [16–19], while others leveraged ultrasonic signals to conduct respiration detection and sleep monitoring [20, 21]; they utilized built-in speakers and microphones in smartphones to play and record ultrasound signals and extract useful information such as respiration patterns from them [22]. In our paper, we will also use ultrasound as the media to detect the respiration rate. Actually, ultrasound is a proper medium for vital sign detection, as it can be easily generated by smartphones, which are taken along by people all the time and are suitable for long-term monitoring. Besides, ultrasound is a mechanical wave; therefore, users do not need to worry about the electromagnetic radiation for long-term monitoring. Because of the above advantages, ultrasound is also considered to be used for other applications, such as activity recognition and human computer interaction [23–25]. However, in existing ultrasound-based monitoring systems, complicated signals such as frequency modulated continuous wave (FMCW) or orthogonal frequency division multiplexing (OFDM) are used as the carrier to conduct object detection, which requires complicated modulation and demodulation signal processing modules. The complexity in signal generation and signal processing leads to large power consumption, which is not suitable for power-limited mobile devices such as smartphones.

In this paper, to overcome the above-mentioned limitation of existing solutions, we propose a smartphone-based respiration rate detection system using a single-frequency ultrasonic signal, which enables simple signal processing and low power consumption. We observed that when a single-frequency cosine ultrasonic signal is transmitted by the speaker and the signals reflected by the human being are collected using the built-in microphone, the signal strength of the received signal can reflect the breath pattern of the person being detected. The reason is that the received signal is a combination of the reflected signal and the line-of-sight signal. The chest movement of the respiration activity changes the distance between the speaker and the microphone and, therefore, changes the phase of the reflected signal, which finally results in the amplitude of the total received signal changing periodically. By analyzing the amplitude of the received signal, the respiration rate can be correctly detected. In our proposed system, to detect the respiration rate, we only need to calculate the amplitude of the signal, without complicated frequency domain analysis. Our method guarantees a high detection accuracy while keeping the analysis relatively simple.

The contributions of our paper are as follows. First, we make an observation that the received signal strength of the single-frequency ultrasound signal can reflect the respiration pattern of the user, by theoretical analysis. Second, we propose a smartphone-based respiration rate detection system utilizing single-frequency ultrasonic signals. We also design a rate detection algorithm to estimate the respiration rate based on the amplitude of the received signal. Third, we implement the system on an Android smartphone and the experimental result shows that our system can achieve accurate respiration rate estimation results under various testing scenarios.

2. Ultrasonic Signal Analysis

In this section, we will analyze the characteristics of the received ultrasonic signal when a single-frequency signal is transmitted by the speaker and reflected by the tester. The analysis result shows that the strength of the received signal reflects the respiration rate; thus, by detecting the signal strength, the respiration rate can be estimated.

2.1. Overview. In this paper, we use the built-in speaker of a smartphone to generate single-frequency ultrasound of 20 kHz. Most off-the-shelf smartphones can generate sound up to 22 kHz using their built-in speakers [20, 21]. The smartphone is placed in front of the tester. The ultrasound signal is reflected by the human body and captured by the built-in microphone of the smartphone. The received signal is mainly composed of two parts. One is the signal directly propagated from the speaker to the microphone. The other is the part that is reflected by the user's moving chest. These two signals have a superposition at the receiving end. Because of the movements of the chest while breathing, the received signal varies. In the following subsection, we will derive the received signal strength of the composed signal. We observed that the signal strength varies along with the respiration. From the amplitude of the received signal, we can extract the respiration waveform and estimate the respiration rate.

2.2. Receiving Signal Analysis. In our system, we use the speaker and microphone of a smartphone to transmit and receive ultrasonic signals. At the transmitter end, the speaker emits a single-frequency cosine signal

$$S(t) = A\cos(2\pi f_s t), \tag{1}$$

where A is the amplitude and f_s is the frequency of the generated ultrasound signal.

The received signal is a superposition of two components: the static signal which propagates directly from the speaker to the microphone and is reflected by the static reflectors and the dynamic one which is caused by the movements of the tester's chest.

For the static component, it contains the line-of-sight signal directly transmitted from the speaker to the microphone and the signals reflected by static objects around. The static reflectors only change the phase of the signal without changing its frequency; therefore, the static component, which is the sum of all static rays, can be written as follows:

$$R_s(t) = A_1\cos(2\pi f_s t + \phi_s), \tag{2}$$

where A_1, f_s are the amplitude and frequency of the sound signal, respectively, and ϕ_s is a constant phase change.

For the dynamic component, the periodical movement of the chest causes a periodical distance change between the smartphone and the chest. Thus, the propagation delay varies because of the chest motion. Therefore, the phase also changes periodically. So, the dynamic component can be written as follows:

$$R_d(t) = A_2\cos(2\pi f_s t + \phi_d(t)), \tag{3}$$

where A_2, f_s are the amplitude and frequency of the sound signal. $\phi_d(t)$ is the periodical phase change caused by the chest movement, which is written as

$$\phi_d(t) = \frac{2\pi f_s (D_0 + D\cos(2\pi f_b t))}{v}, \tag{4}$$

where D_0 is the constant distance of the propagation path, D and f_b are the amplitude and frequency of the chest movement while breathing, and v is the speed of sound. Here, $D_0+D\cos(2\pi f_b t)$ is the distance between the smartphone and the chest while breathing, and $(D_0 + D\cos(2\pi f_b t))/v$ is the propagation delay.

At the receiver end, the received signal is the superposition of the static component and dynamic component:

$$R(t) = R_s(t) + R_d(t) = A_r(t)\cos(2\pi f_s t + \varphi), \tag{5}$$

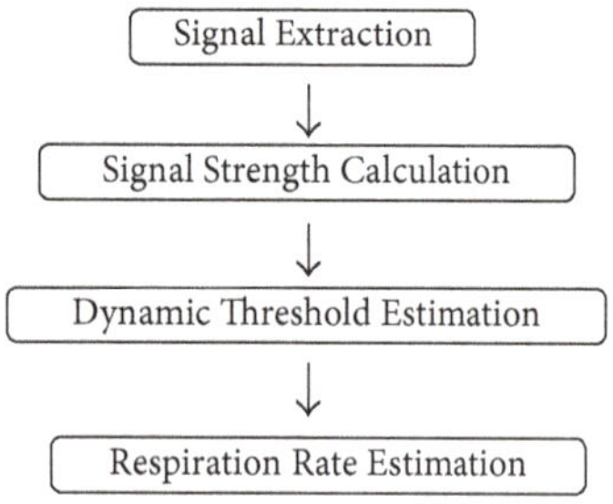

FIGURE 1: Algorithm overview.

where

$$A_r(t) = \sqrt{A_1^2 + A_2^2 + 2A_1A_2\cos(\phi_s - \phi_d(t))}, \quad (6)$$

$$\tan\varphi = \frac{A_1\cos\phi_s + A_2\cos\phi_d(t)}{A_1\sin\phi_s + A_2\sin\phi_d(t)}. \quad (7)$$

Bringing (4) into (6), we have

$$\begin{aligned} A_r(t) &= \sqrt{A_1^2 + A_2^2 + 2A_1A_2\cos\left(\phi_s - \frac{2\pi f_s(D_0 + D\cos(2\pi f_b t))}{v}\right)} \\ &= \sqrt{A_1^2 + A_2^2 + 2A_1A_2\cos\left(\phi_s - \frac{2\pi f_s D_0}{v} - \frac{2\pi f_s D}{v}\cos(2\pi f_b t)\right)}. \end{aligned} \quad (8)$$

In the above expression of $A_r(t)$, $\cos(2\pi f_b t)$ varies with t at a frequency of f_b, which is the breathing rate. So, the amplitude of the received signal $A_r(t)$ varies periodically at the same frequency with breath. Therefore, the signal strength, which is the square of the amplitude, follows the same changing pattern. Therefore, by detecting the received signal strength, we can estimate the chest movement of a person.

3. Respiration Rate Estimation Algorithm

In this section, we present our respiration rate estimation algorithm based on single-frequency ultrasound signals. We aim to run our algorithm and make it work well on smartphones. Considering the limited resources and power supply of smartphones, we try our best to reduce the complexity of our algorithm. We are trying to find the most effective way instead of the most accurate way for signal processing. Our system is composed of four stages: signal extraction, signal strength calculation, dynamic threshold estimation, and respiration rate estimation, as shown in Figure 1.

In the first stage, the smartphone generates an inaudible 20 kHz ultrasound signal, plays it with the built-in speaker, and records the signal using the microphone. The recording process is to sample the received sound signal $R(t)$ with sampling rate F_s, which achieves a discrete signal:

$$X_n = R\left(\frac{n}{F_s}\right), \quad n = 1, 2, 3, \ldots. \quad (9)$$

In the second stage, we calculate the received signal strength. From the analysis in Section 2, we know that the received signal strength changes at the same frequency as the breath. We can estimate the respiration rate based on the received signal strength P_m. Although the audio file is sampled with a high sampling frequency F_s, the received signal strength P_m can be calculated at relatively low frequency F_p, where $F_p = F_s/K$ and K is the coefficient that is used to reduce the sampling rate of signal strength. The signal strength P_m can be defined as the average of the signal strength of K samples in X_n:

$$P_m = \frac{1}{K}\sum_{n=(m-1)*K+1}^{m*K} X_n^2, \quad m = 1, 2, 3, \ldots. \quad (10)$$

Then, we smooth the received signal strength using a moving average filter with a window size of W points. The smoothed signal strength $\overline{P}_m$ is given by

$$\overline{P}_m = \frac{1}{W}\sum_{i=\max\{i-W,1\}}^{m} P_i \quad (m = 1, 2, 3, \ldots). \quad (11)$$

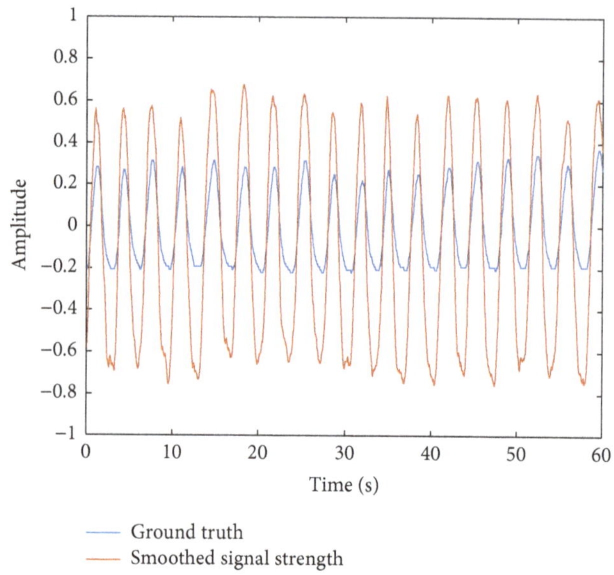

FIGURE 2: Smoothed signal strength compared with ground truth.

Figure 2 shows the smoothed received signal strength $\overline{P}_m$ as well as the ground truth achieved by respiration belt SCHUHFRIED Biofeedback Xpert [26]. The result shows that the smoothed signal strength matches the ground truth well.

With the estimated respiration waveform, we can derive the respiration rate of the tester. To simplify the algorithm, instead of frequency domain analysis, we want to use simple time domain analysis to detect the respiration rate by counting the number of peaks and valleys in the strength signal. To accurately count the peaks and valleys, a threshold is required, and by comparing the signal strength with the threshold, we can get the respiration period and respiration rate. However, in some cases, the signal strength may fluctuate severely because of the changing environment, just as Figure 3 shows. To solve this problem, we conduct dynamic threshold estimation at the third stage before calculating the

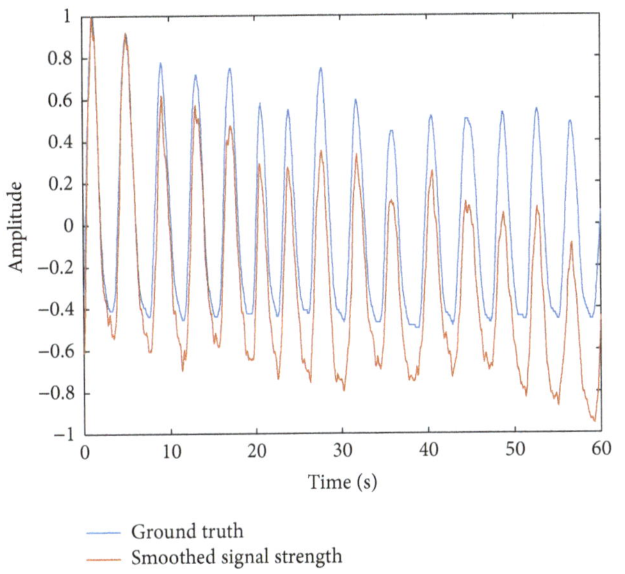

FIGURE 3: Signal strength fluctuation caused by the environment while testing.

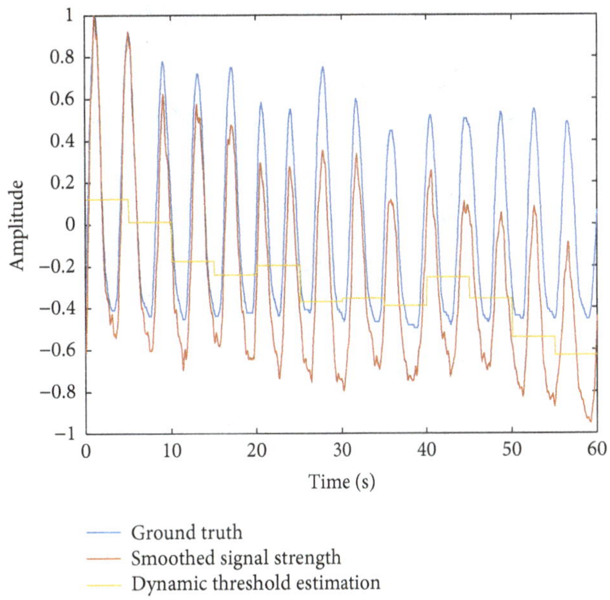

FIGURE 4: Dynamic threshold estimation.

respiration rate in the fourth stage. We calculate the dynamic threshold P_{th} by averaging W_{th} continuous points in $\overline{P}_m$.

$$P_{th}(m) = \frac{1}{W_{th}} \sum_{i=\lfloor m/W_{th} \rfloor * W_{th}+1}^{(\lfloor m/W_{th} \rfloor+1)*W_{th}} \overline{P}_i, \quad m = 1, 2, 3, \ldots. \tag{12}$$

The dynamic threshold P_{th} varies as the signal strength $\overline{P}_m$ changes.

At the last stage, we use the smoothed signal strength $\overline{P}_m$ and the dynamic threshold P_{th} to estimate the respiration rate of the tester. We record the times that the estimated respiration waveform passes through the threshold, and then we get the period of the breathing. Using the average value of several recent periods, we can get the estimated respiration rate. Using this algorithm, we get Figure 4. This figure shows that our algorithm works well and gets a mean estimation error of 0.32 bpm in this example. We also test the case of deep breath; the algorithm still performs well as Figure 5 shows. The mean estimation error under deep breath is 0.35 bpm.

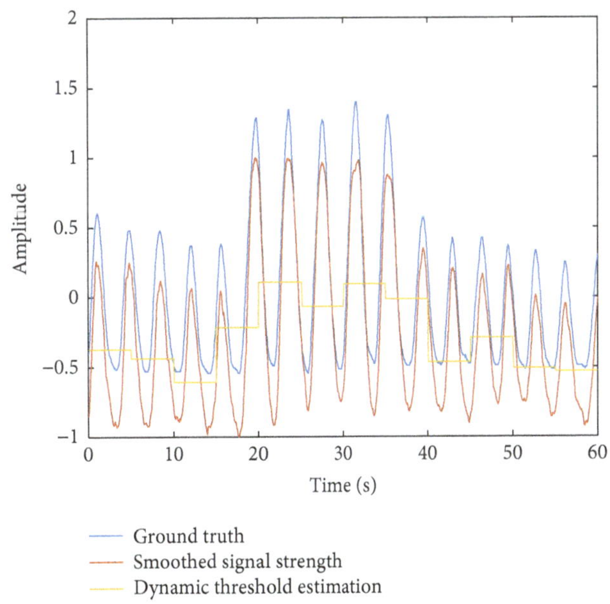

FIGURE 5: Dynamic threshold estimation under deep breath.

4. System Evaluation

4.1. System Implementation and Evaluation Setup. We develop an Android application to implement our algorithm on smartphones and evaluate the performance under various scenarios. We conduct experiments on two smartphones, a Xiaomi MI5 and a Samsung Galaxy S4, which are both based on the Android OS. Our application is developed with a minimum version of Android 4.0.4 and it works well on both smartphones. In our system, to generate ultrasound, we first generate a sound file in Pulse Code Modulation (PCM) format, and then we use AudioTrack in Android to play the generated sound file. The main speaker of the smartphone is used to transmit the ultrasound signal. In the receiving end, we use one microphone to receive the reflected signal, which works on the mono record mode instead of the stereo mode to reduce the complexity of the computation. The ultrasound signal is at the frequency of 20 kHz. The sampling rate of the microphone is 48 kHz which is the maximum sampling rate that most smartphones can support. To achieve the ground truth, we use the SCHUHFRIED Biofeedback Xpert [26] to monitor the respiration rate of the testers. Figure 6 shows the experimental scenario in the office environment.

In the remaining part of this section, without specific instruction, the parameters are set as follows: sampling rate of the smartphone F_s = 48 kHz and coefficient K = 2400. Thus, the sampling rate of P_m is at a frequency of

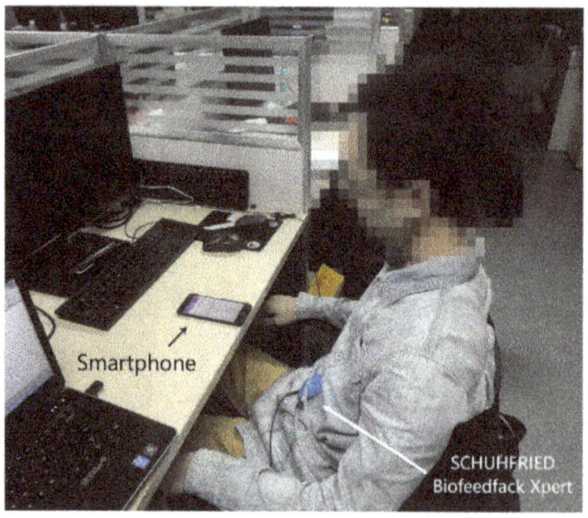

FIGURE 6: Experimental scenario in the office.

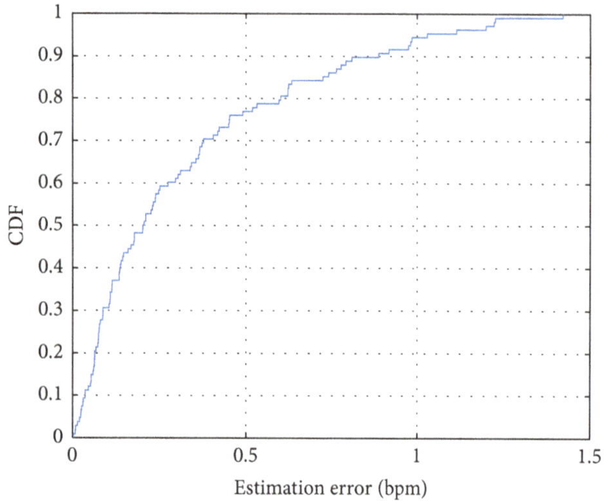

FIGURE 7: CDF of the estimation error.

$F_p = F_s/K = 20$ Hz. The window size of moving average filter $W = 5$. The window size of dynamic threshold estimation $W_{th} = 100$.

4.2. Evaluation Results. We evaluate the overall performance of the system and calculate the cumulative distribution functions (CDF) of estimation errors on the respiration rate, when the distance between the smartphone and the tester is 15 cm. As Figure 7 shows, over 90% of the results have an estimation error under 0.8 bpm. The medium estimation error is 0.2101 bpm while the mean estimation error is 0.4137 bpm. Traditional medical respiration monitoring devices usually allow an error of 5%, considering that a normal breath rate is usually 15–20 bpm, and our estimation error is less than 4%, which is accurate enough for daily use.

Figure 8 shows the respiration rate comparison for a test that lasts for 30 minutes. From this figure, we find that the estimated breathing rate follows the ground truth well. Due to the convenience of the smartphones, they are suitable for our algorithm to do a long-term monitoring.

Compared with Wang et al.'s work [21], we get similar results in estimation error with a much simpler algorithm.

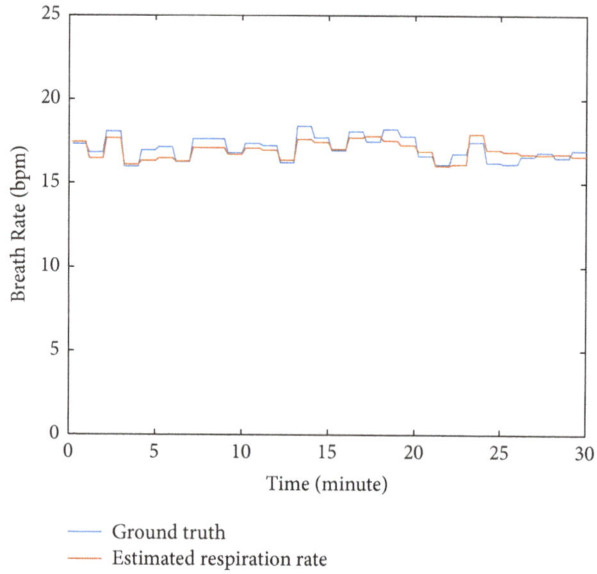

FIGURE 8: Respiration rate estimation result for 30-minute testing.

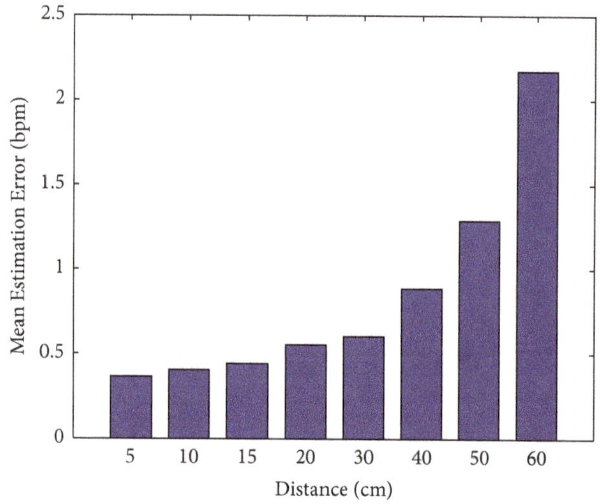

FIGURE 9: Estimation error versus the distance between the user and the smartphone.

They employed sonar phase data to get the breath rate, resulting in high complexity in algorithm. In their work, the estimation error at a distance of 10 cm is about 0.3 bpm, which is very close to our results, but we achieve similar results using a much simpler algorithm.

4.3. Impact of Different Factors

Impact of Distance. Figure 9 shows the impact of distance between the microphone and people's chest. When the distance gets longer, the mean error gets larger. With the increase of distance, the signal attenuation becomes severe and the amount of reflected ultrasound signal becomes less. Thus, the energy change caused by the reflected ultrasound signal becomes more inconspicuous; therefore, the estimation accuracy will be reduced. We can see that our algorithm

Table 1: Estimation error versus the orientation of the smartphone.

Orientation (degree)	Mean estimation error (bpm)
0	0.39
45	0.50
90	0.62

Table 2: Estimation errors on different persons.

Person number	Mean estimation error (bpm)
1	0.41
2	0.45
3	0.42

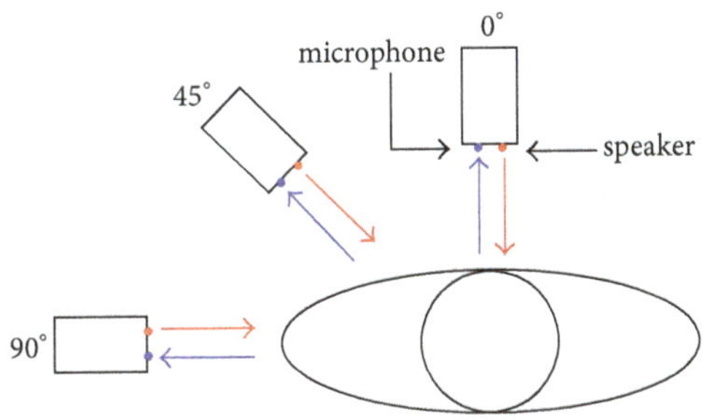

Figure 10: Orientation of the smartphone.

works well when the distance is under 40 cm. The mean estimation error is around 0.5 bpm under a distance of 30 cm and it is acceptable for daily use. When the distance is 40 cm, the mean estimation error is almost 1 bpm.

Impact of Orientation. Table 1 shows how the orientation influences the mean estimation error. In this experiment, we keep the distance at 10 cm and control the angle between the smartphone and the tester as Figure 10 shows. We consider three cases when the angle is equal to 0, 45, and 90 degrees, respectively. At 0 degrees, the person directly faces the smartphone, and we get the best performance at a mean estimation error of 0.39 bpm. At 45 degrees, the chest movements become less clear compared with 0 degrees. So, we get a worse mean estimation error at 0.50 bpm. At 90 degrees, the estimation error of 0.62 bpm is the worst due to the minimum chest movements in these three situations. Despite being much worse than the situation of facing the smartphone, the results at 90 degrees are still acceptable. The system can work in all orientations, because during breathing, the chest moves at both the front and side directions. However, the performance is optimal when the tester directly faces the smartphone.

Results on Different Persons. Table 2 shows whether the tester influences the estimation error a lot. In the experiment, we test three persons (two men and one woman) as shown in Table 3. The experiment is done at 10 cm distance in the office. We can see that there are small differences between different persons. These small differences may be caused by environmental noise, different chest movement length, and experimental error. The mean estimation errors of these three persons are all between 0.4 and 0.5 bpm, which is an acceptable result. Thus, our algorithm works well on different persons.

Impact of Different Smartphones. In this experiment, we use two smartphones, Xiaomi MI5 and Samsung Galaxy S4. From Table 4, we can see that the MI5 has a smaller mean estimation error than the S4. This is because the structures of the two smartphones are slightly different. For MI5, the speaker and microphone are both at the bottom of the phone. This means that when you put it on the desk, the speaker and microphone are directly pointing to the chest of the person. In contrast, the speaker of the S4 is on its back and its microphone is at the bottom. Thus, when we put the S4 on the desk, the volume may be reduced due to the position of the speaker. So, the received signal of MI5 is stronger than that of S4 and MI5 achieves a better performance in the experiment. We now know that the position of the speaker and the microphone does matter. A smartphone with its speaker and microphone at the bottom usually yields a better result. However, even when the speaker is at a relatively bad position, the performance is still acceptable.

Impact of Various Testing Scenarios. Table 5 shows the impact of different scenarios. We test four scenarios including office, dormitory, library, and office with music playing. They achieved mean estimation errors of 0.48, 0.43, 0.31, and 0.78 bpm, respectively. We get the minimum mean estimation error at the library, because in the library, it is quiet and there are a few objects around, which results in the minimum impact of the environment, and thus it achieves the minimum mean estimation error. In the dormitory and office, the situation is similar: crowded room with computer and air conditioners running, even some people talking with others. The noise generated from machines and people does influence the performance of our experiment. Because our algorithm is based on the energy of the signal, a loud noise may cover the signal that we want, resulting in bad performance. We further run experiments in the office with music playing to verify the impact of noise. The mean estimation error with music is 0.78 bpm, which is much larger than the general situation. This shows that a loud noise does make the performance worse. Furthermore, the music has a more severe impact than people's talking. That is because the frequency domain of music is relatively higher than people's voice; therefore, music has a higher influence on the 20 kHz signal that we use to monitor the breath. We did not test our system under the scenario when the tester is running, because all existing works [17, 21] conduct their experiments in a stable scenario. The experiments validate that, even in noisy scenarios, we can get mean estimation errors less than 1 bpm. In general scenarios, the mean estimation error is about 0.5 bpm, which is accurate enough for daily use.

Impact of Different Parameters. Figures 11 and 12 show the reason why we choose the sample frequency of signal strength

TABLE 3: Subjects information.

Person number	Gender	Age	Height (cm)	Weight (kg)
1	Male	23	172	65
2	Male	22	175	50
3	Female	22	160	50

TABLE 4: Estimation errors on different smartphones.

Phone	Mean estimation error (bpm)
Xiaomi MI5	0.37
Samsung Galaxy S4	0.41

TABLE 5: Estimation errors in different testing scenarios.

Scenarios	Mean estimation error (bpm)
Office	0.49
Dormitory	0.44
Library	0.31
Office (with music)	0.78

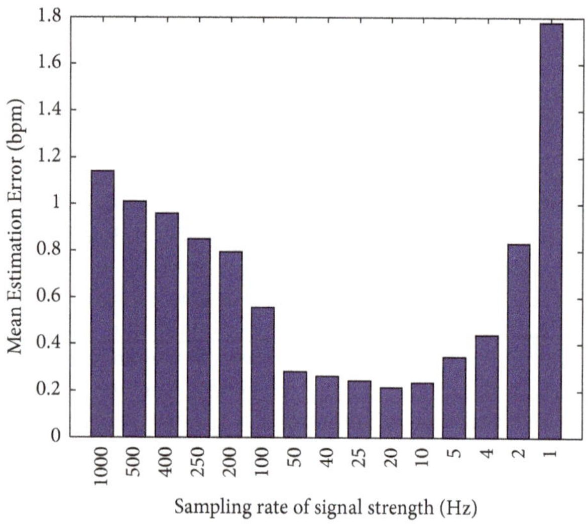

FIGURE 11: Mean estimation error versus sampling rate of signal strength F_p.

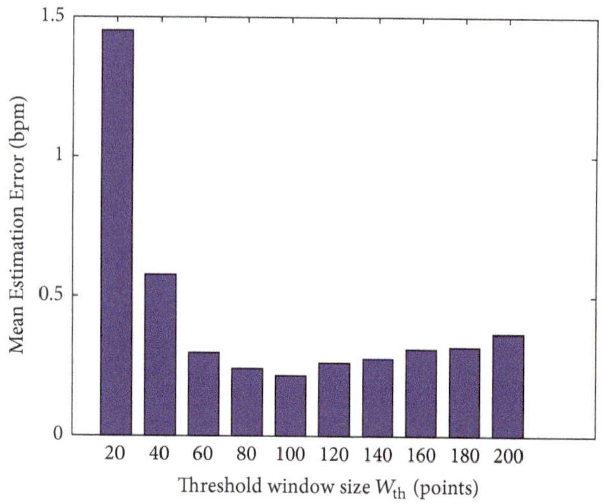

FIGURE 12: Mean estimation error versus threshold window size W_{th}.

F_p equal to 20 Hz and the moving window size W_{th} equal to 100 points in our algorithm. We tried different parameters in our algorithm. For sampling rate F_p, we get the minimum mean estimation error at 20 Hz. Although it is acceptable from 5 Hz to 50 Hz, we choose 20 Hz because the medical device SCHUHFRIED Biofeedback Xpert [26], which we use as our ground truth, also works at a sampling rate of 20 Hz. And for window size W_{th} of dynamic threshold estimation, we get the minimum estimation error at $W_{th} = 100$ points.

5. Conclusion

In this paper, we proposed a smartphone-based respiration rate detection system based on single-frequency ultrasound signals. The proposed system can track the movement of the human chest by observing the signal strength of the recorded ultrasound data. We implemented our system on an Android smartphone and conducted extensive experiments to show the feasibility and accuracy of our system. The results show that this system can achieve accurate respiration rate estimation under various scenarios.

Conflicts of Interest

The authors declare that there are no conflicts of interest regarding the publication of this paper.

Acknowledgments

This work was supported in part by the National Natural Science Foundation of China under Grant no. 61701216, by Guangdong Innovative and Entrepreneurial Research Team Program under contract no. 2016ZT06G587, and by Shenzhen Science, Technology and Innovation Commission Basic Research Project under Grants nos. JCYJ 20160531190935987 and JCYJ 20160531191011045.

References

[1] W. Website, "Who chronic respiratory diseases (crds)," http://www.who.int/respiratory/en/.

[2] T. Young, M. Palta, J. Dempsey, J. Skatrud, S. Weber, and S. Badr, "The occurrence of sleep-disordered breathing among middle-aged adults," *The New England Journal of Medicine*, vol. 328, no. 17, pp. 1230–1235, 1993.

[3] R. B. Berry, R. Brooks, C. E. Gamaldo, S. M. Harding, C. Marcus, and B. Vaughn, "The aasm manual for the scoring of sleep and associated events," in *Rules, Terminology and Technical Specifications*, American Academy of Sleep Medicine, Darien, Ill, USA.

[4] V. respiration monitor belt, Vernier software and technology vernier," http://www.vernier.com/.

[5] O. P. A. nasal mask, Af531 oro-nasal mask philips," http://www.philips.com.hk/healthcare/product/HCNOCTN102/respironics-af531-patient-interface.

[6] H. Abdelnasser, K. A. Harras, and M. Youssef, "Ubibreathe: A ubiquitous non-invasive wifi-based breathing estimator," in *Proceedings of the 16th ACM International Symposium on Mobile Ad Hoc Networking and Computing*, pp. 277–286, ACM, Hangzhou, China, June 2015.

[7] H. Wang, D. Zhang, J. Ma et al., "Human respiration detection with commodity wifi devices," in *Proceedings of the the 2016 ACM International Joint Conference*, pp. 25–36, Heidelberg, Germany, September 2016.

[8] X. Wang, C. Yang, and S. Mao, "PhaseBeat: Exploiting CSI Phase Data for Vital Sign Monitoring with Commodity WiFi Devices," in *Proceedings of the 37th IEEE International Conference on Distributed Computing Systems, ICDCS 2017*, pp. 1230–1239, USA, June 2017.

[9] J. Liu, Y. Wang, Y. Chen, J. Yang, X. Chen, and J. Cheng, "Tracking Vital Signs During Sleep Leveraging Off-the-shelf WiFi," in *Proceedings of the the 16th ACM International Symposium*, pp. 267–276, Hangzhou, China, June 2015.

[10] X. Wang, C. Yang, and S. Mao, "ResBeat: Resilient Breathing Beats Monitoring with Realtime Bimodal CSI Data," in *Proceedings of the 2017 IEEE Global Communications Conference (GLOBECOM 2017)*, pp. 1–6, Singapore, December 2017.

[11] A. Lazaro, D. Girbau, and R. Villarino, "Analysis of vital signs monitoring using an IR-UWB radar," *Progress in Electromagnetics Research*, vol. 100, pp. 265–284, 2010.

[12] J. Salmi and A. F. Molisch, "Propagation parameter estimation, modeling and measurements for ultrawideband mimo radar," *Institute of Electrical and Electronics Engineers. Transactions on Antennas and Propagation*, vol. 59, no. 11, pp. 4257–4267, 2011.

[13] T. Rahman, A. T. Adams, R. V. Ravichandran et al., "DoppleSleep," in *Proceedings of the the 2015 ACM International Joint Conference*, pp. 39–50, Osaka, Japan, September 2015.

[14] S. Suzuki, T. Matsui, H. Imuta et al., "A novel autonomic activation measurement method for stress monitoring: non-contact measurement of heart rate variability using a compact microwave radar," *Medical & biological engineering & computing*, vol. 46, no. 7, pp. 709–714, 2008.

[15] G. G. Senaratne, R. B. Keam, W. L. Sweatman, and G. C. Wake, "Solutions of inverse problems with potential application for breast tumour detection using microwave measurements," *Computational and Mathematical Methods in Medicine. An Interdisciplinary Journal of Mathematical, Theoretical and Clinical Aspects of Medicine*, vol. 8, no. 4, pp. 245–261, 2007.

[16] H. Aly and M. Youssef, "Zephyr demo: Ubiquitous accurate multi-sensor fusion-based respiratory rate estimation using smartphones," in *Proceedings of the IEEE INFOCOM 2016 - IEEE Conference on Computer Communications Workshops (INFOCOM WKSHPS)*, pp. 1-2, San Francisco, CA, USA, April 2016.

[17] Y. Ren, C. Wang, J. Yang, and Y. Chen, "Fine-grained sleep monitoring: Hearing your breathing with smartphones," in *Proceedings of the 34th IEEE Annual Conference on Computer Communications and Networks, IEEE INFOCOM 2015*, pp. 1194–1202, Hong Kong, May 2015.

[18] R.-C. Peng, X.-L. Zhou, W.-H. Lin, and Y.-T. Zhang, "Extraction of heart rate variability from smartphone photoplethysmograms," *Computational and Mathematical Methods in Medicine*, Article ID 516826, 11 pages, 2015.

[19] L. J. Mena, V. G. Felix, R. Ostos et al., "Mobile personal health system for ambulatory blood pressure monitoring," *Computational and Mathematical Methods in Medicine*, vol. 2013, Article ID 598196, 2013.

[20] R. Nandakumar, S. Gollakota, and N. Watson, "Contactless Sleep Apnea Detection on Smartphones," in *Proceedings of the the 13th Annual International Conference*, pp. 45–57, Florence, Italy, May 2015.

[21] X. Wang, R. Huang, and S. Mao, "SonarBeat: Sonar phase for breathing beat monitoring with smartphones," in *Proceedings of the 26th International Conference on Computer Communications and Networks, ICCCN 2017*, Canada, August 2017.

[22] W. Wang, A. X. Liu, and K. Sun, "Device-free gesture tracking using acoustic signals," in *Proceedings of the the 22nd Annual International Conferenceon ON Mobile Computing and Networking*, pp. 82-49, ACM, New York City, New York, October 2016.

[23] W. Ruan, Q. Z. Sheng, L. Yang, T. Gu, P. Xu, and L. Shangguan, "Audiogest: enabling fine-grained hand gesture detection by decoding echo signal," in *Proceedings of the the 2016 ACM International Joint Conference*, pp. 474–485, Heidelberg, Germany, September 2016.

[24] W. Mao, J. He, and L. Qiu, "CAT: High-precision acoustic motion tracking," in *Proceedings of the 22nd Annual International Conference on Mobile Computing and Networking, MobiCom 2016*, pp. 69–81, New York, NY, USA, October 2016.

[25] R. Nandakumar, V. Iyer, D. Tan, and S. Gollakota, "Fingerio: Using active sonar for fine-grained finger tracking," in *Proceedings of the the 2016 CHI Conference*, pp. 1515–1525, Santa Clara, California, USA, May 2016.

[26] S. GmbH, Schuhfried - biofeedback xpert, https://www.schuhfried.com/biofeedback/biofeedback-xpert/.

4

Biomechanical Aspects of Closing Approaches in Postcarotid Endarterectomy

Idit Avrahami, Dafna Raz, and Oranit Bash

Department of Mechanical Engineering and Mechatronics, Ariel University, Ariel, Israel

Correspondence should be addressed to Idit Avrahami; iditavrahami@gmail.com

Guest Editor: Dominique J. Monlezun

The carotid bifurcation tends to develop atherosclerotic stenoses which might interfere with cerebral blood supply. In cases of arterial blockage, the common clinical solution is to remove the plaque via carotid endarterectomy (CEA) surgery. Artery closure after surgery using primary closures along the cutting edge might lead to artery narrowing and restrict blood flow. An alternative approach is patch angioplasty which takes longer time and leads to more during-surgery complications. The present study uses numerical methods with fluid-structure interaction (FSI) to explore and compare the two solutions in terms of hemodynamics and stress and strain fields developed in the artery wall.

1. Introduction

The carotid bifurcation tends to develop atherosclerotic stenosis, which might interfere with cerebral blood supply and can cause coma, hemodynamic disturbance, stroke, and even death. The common clinical solution is to remove the plaque via carotid endarterectomy (CEA) surgery [1], with approximately 100,000 CEAs performed in the United States each year [2–4]. There are few approaches for postsurgery closure [5], and the preferred closure technique still remains an issue of debate [6, 7].

The CEA procedure often associated with intimal hyperplasia [8] or progression of atherosclerosis in the zone of arterial reconstructions [9], attributed to the inflammatory process due to foreign materials in vascular reconstruction [10].

A regular suture (primary closure) is the simplest way to close an arteriotomy, which leads to smaller artery diameter and increased stiffness of the structure. Therefore, routine use of patch graft has been advocated to reduce restenosis, stroke, and death [11, 12]. Patch angioplasty reduces the risk of immediate postoperative complications, results in a larger carotid artery diameter, and significantly lowers vessel restenosis and occlusion rates [13], especially in women [14].

However, the procedure of patch suturing takes longer and thus might increase the risk of stroke and death during surgery. Moreover, a concern is raised about the thrombogenic nature of the conventionally used patches and their protective effect, particularly from late restenosis [15]. Patch angioplasty has also been associated with specific complications, such as patch rupture or expansion and increased risk of infection [16]. In addition, carotid patch was shown to promote irregular neointimal lining with prominent proliferative activity [17].

According to a statistical study by Mannheim et al. [18], a 2-year restenosis-free rate is 97.6% for patch angioplasty vs. 90.9% for primary closure. Nevertheless, overall mortality or morbidity is similar for all closure procedures [13, 19, 20]. Therefore, the use of patch grafts today is targeted selectively only to patients who have very small (<4 mm) or highly constricted and tortuous vessels [7, 20].

There are several types of patches such as prosthetic patches (woven Dacron or PTFE), venous patches, and biomaterial patches (bovine pericardium) [6, 21]. Vein patch is considered the preferred one, thanks to its high compliance and biocompatibility [14]; its main disadvantage is the need for an additional incision to obtain it and occasional deterioration with aneurysmal dilatation and rupture [16]. The advantages of prosthetic patches include immediate availability, avoidance of additional incision, and preservation of vein for future use in other cardiovascular operations. The main shortcomings of the prosthetic patches are higher thrombogenicity, increased risk for infection [22], and higher risks for infections [6]. Bovine pericardium patches recently proved high durability and better long-term survival rates, compared to the other patches [23]. Overall, despite the preference of using biological patches, the clinical results with most available synthetic patches are currently similar enough to prevent clear recommendation of any particular one [6].

Some studies used computational fluid dynamics (CFD) to investigate the blood flow regime in the post-CEA region in patient-specific geometries. Harrison et al. [24] showed that incorporation of a patch indeed increases the artery diameter, but it results with larger areas of low-wall shear stress (WSS) and high-oscillatory shear index (OSI) at the bifurcation, and therefore, its benefit is questionable. Similarly, Guerciotti et al. [25] and Domanin et al. [26] also analyzed WSS, vorticity, time-averaged OSI, and relative residence time (RRT). According to all these studies, cases with primary suture resulted with better hemodynamic parameters and smaller areas of disturbed flow in comparison with patch graft cases. Especially, OSI and RRT values were generally higher in patch graft cases with respect to primary closure, especially for high carotids or when the arteriotomy is mainly at the bulb region.

Although these studies discussed the effect of the closing approach on blood flow hemodynamics to support the clinical decision and to provide a hemodynamic insight into the patch complications, they neglected the biomechanical effect on the arterial tissue and the artificial graft. Excessive stress and cyclic strains are correlated to risk of patch rapture, aneurysms, wall injury, restenosis, irregular neo-intimal lining [27]. Kamenskiy et al. [22] used fluid-structure interaction (FSI) numerical simulations to explore tissue strain and stress in addition to WSS parameters for models with different patch types, widths, and location. In their study, they showed that narrow patches are superior to wide patches, and anterior arteriotomy are superior to lateral arteriotomy.

As far as we know, no study examined the biomechanical aspects of patch angioplasty in comparison to primary suture. In this study, we use FSI numerical models in order to examine the hemodynamics and biomechanical aspects of the patch procedure in comparison with primary suture.

2. Methods

2.1. Cases Studied and Models Geometry. The research study used FSI numerical simulations to explore blood flows and wall dynamics in five different models of the carotid artery region. The examined cases are listed in Table 1.

The geometric specifications of healthy and postsurgery bifurcation models were based on data specified by Tada and Tarbell [28] and Halak et al. [29]. The geometries were idealized to represent a typical model within the framework of the anatomy variance. Geometry dimensions are shown in Figure 1 and detailed in Table 2. The fluid and the structural domains of the model are shown in Figure 2.

Table 1: Cases studied.

Case#1	A healthy carotid bifurcation: a time-dependent simulation of the coupled domains
Case#2	A narrowed postoperation carotid bifurcation with a suture: a time-dependent simulation of the coupled domains
Case#3	A widened carotid bifurcation with a *high* flexibility patch: a time-dependent simulation of the coupled domains
Case#4	A widened carotid bifurcation with a *medium* flexibility patch: a time-dependent simulation of the coupled domains
Case#5	A widened carotid bifurcation with a *low* flexibility patch: a time-dependent simulation of the coupled domains

2.2. Mathematical Model. Blood was assumed homogenous and Newtonian fluid with viscosity of $\mu = 0.0035$ gr/cm·s and density of $\rho = 1.05\,\text{g/cm}^3$. The flow was assumed laminar, and the arterial wall was assumed linearly elastic with an elasticity of $E = 5 \times 10^6\,\text{dyn/cm}^2$ and Poisson's ratio of $\nu = 0.499$ [30]. Small displacement/small-strain formulation was assumed. Patches and suture parameters are listed in Table 3.

The flow and pressure fields in the fluid domain (Ω_f) were calculated by solving the governing equations for the fluid domain for laminar, Newtonian, and incompressible flow in a nongravity field:

$$\left.\begin{aligned} &\nabla \cdot \mathbf{U}_f = 0, \\ &\rho_f\left(\frac{\partial \mathbf{U}_f}{\partial t} + \mathbf{U}_f \cdot \nabla \mathbf{U}_f\right) = -\nabla P + \mu \cdot \nabla^2 \mathbf{U}_f, \end{aligned}\right\} \quad \text{in } \Omega_f, \qquad (1)$$

where P is the static pressure, $\mathbf{U}_f$ is the velocity vector, t is the time, ρ_f is the fluid density, and μ is the dynamic viscosity.

The governing equation for the solid domain (Ω_s) is the Lagrangian momentum conservation equation:

$$\left.\rho_s \frac{\partial^2 \mathbf{d}_s}{\partial t^2} - \nabla \boldsymbol{\sigma}_s = \mathbf{f}\right\} \quad \text{in } \Omega_s, \qquad (2)$$

where $\boldsymbol{\sigma}_s$ is the Cauchy stress tensor, $\mathbf{d}_s$ is the vector of structure displacement, ρ_s is the wall density, and $\mathbf{f}$ represents the body force applied on the structure.

We used artificial boundary conditions (BC) [22, 31] for the fluid domain as follows. Prescribed velocity conditions $\int (\mathbf{U} \cdot \mathbf{n})\, dA = Q(t)$ were imposed at the ICA and the ECA outlets (marked as Γ_f^{ICA} and Γ_f^{ECA} in Figure 2), reflecting a typical physiological waveform [30] of a normal healthy human with a heart rate of 60 bpm, as shown in Figure 3. At the CCA inlet (Γ_f^{in}), stress free conditions were employed, thus the CCA flow was achieved from mass conservation. In addition, a typical physiological time-dependent pressure was imposed, as shown in Figure 4, based on a systolic/diastolic pressure of 120/80 mmHg:

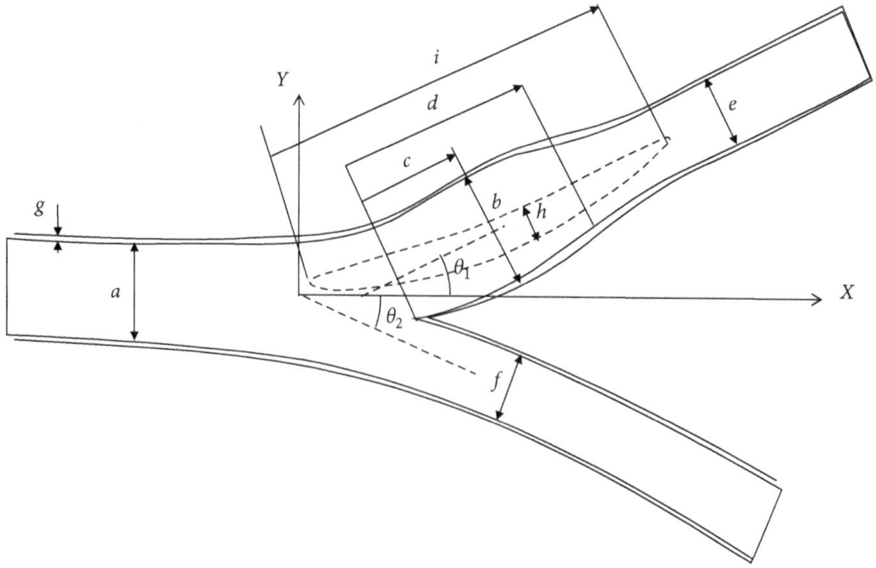

FIGURE 1: Geometry of the artery bifurcation.

TABLE 2: Geometric dimensions of the bifurcation models.

Dimension	Labels in Figure 1	Value (mm)		
		Healthy	Suture	Patch
CCA internal diameter	a	6.6		
Max. bulb internal diameter	b	7.8	7.0	8.4
Max. bulb location	c	7.7		
Bulb length	d	16.0		
ICE internal diameter	e	5.0		
ECA internal diameter	f	4.7		
Wall thickness	g	0.7		
Patch width	h			8.0
Patch length	i			18.0
Patch thickness		0.7	0.7	0.7
ICA bifurcation	θ_1	25°		
ECA bifurcation	θ_2	25°		

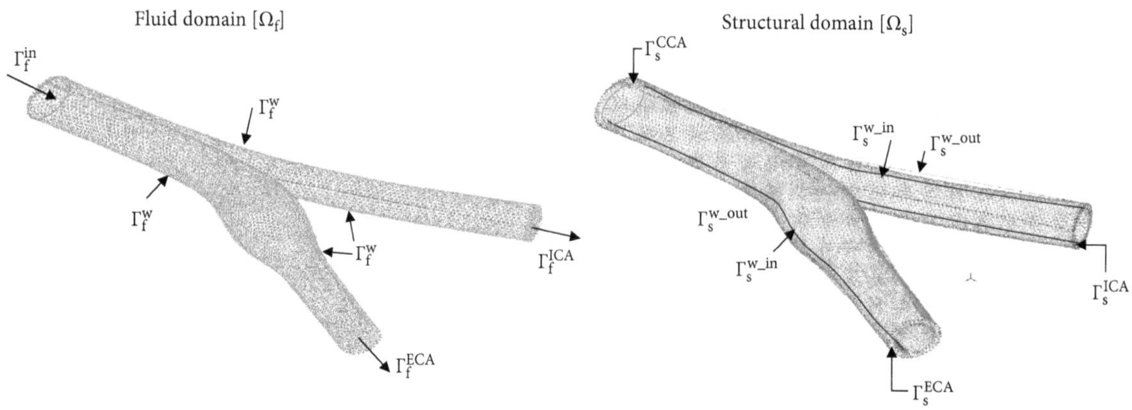

FIGURE 2: The fluid (Ω_f) and Structural (Ω_s) domains and the corresponding boundary conditions (specified in Equations (1)–(3)).

TABLE 3: Material properties.

ν	E (*10^6 dyn/cm^2)	ρ (g/cm^3)	μ (dyn*s/cm^2)	
		1.05	$3.5 * 10^{-2}$	Blood
0.499	5	1.05		Artery
0.499	10	1.05		High flexibility patch
0.499	200	1.05		Medium flexibility patch
0.499	800	1.05		Low flexibility patch
0.37	2000	1.35		Suture

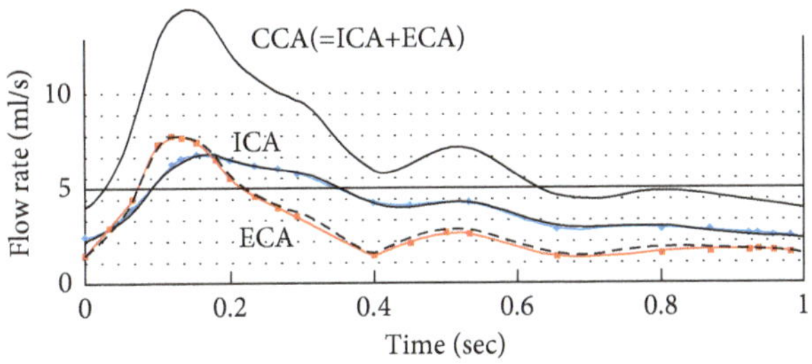

FIGURE 3: Time-dependent flow at the three arteries outlets (CCA, ICA, and ECA) [30].

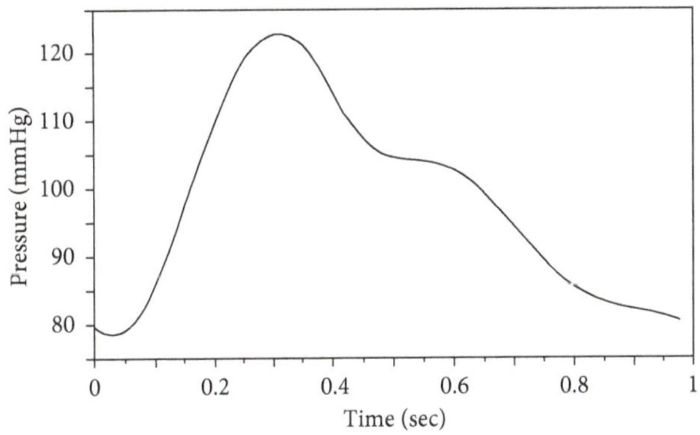

FIGURE 4: Imposed arterial pressure as a function of time [30].

$$\mathbf{n} \cdot \boldsymbol{\tau}_f = 0, \; P = P(t)\} \quad \text{on } \Gamma_f^{in}, \tag{3}$$

where $\mathbf{n} \cdot \boldsymbol{\tau}_f$ are the normal stresses to the surface. BC on the structural domains were fixed ($\dot{\mathbf{d}}_s = 0$) at the edges Γ_s^{CCA}, Γ_s^{ECA}, and Γ_s^{ICA}, and stress-free conditions ($\partial \dot{\mathbf{d}}_s / \partial \mathbf{n} = 0$) were set at the outer faces $\Gamma_s^{w\_out}$.

The BC at the FSI interfaces (Γ_f^w and $\Gamma_s^{w\_in}$) states that (i) displacements of the fluid and solid domain must be compatible, (ii) tractions at these boundaries must be at equilibrium, and (iii) fluid obeys the no-slip\no-penetration conditions. These conditions are given in the following equations:

$$\left.\begin{array}{l} \mathbf{d}_f = \mathbf{d}_s, \; \mathbf{U}_f = \dot{\mathbf{d}}_s, \\ \mathbf{n} \cdot \boldsymbol{\tau}_f = \mathbf{n} \cdot \boldsymbol{\sigma}_s, \end{array}\right\} \quad \text{on } \Gamma_f^w \text{ and } \Gamma_s^{w\_in}, \tag{4}$$

where $\boldsymbol{\sigma}_s$, $\boldsymbol{\tau}_f$, $\mathbf{d}_s$, and $\mathbf{d}_f$ are the wall structure, fluid stress tensors, and wall displacement, respectively.

2.3. Numerical Model. The simulations used the commercial package ADINA (ADINA R&D, Inc., v. 9.0.0) to solve numerically the governing differential equations (Equations (1) and (2)) using the finite elements method (FEM).

The fluid domain (Ω_f) was meshed using 3D 1st order tetrahedral elements, and the structural domain was meshed using triangular 3D 2nd order tetrahedral elements. Several mesh and time-step independence tests were conducted, where the simulations results obtained from several numerical models with different mesh and time steps were compared (Appendix). Based on these tests, the numerical model selected was with ~100,000 fluid elements and ~20,000 structural elements. Each cardiac cycle consisted of 28 time steps of 0.035 sec, and the automatic-time-stepping procedure was used to subdivide the load-step increment when necessary. Three cardiac cycles were computed to obtain results independent of the initial conditions. The results of the third calculated cycle were fully periodic. Convergence is achieved when all mass, velocity component, and energy changes, from iteration to iteration, achieved are less than 10^{-5} root-mean-square error (RMSE).

The three mesh models (of the healthy, suture, and patch) are shown in Figure 5. The suture or the patch elements (marked in red in the figure) were fully connected to the

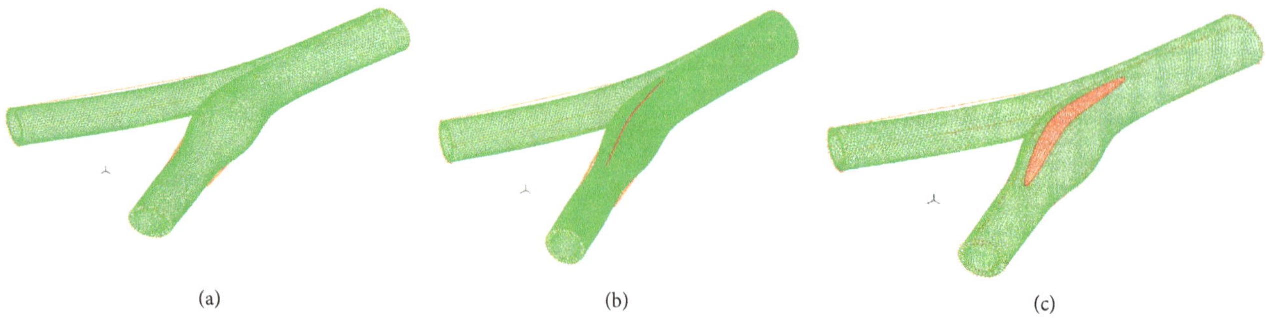

FIGURE 5: Mesh models of the healthy artery (a), suture (b), and patch (c).

arterial elements (marked in green). In the suture model, 817 elements were defined as suture, and in the patch models, 5790 elements were defined as patches. The suture or the patch elements were defined as fully connected to the arterial elements.

The ADINA iterative solver was used for the FSI coupling [32]. The fluid and the structure solvers were solved iteratively until convergence was reached or until it reached 150 iterations. The FSI algorithm included models of linear wall displacements and strains. In order to control the moving mesh under deformations of the flow domains, the arbitrary Lagrangian-Eulerian (ALE) approach was defined on geometric entities. The ALE approach integrates the Eulerian description of the fluid domain with the Lagrangian formulation of the moving mesh using curvature correction [32, 33]. The Newton–Raphson method was used to solve the nodal matrices [34, 35]. A first-order Euler backward implicit time integration method was used for the time marching.

2.4. Examined Parameters. The simulations examined several hemodynamic and biomechanical parameters in order to address the effect of the patch or primary suture in the carotid bifurcation, including effective stress, flow patterns, time-averaged WSS (TAWSS), and OSI. These parameters are known as critical factors in artery occlusion and thrombosis [36].

Wall stresses of the arterial tissue and artificial graft were calculated as the product of blood viscosity and the local velocity gradient in the direction of local surface normal ($\mathbf{n}$):

$$\boldsymbol{\tau}_{\mathrm{w}} = \mu \frac{\partial \mathbf{u}}{\partial \mathbf{n}}. \tag{5}$$

TAWSS and OSI were calculated according to Equations (6) and (7):

$$\mathrm{TAWSS} = \frac{1}{T} \int_0^T \left| \boldsymbol{\tau}_{\mathrm{w}} \right| dt, \tag{6}$$

$$\mathrm{OSI} = 0.5 \left(1 - \frac{\int_0^T \boldsymbol{\tau}_{\mathrm{w}}\, dt}{\int_0^T \left| \boldsymbol{\tau}_{\mathrm{w}} \right| dt} \right), \tag{7}$$

where $\boldsymbol{\tau}_{\mathrm{w}}$ represents the instantaneous WSS vector and T represents the period of the cardiac cycle.

3. Results

3.1. Structural Results. Effective stresses at the wall were calculated at each point according to von Mises criteria. In all cases, elevated effective stresses were found at the bifurcation origin and bifurcation junction (Figures 6 and 7). In the suture and patches with medium and low flexibility cases, there were higher effective stresses (with values above 70 kPA) along the sutures. For the case with high flexibility patch, stress distribution resembled the values of healthy case.

4. Results of Fluid

Figure 8 presents midplane velocity vectors in a magnified view of the bulb for the five models after peak flow at the time $t = 0.385$ sec. In all five cases, a recirculation zone appears in the outer side of the bifurcation bulb. In the healthy case, the recirculation zone is small and most of the flow is unidirectional. In the model of the primary suture, the smaller diameter leads to a sustained velocity stream leaving room to a narrow vortex with less recirculation. In the patch model, although the diameter bulb is wide and the flow has plenty of room, the larger diameter promotes a large vortex that takes over almost the whole bulb space, which interferes with the axial flow stream.

Figure 9 presents time-averaged wall shear stress (TAWSS), and Figure 10 presents the oscillatory shear index (OSI) distribution in the five models. Lower TAWSS and higher OSI values were present mostly in the bulb area, and low values of TAWSS and high values of OSI were rarely found in the primary suture case (in respect with the other cases) because of the smaller diameter in the bulb area.

5. Discussion

High-concentration stresses are found in the suture and low-flexibility patch cases (Figures 6 and 7). The stress concentrations are a result of the discontinuity of the material property of the patch and suture from the arterial wall. High-effective stress values are also found at the bifurcation junction. The values (in the range of 0.5–70 kPa) are in agreement with values reported by Kamenskiy et al. [22]. Concentration of stresses on the artificial graft might imply a higher risk of rupture. High stresses at the arterial wall might lead to atherosclerosis and neointima growth [37–39].

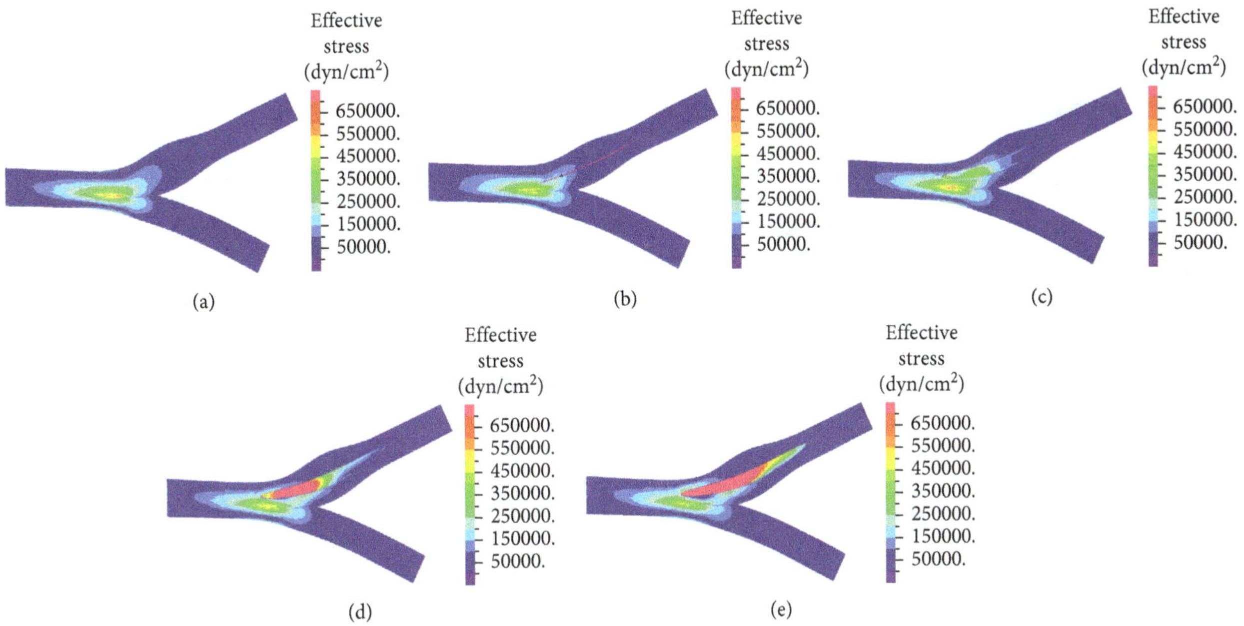

FIGURE 6: Effective stresses—top view in healthy artery (a), suture (b), high flexibility patch (c), medium flexibility patch (d), and low flexibility patch (e).

When there is stress concentration over a larger area (like in the low flexibility patch), the chances of developing restenosis are larger. The high flexibility patch resulted in lower stresses (>35 kPa), both at the artificial graft and the arterial wall. Therefore, in specific cases, when it is decided to prefer a patch over a suture in patients with relatively small arteries, high flexibility patch (such as vein or bovine patch) should result in lower risk for stress-induced restenosis because its properties resemble those of the artery. However, they have a larger risk of forming an aneurysm. Ultimately, the primary suture exhibits effective stress distribution in the artery similar to the healthy artery, indicating that, from our study, the primary suture has less potential effect on the arterial tissue.

From examination of flow patterns in the different models (Figure 8), we conclude that some vortical flow patterns are found in all cases but in different sizes. In a primary suture model, the flow in the bulb is mostly unidirectional and only a negligible vortex is found. In the patch and in the healthy models, the large diameter bulb promotes a large vortex that dominants the flow. According to Gimbrone et al. [40], vortices and disruptions in the flow are factors that cause a decrease in endothelial function and eventually might increase risk for stenosis or other vascular diseases. In addition, these vortices might disturb the flow and lead to thrombus formation and thus increase the risk for stroke.

The results of TAWSS (Figure 9) and OSI (Figure 10) reveal that WSS values in the bulb region are similar to the healthy case and the patches cases, while in the suture case, the values are higher. It can be concluded that bulb diameter is the cause for variations in WSS. In regions with larger diameter, the velocities are smaller, and therefore, WSS are smaller and OSI are higher. Low WSS and high OSI are correlated with plaque deposition, artery occlusion, and endothelium dysfunction [41]. Therefore, from our study, the suture case which has higher WSS and lower OSI is also preferred by this parameter. These results agree with Domanin et al. [26] that showed higher values of OSI and RRT in patch graft vs. direct suture cases in patient-specific simulations, and with Kamenskiy et al. [22] that showed that artery with a narrow patch showed significant improvement in hemodynamics in comparison with wider patches.

The influence of fluid shear forces on structure dynamics is relatively small. WSS is negligible (<70 dyn/cm^2) compared to the effected stresses due to hydrostatic pressure (<70,000 dyn/cm^2). Therefore, when a simplified simulation is needed for estimation of local effective stresses, it may be valid to consider the structural domain separately from the fluid domain.

The study assumes a relatively simplified model of a specific anatomy, with linearly elastic material and Newtonian fluid, and does not consider possible physiological or anatomic variations between healthy and post-CEA [29]. The actual values are highly geometry-dependent and therefore, patient-specific. Moreover, in reality, during the CEA procedure, a part of the media layer is removed, with only adventitia layer remaining. This might further increase the effective stress in the bifurcation wall in relation with the healthy case and increase the risk for rapture or aneurysm. In addition, the results and conclusions discuss only the mechanical aspects and did not take the clinical aspect as argument.

Another limitation relates to the time and space discretization parameters. Note that a time-step interval of 0.035 sec is relatively elevated, in respect of systolic interval (of 0.4 sec), and may limit capturing the systolic fluid dynamics. To minimize this limitation, an automatic-time-stepping

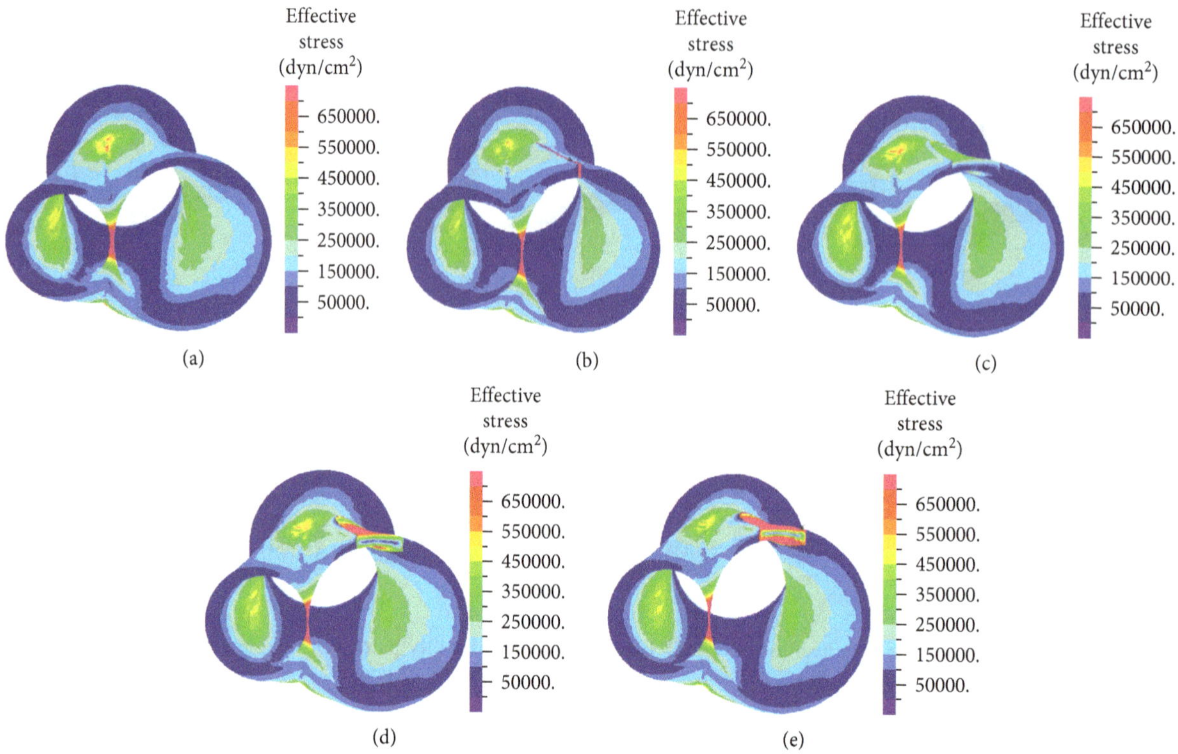

FIGURE 7: Effective stresses—cut view in healthy artery (a), suture (b), high flexibility patch (c), medium flexibility patch (d), and low flexibility patch (e).

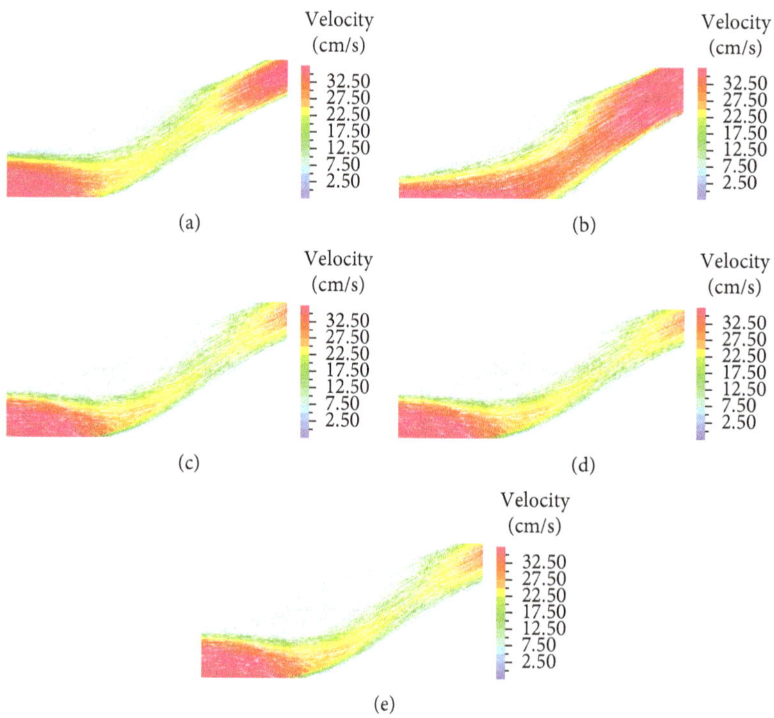

FIGURE 8: Velocity vectors—magnified view in healthy artery (a), suture (b), high flexibility patch (c), medium flexibility patch (d), and low flexibility patch (e).

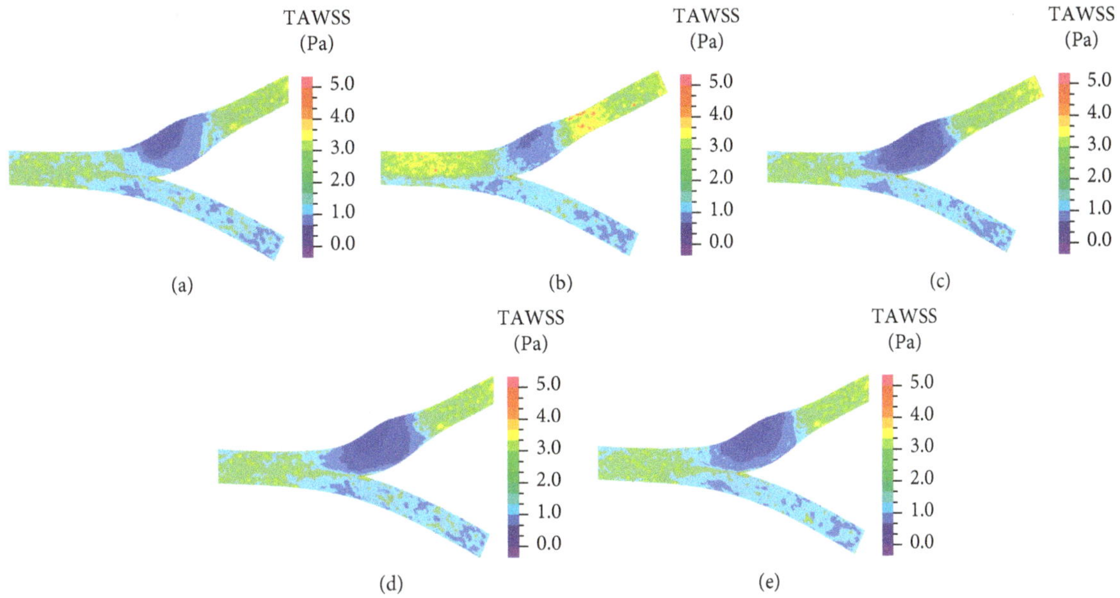

FIGURE 9: Time-averaged WSS in healthy artery (a), suture (b), high flexibility patch (c), medium flexibility patch (d), and low flexibility patch (e).

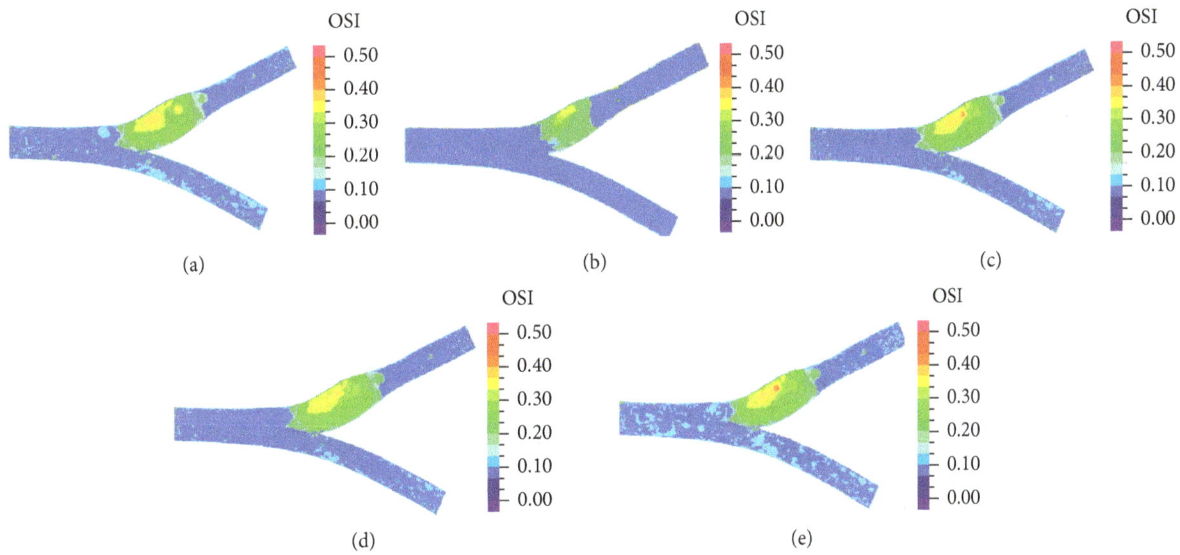

FIGURE 10: OSI in healthy artery (a), suture (b), high flexibility patch (c), medium flexibility patch (d), and low flexibility patch (e).

procedure was used to subdivide the load-step increment when necessary. According to our mesh and time discretization refinement studies (Appendix), the discretization errors in the calculation of WSS are up to 10%.

However, the model results are in agreement with clinical and previous studies [25, 26, 42], and the comparative results between the different models are clear and distinct. Most guidelines suggest to prefer primary suture for narrow ICA (>4 mm), and the current model assumes ICA with 5 mm, showing preferred performance with primary patch. This study may delineate the dominant parameters affecting the combined hemodynamic and biomechanics of the patches versus suture approach. Thus, its results may shed a light on the controversy between physicians regarding the preferred approach and explain the reason for the nonsignificant advantage of the patch procedure in CEA.

6. Conclusion

In this manuscript, we examined the biomechanical aspects of patch angioplasty in comparison with primary suture. The examined parameters included elevated stress, in addition to the previously examined hemodynamic (WSS) parameters. Based on our results of elevated stress and OSI values and low TAWSS values, primary suture has shown better performance in our study than patch, and the high flexibility patch has shown better performance compared to lower flexibility patch.

Conflicts of Interest

The authors declare that they have no conflicts of interest.

Acknowledgments

The research was partly funded by the Ariel Research Authority.

Appendix

A. Model Validation

A.1. Time-Step Independence Tests. Mesh and time-step independence tests were conducted to validate the numerical model. To evaluate the optimal time-step size for the transient simulations, FSI simulations of the healthy case with the mesh of 100,000 elements were performed using 28 steps of 0.035 sec and 98 steps of 0.01 sec. The resulted maximal values in the domains for the two cases are detailed in Table 4.

To evaluate the discretization error, we calculated the relative difference (ERR) between the maximal value in the course and the fine time resolution, as follows:

$$\mathrm{ERR}_i\,(\%) = \left(\frac{(X)_{\mathrm{fine}} - (X)_{\mathrm{course}}}{(X)_{\mathrm{fine}}}\right) * 100\,(\%), \qquad (\mathrm{A.1})$$

where X is the maximal value in the domain (velocity and shear stress in the fluid domain and effective stress in the structural domain). The resulted errors are detailed in the table. The results show that the time step of $dt = 0.035$ sec is sufficient, with ERR < 2% in the three critical parameters. Therefore, for the transient analyses in this study, 28 steps were set per cycle.

Note that only maximum values were examined (in space and time), thus instantaneous differences at a given instance (e.g., systole) were not examined. Note that this time-step interval is relatively elevated, in respect of systolic interval (of 0.4 sec), and may limit capturing the systolic fluid dynamics. To minimize this limitation, an automatic-time-stepping procedure was used to subdivide the load-step increment when necessary.

A.2. Mesh Independence Tests. To evaluate the optimal mesh resolution, three models of the healthy base case with different mesh resolutions were built (with 50,000–500,000 elements). The models were simulated during a period of one cardiac cycle, each with 28 time steps. The resulted maximal values in the domains of each model mesh are listed in Table 5, together with their error relative to the finest mesh (Equation (A.1)). Based on these results, mesh resolutions of 100,000 elements were found suitable for our model with ERR ≤ 10% of the finest mesh.

TABLE 4: Resulted values with different time steps (error in % is indicated in parentheses).

Time step (sec)	0.035	0.01
Max. velocity magnitude (cm/s)	58.5 (1.8%)	59.5
Max. shear stress (dyn/cm^2)	122.1 (1.7%)	124.2
Max. effective stress (dyn/cm^2)	2.12×10^7(1.8%)	2.15×10^7

TABLE 5: Maximum values for different mesh resolutions (error in % is indicated in parentheses).

No. of elements	51,544	100,268	484,401
Max. velocity magnitude (cm/s)	57.5 (5.7%)	58.5 (2.7%)	60.1
Max. shear stress (dyn/cm^2)	94.7 (28.1%)	122.1 (3.5%)	126.4
Max. effective stress (dyn/cm^2)	1.64×10^7 (28.9%)	2.12×10^7 (8.5%)	2.21×10^7

References

[1] N. A. S. C. E. T. Collaborators*, "Beneficial effect of carotid endarterectomy in symptomatic patients with high-grade carotid stenosis," *New England Journal of Medicine*, vol. 325, no. 7, pp. 445–453, 1991.

[2] J. H. Lichtman, M. R. Jones, E. C. Leifheit et al., "Carotid endarterectomy and carotid artery stenting in the US Medicare population, 1999-2014," *JAMA*, vol. 318, no. 11, pp. 1035–1046, 2017.

[3] D. C. Hsia, L. M. Moscoe, and W. M. Krushat, "Epidemiology of carotid endarterectomy among Medicare beneficiaries: 1985–1996 update," *Stroke*, vol. 29, no. 2, pp. 346–350, 1998.

[4] R. W. Hobson, W. C. Mackey, E. Ascher et al., "Management of atherosclerotic carotid artery disease: clinical practice guidelines of the Society for Vascular Surgery," *Journal of Vascular Surgery*, vol. 48, no. 2, pp. 480–486, 2008.

[5] K. I. Paraskevas, V. Robertson, A. N. Saratzis, and A. R. Naylor, "An updated systematic review and meta-analysis of outcomes following eversion vs. conventional carotid endarterectomy in randomised controlled trials and observational studies," *European Journal of Vascular and Endovascular Surgery*, vol. 67, no. 5, p. 1632, 2018.

[6] D. Treat-Jacobson, K. Fitzgerald, A. Foley, C. AGACNP, and D. Kohlman-Trigoboff, "Society for Vascular Nursing Carotid Endarterectomy (CEA) Updated Nursing Clinical Practice Guideline," *Journal of Vascular Nursing*, vol. 35, no. 2, pp. 90–111, 2017.

[7] J. Byrne, P. Feustel, and R. C. Darling, "Primary closure, routine patching, and eversion endarterectomy: what is the current state of the literature supporting use of these techniques?," *Seminars in Vascular Surgery*, vol. 20, no. 4, pp. 226–235, 2007.

[8] A. Szabó, E. Brazda, E. Dósa, A. Apor, Z. Szabolcs, and L. Entz, "Long-term restenosis rate of eversion endarterectomy on the internal carotid artery," *European Journal of Vascular and Endovascular Surgery*, vol. 27, no. 5, pp. 537–539, 2004.

[9] G. P. Clagett, M. Robinowitz, J. R. Youkey et al., "Morphogenesis and clinicopathologic characteristics of recurrent carotid disease," *Journal of Vascular Surgery*, vol. 3, no. 1, pp. 10–23, 1986.

[10] C. Lattimer and K. Burnand, "Recurrent carotid stenosis after carotid endarterectomy," *British Journal of Surgery*, vol. 84, no. 9, pp. 1206–1219, 1997.

[11] R. Bond, K. Rerkasem, A. Naylor, A. Aburahma, and P. Rothwell, "Systematic review of randomized controlled trials of patch angioplasty versus primary closure and different types of patch materials during carotid endarterectomy," *Journal of Vascular Surgery*, vol. 40, no. 6, pp. 1126–1135, 2004.

[12] K. J. Ho, L. L. Nguyen, and M. T. Menard, "Intermediate-term outcome of carotid endarterectomy with bovine pericardial patch closure compared with Dacron patch and primary closure," *Journal of Vascular Surgery*, vol. 55, no. 3, pp. 708–714, 2012.

[13] V. Maertens, H. Maertens, M. Kint, C. Coucke, and Y. Blomme, "Complication rate after carotid endarterectomy comparing patch angioplasty and primary closure," *Annals of Vascular Surgery*, vol. 30, pp. 248–252, 2016.

[14] A. F. AbuRahma, P. A. Robinson, S. Saiedy, J. H. Khan, and J. P. Boland, "Prospective randomized trial of carotid endarterectomy with primary closure and patch angioplasty with saphenous vein, jugular vein, and polytetrafluoroethylene: long-term follow-up," *Journal of Vascular Surgery*, vol. 27, no. 2, pp. 222–234, 1998.

[15] A. F. AbuRahma, P. A. Robinson, R. S. Hannay, J. Hudson, and L. Cutlip, "Prospective controlled study of carotid endarterectomy with Hemashield patch: is it thrombogenic?," *Vascular and Endovascular Surgery*, vol. 35, no. 3, pp. 167–174, 2001.

[16] J. P. Archie, "Carotid endarterectomy saphenous vein patch rupture revisited: selective use on the basis of vein diameter," *Journal of Vascular Surgery*, vol. 24, no. 3, pp. 346–352, 1996.

[17] M. Ito, Y. Niiya, H. Uchino, N. Nakayama, S. Mabuchi, and K. Houkin, "Irregular neointimal lining with prominent proliferative activity after carotid patch angioplasty: an autopsy case report," *World Neurosurgery*, vol. 82, no. 1–2, pp. 240.e1–240.e6, 2014.

[18] D. Mannheim, B. Weller, E. Vahadim, and R. Karmeli, "Carotid endarterectomy with a polyurethane patch versus primary closure: a prospective randomized study," *Journal of Vascular Surgery*, vol. 41, no. 3, pp. 403–407, 2005.

[19] P. G. Al-Rawi, C. L. Turner, V. Waran, I. Ng, and P. J. Kirkpatrick, "A randomized trial of synthetic patch versus direct primary closure in carotid endarterectomy," *Neurosurgery*, vol. 59, no. 4, pp. 822–829, 2006.

[20] K. Rerkasem and P. M. Rothwell, "Patch angioplasty versus primary closure for carotid endarterectomy," *Cochrane Database of Systematic Reviews*, vol. 7, no. 4, 2009.

[21] A. Muto, T. Nishibe, H. Dardik, and A. Dardik, "Patches for carotid artery endarterectomy: current materials and prospects," *Journal of Vascular Surgery*, vol. 50, no. 1, pp. 206–213, 2009.

[22] A. V. Kamenskiy, J. N. MacTaggart, I. I. Pipinos, P. K. Gupta, and Y. A. Dzenis, "Hemodynamically motivated choice of patch angioplasty for the performance of carotid endarterectomy," *Annals of Biomedical Engineering*, vol. 41, no. 2, pp. 263–278, 2013.

[23] W. A. Oldenburg, T. Almerey, M. Selim, H. Farres, and A. G. Hakaim, "Durability of carotid endarterectomy with bovine pericardial patch," *Annals of Vascular Surgery*, vol. 50, pp. 218–224, 2018.

[24] G. J. Harrison, T. V. How, R. J. Poole et al., "Closure technique after carotid endarterectomy influences local hemodynamics," *Journal of Vascular Surgery*, vol. 60, no. 2, pp. 418–427, 2014.

[25] B. Guerciotti, C. Vergara, L. Azzimonti et al., "Computational study of the fluid-dynamics in carotids before and after endarterectomy," *Journal of Biomechanics*, vol. 49, no. 1, pp. 26–38, 2016.

[26] M. Domanin, D. Bissacco, D. Le Van, and C. Vergara, "Computational fluid dynamic comparison between patch-based and primary closure techniques after carotid endarterectomy," *Journal of Vascular Surgery*, vol. 67, no. 3, pp. 887–897, 2018.

[27] J. Ando and K. Yamamoto, "Effects of shear stress and stretch on endothelial function," *Antioxidants & Redox Signaling*, vol. 15, no. 5, pp. 1389–1403, 2011.

[28] S. Tada and J. M. Tarbell, "Oxygen mass transport in a compliant carotid bifurcation model," *Annals of Biomedical Engineering*, vol. 34, no. 9, pp. 1389–1399, 2006.

[29] M. Halak, S. Fajer, H. Ben-Meir, Z. Loberman, B. Weller, and R. Karmeli, "Neck irradiation: a risk factor for occlusive carotid artery disease," *European Journal of Vascular and Endovascular Surgery*, vol. 23, no. 4, pp. 299–302, 2002.

[30] S. Tada and J. Tarbell, "A computational study of flow in a compliant carotid bifurcation–stress phase angle correlation with shear stress," *Annals of Biomedical Engineering*, vol. 33, no. 9, pp. 1202–1212, 2005.

[31] L. Formaggia, A. Quarteroni, and A. Veneziani, *Cardiovascular Mathematics: Modeling and Simulation of the Circulatory System*, Vol. 1, Springer Science & Business Media, Berlin, Germany, 2010.

[32] I. ADINA R&D, *ADINA Theory and Modeling Guide*, ADINA, Watertown, MA, USA, 2000.

[33] K.-J. Bathe, *Finite Element Procedures*, Prentice Hall, Upper Saddle River, NJ, USA, 2006.

[34] H. Zhang and K.-J. Bathe, *Direct and Iterative Computing of Fluid Flows Fully Coupled with Structures*, Computational Fluid and Solid Mechanics, 2001.

[35] M. G. Doyle, S. Tavoularis, and Y. Bourgault, "Application of Parallel Processing to the Simulation of Heart Mechanics," in *High Performance Computing Systems and Applications*, Springer, Berlin, Germany, 2010.

[36] T. Meirson, E. Orion, and I. Avrahami, "Numerical analysis ofvenous external scaffolding technology for saphenous vein grafts," *Journal of Biomechanics*, vol. 48, no. 10, pp. 2090-2095, 2015.

[37] M. J. Thubrikar and F. Robicsek, "Pressure-induced arterial wall stress and atherosclerosis," *Annals of Thoracic Surgery*, vol. 59, no. 6, pp. 1594–1603, 1995.

[38] R. S. Salzar, M. J. Thubrikar, and R. T. Eppink, "Pressure-induced mechanical stress in the carotid artery bifurcation: a possible correlation to atherosclerosis," *Journal of Biomechanics*, vol. 28, no. 11, pp. 1333–1340, 1995.

[39] M. J. Thubrikar, S. K. Roskelley, and R. T. Eppink, "Study of stress concentration in the walls of the bovine coronary arterial branch," *Journal of Biomechanics*, vol. 23, no. 1, pp. 1519–1726, 1990.

[40] M. A. Gimbrone, J. N. Topper, T. Nagel, K. R. Anderson, and G. Garcia-cardeña, "Endothelial dysfunction, hemodynamic forces, and atherogenesis," *Annals of the New York Academy of Sciences*, vol. 902, no. 1, pp. 230–240, 2000.

[41] C. Caro, J. Fitz-Gerald, and R. Schroter, "Atheroma and arterial wall shear observation, correlation and proposal of a shear dependent mass transfer mechanism for atherogenesis," in *Proceedings of the Royal Society of London. Series B. Biological Sciences*, vol. 177, 1046, pp. 109–133, 1971.

[42] M. Domanin, A. Buora, F. Scardulla et al., "Computational fluid-dynamic analysis after carotid endarterectomy: patch graft versus direct suture closure," *Annals of Vascular Surgery*, vol. 44, pp. 325–335, 2017.

5

Unsupervised Scoliosis Diagnosis via a Joint Recognition Method with Multifeature Descriptors and Centroids Extraction

Liyuan Zhang, Jiashi Zhao, Huamin Yang, Zhengang Jiang, and Qingliang Li

School of Computer Science and Technology, Changchun University of Science and Technology, No. 7089, Weixing Road, Changchun, China

Correspondence should be addressed to Jiashi Zhao; zhaojiashi@cust.edu.cn and Huamin Yang; yhm@cust.edu.cn

Academic Editor: David A. Winkler

To solve the problem of scoliosis recognition without a labeled dataset, an unsupervised method is proposed by combining the cascade gentle AdaBoost (CGAdaBoost) classifier and distance regularized level set evolution (DRLSE). The main idea of the proposed method is to establish the relationship between individual vertebrae and the whole spine with vertebral centroids. Scoliosis recognition can be transferred into automatic vertebral detection and segmentation processes, which can avoid the manual data-labeling processing. In the CGAdaBoost classifier, diversified vertebrae images and multifeature descriptors are considered to generate more discriminative features, thus improving the vertebral detection accuracy. After that, the detected bounding box represents an appropriate initial contour of DRLSE to make the vertebral segmentation more accurate. It is helpful for the elimination of initialization sensitivity and quick convergence of vertebra boundaries. Meanwhile, vertebral centroids are extracted to connect the whole spine, thereby describing the spinal curvature. Different parts of the spine are determined as abnormal or normal in accordance with medical prior knowledge. The experimental results demonstrate that the proposed method cannot only effectively identify scoliosis with unlabeled spine CT images but also have superiority against other state-of-the-art methods.

1. Introduction

Scoliosis is a common spinal abnormality, and it seriously endangers the people's health [1]. Scoliosis recognition is an important premise for preventing the spinal curve from getting worse. Computer-aided diagnosis (CAD) [2] has been a powerful tool to identity scoliosis by analyzing medical imaging. In scoliosis recognition, scoliosis curvature [3] is the most valuable spinal parameter, which can provide the decision value of the normal or abnormal condition. Due to unclear boundaries and degenerative disorders of CT spine images, it is difficult to extract effective scoliosis features from available diagnostic images. A specially designed computer-aided method for accurate scoliosis recognition has a significant research meaningful.

Until recently, many methods have been developed for diagnosing the scoliotic deformity [4–6]. Zhang et al. [7] proposed a semiautomatic scoliosis measurement method to reduce the assessment variability. Hough transform and snake model are integrated together with a shape prior, thus improving the performance and reliability. Little user judgments are still needed. Zukić et al. [8] adopted the Viola–Jones algorithm with candidate filtering to identify scoliosis. Geometric diagnostic features are deduced by detecting vertebral centers. This method extracts the key pathological features, which can further increase robustness of the algorithm, but the optimal parameter largely depends on a manually segmented dataset. Korez et al. [9] presented an automated vertebral detection and segmentation framework with interpolation theory and shape-constrained deformable model. All local optima that correspond to candidate vertebral locations are detected, thereby preserving the vertebral shape. Nonetheless, when severe disorders occur in the CT spine image, the segmentation accuracy will be decreased. Pinheiro et al. [10] proposed a novel computerized methodology with genetic algorithm optimization to evaluate the scoliotic deformity. The ellipse that best fits to the spine curve is introduced. This method can reproduce scoliotic curvatures using the geometric parameters of the underlying ellipses.

To intelligently diagnose scoliosis using machine learning, Glocker et al. [11] developed an approach combining the supervised classification forest and dense probabilistic centroid estimation. Pathological vertebrae images are considered into the discriminative centroid classier. Consequently, the minimum of an estimated centroid location error of 4.4 mm and the detection rate of 86% are obtained. In [12], the authors developed a computer-aided Cobb angle method using deep neural network (DNN) for scoliosis assessment. Enough vertebral patches are needed in the training dataset to automatically determine the slope of the vertebrae. This solution shows a promising result owing to more training data. In [13], a novel BoostNet architecture was designed to estimate vertebral landmarks for adolescent idiopathic scoliosis assessment. Convolutional neural network with statistical theory is used as robust feature extraction. The effectiveness of BoostNet is verified on plenty of spinal X-ray images. In practice, a large amount of high-quality labeled data are crucial to build a better classifier, while the data-labeling processing is very expensive and time-consuming. Additionally, the neighboring vertebrae have similar morphological appearance, which makes them difficult to be distinguished. Therefore, the feature mining of unlabeled CT images is a significant challenge in the scoliosis recognition without manual annotation.

To address this problem, this paper proposes a scoliosis recognition method in the unsupervised setting with unlabeled CT images. The cascade gentle AdaBoost (CGAdaBoost) classifier with multifeature descriptors and distance regularized level set evolution (DRLSE) model are combined into the centroids method. The main contributions of this paper are presented as follows. First, the relationship between individual vertebrae and the whole spine is established using vertebral centroids, which is beneficial to reduce the data-labeling burden on medical staff. Second, three different descriptors are fully combined to achieve more effective features for the CGAdaBoost classifier. Moreover, detected bounding boxes are used as an initial contour of DRLSE to segment vertebral bodies without manual interaction. Our work can provide a feasible and effective scoliosis recognition method for medical intelligence diagnosis.

The remainder of this paper is organized as follows. In Section 2, the gentle AdaBoost classifier and edge-based level set method are briefly introduced. Section 3 describes the proposed method in detail. Section 4 focuses on experimental results and discussion, followed by the conclusion and future work in Section 5.

2. Related Basic Knowledge

In general, the detection and segmentation of the object are important processes of the recognition task. To achieve a reliable detection result, learning-based technique has been extensively adopted to detect the vertebrae [6, 14–17]. Besides, the vertebral body segmentation exerts a tremendous influence on the extracted centroid result. Level set model is an effective contour evolution method using the image gradient to converge on the object boundary.

2.1. Gentle AdaBoost Classifier. Gentle AdaBoost [17] is a type of probability detector with strong robustness. In the aspect of detection accuracy, gentle AdaBoost outperforms discrete and real AdaBoost classifiers owing to a unique weighting update way. Furthermore, a small number of features are only required in the gentle AdaBoost classifier. There is a lower computational complexity. Using the small decision tree as weak classifiers is to improve the generalization ability. The training process of the gentle AdaBoost classifier is described as follows:

Step 1. There is a given training dataset $S = \{(x_1, y_1), \ldots, (x_n, y_n)\}$. $y_i = 0$ and $y_i = 1$ represent negative and positive samples, respectively.

Step 2. Initialize the weight w_i of the training sample, that is, the initial probability distribution of the sample i.

$$w_i = \begin{cases} \dfrac{1}{2m}, & y_i = 0, \\ \dfrac{1}{2l}, & y_i = 1, \end{cases} \tag{1}$$

where m and l denote numbers of negative and positive samples, respectively.

Step 3. Perform t-stage ($t = 1, \ldots, T$) training on the classifier. T is the maximum iteration number. Set $w_{t,i} \longleftarrow w_{t,i}/\sum_{j=1}^{n} w_{t,j}$, and renormalize weight such that w_t is a probability distribution. For each feature j, the weak classifier h_j is trained according to the weight distribution w_t. The minimum error ε_t is got by constructing the error ε_j with weight w_t.

$$\varepsilon_j = \sum_i w_i \mid h_j(x_i - y_i). \tag{2}$$

Step 4. Update the weight $w_{t+1,i} \longleftarrow w_{t,i}\beta_t^{1-e_i}$ and $\beta_t = \varepsilon_t/(1-\varepsilon_t)$. If x_i is classified correctly, e_i will be equal to zero. Otherwise, e_i is equal to one.

Step 5. The final strong classifier is output as follows:

$$h(x) = \begin{cases} 1, & \sum_{t=1}^{T} \alpha_t h_t(x) \geq \dfrac{1}{2} \sum_{t=1}^{T} \alpha_t, \\ 0, & \text{otherwise}, \end{cases} \tag{3}$$

where the coefficient α_t is set as $\alpha_t = \log 1/\beta_t$.

The weight of each sample is constantly adjusted to form the strongest classifier, thus improving the performance of the classifier and avoiding the overfitting. Hence, the gentle AdaBoost classifier with high efficiency is suitable for the vertebral detection.

2.2. Edge-Based Level Set Model. Level set based on edge information [18] constructs an edge detecting function to drive the evolving contour to the desired boundary. Given the initial contour $C = \{(x, y) \in \Omega | \phi(x, y) = 0\}$, $\phi : \Omega \longrightarrow \mathfrak{R}$ is a level set function (LSF) defined on a domain Ω. For the edge distribution of the image, an edge indicator function g is defined by

$$g \triangleq \frac{1}{1 + |\nabla G_\sigma * I|^2}, \tag{4}$$

where I represents an image. G_σ is a Gaussian kernel with standard deviation σ. $*$ denotes a convolution operation to reduce the image noise. The function $g \in [0, 1]$ should take smaller values at object boundaries than other locations. The basic energy function for ϕ can be defined as

$$\varepsilon(\phi) = \text{length}(\phi) + \text{area}(\phi), \tag{5}$$

where length(ϕ) is the length term of the initial contour. Area(ϕ) represents the energy of the area inside the contour C. Here, both the length term and area term rely on the edge information of the image. These two energy terms can be given as follows:

$$\begin{aligned} \text{length}(\phi = 0) &= \int_\Omega g\delta(\phi)|\nabla\phi|\, dx, \\ \text{area}(\phi \geq 0) &= \int_\Omega gH(\phi)\, dx, \end{aligned} \tag{6}$$

where

$$H(x) = \begin{cases} 1 & \text{if } x > 0, \\ 0 & \text{if } x < 0. \end{cases} \tag{7}$$

$H(x)$ is the Heaviside function. And $\delta(x)$ is a Dirac delta smoothing function defined by

$$\delta(x) = \begin{cases} \frac{1}{2\varepsilon}\left[1 + \cos\left(\frac{\pi x}{\varepsilon}\right)\right], & |x| < \varepsilon, \\ 0, & |x| > \varepsilon. \end{cases} \tag{8}$$

3. The Proposed Method

The motivation of this paper is to accurately recognize scoliosis from unlabeled CT images. For convenience, the proposed method is called as CGAdaBoost-DRLSE. Figure 1 depicts an overview of the CGAdaBoost-DRLSE method. The proposed method is comprised of three main stages. The first stage is the automated vertebrae detection, including training and testing processes. Diversified training samples and multifeature descriptors are considered to train the CGAdaBoost classifier and detect all vertebral bodies in the CT spine image. The second stage is the vertebral region segmentation. DRLSE strongly depends on a relatively accurate initialization. Consequently, the detected bounding boxes are used as the initial contours of DRLSE. Acquired segmentation results further serve for vertebral centroids extraction. Vertebral centroids are extracted to connect the whole spine, which can make scoliotic deformity recognition more accurate. After spinal curve fitting, calculating the angle between two tangents is to measure the spinal curvature angle, thereby diagnosing spinal abnormalities.

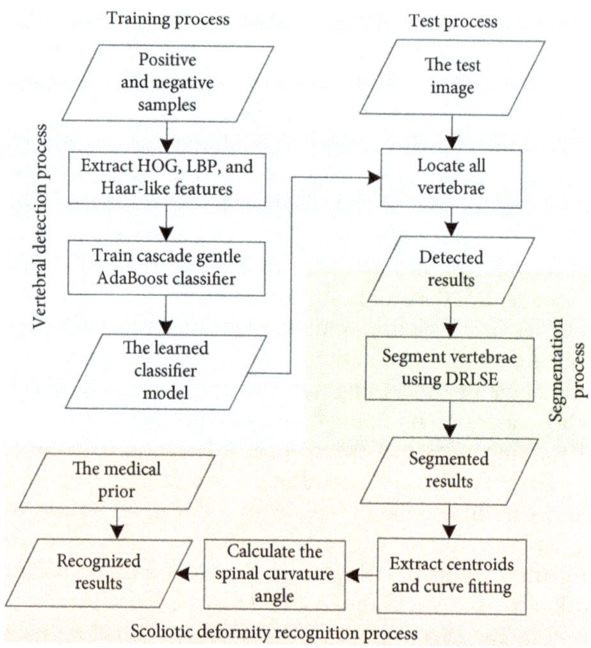

FIGURE 1: The overview of the CGAdaBoost-DRLSE method.

3.1. Vertebrae Detection. We make the first attempt to use the cascade gentle AdaBoost detector with multifeature fusion to detect vertebrae. The classifier with the cascade structure is essentially a degenerated decision tree, which arranges a series of strong AdaBoost classifiers from simple to complex [19]. By continuously training, each strong classifier will have a higher detection rate and lower false-positive rate. Figure 2 illustrates the basic schematic of a cascade classifier.

In the training process, only positive samples of the previous classifier will be transmitted into the next classifier to continue learning. Then, some subwindows belonging to positive samples in each classifier are output as the detected vertebrae. On the contrary, subwindows belonging to negative samples will be rejected directly. Obviously, the cascade classifier can overcome the problem of the imbalanced sample and significantly improve the efficiency of the detector. The training process of the CGAdaBoost classifier is briefly described, as shown in Algorithm 1.

3.2. Multifeature Fusion. Currently, Haar-like [20], local binary pattern (LBP) [21], and histogram of oriented gradients (HOG) [22] descriptors have become the most useful digital image feature methods in object recognition. Among them, Haar-like descriptor is used to describe the image intensity differences. LBP is suitable for describing the local texture feature of the image. HOG descriptor can better describe appearance and shape of the object, that is, the local gradient or the distribution of the edge direction.

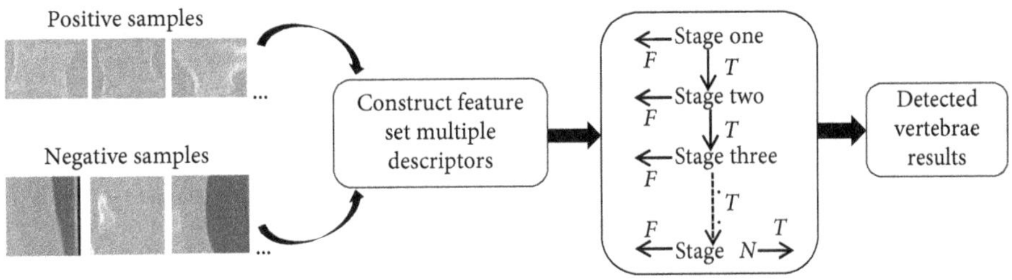

FIGURE 2: Schematic of the cascade gentle AdaBoost detector.

```
Input: the cascade layer number K, the minimum detection rate d (D_0 = 1), the maximum false-positive rate f (F_0 = 1), the positive
sample set P, the negative sample set N, and the final false-positive rate F_t.
Output: subwindows of the image belonging to positive samples
  for i = 0:K, F_i > F_t do
    ++i; n_i = 0; F_i = F_{i-1};
    if F_i > f × F_{i-1} then
      ++n_i; train strong classifier with n_i weak classifier using AdaBoost on P and N;
      Calculate the detection rate D_i and false-positive rate F_i of the current classifier;
      if d_i < d × D_{i-1} then
        Reduce the threshold of the ith layer strong classifier;
        Calculate detection rate D_i and false-positive rate F_i of the current classifier;
      end if
    end if
  N = Φ; use the current classifier to detect testing images and put the negative image into N; output and number subwindows
  belonging to positive samples.
end for
return the detected results
```

ALGORITHM 1: The training process of the cascade gentle AdaBoost classifier.

Although each single descriptor is highly efficient, the extracted features are difficult to accurately distinguish vertebral and nonvertebral regions from spine CT images with low contrast. Therefore, to make full use of the advantage of each feature descriptor, we present a multifeature fusion way. HOG, LBP, and Haar-like features will be combined together to construct a feature vector, thus generating the optimal feature set. Before fusing, these features will be normalized for the facility of computation. The final feature set F_{final} is formally expressed as

$$F_{\text{final}} = F_{\text{Haar}} + F_{\text{LBP}} + F_{\text{HOG}}. \tag{9}$$

Haar-like adopts the black-and-white feature template to perform sliding detection on the image. The integral graph is employed to realize the fast summation of subregions. And the sum of the pixel in the white region subtracts the sum of the pixel in the black region as feature value F_{Haar}, which can be described as

$$F_{\text{Haar}} = \text{sum}\left(R_{\text{white}}\right) - \text{sum}\left(R_{\text{black}}\right). \tag{10}$$

For the LBP descriptor, the image is divided into several subregions. Then, the LBP feature of each pixel in the subregion is extracted. The statistical histogram of each subregion constitutes the texture feature vector of the whole image. A mathematical description of the LBP can be given as follows:

$$F_{\text{LBP}} = \sum_{p=0}^{P-1} 2^P s\left(f_p - f_c\right) = \begin{cases} \sum_{p=0}^{P-1} 2^P, & \text{if } f_p \geq f_c, \\ 0, & \text{else}, \end{cases} \tag{11}$$

where p is the pth pixel of the subregion. f_p and f_c represent the gray value of the neighboring pixel and central pixel, respectively.

HOG descriptor also divides the image into several small connected regions. The gradient direction (or edge direction) histogram of the pixels in each region is calculated, thereby combining these histograms as a feature vector. The gradient vector of the HOG descriptor can be obtained as

$$F_{\text{HOG}} = \nabla f(x, y) = \begin{bmatrix} \dfrac{\partial f(x, y)}{\partial x} \\ \dfrac{\partial f(x, y)}{\partial y} \end{bmatrix} = \begin{bmatrix} f_x \\ f_y \end{bmatrix}, \tag{12}$$

where $f(x, y)$ denotes a positive sample; f_x and f_y represent the gradient vectors of x and y directions, respectively.

3.3. Vertebral Body Segmentation.

3.3. Vertebral Body Segmentation. Vertebrae segmentation is vitally important to extract centroids for scoliosis recognition. DRLSE [23] is an edge-based level set evolution method. A distance regularization term is introduced into the energy function as a penalty term. DRLSE can produce a better effect of edge detection and avoid the reinitialization process. Aiming at the low contrast and high noise of CT image, it is necessary to employ a contrast limited adaptive histogram equalization (CLAHE) method [24] to enhance the vertebral region before segmenting.

To further reduce manual intervention, detected bounding boxes are viewed as the initial contour C of DRLSE. The energy function is minimized by solving the following gradient flow equation:

$$\frac{\partial \phi}{\partial t} = \mu \operatorname{div}\left(d_p(|\nabla \phi|)\nabla \phi\right) + \lambda \delta_\varepsilon(\phi)\operatorname{div}\left(g\frac{\nabla \phi}{|\nabla \phi|}\right) + \alpha g \delta_\varepsilon(\phi), \tag{13}$$

where the first term is the distance regularization term, and $\mu > 0$ is the coefficient of the distance regularization term; the second term stands for the gradient flow of the length term, and λ denotes the weighted coefficient of the length term; the third term is the gradient flow of the area term, and α is the weighted coefficient of the area term; and the $\operatorname{div}(\cdot)$ is the divergence operator.

In (13), the definition of d_p is given by

$$d_p(s) = \frac{p'(s)}{s}, \tag{14}$$

where the potential function $p(\cdot)$ should have minimum points at $s = 0$ and $s = 1$. The flow in (13) has a diffusion effect on the level set function ϕ. This diffusion is termed as forward-and-backward diffusion, which adaptively increases or decreases $\nabla\phi$ to maintain the desired shape of the function ϕ, thus avoiding the effect of bad edges.

A preferable potential function can keep the sign distance function smoothing in the distance regularization term, which can be expressed as

$$p(s) = \begin{cases} \dfrac{1}{2\pi^2}(1-\cos(2\pi s)), & \text{if } s \le 1, \\ \dfrac{1}{2}(s-1)^2, & \text{if } s \ge 1. \end{cases} \tag{15}$$

As a consequent, the vertebral body region is obtained by evolving iteratively. It is noticed that the combination of detection and segmentation solves not only the sensitivity of initialization, but also quickly and accurately converges on vertebral boundaries.

3.4. Scoliosis Recognition with the Medical Prior Knowledge. To establish the relationship between the individual vertebrae and the whole spine, we extract vertebral centroids as a shared feature from the segmented result. This work highlights the importance of the vertebral centroid. Let $I(x, y)$ be a segmented binary mask, where its size is $m \times n$. The centroid coordinate (k, l) of each vertebral body can be expressed as

$$\begin{aligned} k &= \frac{\sum_x^m \sum_y^n x I(x,y)}{\sum_x^m \sum_y^n I(x,y)}, \\ l &= \frac{\sum_x^m \sum_y^n y I(x,y)}{\sum_x^m \sum_y^n I(x,y)}, \end{aligned} \tag{16}$$

where $I(x, y) = \begin{cases} 1, & (x, y) \in \text{object}, \\ 0, & (x, y) \in \text{background}. \end{cases}$ All vertebral centroids are computed by solving (16) iteratively.

After extracting vertebral centroids, the least squares method [25] is adopted to construct the spinal curve. In Figure 3, suppose that the spinal curve $y = f(x)$ has a continuous derivative. And the points A and P represent anterior and posterior vertebral centroids, respectively. Δs denotes the length of arc AP, and φ is the angle between two tangents of the curve. Therefore, the curvature of y at the point (x, y) is given by

$$k = \left|\frac{d\varphi}{ds}\right| = \frac{|y''|}{\left(1+y'^2\right)^{3/2}}, \tag{17}$$

where $d\varphi = (y''/1 + y'^2)dx$. In centroid measurement method, the spinal curvature angle is defined as the angle between two tangents at centroid points of two terminal vertebrae, which can be calculated as follows:

$$\varphi = \int_a^b \frac{y''}{1+y'^2}\,dx. \tag{18}$$

According to the prior medical treatment, viewed from the coronal, the spinal curve looks like a straight line. Generally, if the curvature angle is greater than ten degrees, the spine will be diagnosed as scoliosis:

$$\text{diagnosis result 1} = \begin{cases} \text{scoliosis}, & \text{if } \varphi \ge 10^\circ, \\ \text{normal}, & \text{if } \varphi < 10^\circ. \end{cases} \tag{19}$$

In addition, viewed from the sagittal, the curve of the spine is in "S" shape. There are three normal curvatures of spine, including cervical lordosis (35° to 45°), thoracic kyphosis (20° to 45°), and lumbar lordosis (40° to 60°) [26]:

$$\text{diagnosis result 2} = \begin{cases} \text{cervical lordosis}, & \text{if } \varphi \notin [35^\circ, 45^\circ], \\ \text{thoracic kyphosis}, & \text{if } \varphi \notin [20^\circ, 45^\circ], \\ \text{lumbar lordosis}, & \text{if } \varphi \notin [40^\circ, 60^\circ] \end{cases} \tag{20}$$

The measurement from the coronal view is focused on the diagnosis of scoliosis, while the measurement from the sagittal view refers to the diagnosis of lumbar lordosis, thoracic kyphosis, and cervical lordosis.

4. Experimental Results and Discussion

4.1. Data Description and Experimental Platform. To verify the effectiveness and feasibility of the proposed method, a variety of experiments are conducted on about 500 spine CT images to automatically recognize scoliotic deformity. These images are from 20 subjects (11 males and 9 females;

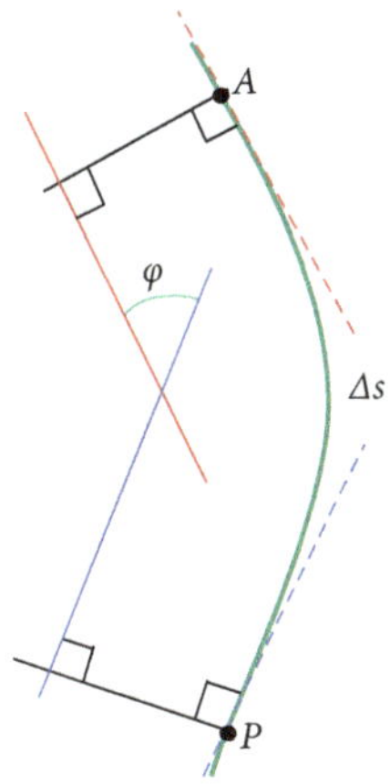

FIGURE 3: The schematic diagram of the curvature angle.

age range 18–56 years) of available spine CT volumes on the publicity platform *SpineWeb* [27]. The volume size is $512 \times 512 \times (100 - 240)$. The view of the spine image is limited to 5–20 vertebrae. Each dataset may include high-grade scoliosis, kyphosis, and fractures, which is along with ground-truth centroid of each vertebra. All experiments are implemented in the Matlab 2014a platform, which run on Microsoft Windows 7 64-bit operating system with 3.20 GHz Intel® Xeon® CPU, 8 Gbyte RAM.

4.2. The Analysis of Experimental Results. In the vertebral detection experiment, we only select the vertebral bodies as positive samples without considering the spinal cord, ribs, and sacrum. The positive and negative samples are created by the Training Image Labeler Toolkit of Matlab 2014a, resulting in 520 positive samples and 1058 negative samples. Figure 4 shows a part of positive and negative samples.

The positive samples contain various parts (cervical, thoracic, and lumbar) of the whole spine. Additionally, vertebrae images with different views (sagittal and coronal), arbitrary contrasts, and lesions are also considered as positive samples. We select nonvertebral regions from CT images as negative samples. As far as possible to increase the distinguishability of interclass samples, diverse features will be provided for classifying.

For feature extraction, Figures 5 and 6 show extracted results of HOG, LBP, and Haar-like from the same positive sample. Here, the 8×8 cell constitutes a HOG block. From Figure 5(c), it can be seen that the HOG feature better describes the edge gradient and shape information of the vertebrae.

Figure 6(a) illustrates the texture image via LBP. The vertebral edge can be visualized clearly. However, Haar-like is an effective method to reflect the change of image gray. In Figure 6(b), two edge features, two center-surround features, and four line features are extracted as the final Haar-like feature. Furthermore, we have taken into account the complete feature set for the classifier as only combining HOG, LBP, and Haar-like descriptors. To do this, the intensity, appearance, and shape of the individual vertebrae can be fully expressed.

In the training process, the gentle AdaBoost classifier built a powerful classifier with high accuracy through several simple weak classifiers. Figure 7 shows the processes of training weak classifiers and the optimal classifier. Red dots and green boxes represent two classes. After numerous iterations, red dots and green boxes are classified into two regions (white region and black region). Figures 7(a) and 7(b) display weak classifiers with different false-positive rates. Actually, the construction of the optimal classifier is to find the appropriate classifier parameter. Hence, the classifier has the lowest false-positive rate of 0.03 for all samples, which is as shown in Figure 7(c).

The CGAdaBoost classifier captured the shape and pathological features of the vertebrae. At the same time, the vertebrae of sagittal and coronal views also are detected from CT spine images. We obtain the optimal parameter of the classifier after several experiments. To reduce the loss of the vertebrae, the true-positive rate should be set to a larger value. Likewise, the smaller the value of the false-positive rate is, the less the number of the falsely detected vertebrae is. As a result, TruePositiveRate (true-positive rate) is set to 0.9, and FalseAlarmRate (false-positive rate) is set to 0.03. The number of the training stage (NumCascadeStages) is set to 10 according to the total number of samples. In our implementation, the subwindow size is experimentally set to 90×80. Only in this way, the initial contour of the DRLSE method is closer to the edge of the vertebrae such that the final segmentation results are more accurate to serve for vertebral centroids extraction.

Various detection results with different feature descriptors on sagittal and coronal planes are illustrated in Figure 8. When the single HOG feature is used to detect the vertebrae on the coronal view, there are undetected vertebrae (red circle) and false detection vertebrae (green circle) in Figure 8(a). In contrast, Figure 8(b) shows better the detected result with multifeature fusion. To demonstrate the diversity of our result, the detected result with multifeature on the high-contrast coronal image is shown in Figure 8(c). It can be seen from Figure 8(d) that in spite of vertebrae 7 having a complex shape deformation, it can still be detected. Figure 9 shows the detected result after enhancing the image of CLAHE. It is evident that the enhanced CT image can

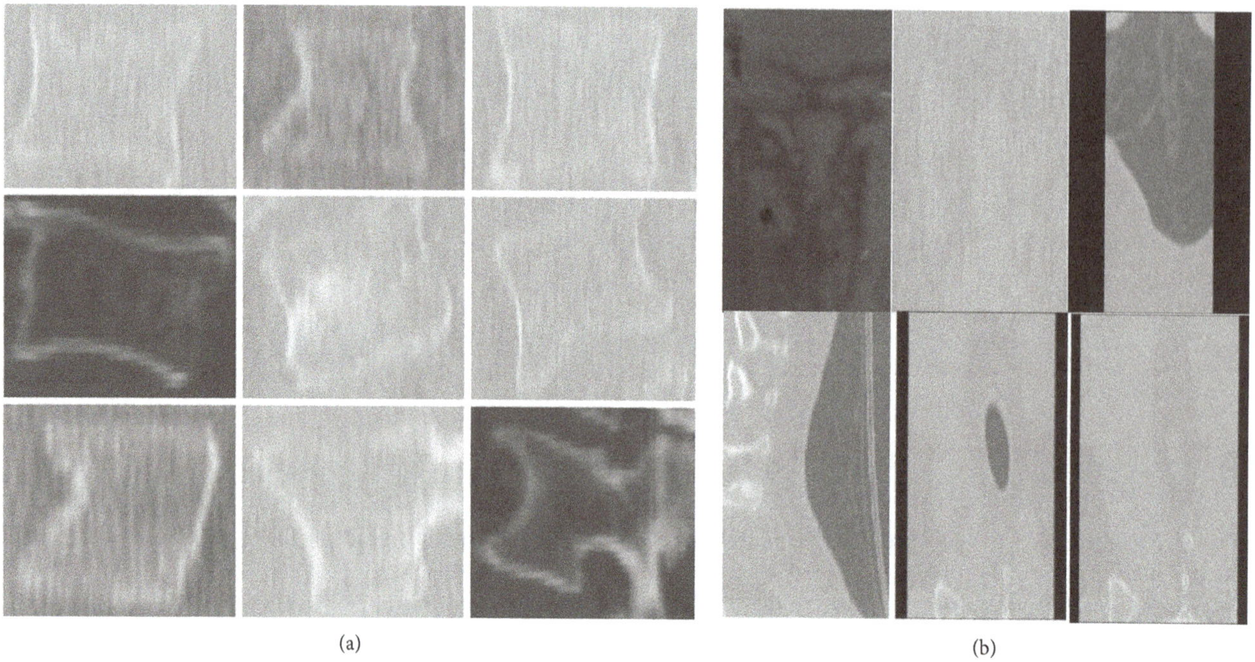

Figure 4: Parts of training samples: (a) positive samples; (b) negative samples.

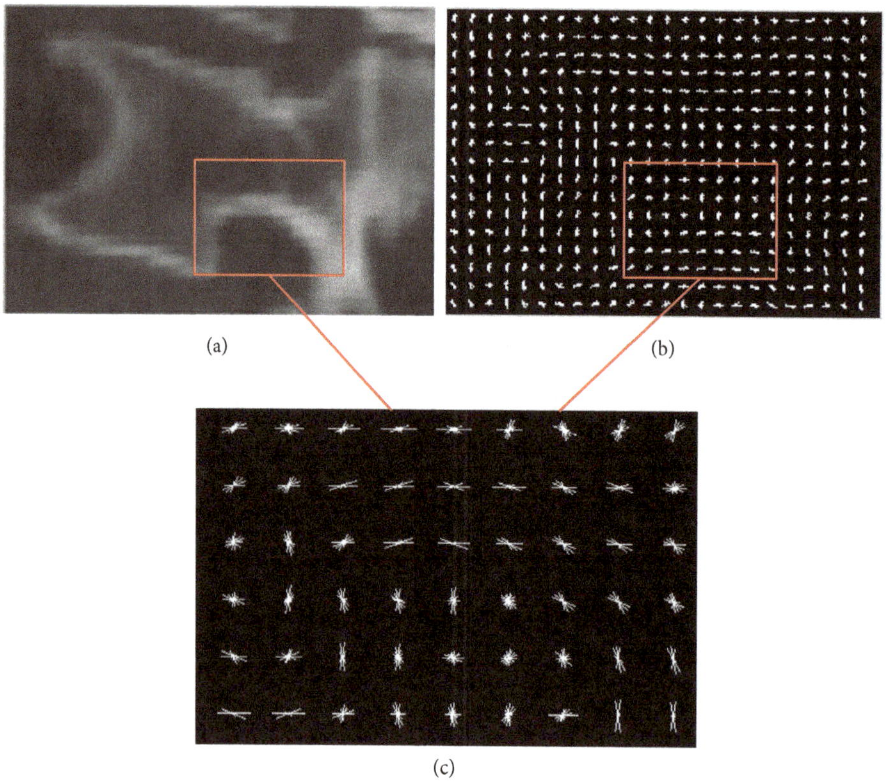

Figure 5: The result of the HOG descriptor: (a) CT vertebral image; (b) the visualization of HOG; (c) local amplification result.

make detection more accurate and provide suitable initial contours for the subsequent DRLSE model. It is worth noting that the scanning order of the testing process is from left to right and from top to bottom on the test image. Then, the scanned subwindow belonging to the vertebrae will be labeled using numbers. The aim of numbering is to facilitate different parts extraction from the whole spine.

We applied the DRLSE method to segment the vertebrae from spine CT images without any user intervention. On the basis of vertebral detected results, located bounding boxes

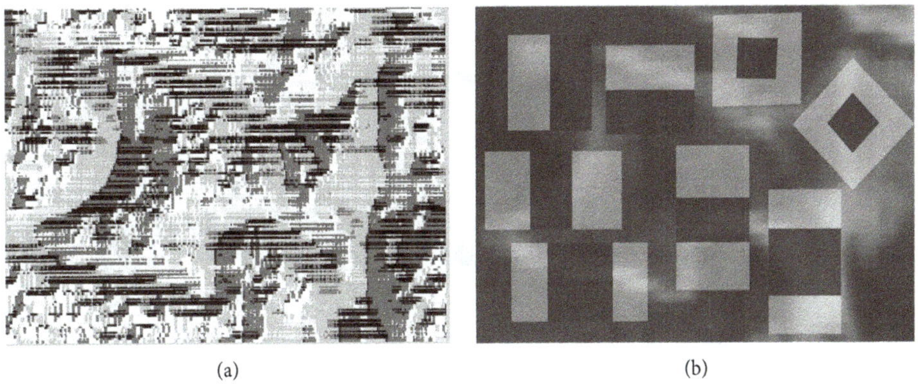

FIGURE 6: Results of LBP and Haar-like descriptors: (a) the visualization of LBP; (b) Haar-like feature with edge features, center-surround features, and line features.

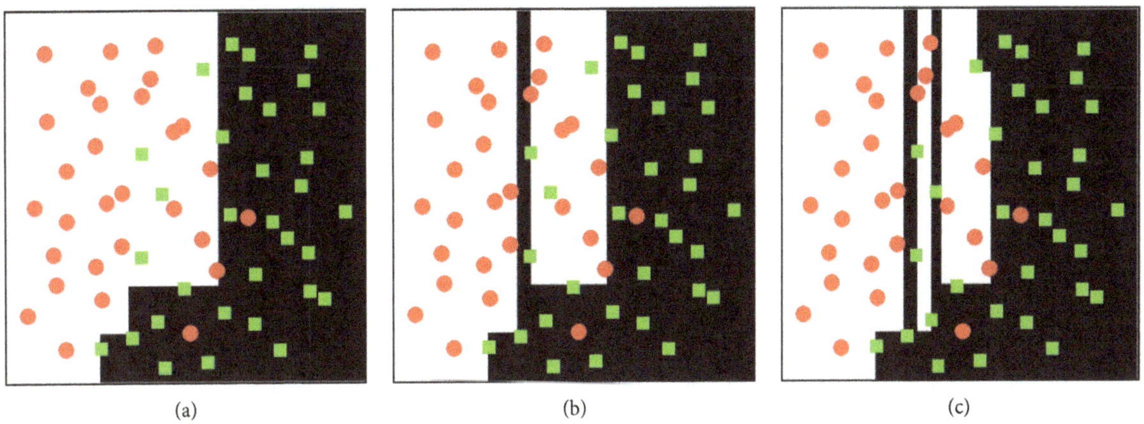

FIGURE 7: The training process of the optimal classifier: (a) the weak classifier with false-positive rate 0.1; (b) the weak classifier with false-positive rate 0.08; (c) the strong classifier with false-positive rate 0.03.

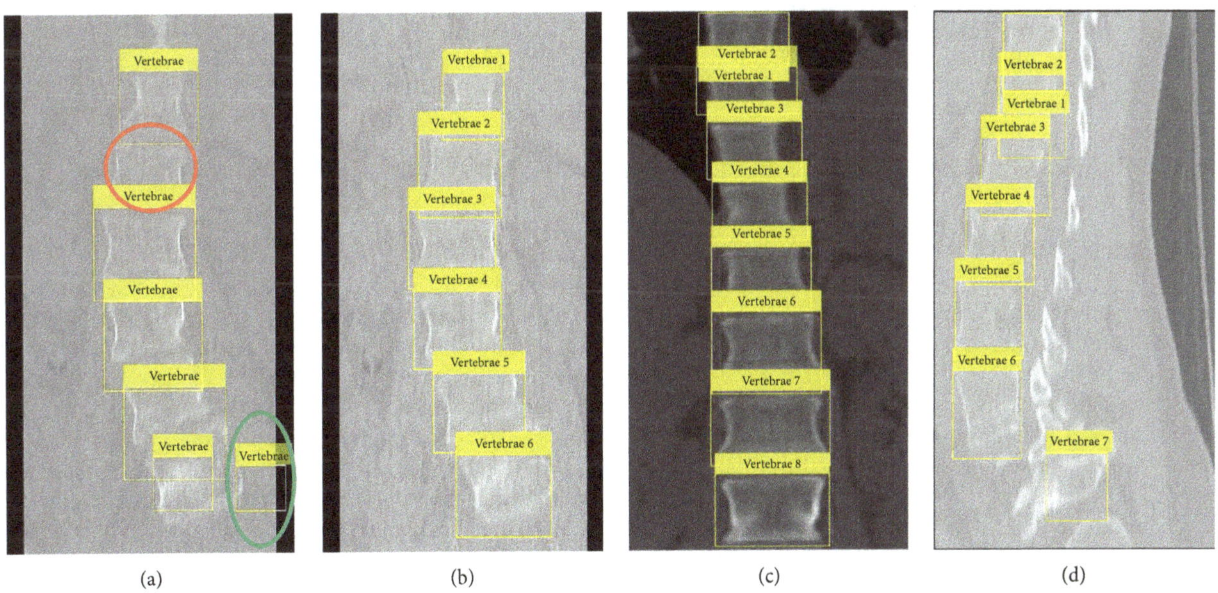

FIGURE 8: Detection results of the CGAdaBoost classifier with FalseAlarmRate 0.03 and TruePositiveRate 0.9: (a) the detection result with single feature on the coronal view (a red circle for the undetected vertebrae and a green circle for the false detected vertebrae); (b) the detection result with multifeature on the coronal view; (c) the detection result with multifeature on the high-contrast coronal view; (d) the detection result with multifeature on the sagittal view.

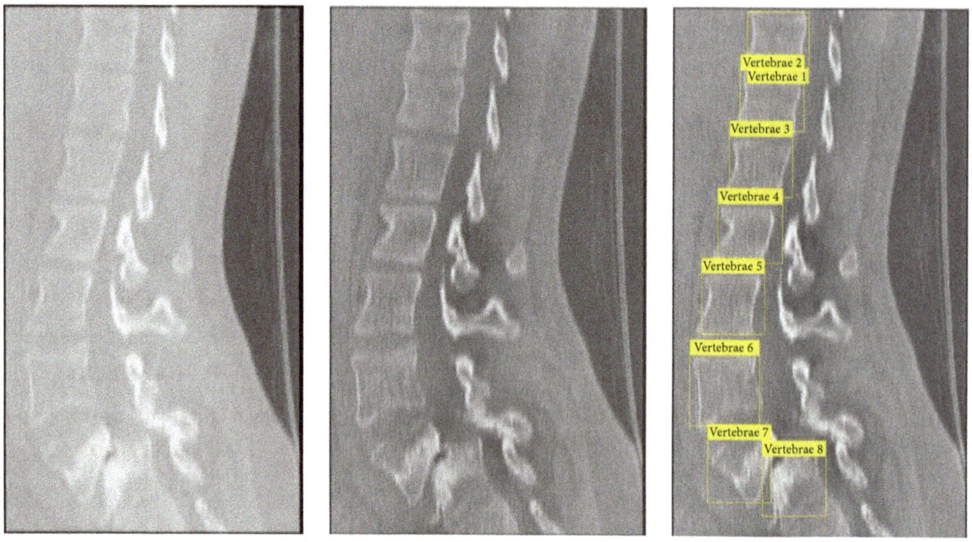

(a) (b) (c)

Figure 9: Another detected result: (a) original CT image; (b) the enhanced image by the CLAHE method; (c) the final detected result.

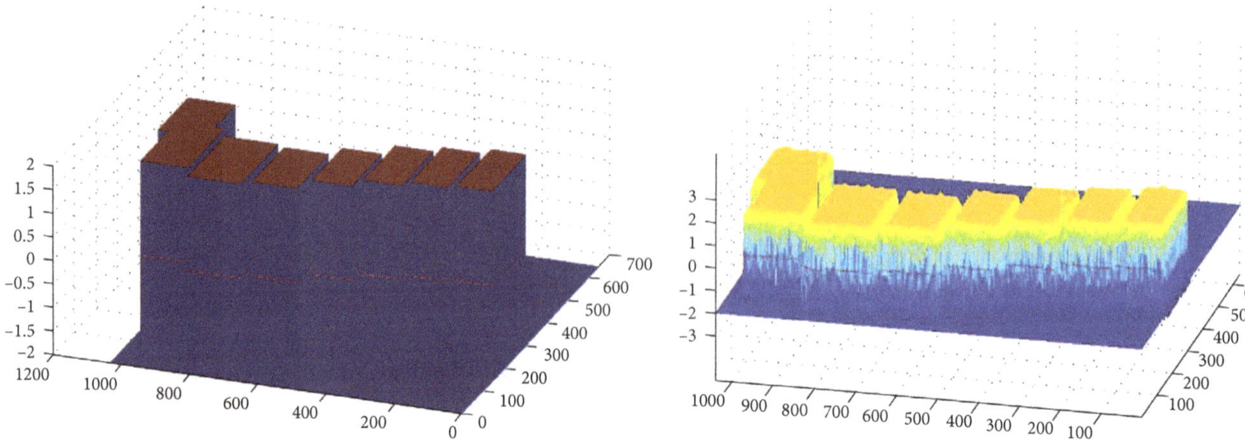

(a) (b)

Figure 10: The evolution result of DRLSE: (a) the initial level set function; (b) the final level set function.

are regarded as the initial contour C to eliminate the sensibility of initialization of DRLSE. In the segmentation experiment, we set the appropriate coefficient $\mu = 0.04$ of the distance regularization term. The weighted coefficients of the length term λ and area term α are set to 5 and 1.5, respectively. Scale parameter ε in Gaussian kernel is set to 1.5. To visualize the curve evolution process, Figure 10 displays the evolution results of the initial and the final level set functions. The red curve represents a zero level set function.

By adjusting appropriate parameters, Figures 11 and 12 show two spinal segmentation results. Figure 11(a) presents the evolutionary result of 100 iterations on the sagittal view. Figure 11(b) shows the evolution result of 200 iterations. Figure 11(c) shows the binary segmentation result after 200 iterations. Figure 12 displays the segmented result on the coronal view of the spine CT image. Obviously, it can be observed that the initial contour can quickly converge on vertebra boundaries to improve the segmentation accuracy. In addition, less iteration is needed to obtain the final converged result, thereby avoiding a larger displacement of the initial contour.

The final goal of our method is to represent the detailed shape of the spinal curve using the centroid method. It should be pointed out that we would select the appropriate CT slice of the spine image in order to extract vertebral centroids. Figure 13 shows a comparison result between the extracted centroid and ground-truth centroid. It can be seen that all extracted centroids of CGAdaBoost-DRLSE with multifeature fusion are nearly consistent with the ground truth. Only less extracted centroid has a larger error because of the severe vertebral fracture, which is seen in Figure 13(b).

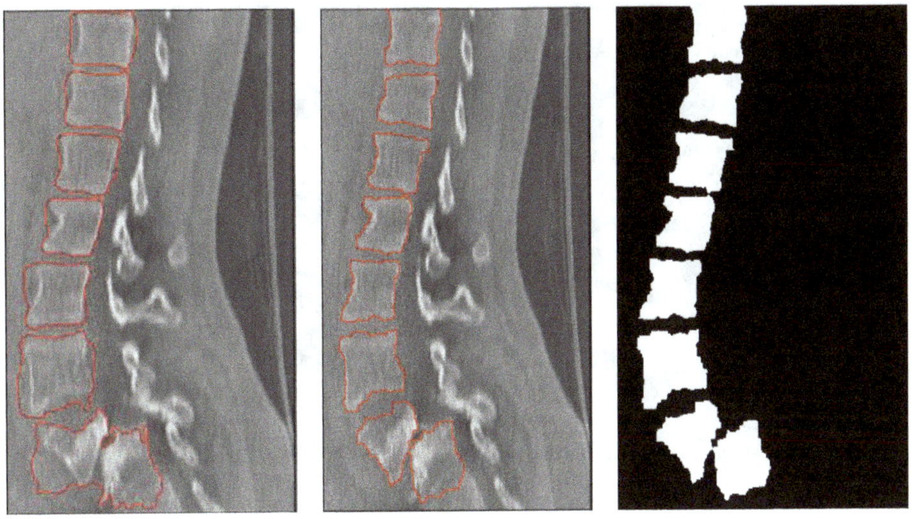

(a) (b) (c)

FIGURE 11: The segmentation result on the sagittal view: (a) the contour with 100 iterations; (b) the final contour with 200 iterations; (c) the final segmentation result.

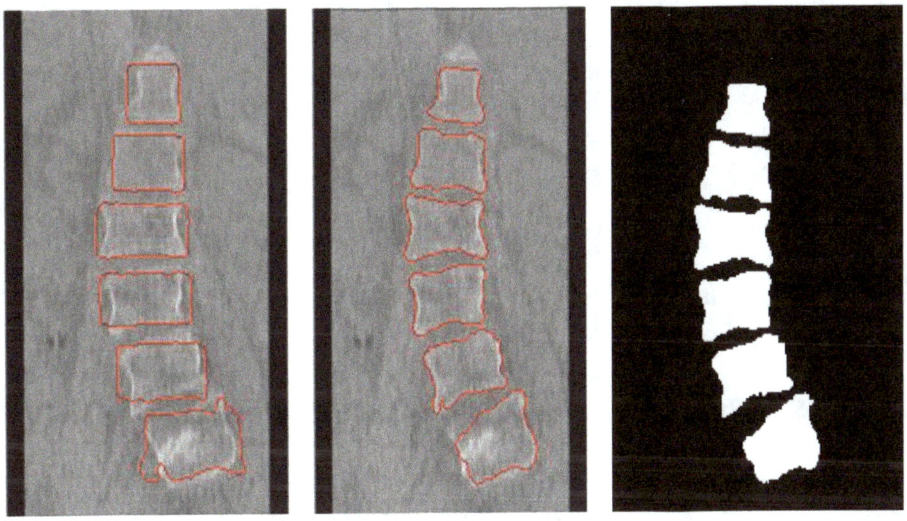

(a) (b) (c)

FIGURE 12: The segmentation result on the coronal view: (a) the contour with 100 iterations; (b) the final contour with 200 iterations; (c) the final binary segmentation result.

Furthermore, we extract the lumbar vertebrae, cervical vertebrae, and thoracic vertebrae from the whole spine. The corresponding spinal curve is calculated by the least squares method. Figure 14 displays the curve fitting results of various parts of the whole spine. The curve of the lumbar (i.e., from L1 to L5) on the sagittal view is presented in Figure 14(a). The curve of the lumbar (i.e., from T12 to L5) on the coronal view is given in Figure 14(b). Figures 14(c) and 14(d) show the curves of the lumbar on the coronal view. The severe scoliosis curve of the thoracic (i.e., from T1 to T12) on the coronal view is shown in Figure 14(e). The curve of the cervical (i.e., from C1 to C7) on the coronal view is shown in Figure 14(f). Specially, from Figure 14(c), we can see that, even if there are some distorted vertebrae (indicated by the arrow), it has no influence on the whole spinal curvature.

After curve fitting, the angle between two tangents to the curve is determined as the spinal curvature angle. Table 1 lists a part of spinal abnormalities diagnostic results in Figure 13. Comparing the spinal curvature angle with the prior medical treatment, the spine is decided as normal or abnormal. By analyzing the above results, our method shows promising scoliosis recognition results and also can be applicable to other image patterns. Meanwhile, this method also diagnoses lordosis and kyphosis by observing the spinal curvature from a sagittal view. The resulting curvature of the spine is not affected by bone lesions, without the need to manually identify the endplate vertebrae.

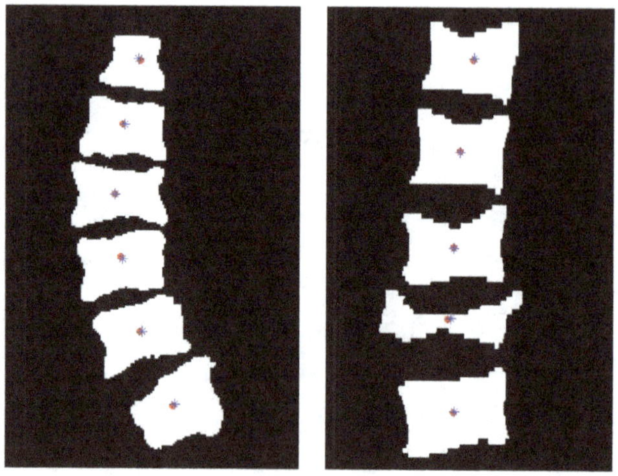

(a) (b)

FIGURE 13: Comparison results of the extracted centroids (red solid point) with the ground-truth centroids (blue star): (a) the result without the vertebral pathology case; (b) the result with the severe vertebral fracture case.

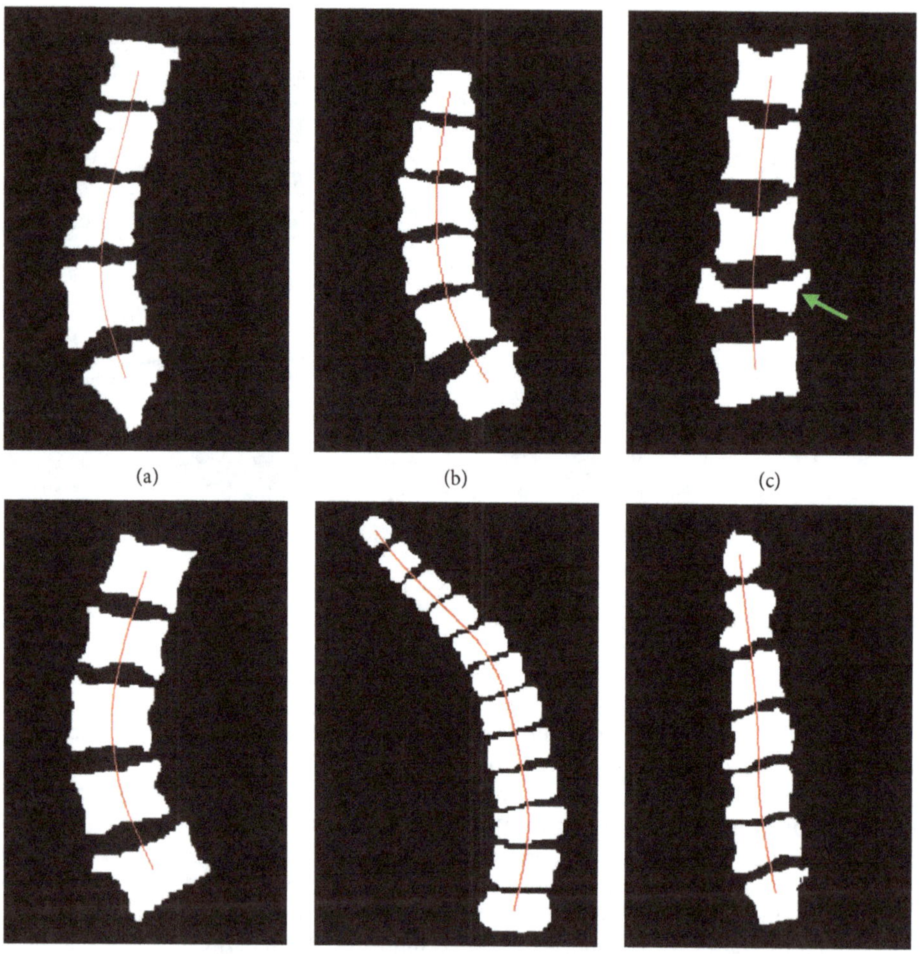

(a) (b) (c)

(d) (e) (f)

FIGURE 14: Curve fitting results of various parts in the whole spine: (a) sagittal lumbar curvature; (b)–(d) coronal lumbar curvature; (e) coronal thoracic curvature; (f) coronal cervical curvature.

TABLE 1: A part of diagnosis results.

Binary mask	The curvature angle	The diagnosis result
Figure 14(a)	Sagittal lumbar 40.4°	Normal
Figure 14(b)	Coronal lumbar 30.5°	Abnormal
Figure 14(c)	Coronal lumbar 3.2°	Normal
Figure 14(d)	Sagittal lumbar 44.6°	Normal
Figure 14(e)	Coronal thoracic 55.2°	Abnormal
Figure 14(f)	Coronal cervical 2.6°	Normal

4.3. The Assessment of Performance. To evaluate the performance of the cascade gentle AdaBoost classifier, the receiver operator characteristic (ROC) is used as an evaluation criterion. ROC intuitively shows the compromise between true-positive rate and false-positive rate for the classification model. Figure 15 compares the ROC curve of CGAdaBoost with a single-feature and multifeature fusion. It is obvious that the ROC of the multifeature fusion method is higher than a single-feature method. This further reveals that the detected vertebrae are more accurate by the CGAdaBoost classifier with multifeature fusion.

Additionally, to verify the credibility of the spinal curve fitting, using two evaluation criteria is to comprehensively assess the quality of curve fitting. One is the coefficient of determination (R^2). It is a statistic of goodness of fit. The value range is from 0 to 1. And the larger the value is, the better the fitting effect is. The other is root mean squared error (RMSE) that is the relative deviation between the predicted value and the real value. The mathematical expressions of two criteria are given as follows:

$$R^2 = 1 - \frac{\sum_{i=1}^{n} (v_i' - v_i)^2}{\sum_{i=1}^{n} (v_i - \overline{v})^2},$$
$$\text{RMSE} = \sqrt{\frac{1}{n} \sum_{i=1}^{n} (v_i - v_i')^2}. \quad (21)$$

where v and v' indicate the real value and the fitting value, respectively, and $\overline{v}$ is the mean value of n samples. By calculating the values of R^2 and RMSE, the results of different methods on twenty spine CT volumes are displayed in Figures 16 and 17, respectively.

In Figure 16, we can see that no matter which method, it gains relatively higher R^2 on each volume. The R^2 of multifeature fusion is closer to one. It reflects that the spine curve fitting has better goodness of fit. Meanwhile, from Figure 17, we clearly know that multifeature fusion outperforms a single-feature method in terms of RMSE on each dataset. It is because that better spine curve fitting may depend on accurate vertebral centroids. This also confirms the effectiveness and quality of curve fitting in the proposed method.

Furthermore, the detection accuracy rate and centroid location error are employed to further assess the proposed method. From the total 231 vertebrae, CGAdaBoost-DRLSE with single feature detects 227 vertebrae, resulting in the detection accuracy rate of 98%. By contrast, the multifeature fusion successfully detects 229 vertebrae and has the detection accuracy rate of about 99%. The centroid location error is computed using the Euclidean distance between extracted centroids and ground-truth centroids. The single-feature method achieves an average centroid location error of 1.51 mm. The multifeature fusion method has the mean centroid location error of 0.87 mm. Two related methods reported by Korez et al. [9, 11] use the same database as our method for fair comparison. The performance comparison result is summarized in Table 2.

In Table 2, the proposed method is superior to other competitive methods in terms of the detection accuracy rate.

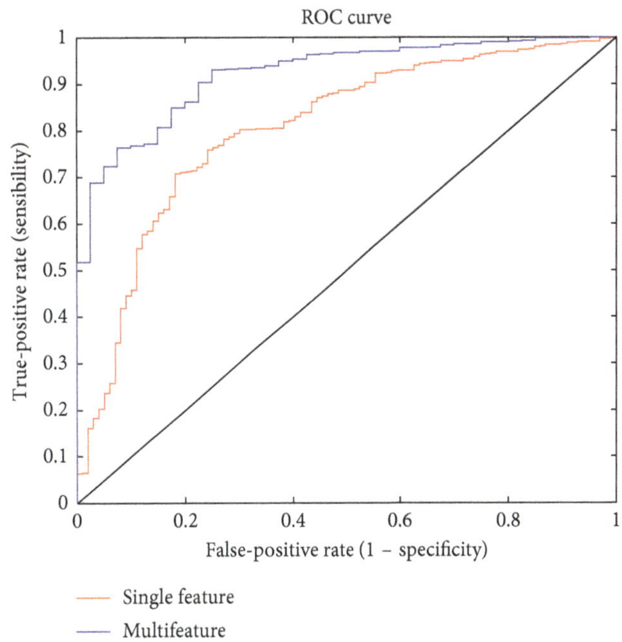

FIGURE 15: The ROC curve on different methods.

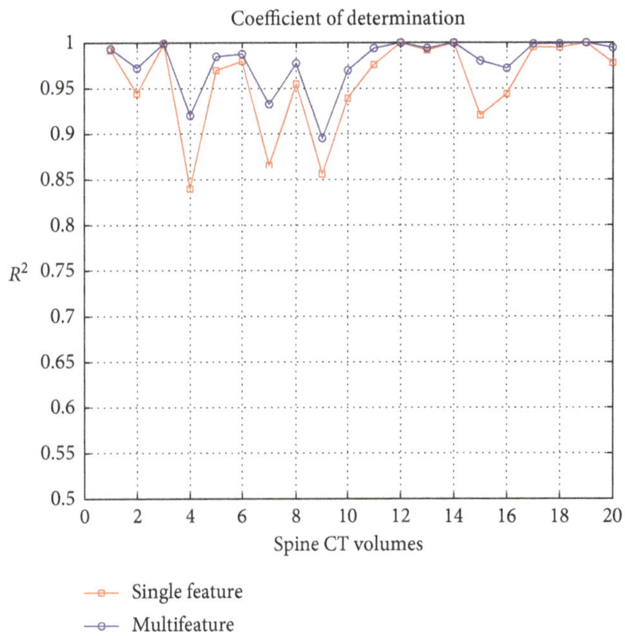

FIGURE 16: R^2 of different methods on twenty subjects.

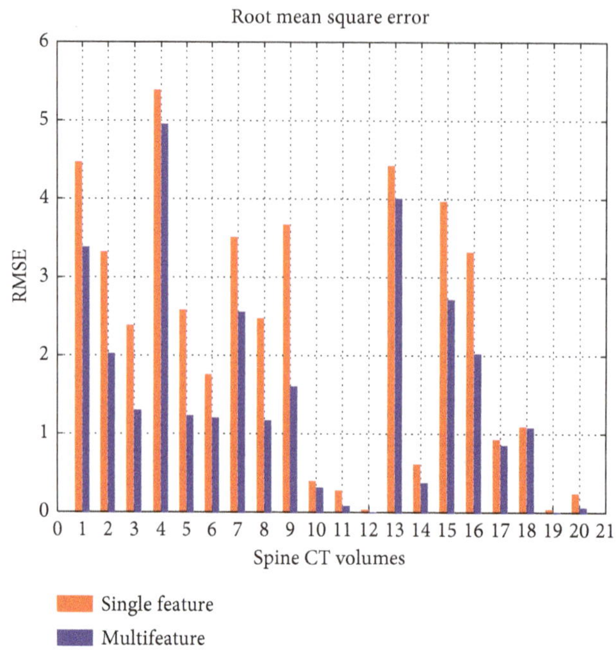

Figure 17: RMSE of different methods on twenty subjects.

Table 2: The performance comparison of our methods and other related works.

Evaluation criteria	Methods			
	Supervised classification forests [11]	Interpolation theory + shape-constrained [9]	Single-feature + CGAdaBoost-DRLSE	Multifeature + CGAdaBoost-DRLSE
Detection accuracy rate	86%	97%	98%	99%
Centroid location error	4.4 mm	1.1 mm	1.51 mm	0.87 mm

This is because that vertebral shape variations and pathological features are fully fused. Thus, effective vertebrae features are trained to improve the detection accuracy. On account of inadequate estimation using the dense probabilistic centroid estimation method in [11], our method potentially produces less centroid location error. By compared with interpolation theory and shape-constrained method [9], we found that CGAdaBoost-DRLSE with multifeature is slightly superior to the result of [9], which is most probably due to incorrect segmentation of some smaller vertebrae. Overall, the proposed method can automatically detect and segment vertebrae from CT images. In particular, for a relatively small number of subjects with some pathological vertebra, it can provide accurate spinal curvature to recognize scoliosis deformity from unlabeled CT images. Moreover, any manual landmark is not required. These unknown CT images can be labeled as abnormal or normal through this unsupervised method.

5. Conclusion

This paper proposes an unsupervised scoliosis recognition method with unlabeled CT images to improve the accuracy. The CGAdaBoost-DRLSE method consists of vertebral bodies' detection, segmentation, and centroids extraction. Firstly, diversified training samples and multifeature descriptors are considered to achieve better detection results in the cascade gentle AdaBoost classifier. Then, located bounding boxes represent the initial contour of DRLSE to eliminate the sensitivity of initialization and quickly converge on vertebral boundaries. Finally, vertebral centroid extraction and curve fitting are performed to compute the spinal curvature angle, thereby recognizing scoliosis with the prior medical treatment. Experimental results have demonstrated that the proposed method can effectively and accurately diagnose scoliosis deformity and reduce the need for manual landmark. Besides, the proposed method also is suitable for clinical work with acceptable results and serves as a quick guideline for non-experts. In future work, we will extend the proposed method to the three-dimensional case by introducing spatial information of CT spine volume and classify various vertebral fractures by the designed multiclass classifier.

Conflicts of Interest

The authors declare that there are no conflicts of interest about this paper.

Acknowledgments

This work was supported by the Science and Technology Development Program of Jilin Province, China (nos. 20150307030GX, 2015Y059, and 20160204048GX), International Science and Technology Cooperation Program of China (Grant no. 2015DFA11180), Science Foundation for Young Scholars of Changchun University of Science and Technology (no.XQNJJ-2016-08), and National Key Research and Development Program of China (no.2017YFC0108303).

References

[1] W. Skalli, C. Vergari, E. Ebermeyer et al., "Early detection of progressive adolescent idiopathic scoliosis: a severity index," *Spine*, vol. 11, no. 42, pp. 823–830, 2017.

[2] T. Kooi, G. Litjens, B. van Ginneken et al., "Large scale deep learning for computer aided detection of mammographic lesions," *Medical Image Analysis*, vol. 35, pp. 303–312, 2017.

[3] T. Vrtovec, F. Pernuš, and B. Likar, "A review of methods for quantitative evaluation of spinal curvature," *European Spine Journal*, vol. 18, no. 5, pp. 593–607, 2009.

[4] Y. L. Chen, W. J. Chen, and W. K. Chiou, "An alternative method for measuring scoliosis curvature," *Orthopedics*, vol. 30, no. 10, pp. 828–831, 2007.

[5] M. C. Tanure, A. P. Pinheiro, and A. S. Oliveira, "Reliability assessment of Cobb angle measurements using manual and digital methods," *Spine Journal*, vol. 10, no. 9, pp. 769–774, 2010.

[6] A. Suzani, A. Rasoulian, A. Seitel, S. Fels, R. N. Rohling, and P. Abolmaesumi, "Deep learning for automatic localization, identification, and segmentation of vertebral bodies in volumetric MR images," *SPIE Medical Imaging, International Society for Optics and Photonics*, article 941514, vol. 9415, article 941514, 2015.

[7] J. Zhang, E. Lou, X. Shi et al., "Computer-aided assessment of scoliosis on posteroanterior radiographs," *Medical and Biological Engineering and Computing*, vol. 2, no. 48, pp. 185–195, 2010.

[8] D. Zukić, A. Vlasák, J. Egger, D. Hořínek, C. Nimsky, and A. Kolb, "Robust detection and segmentation for diagnosis of vertebral diseases using routine MR images," *Computer Graphics Forum*, vol. 33, no. 6, pp. 190–204, 2014.

[9] R. Korez, B. Ibragimov, B. Likar, F. Pernus, and T. Vrtovec, "A framework for automated spine and vertebrae interpolation-based detection and model-based segmentation," *IEEE Transactions on Medical Imaging*, vol. 34, no. 8, pp. 1649–1662, 2015.

[10] A. P. Pinheiro, J. C. Coelho, A. C. P. Veiga, and T. Vrtovec, "A computerized method for evaluating scoliotic deformities using elliptical pattern recognition in X-ray spine images," *Computer methods and programs in biomedicine*, vol. 161, pp. 85–92, 2018.

[11] B. Glocker, D. Zikic, E. Konukoglu, D. R. Haynor, and A. Criminisi, "Vertebrae localization in pathological spine CT via dense classification from sparse annotations," *Medical image computing and computer-assisted intervention*, vol. 16, pp. 262–270, 2013.

[12] J. H. Zhang, H. J. Li, L. Lv, and Y. Zhang, "Computer-aided Cobb measurement based on automatic detection of vertebral slopes using deep neural network," *International Journal of Biomedical Imaging*, vol. 2017, Article ID 9083916, 6 pages, 2017.

[13] H. Wu, C. Bailey, P. Rasoulinejad, and S. Li, "Automatic landmark estimation for adolescent idiopathic scoliosis assessment using BoostNet," *International Conference on Medical Image Computing and Computer-Assisted Intervention*, vol. 2017, pp. 127–135, Springer, Cham, Switzerland, 2017.

[14] B. M. Kelm, M. Wels, S. K. Zhou et al., "Spine detection in CT and MR using iterated marginal space learning," *Medical Image Analysis*, vol. 17, no. 8, pp. 1283–1292, 2013.

[15] Y. L. Cai, L. Mark, T. L. David, A. Kornecki, A. Lum, and S. Li, "Multi-modal vertebrae recognition using transformed deep convolution network," *Computerized Medical Imaging and Graphics*, vol. 51, pp. 11–19, 2016.

[16] M. Rak and K. D. Tönnies, *A Learning-Free Approach to Whole Spine Vertebra Localization in MRI*, MICCAI, vol. 2016, pp. 283–290, Springer, Cham, Switzerland, 2016.

[17] N. Hamdi, K. Auhmani, M. M. R. Hassani et al., "An efficient gentle Adaboost-based approach for mammograms classification," *Journal of Theoretical and Applied Information Technology*, vol. 81, no. 1, pp. 138–143, 2015.

[18] J. Tang and X. Jiang, "A variational level set approach based on local entropy for image segmentation and bias field correction," *Computational and Mathematical Methods in Medicine*, vol. 2017, Article ID 9174275, 15 pages, 2017.

[19] V. Uc-Cetina, C. Brito-Loeza, and H. Ruiz-Piña, "Chagas parasite detection in blood images using AdaBoost," *Computational and Mathematical Methods in Medicine*, vol. 2015, Article ID 139681, 13 pages, 2015.

[20] P. Klęsk, A. Godziuk, M. Kapruziak, and B. Olechet, "Fast analysis of C-Scans from ground penetrating radar via 3-D Haar-like features with application to landmine detection," *IEEE Transactions on Geoscience and Remote Sensing*, vol. 53, no. 7, pp. 3996–4009, 2015.

[21] Z. Guo, L. Zhang, and D. Zhang, "A completed modeling of local binary pattern operator for texture classification," *IEEE Transactions on Image Processing*, vol. 19, no. 6, pp. 1657–1663, 2010.

[22] P. Y. Chen, C. C. Huang, C. Y. Lien, and Y.-H. Tsai, "An efficient hardware implementation of HOG feature extraction for human detection," *IEEE Transactions on Intelligent Transportation Systems*, vol. 15, no. 2, pp. 656–662, 2014.

[23] C. Li, C. Xu, C. Gui et al., "Distance regularized level set evolution and its application to image segmentation," *IEEE Transactions on Image Processing*, vol. 19, no. 12, pp. 3243–3254, 2010.

[24] V. Magudeeswaran and J. F. Singh, "Contrast limited fuzzy adaptive histogram equalization for enhancement of brain images," *International Journal of Imaging Systems and Technology*, vol. 27, no. 1, pp. 98–103, 2017.

[25] T. Van Dao, S. Chaitusaney, and H. T. N. Nguyen, "Linear least-squares method for conservation voltage reduction in distribution systems with photovoltaic inverters," *IEEE Transactions on Smart Grid*, vol. 8, no. 3, pp. 1252–1263, 2017.

[26] K. Takizawa, T. E. Tezduyar, R. Kolesar, C. Boswell, T. Kanai, and K. Montel, "Multiscale methods for gore curvature calculations from FSI modeling of spacecraft parachutes," *Computational Mechanics*, vol. 54, no. 6, pp. 1461–1476, 2014.

[27] The publicity platform SpineWeb, http://spineweb.digitalimaginggroup.ca/dataset.html.

6

Automatic Optic Disc Detection in Color Retinal Images by Local Feature Spectrum Analysis

Wei Zhou,[1] **Hao Wu**,[2] **Chengdong Wu**,[3] **Xiaosheng Yu**,[3] **and Yugen Yi**[4]

[1]*College of Information Science and Engineering, Northeastern University, Shenyang, China*
[2]*Engineering Faculty, University of Sydney, NSW 2006, Austria*
[3]*Faculty of Robot Science and Engineering, Northeastern University, Shenyang, China*
[4]*School of Software, Jiangxi Normal University, Nanchang, China*

Correspondence should be addressed to Wei Zhou; zhouweineu@outlook.com and Chengdong Wu; wuchengdong@ise.neu.edu.cn

Academic Editor: Po-Hsiang Tsui

The optic disc is a key anatomical structure in retinal images. The ability to detect optic discs in retinal images plays an important role in automated screening systems. Inspired by the fact that humans can find optic discs in retinal images by observing some local features, we propose a local feature spectrum analysis (LFSA) that eliminates the influence caused by the variable spatial positions of local features. In LFSA, a dictionary of local features is used to reconstruct new optic disc candidate images, and the utilization frequencies of every atom in the dictionary are considered as a type of "spectrum" that can be used for classification. We also employ the sparse dictionary selection approach to construct a compact and representative dictionary. Unlike previous approaches, LFSA does not require the segmentation of vessels, and its method of considering the varying information in the retinal images is both simple and robust, making it well-suited for automated screening systems. Experimental results on the largest publicly available dataset indicate the effectiveness of our proposed approach.

1. Introduction

Computer based retinal image analysis was first implemented in 1974 [1] and it is now becoming a mainstream technique for quick and accurate detection of retinal diseases such as diabetic retinopathy (DR) and glaucoma [2]. Several important anatomical features appear in the fundus images, such as the retinal blood vessels, the optic disc, and the fovea, as shown in Figure 1. Among these anatomical features, the optic disc appears as a bright, circular-shaped anatomical structure on which the retinal blood vessel network converges (see Figure 1). The position and radius of the optic disc can be used as the references for approximating fovea detection. In addition, the size and shape of the optic disc outer boundaries are useful for diagnosing glaucoma [3]. Therefore, the optic disc is an important anatomical feature in the retinal images, and its detection is a prerequisite for developing automatic screening systems.

Recently, many optic disc detection studies have been presented. Some of them are reviewed below.

The existing works mainly use features such as intensity, appearance, and shape to locate the position of the optic disc. Sinthanayothin et al. [4, 5] proposed optic disc detection methods based on the highest variability in the intensity of adjacent pixels. However, these algorithms do not work well when large areas of bright lesions exist that are similar to the optic disc. Shape is also a useful feature for describing optic discs. Park et al. [6] employed both the round-shape and brightness to detect the optic disc in retinal images. This approach first selects several rounded areas with high intensity variations. Then, the Hough transform is used to estimate the optic disc contour. Finally, the optic disc location is determined by the circle with the highest intensity. However, diseased areas and noisy spots within the retinal images may also present as bright circular lesions that appear similar to the optic disc, which weakens the detection performance. Moreover, optic discs have large variances (e.g., in shape, color, and size), especially in the presence of retinopathies. Hence, localizing the optic disc using only these properties is insufficient [7]. To improve the optic disc detection

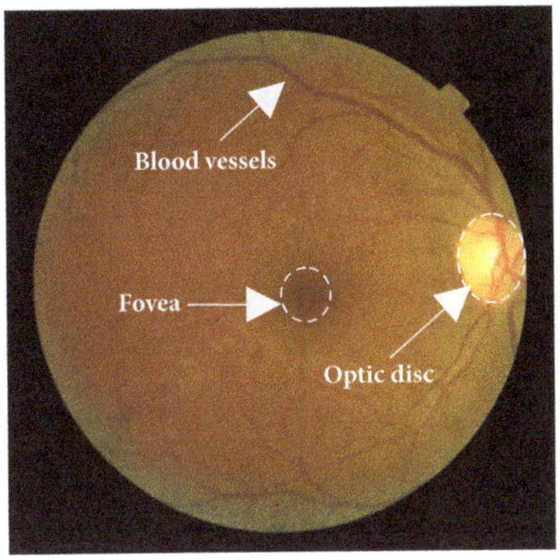

FIGURE 1: Some anatomical features in retinal fundus images.

accuracy, some template matching-based approaches have been proposed to approximately locate the optic disc [8–10]. However, the range of color and contrast variations and the possible presence of pathologies that present as bright spots make it difficult to find a suitable template for matching optic discs. In addition, most of the recently proposed techniques attempt to use information provided by the retinal vasculature to identify the optic disc. In [11], optic disc localization was based on the major blood vessels and their intersection. Hoover et al. [12] identified optic discs using geometrical relationships between the optic disc and blood vessels. Youssif et al. [13] used the vessels' directional matched filter (VDMF) and the outgoing vessels to locate the center of the optic disc. Zhang et al. [14] proposed an optic disc center detection method that used multiscale Gaussian filtering and a VDMF. The optic disc center is determined by the pixel with the smallest difference from its surrounding pixels. The use of a vascular tree can improve the performance of optic disc detection, especially when the optic disc is not clearly visible due to poor contrast or retinal pathologies within the images. However, the accurate segmentation of the vascular tree is a complex and time-consuming task. Apart from the above-mentioned approaches, some machine learning techniques have been incorporated into optic disc localization. Perez et al. [15] utilized a cascade classifier to detect the optic disc. In [15], the cascade classifiers are trained using Haar features extracted from the segmented optic disc and nonoptic disc images. The main drawback of this approach is its computational complexity. Li et al. [16] combined Principle Component Analysis (PCA) with active shape techniques to determine the center of the optic disc by the minimum distance between the original retinal image and its projection onto "disk space". However, it is difficult to choose a suitable shape model to detect the various disc shapes resulting from many pathological changes.

Although the above-mentioned works have achieved good performances, each method has its own limitations. In summary, the complex vessel structures and its occlusion position variations on the optic disc are two main challenging issues. In other words, we need to detect the optic disc from a complex scene.

In pattern recognition and computer vision, the idea of global frequency with local spatial constraints is widely employed to the complex scene classification [17]. Inspired by it, Li et al. used intermediate representations before classifying scenes [18]. In their approach, they model an image as a collection of local patches. Each patch is obtained from a large vocabulary of codewords. Their goal is to learn a model that best represents the distribution of the codewords of a particular image. Furthermore, it is easier to understand a complex scene by going through a generative model.

Optic disc always varies in terms of appearance, size, and location in different retinal images. Humans can recognize images containing optic discs effortlessly by observing parts of the local features of a retinal image, but machines seem to have a hard time doing the same task. Inspired by this, we propose a novel feature extraction method named local feature spectrum analysis (LFSA), which eliminates the influence caused by the variance in the spatial positions of local features. The local feature spectrum is stable for most optic discs. Consequently, we need only a few training samples to achieve a satisfactory detection performance. A flow chart of our approach is depicted in Figure 2.

The remainder of this paper is organized as follows: Section 2 describes the process of candidate extraction. Section 3 presents the proposed optic disc detection approach. Experimental results are depicted in Section 4 and Section 5 concludes the paper.

2. Candidate Extraction

By utilizing some basic image processing techniques, we can acquire a series of candidates and some of them contain the optic disc. And then, the optic disc will be finally identified using the proposed LFSA approach. In this section, the candidate extraction is given briefly.

There are many approaches that have been proposed for candidate extraction such as intensity-based thresholding [4, 5] and template matching [6, 8–10]. Here, we employ our previous study [19] to extract the optic disc candidates which uses a two-step reconstruction based saliency detection approach. It is particularly suitable for those retinal fundus images with uneven illumination and poor contrast. Meanwhile, the optic disc often appears as the area with the highest contrast compared to the background in the red channel [20, 21]. Therefore, the red channel image is employed in this paper.

Some of the optic disc and nonoptic disc candidate example images are shown in Figure 3, in which the first row of images is the extracted optic disc images and the second row of images is the nonoptic disc images.

3. The Proposed Method

The proposed optic disc detection approach consists of the following four steps. First, local features are extracted from the retinal fundus images. Second, with the use of the extracted local features, an optimal dictionary is generated. Third, the frequencies of every atom in the dictionary used for new image reconstruction are considered as a kind of

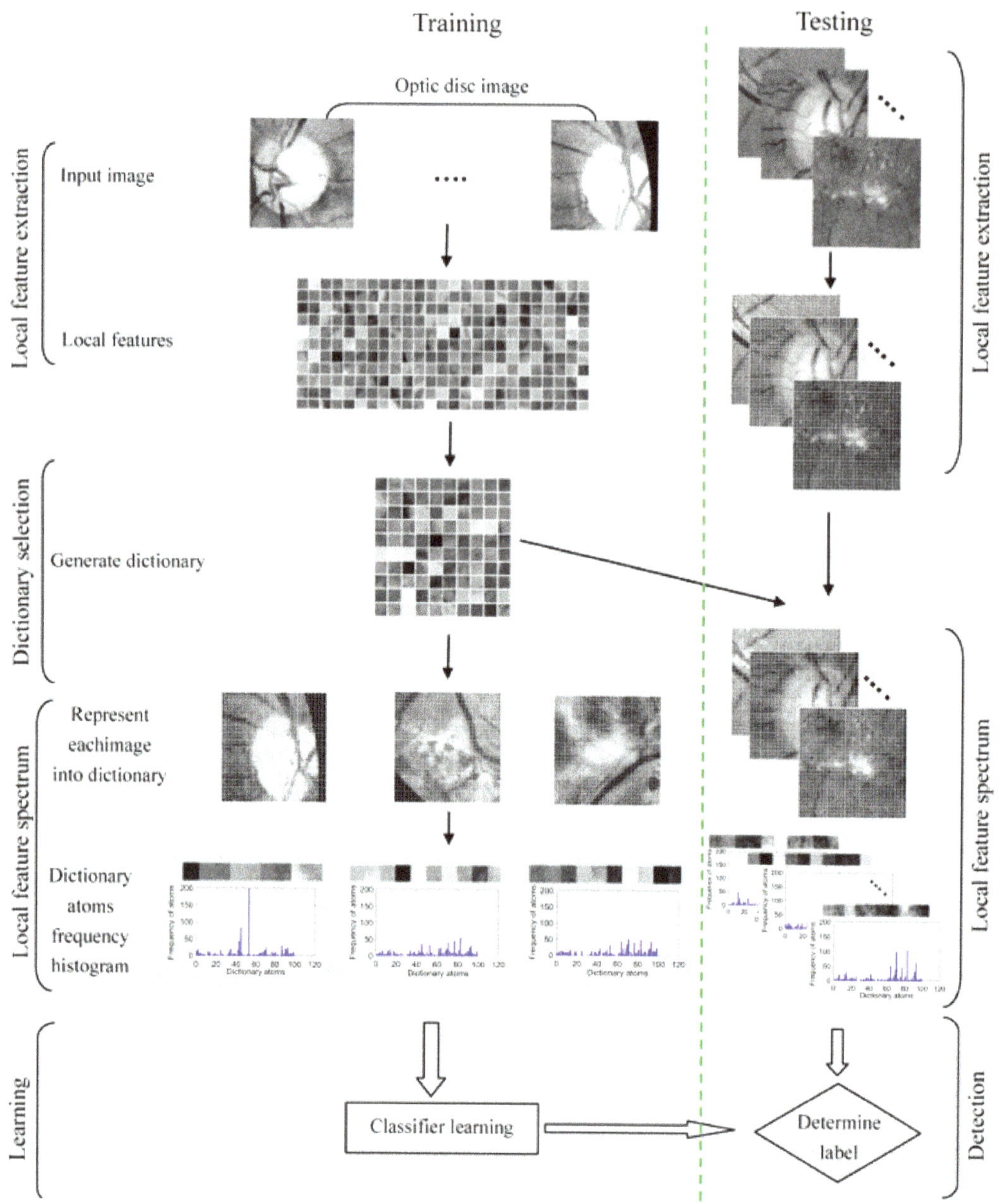

FIGURE 2: The flow chart of our work.

"spectrum". Finally, we use the generated "spectrum" feature for optic disc classification. Each step of the proposed method will be described in detail as follows.

3.1. Local Feature Extraction. After completing candidate extraction, we obtain a series of subimages (candidates) whose size is 300 × 300 pixels. Then, we define small image patch sizes of $M \times M$ pixels as the local features. Each retinal image is partitioned into n nonoverlapping patches where $n = (300/M)^2$. Each patch is ordered lexicographically and denoted as $r_i \in \Re^{d\times 1} (i = 1, 2, \ldots, n)$ where $d = M \times M$. Subsequently, the full set of the local features of a retinal image can be denoted as $R = [r_1, r_2, \ldots, r_n] \in \Re^{d\times n}$. Here, the parameter M ranges among the values in the set $\{5, 10, 30, 50, 100\}$. The corresponding segmentation results are depicted in Figure 4.

3.2. Dictionary Selection. A dictionary used to reconstruct a whole image can be acquired by using local features. Since the dictionary can directly determine the reconstruction error and the final classification performance, how to select an optimal dictionary is a key issue.

Traditional intermediate representation based models [17, 18] obtain a dictionary of local features by clustering local image descriptors extracted from images. However, clustering methods have several limitations. For example, the number of clusters k needs to be specified in advance, which is considered to be one of the greatest drawbacks

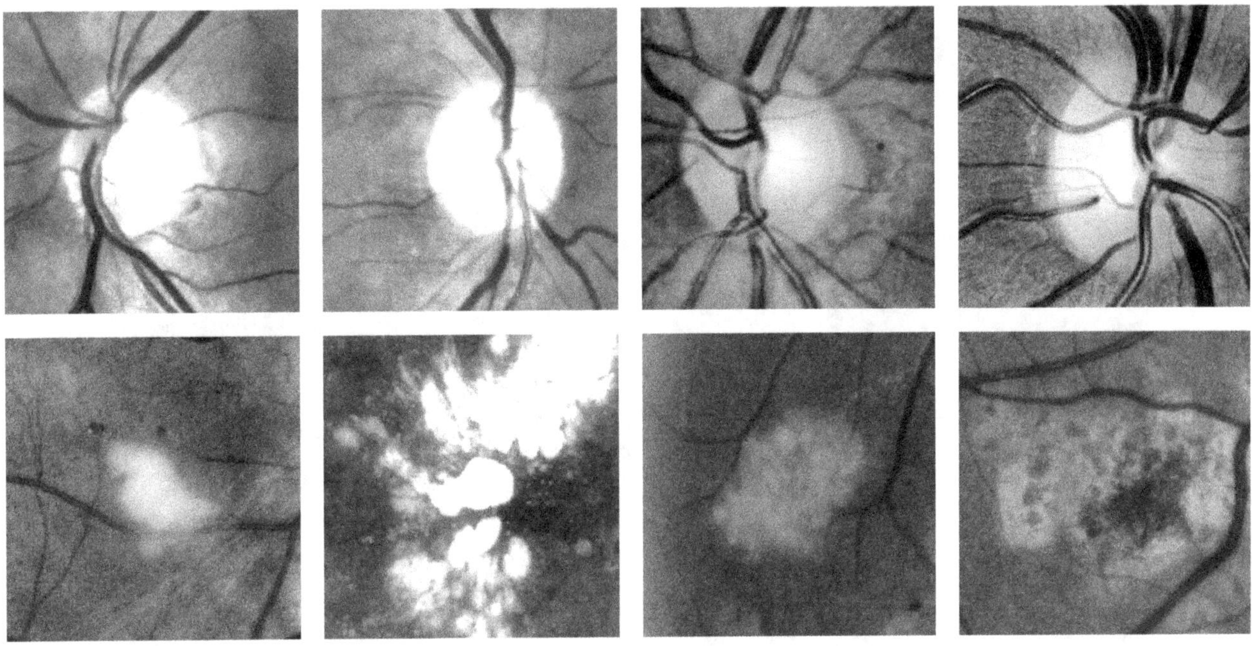

FIGURE 3: Examples of extracted optic disc regions (the first row) and nonoptic disc regions (the second row).

of these algorithms. Furthermore, the algorithms prefer clusters of approximately similar size, because they always assign an object to the nearest centroid. This often leads to incorrectly defined borders between clusters. Additionally, the clustering process does not consider representational ability; consequently, some clusters will be imported that are useless for image reconstruction tasks. All these limitations affect the classification performance.

In this subsection, with the aim of representing all candidates fully, we employ the sparse dictionary selection approach proposed by Cong et al. [22] to construct an optimal dictionary. Each sample is given a weight that reflects its representational ability. The most representative samples are selected as dictionary atoms [23–28].

After image partitioning, we obtain many equal-sized patches from all the training images. Let $X = [x_1, x_2, ..., x_N] \in \Re^{d\times N}$ denote the data matrix that includes all patches. Here, $x_j \in \Re^{d\times 1}$ denotes the j-th sample and N is the number of samples. Our goal is to select an optimal subset to form a dictionary $X' = [x'_1, x'_2, ..., x'_M] \in \Re^{d\times Dsize}$ such that the set X is well represented by the X' with the smallest $Dsize$. This dictionary selection task can be achieved by solving the following optimization problem:

$$\min_{B} \frac{1}{2} \|X - XB\|_F^2 + \lambda \|B\|_{2,1}, \tag{1}$$

where $B \in \Re^{N\times N}$. Here, $\|B\|_F := \sum_{i,j} B_{ij}^2$ is defined as the Frobenius norm of B and $\|B\|_{2,1} := \sum_{i=1}^{N} \|B_{i.}\|_2$ is defined as the $l_{2,1}$-norm of B, where $B_{i.}$ denotes the i-th row of B.

The objective function in (1) consists of two terms, the first term is the representation error, and the second term is the regularization term. The $l_{2,1}$-norm is a general version of the l_1-norm when B is a vector, i.e., $\|B\|_{2,1} = \|B\|_1$. In addition, by constructing a new vector $b \in \Re^N$ with $b_i = \|B_{i.}\|_2$, $\|B\|_{2,1}$ is equivalent to $\|b\|_1$. From this point, it is easy to conclude that minimizing (1) will lead to row sparsity in the representative matrix B; i.e., B usually contains some zero rows because the corresponding samples in X were not selected as dictionary atoms [22]. Here, λ is a parameter used to balance the first term and the second term. A larger λ leads to more zero rows in B.

As shown in (1), the objective function involves the $l_{2,1}$-norm, which is nonsmooth and cannot be solved by a closed form [22]. Therefore, in this paper, we adopt the Nesterov [29] algorithm to solve the optimization problem. After obtaining the optimal B, we compute each sample's weight using $\|B_{i.}\|_2$ $(i = 1, 2, ..., N)$ and sort all the weight values in descending order. Finally, we select the top $Dsize$ ranked dictionary atoms to construct the optimal dictionary V.

Here, using the patch size parameter M=10 and the number of dictionary atoms $Dsize$=100 as an example, the optimal dictionary $V \in \Re^{Dsize\times N}$ can be obtained by the above-mentioned dictionary selection approach. Figures 5(a) and 5(b) depict the segmented image patches and the selected optimal dictionary, respectively.

3.3. Local Feature Spectrum. Based on the obtained dictionary V, we can represent an image as a "bag" of "dictionary atoms", disregarding the atoms' order. For example, for each atom w in V, the local feature spectrum analysis model estimates the frequencies of atoms in an image as follows:

$$LFSA(w) = \sum_{i=1}^{n} \begin{cases} 1 & if\ w = \arg\min_{v\in V} (D(v, r_i)); \\ 0 & otherwise, \end{cases} \tag{2}$$

where n is the number of local features in an image, r_i is the i-th local feature, and $D(v, r_i)$ is the distance between a

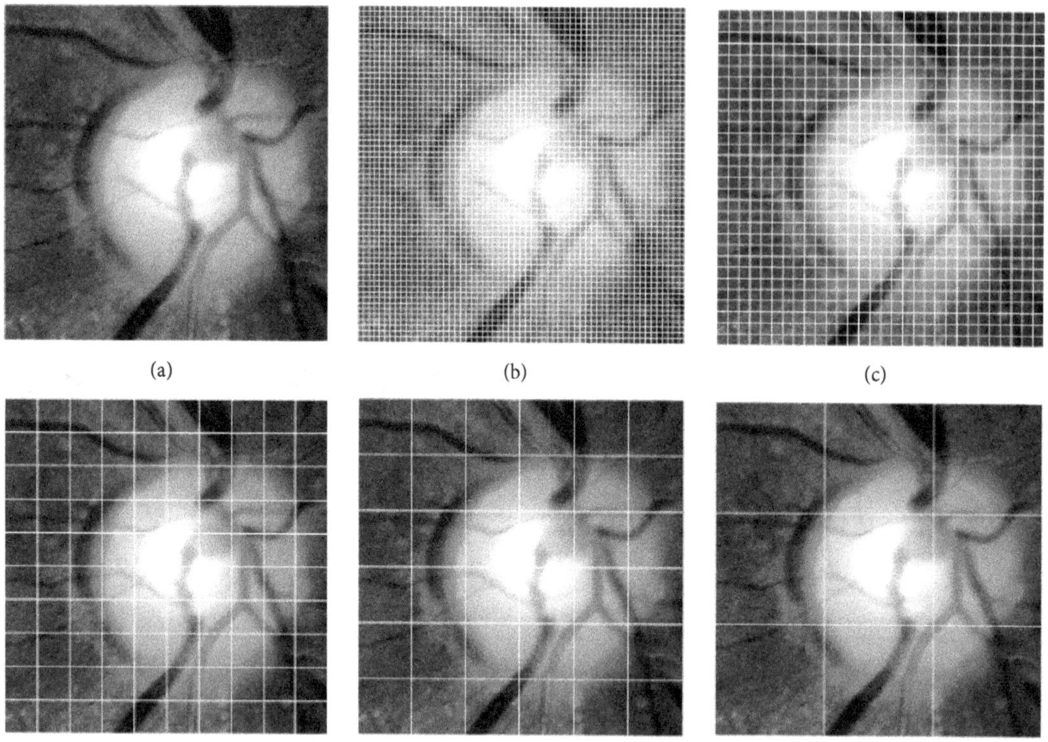

FIGURE 4: Image partition results using different patch sizes. (a) Original image; (b-e) different patch sizes (from left to right: 5×5, 10×10, 30×30, 50×50, and 100×100 pixels).

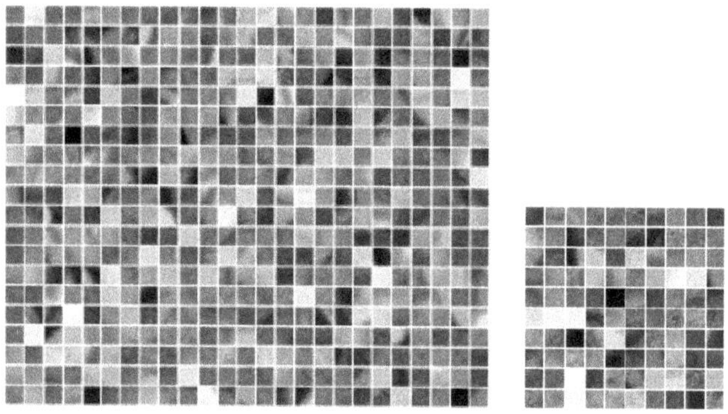

FIGURE 5: (a) A portion of the example patches used to construct the dictionary; (b) the selected dictionary.

dictionary atom v and a local feature r_i. The atom with the smallest distance calculated by $D(v, r_i)$ is used to represent the current patch.

Figure 6 gives the details of local feature spectrum extraction. First, given several candidate region images obtained by our previous approach (see Figures 6(a1)–6(a4)), we partition them into a series of patches, regarded as local features (see Figures 6(b1)–6(b4)). Second, for each patch, the atom with the smallest distance to the current patch is selected to reconstruct it. This process is repeated until all the patches have been reconstructed. Then, we can obtain the full reconstructed images (see Figures 6(c1)–6(c4)) and the corresponding error images, which are calculated by subtracting the original image (see Figures 6(a1)–6(a4)) from the reconstructed image (see Figures 6(c1)–6(c4)) as shown in Figures 6(d1)–6(d4). Finally, we describe an image by the frequency over dictionary atoms as shown in Figures 6(e1)–6(e4).

3.4. Classification. In this subsection, we employ the generated local feature spectra (see Figures 6(e1)–6(e4)) as the features for optic disc classification. Two widely used classifiers including k-Nearest Neighbor (kNN) [30] and

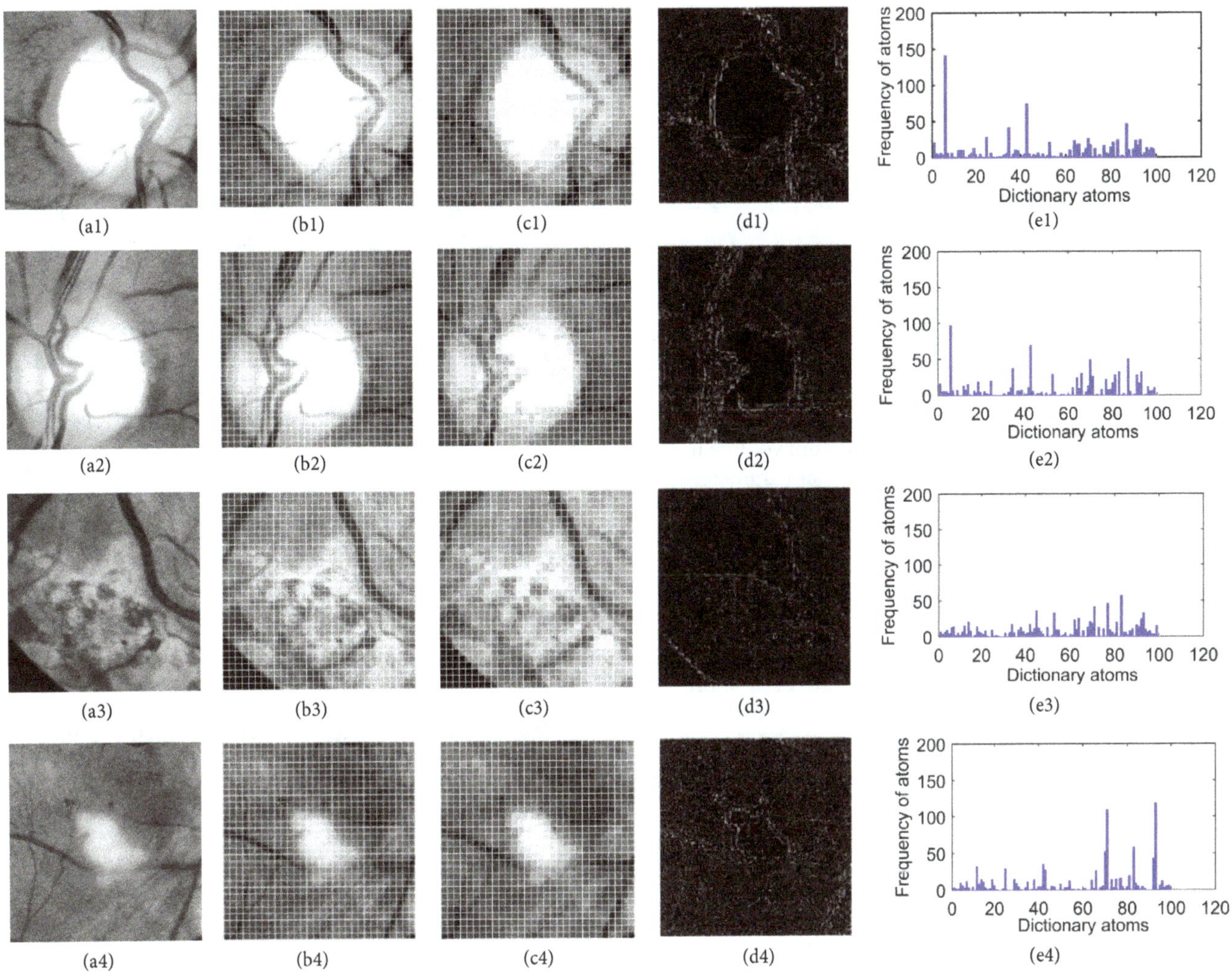

FIGURE 6: Local feature spectrum analysis. (a1–a4) Original optic disc and nonoptic disc images; (b1–b4) patch segmentation with 10 × 10 pixels; (c1–c4) reconstruction results of original images; (d1–d4) the error images between the original images (a1–a4) and the reconstructed images (c1–c4); (e1–e4) local feature spectrum analysis results for (a1–a4), respectively.

the support vector machine (SVM) [31] are utilized in this paper.

First, the obtained local feature spectra of all the candidates can be expressed as a set of *Dsize* dimensional features:

$$F = \{f_1, f_2, \ldots, f_i, \ldots, f_S\}, \quad (3)$$

where $f_i \in \mathfrak{R}^{Dsize \times 1}$ denotes the i-th candidate and S is the number of candidates.

Then, each of these features is normalized to a vector with zero mean and unit variance by applying

$$f' = \frac{f - \mu}{\sigma}, \quad (4)$$

where μ and σ are the mean and standard deviation vector of the feature, respectively.

Finally, with the use of these normalized spectrum features, we can train the classifier for distinguishing the optic disc candidates from the nonoptic disc candidates, achieving classification.

However, one main problem occurs when performing binary classification for computer-aided diagnosis of medical images. That is, the object candidate (positive samples) features we are trying to identify are similar, whereas the features of nonobject candidates (negative samples) vary widely. For example, in our study, the local feature spectra of optic disc candidates are stable and similar, whereas the local feature spectra of nonoptic disc candidates are varied, as shown in Figure 6. When training a binary classifier, it is desirable that the negative samples should appear in sufficient quantity and species. Therefore, the training set should contain most types of nonobject candidates. Otherwise, the trained model will not achieve a good classification performance. Meanwhile, the number of positive samples should approximately equal the number of negative samples to avoid the class imbalance problem, which also substantially reduces classification accuracy [32]. However, collecting all types of nonobject candidates is difficult. Therefore, we employ a one-class classifier to identify objects of a specific class among all the objects by learning from a training set containing only one-class object.

Table 1: Descriptive statistics of the Messidor database.

	Resolution	Coverage of the retina	Number of images
Type	1440 × 960	45°	151
Type	2240 × 1488	45°	881
Type	2304 × 1536	45°	168

As an unsupervised machine learning algorithm, one-class SVM (also called support vector data description or support vector domain description (SVDD)) was proposed by Schölkopf et al. [33] to estimate high-dimensional distributions. SVDD first maps the data from the original space to the feature space using a nonlinear transformation and then finds the hypersphere with the minimum volume in the feature space. Parameters such as the kernel parameter and slack factor in SVM and SVDD or the value of k in kNN are determined using cross validation [34].

4. Experimental Results and Analysis

4.1. Database. In this section, we employ the publicly available Messidor database to evaluate the proposed approach [35]. The Messidor database contains 1,200 color fundus images with three different resolutions: 1,440 × 960, 2,240 × 1,488, and 2,304 × 1536 pixels. Each image has a reference standard marking every optic disc and agreed upon by the consensus of 4 experts. All the images in Messidor were acquired by three ophthalmologic departments using a color video 3CCD camera on a Topcon TRC NW6 non-mydriatic retinograph with a 45-degree field of view. Table 1 lists the detailed information of this database. In our experiment, all the retinal images are resized to 1,440 ×960 pixels using bilinear interpolation.

From the Messidor database, after performing candidate extraction, we obtained 1,200 optic disc candidates and 10,985 nonoptic disc candidates. In our experiments, we selected all the optic disc samples and the same number of nonoptic disc samples for training and testing. Classification accuracy is used as the evaluation criterion in Sections 4.2, 4.3, and 4.4. However, we adopt detection accuracy as the evaluation criterion in the last experiment in Section 4.5. From all the selected samples, we randomly selected 70% for training and used the remaining samples for testing. This random sample selection was repeated 10 times, and we report the average classification accuracy as the final result.

4.2. Parameter Selection. The proposed method has two parameters: patch size (M) and the number of dictionary atoms ($Dsize$), and determining their values appropriately is important in the proposed approach. Here, the patch sizes are set to 5×5, 10×10, 30×30, 50×50, and 100×100. The number of dictionary atoms varies with different patch sizes. For example, we tuned the value of the $Dsize$ parameter by searching the grids $\{5, 10, 15, 20, 25\}$ and $\{10, 30, 50, 70, 90, 100\}$ for patches with 5 × 5 and 10 × 10 pixels, respectively. We tuned $Dsize$ on the grid $\{50, 150, 300, 500, 700, 1000\}$ by searching for patches with 30 × 30, 50 × 50, and 100 × 100 pixels.

Table 2: The average classification accuracy rates (%), standard deviations, (%) and reconstruction errors (R_error) of the proposed approach with M=5.

Dsize	5	10	15	20	25
SVM + random	92.01 ±2.15	94.23 ±2.47	95.57 ±2.28	95.04 ±2.63	94.80 ±2.89
SVM + K-means	93.67 ±2.44	94.88 ±3.01	96.23 ±2.50	97.67 ±2.37	96.35 ±2.89
SVM + ours	96.25 ±2.42	98.00 ±1.96	98.50 ±1.82	99.00 ±2.29	98.80 ±2.33
kNN + random	91.36 ±2.17	92.75 ±2.46	94.37 ±2.86	94.50 ±2.05	94.21 ±2.49
kNN + K-means	92.69 ±3.26	93.97 ±2.87	95.61 ±2.54	95.11 ±2.63	94.93 ±2.37
kNN + ours	94.50 ±2.79	97.25 ±2.33	98.00 ±2.13	98.50 ±2.35	98.31 ±2.64
R_error (random)	0.427	0.410	0.407	0.398	0.378
R_error (K-means)	0.329	0.318	0.312	0.301	0.296
R_error (ours)	0.118	0.116	0.108	0.105	0.094

We compared three dictionary selection methods, including K-means clustering, sparse dictionary selection, and random selection, under different patch sizes and atom numbers. The kNN and SVM algorithms are employed for classification, and classification accuracy is adopted as the evaluation criterion. We also report the average reconstruction error, which is defined as the mean value of the mean square error between the original images and the reconstructed images. The reconstruction error can be calculated as follows:

$$R\_error = \frac{1}{T}\sum_{t=1}^{T} \left\| I_t - I_t^R \right\|_2 \quad (5)$$

where I_t is the t-th original image, I_t^R is the t-th reconstructed image, and T is the total number of reconstructed images. A smaller error represents a better reconstruction, but the experimental results show that the minimum error does not equate the highest classification accuracy. Therefore, the best number of dictionary atoms will be found via experiment.

Tables 2–6 show the average classification accuracies obtained by two different classifiers (SVM and kNN) with three different dictionary selection methods (random selection, K-means clustering, and our sparse dictionary selection) and the corresponding reconstruction errors. For example, we denote "SVM + random" as the combination of the SVM classifier with random sample selection dictionary construction for classification, and R_error (random) denotes the reconstruction error using a dictionary constructed by random sample selection.

In each table, we fixed the value of M. In Table 2, for example, the classification performance obtained by the proposed approach is not good when $Dsize$ is small. However, as $Dsize$ increases, the classification accuracy improves. The best performance was achieved with $Dsize$=20 using SVM and

TABLE 3: The average classification accuracy rates (%), standard deviations (%), and reconstruction errors (*R_error*) of the proposed approach with *M*=10.

Dsize	10	30	50	70	90	100
SVM + random	94.07 ±2.96	94.79 ±2.78	95.14 ±2.70	96.88 ±2.87	96.62 ±2.34	96.51 ±3.06
SVM + *K*-means	95.63 ±2.45	95.77 ±2.62	96.21 ±2.33	97.96 ±2.91	98.71 ±1.99	97.60 ±2.98
SVM + ours	98.50 ±2.10	98.75 ±1.32	99.20 ±1.93	99.50 ±2.06	99.75 ±1.79	99.65 ±1.96
*k*NN + random	93.89 ±3.12	94.20 ±2.09	95.53 ±2.96	96.34 ±1.98	96.55 ±2.37	96.43 ±2.55
*k*NN + *K*-means	94.22 ±2.93	95.23 ±2.68	96.60 ±2.45	97.59 ±2.85	98.21 ±2.14	97.83 ±2.22
*k*NN + ours	97.25 ±2.13	97.25 ±2.27	98.50 ±2.35	98.64 ±2.13	98.80 ±2.25	98.58 ±2.38
R_error (random)	0.449	0.437	0.423	0.415	0.408	0.396
R_error (*K*-means)	0.334	0.326	0.320	0.316	0.306	0.297
R_error (ours)	0.181	0.176	0.162	0.153	0.140	0.138

TABLE 4: The average classification accuracy rates (%), standard deviations (%), and reconstruction errors (*R_error*) of the proposed approach with *M*=30.

Dsize	50	150	300	500	700	1000
SVM + random	92.95 ±3.64	95.79 ±3.86	95.09 ±2.14	94.36 ±2.75	94.17 ±2.49	94.02 ±2.16
SVM + *K*-means	93.83 ±3.33	97.02 ±3.64	97.36 ±2.88	96.24 ±2.69	96.03 ±2.72	95.83 ±2.99
SVM + ours	97.75 ±3.10	98.72 ±3.29	97.75 ±2.67	97.00 ±2.71	96.50 ±2.84	96.50 ±2.76
*k*NN + random	92.14 ±3.69	96.86 ±3.91	93.04 ±2.80	92.96 ±3.87	92.36 ±3.60	90.05 ±3.49
*k*NN + *K*-means	92.96 ±3.59	97.50 ±3.55	93.76 ±2.73	93.45 ±3.93	92.61 ±3.47	91.34 ±3.91
*k*NN + ours	96.50 ±3.17	98.00 ±3.06	94.00 ±2.89	93.75 ±3.04	92.75 ±3.21	91.75 ±3.09
R_error (random)	0.463	0.448	0.440	0.436	0.429	0.418
R_error (*K*-means)	0.356	0.349	0.336	0.328	0.319	0.311
R_error (ours)	0.183	0.178	0.174	0.169	0.168	0.162

sparse dictionary selection. Subsequently, the classification accuracy begins to drop, whereas the reconstruction error continues to drop as *Dsize* increases. This result occurs because as the number of dictionary atoms selected for reconstruction becomes larger, the dictionary atoms are more sufficient, leading to a smaller reconstruction error. However, too many atoms introduce noise and cause instability in the local feature spectrum. Thus, although *R_error* always falls as *Dsize* increases, the classification accuracy peaks at a certain *Dsize*.

TABLE 5: The average classification accuracy rates (%), standard deviations (%), and reconstruction errors (*R_error*) of the proposed approach with *M*=50.

Dsize	50	150	300	500	700	1000
SVM + random	91.56 ±2.98	94.76 ±3.77	93.49 ±3.16	94.27 ±2.75	93.17 ±3.73	93.01 ±2.39
SVM + *K*-means	92.98 ±2.75	95.32 ±3.39	94.76 ±3.64	95.19 ±2.63	94.24 ±3.85	93.94 ±2.73
SVM + ours	94.75 ±2.53	95.48 ±3.46	96.50 ±3.10	95.45 ±2.12	94.42 ±3.44	94.40 ±2.53
*k*NN + random	90.13 ±3.17	91.83 ±3.94	94.15 ±3.19	91.37 ±2.34	91.01 ±3.73	90.07 ±2.39
*k*NN + *K*-means	90.87 ±3.61	92.09 ±3.46	95.09 ±3.90	91.88 ±2.65	90.35 ±3.94	90.78 ±2.31
*k*NN + ours	91.95 ±3.11	92.25 ±3.77	96.25 ±3.56	92.00 ±2.89	91.75 ±3.13	91.25 ±2.71
R_error (random)	0.472	0.465	0.459	0.447	0.436	0.429
R_error (*K*-means)	0.375	0.367	0.358	0.344	0.329	0.320
R_error (ours)	0.194	0.187	0.184	0.179	0.177	0.174

TABLE 6: The average classification accuracy rates (%), standard deviations (%), and reconstruction errors (*R_error*) of the proposed approach with *M*=100.

Dsize	50	150	300	500	700	1000
SVM + random	91.51 ±3.76	93.16 ±3.49	93.68 ±3.49	94.03 ±3.18	93.11 ±3.71	91.85 ±2.54
SVM + *K*-means	92.56 ±3.61	94.03 ±3.55	94.36 ±3.72	94.87 ±3.54	93.96 ±3.34	92.03 ±2.72
SVM + ours	94.15 ±3.15	94.25 ±3.19	94.75 ±3.47	95.01 ±3.06	94.00 ±3.35	92.25 ±2.94
*k*NN + random	90.07 ±3.89	90.83 ±3.36	92.07 ±3.82	91.36 ±2.44	90.36 ±3.79	89.06 ±3.99
*k*NN + *K*-means	91.11 ±3.13	91.09 ±3.21	92.31 ±3.64	92.89 ±2.30	90.87 ±3.81	89.47 ±3.74
*k*NN + ours	92.50 ±3.34	91.30 ±3.51	92.50 ±3.04	93.23 ±2.98	91.00 ±3.33	89.75 ±3.36
R_error (random)	0.486	0.479	0.468	0.456	0.439	0.435
R_error (*K*-means)	0.384	0.380	0.371	0.362	0.358	0.349
R_error (ours)	0.225	0.217	0.213	0.209	0.207	0.205

The five tables correspond to five different values of M in the set $\{5, 10, 30, 50, 100\}$. The best scores corresponding to each M are 99% (*Dsize*=20), 99.75% (*Dsize*=90), 98.72% (*Dsize*=150), 96.50% (*Dsize*=300), and 95.01% (*Dsize*=500) using SVM and our dictionary selection approach. Our method's best classification accuracy is 99.75% when *Dsize*=10 and *M*=90. This result demonstrates that the local feature spectrum with *Dsize*=10 and *M*=90 can best reveal the class of an optic disc candidate. Humans tend to find the optic disc by observing local image features with a certain size. When the feature size is too small, all types of images can be represented by the same local features, making the spectra for different classes of candidates inseparable. In contrast, when the feature size is too large, each feature becomes quite specific, which is also unsuitable for reconstructing and recognizing candidate images.

In addition, compared to the dictionary constructed by random selection and K-means clustering, the sparse dictionary selection approach performs well for image reconstruction and supports a better classification performance. These results may occur for several reasons. First, building a dictionary by selecting candidates randomly is a simple idea; however, it carries a large risk of including noisy candidates, which affects the reconstruction. Second, in K-means clustering, the cluster number is preset and the method does not consider the representative ability of each cluster. Consequently, some useless clusters will be imported for the image classification task. Unlike these two methods, the sparse dictionary selection approach can select the optimal dictionary to reconstruct the candidates. Then, we obtain a dictionary with the minimal number of atoms and discard redundant and noisy samples. Therefore, this approach increases both the accuracy and the computational efficiency.

In all the experimental results, SVM performs better than kNN. In particular, when M=10 and $Dsize$=90, our approach achieves its best classification performance of 99.75% using "SVM+ours". In the following experiments, we will regard M=10 and $Dsize$=90 as the optimal parameters.

4.3. One-Class Classification. As mentioned in Section 3.4, in practice, the number of nonobjective candidates may be insufficient. Therefore, in this subsection, we compare the performances of an unsupervised one-class SVDD classifier and a supervised binary SVM classifier. The patch size and the number of dictionary atoms are fixed to M=10 and $Dsize$=90, respectively. The training samples are randomly selected ten times in the training set, while the testing set remains unchanged. The number of optic disc samples varies from 10 to 840 and the number of nonoptic disc samples is equal to the number of optic disc samples to maintain class balance. As an unsupervised classifier, SVDD does not require nonoptic disc samples. The average classification accuracy is shown in Figure 7. Both the mean value and the standard variance are provided to indicate the classification performances.

As shown in Figure 7, when there are fewer than 400 nonoptic disc samples, SVDD outperforms SVM. For SVDD, the curve tends to become stable after the number of samples exceeds 500, and it reaches its highest accuracy (96%) at 840 samples. For SVM, the curve tends to become stable after the number of samples exceeds 800, and it reaches its highest accuracy (99.75%) after the number of samples exceeds 800. The standard variances of both methods decrease as the number of training samples increase, which means that models obtained using more samples also have more stable prediction and detection performances.

Based on these observations, we can conclude that, on one hand, when only a few positive samples (optic discs) are available, SVDD performs better than SVM. On the other hand, when sufficient negative samples (nonoptic discs) exist,

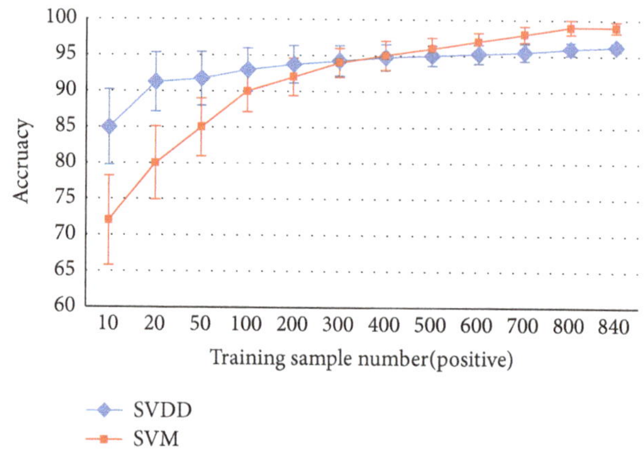

FIGURE 7: The classification accuracies of SVM and SVDD using different numbers of training samples.

TABLE 7: The average classification accuracy rate (%) of the baseline method and our approach using different classifiers.

Classifier	Baseline	Our approach
SVM	93.50	99.75
kNN	91.50	98.80

the SVM model is more accurate and stable. In practice, when negative samples are hard to obtain, we should choose the unsupervised one-class classifier SVDD. Alternatively, if higher accuracy is required, we need to acquire enough and sufficient training samples and the supervised binary classifier SVM. For this paper, we can acquire sufficient labeled samples. Therefore, we compare the detection results using SVM with those of other state-of-the-art methods.

4.4. Comparison with the Baseline Approach. Here, we regard the classification based on the original gray feature as the baseline approach. That is, we directly utilize the candidates obtained by Section 2 as the input samples for training and testing. The corresponding classification results with different classifiers are listed in Table 7. As Table 7 shows, our approach achieves better performances than the baseline algorithm using both classifiers, which indicates that the local feature spectrum analysis process not only reduces the number of input features but also improves classifier performance.

4.5. Comparison with State-of-the-Art Approaches. In the last experiment, to compare the proposed approach with other state-of-the-art approaches, we adopt the accuracy of optic disc detection as the evaluation criterion [30–32]. When the detected optic disc center is within the circumference of the optic disc in the reference standard given in the database, then the detection is considered to be successful. Table 8 shows the comparison results on the Messidor dataset.

The comparative results listed in Table 8 show that the proposed approach is more reliable than other tested approaches in terms of detection accuracy, indicating the effectiveness of the proposed approach. Specially, traditional approaches [36, 37] including our approach outperform the

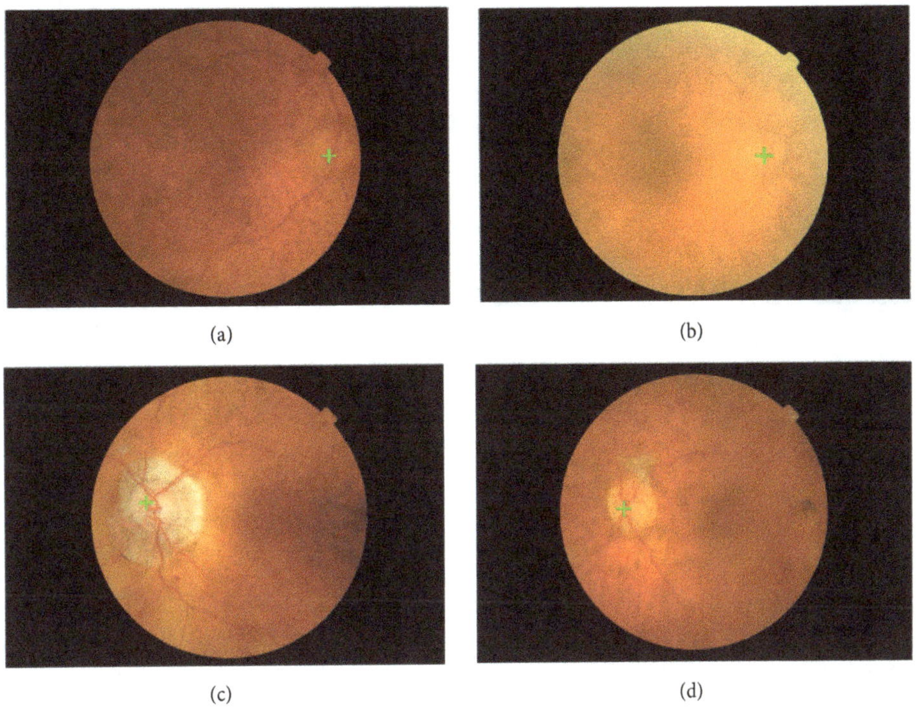

FIGURE 8: Results of the proposed method. The green cross represents the detected optic disc center. (a) Uneven optic disc brightness with blood vessels; (b) blurred optic disc appearance; (c) optic disc appearance with peripapillary atrophy; (d) myelinated nerve fibers connected to optic disc margin.

TABLE 8: Successful detection rate (%) of different methods for optic disc detection on the Messidor dataset.

Methods	Successful detection rate (%)
Ahmed et al., [36]	97.80
Aquino et al., [37]	98.83
Al-Bander et al., [38]	97.00
Ours	99.83

deep neural network (DNNs) based approach [32]. The main reason is that there is still an improvement room for the novel techniques, especially for small datasets.

Figure 8 shows some examples where the proposed approach successfully detected the optic disc center. The center of the candidate subimage is considered as the center of optic disc and marked with green cross. As shown by Figure 8, the proposed approach is robust to the influence caused by the variations in the optic disc's appearance, size, or location (see Figures 8(a), 8(c), and 8(d)). And it is also robust to uneven illumination, which results in the optic disc appearing indistinct and blurred (see Figure 8(b)). Overall, our approach uses only the local feature spectrum without requiring any other information, such as blood vessels or predesigned templates. Therefore, the proposed approach is both simple and efficient for optic disc detection, especially in abnormal retinal images.

The Messidor database includes two images in which our proposed approach failed to locate the optic disc, as shown in Figure 9. The detected locations of the proposed approach are marked by green crosses, while the ground truth locations are marked by black crosses. In these cases, extra-large and bright lesions overshadow the optic discs, making them invisible. The disc areas are not included in the intensity image (see Figures 9(a) and 9(b)) which should be considered in future work.

5. Conclusions

Complex structures and their spatial variability reduce the recognition rate of optic discs. Inspired by the fact that humans can find optic discs in images by observing local features, we propose using local feature spectrum analysis to eliminate the influence caused by the variability of spatial locations. In the proposed approach, each candidate is reconstructed using a set of local features, and the frequencies of all the local features are considered as a type of "spectrum" suitable for the recognition tasks. To find an optimal set of local features, we employ the sparse dictionary selection approach, which aims to find the most representative atoms. We also discuss the common class imbalance problem in medical dataset. When there are insufficient numbers of negative samples, a one-class classifier performs better, whereas when sufficient samples are available, a binary classifier achieves more precise classification performances. The proposed approach is tested on a publicly available Messidor database. The experimental results indicate that the proposed approach yields a better performance than other state-of-the-art approaches.

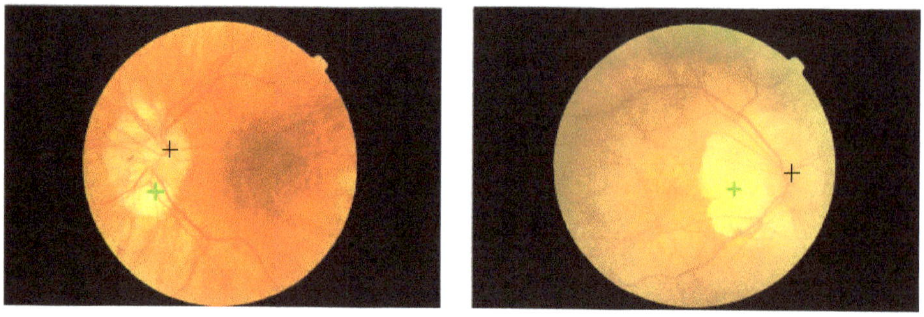

(a) (b)

FIGURE 9: Optic disc detection failure examples. (a-b) Extra-large and bright lesions overshadow the optic disc, making it invisible.

Conflicts of Interest

All the authors declare that they have no conflicts of interest

Acknowledgments

This work is supported by National Key R&D Program of China (no. 2017YFB1300900); National Natural Science Foundation of China (nos. U17132166, 1602222, 61471110, 61603415, and 61602221); Research Fund of Shenyang (no. 17-87-0-00); the Ph.D. Programs Foundation of Liaoning Province (201601019); the Fundamental Research Funds for the Central Universities (N172604004)

References

[1] S. Yamamoto and H. Yokouchi, "Automatic diagnosis of color fundus photographs using TV camera system," *The Journal of the Institute of Television Engineers of Japan*, vol. 29, no. 5, pp. 389–395, 1975.

[2] W. Zhou, C. Wu, D. Chen, Y. Yi, and W. Du, "Automatic microaneurysm detection using the sparse principal component analysis-based unsupervised classification method," *IEEE Access*, vol. 5, pp. 2563–2572, 2017.

[3] E. M. Hoffmann, L. M. Zangwill, J. G. Crowston, and R. N. Weinreb, "Optic disk size and glaucoma," *Survey of Ophthalmology*, vol. 52, no. 1, pp. 32–49, 2007.

[4] C. Sinthanayothin, J. F. Boyce, H. L. Cook, and T. H. Williamson, "Automated localisation of the optic disc, fovea, and retinal blood vessels from digital colour fundus images," *British Journal of Ophthalmology*, vol. 83, no. 8, pp. 902–910, 1999.

[5] C. Sinthanayothin, *Image analysis for automatic diagnosis of diabetic retinopathy [Ph.D. thesis]*, University of London (King's College London), London, U.K, 1999.

[6] M. Park, J. S. Jin, and S. Luo, "Locating the optic disc in retinal images," in *Proceedings of the International Conference on Computer Graphics, Imaging and Visualisation (CGIV '06)*, pp. 141–145, IEEE, Sydney, Australia, July 2006.

[7] E. Trucco and P. J. Kamat, "Locating the optic disk in retinal images via plausible detection and constraint satisfaction," in *Proceedings of the 2004 International Conference on Image Processing, ICIP 2004*, pp. 155–158, Singapore, Singapore, October 2004.

[8] A. Osareh, M. Mirmehdi, B. Thomas, and R. Markham, "Classification and localisation of diabetic-related eye disease," in *ECCV 2002: Computer Vision — ECCV 2002*, vol. 2353 of *Lecture Notes in Computer Science*, pp. 502–516, Springer, Berlin, Germany, 2002.

[9] H. Yu, E. S. Barriga, C. Agurto et al., "Fast localization and segmentation of optic disk in retinal images using directional matched filtering and level sets," *IEEE Transactions on Information Technology in Biomedicine*, vol. 16, no. 4, pp. 644–657, 2012.

[10] M. Lalonde, M. Beaulieu, and L. Gagnon, "Fast and robust optic disc detection using pyramidal decomposition and hausdorff-based template matching," *IEEE Transactions on Medical Imaging*, vol. 20, no. 11, pp. 1193–1200, 2001.

[11] S. Ravishankar, A. Jain, and A. Mittal, "Automated feature extraction for early detection of diabetic retinopathy in fundus images," in *Proceedings of the IEEE Computer Society Conference on Computer Vision and Pattern Recognition Workshops (CVPR '09)*, pp. 210–217, Miami, Fla, USA, June 2009.

[12] A. Hoover and M. Goldbaum, "Locating the optic nerve in a retinal image using the fuzzy convergence of the blood vessels," *IEEE Transactions on Medical Imaging*, vol. 22, no. 8, pp. 951–958, 2003.

[13] A. A.-H. A.-R. Youssif, A. Z. Ghalwash, and A. A. S. A.-R. Ghoneim, "Optic disc detection from normalized digital fundus images by means of a vessels' direction matched filter," *IEEE Transactions on Medical Imaging*, vol. 27, no. 1, pp. 11–18, 2008.

[14] B. Zhang and F. Karray, "Optic disc and fovea detection via multi-scale matched filters and a Vessels' Directional Matched Filter," in *Proceedings of the IEEE 2010 International Conference on Autonomous and Intelligent Systems, AIS 2010*, Povoa de Varzim, Portugal, June 2010.

[15] C. A. Perez, D. A. Schulz, C. M. Aravena, C. I. Perez, and V. T. Juan, "A new method for online retinal optic-disc detection based on cascade classifiers," in *Proceedings of the IEEE International Conference on Systems, Man, and Cybernetics (SMC '13)*, pp. 4300–4304, IEEE, Manchester, UK, October 2013.

[16] H. Li and O. Chutatape, "Automatic location of optic disk in retinal images," in *Proceedings of the 2001 International Conference on Image Processing*, pp. 837–840, Thessaloniki, Greece.

[17] A. Oliva and A. Torralba, "Modeling the shape of the scene: a holistic representation of the spatial envelope," *International Journal of Computer Vision*, vol. 42, no. 3, pp. 145–175, 2001.

[18] F.-F. Li and P. Perona, "A bayesian hierarchical model for learning natural scene categories," in *Proceedings of the IEEE Computer Society Conference on Computer Vision and Pattern Recognition (CVPR '05)*, vol. 2, pp. 524–531, San Diego, Calif, USA, June 2005.

[19] W. Zhou, C. Wu, Y. Gao, and X. Yu, "Automatic optic disc boundary extraction based on saliency object detection and modified local intensity clustering model in retinal images," *IEICE Transactions on Fundamentals of Electronics, Communications and Computer Sciences*, vol. E100A, no. 9, pp. 2069–2072, 2017.

[20] M. Esmaeili, H. Rabbani, and A. M. Dehnavi, "Automatic optic disk boundary extraction by the use of curvelet transform and deformable variational level set model," *Pattern Recognition*, vol. 45, no. 7, pp. 2832–2842, 2012.

[21] C. I. Sánchez, R. Hornero, M. I. López, M. Aboy, J. Poza, and D. Abásolo, "A novel automatic image processing algorithm for detection of hard exudates based on retinal image analysis," *Medical Engineering & Physics*, vol. 30, no. 3, pp. 350–357, 2008.

[22] Y. Cong, J. Yuan, and J. Liu, "Sparse reconstruction cost for abnormal event detection," in *Proceedings of the IEEE Conference on Computer Vision and Pattern Recognition (CVPR '11)*, pp. 3449–3456, Providence, RI, USA, June 2011.

[23] W. Zhou, C. Wu, Y. Yi, and G. Luo, "Structure Preserving Non-negative Feature Self-Representation for Unsupervised Feature Selection," *IEEE Access*, vol. 5, pp. 8792–8803, 2017.

[24] H. Dai, W. Chen, J. Jia, J. Liu, and Z. Zhang, "Exponential synchronization of complex dynamical networks with time-varying inner coupling via event-triggered communication," *Neurocomputing*, vol. 245, pp. 124–132, 2017.

[25] W. Zhou, C. Wu, D. Chen, Z. Wang, Y. Yi, and W. Du, "Automatic microaneurysms detection based on multifeature fusion dictionary learning," *Computational and Mathematical Methods in Medicine*, vol. 2017, Article ID 2483137, pp. 1–11, 2017.

[26] Y. Yi, S. Qiao, W. Zhou, C. Zheng, Q. Liu, and J. Wang, "Adaptive multiple graph regularized semi-supervised extreme learning machine," *Soft Computing*, vol. 22, no. 11, pp. 3545–3562, 2018.

[27] W. Zhou, C. Wu, Y. Yi, and W. Du, "Automatic detection of exudates in digital color fundus images using superpixel multi-feature classification," *IEEE Access*, vol. 5, pp. 17077–17088, 2017.

[28] Y. Yi, W. Zhou, C. Bi, G. Luo, Y. Cao, and Y. Shi, "Inner product regularized nonnegative self representation for image classification and clustering," *IEEE Access*, vol. 5, pp. 14165–14176, 2017.

[29] Y. Nesterov, "Gradient methods for minimizing composite objective function," Tech. Rep., CORE, 2007.

[30] T. Hastie and R. Tibshirani, "Discriminant adaptive nearest neighbor classification," *IEEE Transactions on Pattern Analysis & Machine Intelligence*, vol. 18, no. 6, pp. 607–616, 2002.

[31] T. Furey, N. Cristianini, N. Duffy, D. W. Bednarski, M. Schummer, and D. Haussler, "Support vector machine classification and validation of cancer tissue samples using microarray expression data," *Bioinformatics*, vol. 16, no. 10, pp. 906–914, 2000.

[32] M. M. Rahman and D. N. Davis, "Addressing the class imbalance problem in medical datasets," *International Journal of Machine Learning & Computing*, vol. 3, no. 2, p. 224, 2013.

[33] B. Schölkopf, J. C. Platt, J. Shawe-Taylor, A. J. Smola, and R. C. Williamson, "Estimating the support of a high-dimensional distribution," *Neural Computation*, vol. 13, no. 7, pp. 1443–1471, 2001.

[34] J. Bao, Y. Zhang, X. Su, and R. Zheng, "Unpaved road detection based on spatial fuzzy clustering algorithm," *EURASIP Journal on Image and Video Processing*, vol. 26, 2018.

[35] MESSIDOR database: http://messidor.crihan.fr/index-en.php.

[36] M. I. Ahmed and M. A. Amin, "High speed detection of optical disc in retinal fundus image," *Signal, Image and Video Processing*, vol. 9, no. 1, pp. 77–85, 2015.

[37] A. Aquino, M. E. Gegundez, and D. Marin, "Automated optic disc detection in retinal images of patients with diabetic retinopathy and risk of macular edema," *International Journal of Biological & Life Sciences*, vol. 8, no. 2, pp. 87–92, 2012.

[38] B. Al-Bander, W. Al-Nuaimy, B. M. Williams, and Y. Zheng, "Multiscale sequential convolutional neural networks for simultaneous detection of fovea and optic disc," *Biomedical Signal Processing and Control*, vol. 40, pp. 91–101, 2018.

7

EEG-Based Automatic Sleep Staging using Ontology and Weighting Feature Analysis

Bingtao Zhang[1,2] **Tao Lei,**[3] **Hong Liu,**[4] **and Hanshu Cai**[1]

[1]*School of Information Science and Engineering, Lanzhou University, Lanzhou, China*
[2]*School of Electronic and Information Engineering, Lanzhou Jiaotong University, Lanzhou, China*
[3]*College of Electronical and Information Engineering, Shaanxi University of Science and Technology, Xi'an, China*
[4]*School of Information Science and Engineering, Shandong Normal University, Jinan, China*

Correspondence should be addressed to Hanshu Cai; caihsh13@lzu.edu.cn

Academic Editor: Raul Alcaraz

Sleep staging is considered as an effective indicator for auxiliary diagnosis of sleep diseases and related psychiatric diseases, so it attracts a lot of attention from sleep researchers. Nevertheless, sleep staging based on visual inspection of tradition is subjective, time-consuming, and error-prone due to the large bulk of data which have to be processed. Therefore, automatic sleep staging is essential in order to solve these problems. In this article, an electroencephalogram- (EEG-) based scheme that is able to automatically classify sleep stages is proposed. Firstly, EEG data are preprocessed to remove artifacts, extract features, and normalization. Secondly, the normalized features and other context information are stored using an ontology-based model (OBM). Thirdly, an improved method of self-adaptive correlation analysis is designed to select the most effective EEG features. Based on these EEG features and weighting features analysis, the improved random forest (RF) is considered as the classifier to achieve the classification of sleep stages. To investigate the classification ability of the proposed method, several sets of experiments are designed and conducted to classify the sleep stages into two, three, four, and five states. The accuracy of five-state classification is 89.37%, which is improved compared to the accuracy using unimproved RF (84.37%) or previously reported classifiers. In addition, a set of controlled experiments is executed to verify the effect of the number of sleep segments (epochs) on the classification, and the results demonstrate that the proposed scheme is less affected by the sleep segments.

1. Introduction

Sleep [1] is one basic need for the human, both in terms of physical and mental health. However, nowadays many countries around the world have been into aging society, there have been significant and growing incidences of sleep-related conditions such as obstructive sleep apnea (OSA), insomnia [2], and narcolepsy [3]. Generally, these sleep-related conditions may cause depression [4, 5], irritation, anxiety, or even death, so on and so forth, which seriously affect the quality of life (QoL) for those suffering from such conditions [6].

Accurate and effective identification of sleep staging is an important step in assisting the diagnosis of sleep-related disorders. Traditionally, all night polysomnographic (PSG) recordings are visually scored by well-trained experts based on Rechtschaffen and Kales's (R&K) recommendations [7] or a new guideline developed by the American Academy of Sleep Medicine (AASM) [8]. PSG is a comprehensive record of concurrent physiological signal during sleep, and this record principally includes EEG, electrooculogram (EOG), chin electromyogram (EMG), electrocardiogram (ECG), oxygen saturation (SpO_2), respiration (Resp), and rectal body temperature. The most important signal for sleep staging is EEG from PSG because the clinically acceptable sleep staging is mainly determined by EEG [9, 10]. Thus this study extracts features derived from EEG data. The typical recording period of EEG segment is divided into 20 or 30 seconds prior to being classified as different sleep stages. We use a fixed 30s segment in this study. These different sleep stages and their relationship with EEG signals are briefly described as follows [1, 6, 7, 11, 12]:

(1) *Wakefulness (WA)*: it is an awakening state of subject before falling asleep. In this stage, the EEG signal rapidly changes, and alpha waves (8–13 Hz) are most significant.

(2) *Nonrapid eye movement (NREM) sleep stage 1 (NREM1)*: this first stage of sleep is also called the drowsiness state. During this stage, brain activity slows down and theta waves (4–8 Hz) are more prominent. Meanwhile, the eyes begin to roll slightly.

(3) *NREM sleep stage 2 (NREM2)*: in this stage, the EEG amplitude becomes higher, sleep spindles (11–15 Hz), and K-complexes appear. Meanwhile, the eyes stop moving.

(4) *NREM sleep stage 3 (NREM3)*: this sleep stage is also called the deep sleep. In this stage, 20% to 50% of EEG signals are delta waves (2–4 Hz) and the rest are theta waves. The people may experience sleepwalking, night terrors, and bedwetting during this period.

(5) *NREM sleep stage 4 (NREM4)*: this stage is a continuation of deep sleep. And it also called the slow wave sleep. In this stage, more than 50% of EEG frequency varies between 0.5 Hz and 2 Hz (delta wave).

(6) *Rapid eye movement (REM)*: the eyes remain closed but move rapidly. Beta waves are more predominant and frequency is greater than 13 Hz.

In addition, there are five different combination ways (CWs) for the above six sleep stages according to the R&K standard (Table 1) [10].

Sleep staging on the basis of visual EEG interpretation is problematic for many reasons [9, 13]. The accuracy of sleep staging depends on subjective experience of experts. Furthermore, research shows that the classification consistency among experts is less than 85%. Meanwhile, the analysis of the huge bulks of data from patients makes sleep staging by human experts become onerous and subject to misclassification due to fatigue. In clinical settings, there is a certain lag in sleep staging based on visual inspection, due to its time-consuming process. To solve these problems, it is obvious that there is a great need of an automatic sleep-staging scheme.

With the continuous development of new technology, the ability of EEG to automatically identify sleep staging is increased because a large number of features extracted from EEG data. A huge number of features enhance classification accuracy of sleep staging. However, to manage a large number of unstructured features is a very complex task for data management professionals. Hence, we believe that as an essential condition for an automatic sleep-staging scheme, automatic management and representation of EEG features is necessary. In this situation, ontology techniques exhibit several benefits in terms of information representation, and it has the major advantages as follows [14]: (1) it can realize the effectiveness, specification, and conciseness of information representation; (2) it can manage a large number of features in a hierarchical manner.

The remainder of this paper is organized as follows: related research is considered in Section 2 with data description introduced in Section 3. Section 4 describes the proposed EEG-based automatic sleep-staging scheme. In Section 5, we set out experimentation with the results and a discussion. Finally, the conclusion and future work are presented in Section 6.

Table 1: Different combination criterions of six sleep stages.

Combination criterion	Sleep stages
Six-state	WA, NREM1, NREM2, NREM3, NREM4, REM
Five-state	WA, NREM1, NREM2, SWS, REM
Four-state	WA, light sleep, SWS, REM
Three-state	WA, NREM, REM
Two-state	WA, sleep

Note. Light sleep: NREM1 and NREM2 are pooled into a single stage. SWS: NREM3 and NREM4, are pooled into a single stage. Sleep: NREM1, NREM2, NREM3, NREM4 and REM are pooled into a single stage.

2. Related Research

Automatic sleep staging has become a hot issue in sleep research field; some studies emphasize the importance of selecting a suitable classifier to improve classification accuracy. Such research involves support vector machines [6, 15], neural networks [12, 16, 17], random forest [18], decision trees [19], Hidden Markov Models [20], and visibility graphs [13, 21, 22]. Sharma et al. [6] have developed a single-channel EEG-based sleep stages classification system using a novel three-band time-frequency localized (TBTFL) wavelet filter bank (FB). The system achieved classification accuracies from 91.5% to 98.3% for five CWs. In [12], the power spectral density (PSD) features extracted from multichannel neural data were fed into a multilayer feedforward network classifier for five-state sleep classification, and the classification accuracy of 90.1% was obtained. Hsu et al. [17] present a recurrent neural classifier for automatically classifying sleep stages based on energy features from EEG signal. Zhu et al. [13] analyzed sleep stages based on visibility graphs. The representative graph features were extracted and then were forwarded to a classifier to distinguish different sleep stages. The datasets used in the research were from the Sleep-EDF database, and the classification accuracy of 87% was reported. Diykh et al. [21, 22] proposed a novel method of sleep stages classification based on complex networks, and its method is mainly composed of two steps: (1) Each EEG segment is partitioned into subsegments. The size of a subsegment is determined empirically. (2) Statistical features are extracted, sorted into different feature sets, and then forwarded to the structural graph similarity and the K-means (SGSKM) to classify EEG sleep stages.

To better introduce the research status of automatic sleep staging based on EEG, the following is an overview of some representative studies in recent years. Acharya et al. [1] gave a comprehensive and comparative analysis of 29 nonlinear dynamic measures for EEG-based sleep stages classification. An iterative filtering-based decomposition is presented for automatic classification of sleep stages using EEG signals [10]. In [11], a number of higher order spectra- (HOS-) based features were extracted from unique bispectrum and bicoherence during the different sleep stages, and then these features were fed to a Gaussian mixture model (GMM) classifier for automatic sleep staging. Liang et al. [9] proposed an automatic sleep-scoring method combining multiscale entropy (MSE) and autoregressive (AR) models. Berthomier et al. [23] proposed an approach which uses an iterative adaptive fuzzy logic system to iterative update sleep stage pattern definitions. The above three studies have achieved an average accuracy for five-state sleep stages which are 83.6%, 83.49%, and 71.2%. An intelligent system for sleep stages classification (ISSSC) was designed and implemented [24]. The system consists of four principal modules: signals preprocessing, machine learning, inference, and inference corrector module.

Though many research results of sleep stages based on EEG have been reported in the literature, these studies have differing advantages and limitations. None of them have been pervasively applicable, maybe it is because of a lack of effective mechanism for presentation and management of the large number of EEG features. In our opinion, two important problems should to be solved in order to get an acceptable scheme of automatic sleep staging: (1) the scheme should be able to express and manage related information of sleep staging (such as a large number of unstructured EEG features) using a few simple terms which are readable by humans and computers. Simultaneously, the related information of sleep staging can be processed by computers easier and faster; (2) the scheme should be able to provide a valid analysis and classification method for EEG signals and sleep staging.

OBM [25] as a tool has been applied in many fields of information representation. Such as affective computing [26–28], attack detection [29], agriculture [30], and intelligent transport [31]. In addition, OBM has also been applied in expert system; for example, Bau et al. [32] construct a clinical decision support system (CDSS) for undergoing surgery based on ontology and rules reasoning in the setting of hospitalized diabetic patients. Wang et al. [33] proposes an assisting support system based on ontology, which is used to embody the relations among knowledge points. However, these studies rarely involve the use of ontology to describe and manage physiological features.

The purpose of this study is to improve the accuracy of sleep staging and fill the gap in the field of massive EEG feature representation. So based on inspiration of above research, this study proposes an automatic sleep-staging scheme based on EEG, which employs OBM to represent and organize related information of sleep staging and weighting feature analysis to improve the existing classification algorithm. Before ending this section, the main novelties of this paper are presented as follows:

(1) An ontology-based model (OBM) is designed and applied to manage a high volume of sleep EEG features and context information.

(2) An improved method of self-adaptive correlation analysis is designed to select the most effective EEG features.

(3) Based on weighting features analysis, the improved RF is considered as the classifier to achieve the classification of sleep stages.

(4) The effectiveness of the proposed method is validated by several sets of experiments.

3. Data Description

The raw physiological data are obtained from the Sleep-EDF database (EXPANDED) [34, 35], which contains 61 data recordings taken from 42 Caucasian subjects. The first 39 data records (SC∗PSG.edf files) are obtained from 20 healthy volunteers without any sleep-related disease. At the time of recordings, the demographic range was 25 to 34 years with the population consisting of 10 males and 10 females. The remaining 22 data records (ST∗PSG.edf files) are collected from 22 participants consisting of 7 males and 15 females aged from 18 to 79; all 22 subjects experienced mild difficulty in falling asleep.

The database includes dual-channel EEG from Fpz-Cz and Pz-Oz with a sampling rate of 100 Hz. Meanwhile, this database also contains other physiological signals, such as EOG, EMG, oronasal respiration, and so on. In this study, we only use two-channel EEG data (Fpz-Cz and Pz-Oz) to realize automatic sleep staging because of EEG is derived from the central nervous system, and it is most effective for the sleep stages classification.

To ensure the equality of 61 data records, 840 sleep segments (the length of each segment is 30 seconds) are randomly selected from each data record. So the total numbers of sleep segments are 51240. Specific database usage is shown in Table 2.

4. Methods

4.1. Our Work. Figure 1 provides a structure diagram of our proposed scheme. The simplified execution process is shown as follows:

(1) Raw physiological data are collected during sleep. The EEG in raw physiological data cannot be used directly for sleep staging, so first of all is EEG preprocessing. Meanwhile, other sleep-related information is directly mapped to "EEG-Sleep ontology."

(2) A large number of preprocessed EEG data are also mapped to "EEG-Sleep ontology." It means that these EEG features together with sleep-related information are used to form the "EEG-Sleep ontology."

(3) By correlation analysis of the information stored in "EEG-Sleep ontology," the subsets of EEG features that are most relevant to the sleep staging are obtained.

(4) Next, the classification mechanism is introduced, and then an analysis strategy of the weighting feature is proposed to improve the classification algorithm.

(5) The improved classification algorithm is used to infer different sleep stages. Finally, the results of sleep staging will be used to assist physicians in the diagnosis of sleep disorders.

TABLE 2: Specific database usage in this study (unit: epoch or segment).

		WA	NREM1	NREM2	SWS	REM	Total
SC∗PSG.edf	Females	1397	1476	7557	3864	2540	16834
	Males	1292	919	7675	3529	2523	15938
ST∗PSG.edf	Females	924	1163	5774	2925	1886	12672
	Males	529	554	2809	1282	622	5796
	Total	4142	4112	23815	11600	7571	51240

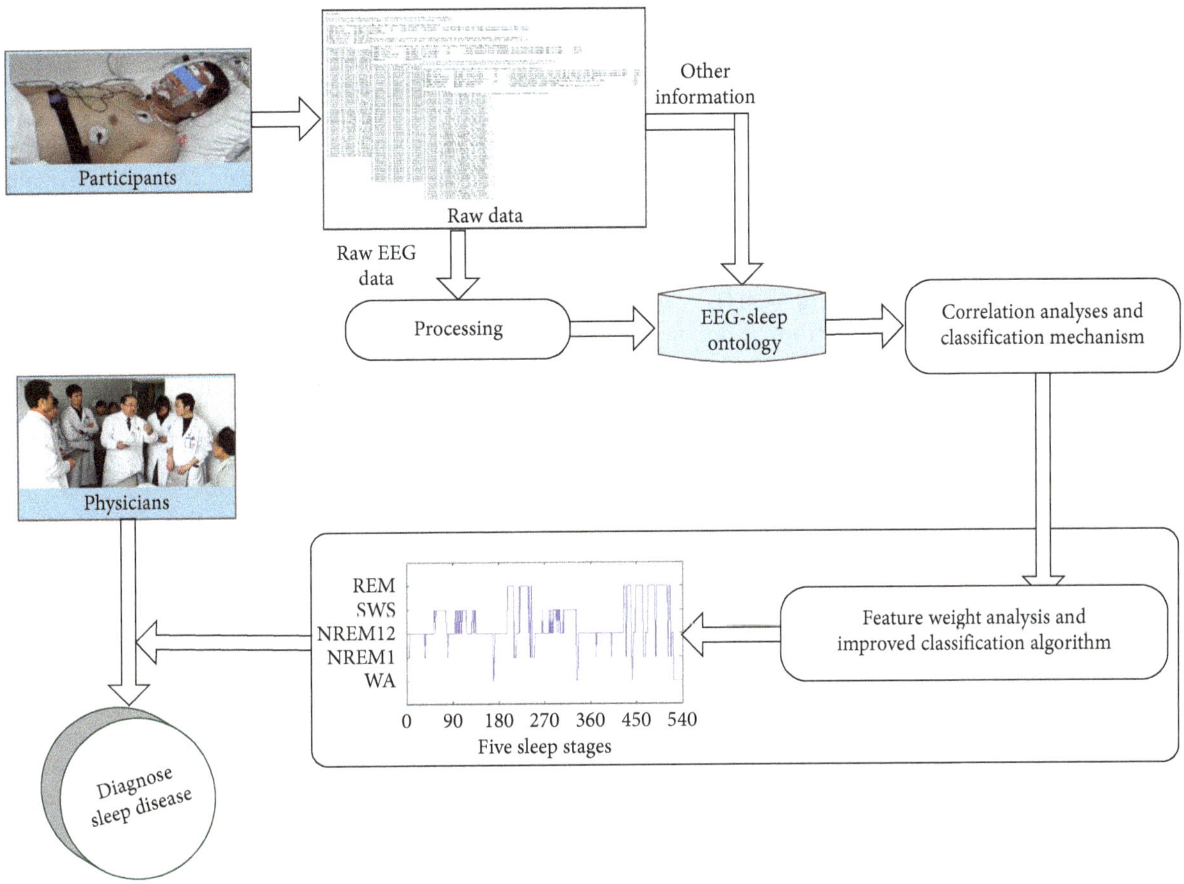

FIGURE 1: The structure diagram of our proposed scheme.

4.2. Data Preprocessing. Usually, for the accurate analysis of data, it is necessary to preprocess the raw EEG data. Research [17] shows that the frequency range associated with sleep stages occur in 0.5 Hz to 30 Hz. Therefore, a 0.5 high-pass filter removed the low-frequency drift, and a 30 Hz low-pass filter removed the high frequency noise. Ocular artifacts appeared in the frequency ranges from 0 to 16, thus overlapping with other frequency bands, such as theta (4–8 Hz), alpha (8–13 Hz), and so on. So this study uses a FastICA [36, 37] to denoise overlapping frequency bands.

Traditionally, EEG features are extracted using the linear analysis method, which can be used to simulate a time series of signals with a specific mathematical formula in several nonoverlapping frequency bands. Recently, various nonlinear methods have also been widely used to extract EEG features. Because the EEG signals are derived from the central nervous system it is necessary to combine linear and nonlinear methods for EEG signal processing.

4.2.1. Linear Method. In this study, 4 linear features of EEG signal are extracted from delta (0.5–2 Hz), sawtooth (2–4 Hz), theta (4–8 Hz), alpha (8–13 Hz), beta (13–30 Hz) bands: absolute power, relative power, center frequency, and max power. In addition, Hjorth parameters (Activity, Mobility, and Complexity), Skewness, and Kurtosis are also extracted from EEG signals in the full-frequency range (0.5–30 Hz) to dig the statistical properties and asymmetry of EEG. Among them, Activity is calculated as follows:

$$\text{Act} = \frac{1}{n-1}\sum_{i=1}^{n}(x(i)-\mu)^2. \quad (1)$$

Mobility is calculated as follows:

$$\text{Mob} = \sqrt{\frac{\left((1/(n-2))\sum_{i=2}^{n}(x_d(i)-\mu_d)^2\right)}{(1/(n-1))\sum_{i=1}^{n}(x(i)-\mu)^2}}. \quad (2)$$

Complexity is calculated as follows:

$$\mathrm{Cpx} = \sqrt{\frac{\left((1/(n-3))\sum_{i=3}^{n}\left(x_{\mathrm{dd}}(i)-\mu_{\mathrm{dd}}\right)^2\right)}{(1/(n-1))\sum_{i=1}^{n}(x(i)-\mu)^2}}. \quad (3)$$

Skewness is calculated as follows:

$$\mathrm{Skew} = \frac{(1/n)\sum_{i=1}^{n}(x(i)-\mu)^3}{(\sqrt{\mathrm{var}})^3}. \quad (4)$$

Kurtosis is calculated as follows:

$$\mathrm{Kurt} = \frac{(1/n)\sum_{i=1}^{n}(x(i)-\mu)^4}{\mathrm{var}^2}. \quad (5)$$

4.2.2. Nonlinear Method. Considering the effectiveness and stability of this scheme, 5 nonlinear features are extracted from the EEG signal in frequency range (0.5–30 Hz), which includes Shannon entropy, Spectral entropy, Kolmogorov entropy, Max lyapunov exponent, and C0-complexity. In these five EEG features, Shannon entropy ShE(X) is used to describe the amount of information contained in the EEG signal. Spectral entropy SpE(X) is used to describe the complexity of the EEG signal. Kolmogorov entropy (KoE) is used to describe the degree of loss rate of the EEG signal. Lyapunov exponent is used to describe the sensitivity to the initial conditions, and Max lyapunov ($\lambda_{\max}$) exponent is the most important one of Lyapunov exponents. C0-complexity is used to describe the irregular rate of EEG signal. Among them, Shannon entropy is calculated as follows [38]:

$$\mathrm{ShE}(X) = -\sum_{i=1}^{n} p(x)\log p(x). \quad (6)$$

Spectral entropy is calculated as follows:

$$p(i) = \frac{P(w)}{P_{xx,\mathrm{tot}}},$$
$$\mathrm{SpE}(X) = -\sum_{i=1}^{n} p(i)\log p(i). \quad (7)$$

Kolmogorov entropy is calculated as follows:

$$\mathrm{KoE} = -\lim_{\Delta t\to 0}\lim_{\varepsilon\to 0}\lim_{n\to\infty}\sum_{i=0}^{m} p(i_0,\ldots,i_n)\log p(i_0,\ldots,i_n). \quad (8)$$

Max lyapunov exponent is calculated as follows:

$$L'(t_i) = \min\left|Y(t_i) - Y_0(t_i)\right|,$$
$$\lambda\max = \frac{1}{t_n - t_0}\sum_{i=1}^{n}\log\frac{L'(t_i)}{L(t_{i-1})}. \quad (9)$$

C0-complexity is calculated as follows:

$$A_0 = \sum_{i=0}^{n}|x(i)|^2,$$
$$A_1 = \sum_{i=0}^{n}|x(i)-y(i)|^2, \quad (10)$$
$$\mathrm{C0\text{-}complexity} = \frac{A_1}{A_0}.$$

The above EEG features are widely acknowledged in the human sleep research. Finally, 50 linear features and 10 nonlinear features are extracted in total.

In order to avoid the effect of EEG features and abnormal values on classification accuracy, each EEG feature is normalized in the range [0, 1] using the Min-Max normalization [39].

$$X_{\mathrm{norm}} = \frac{X-\min}{\max-\min}, \quad (11)$$

where X represents the initial value of the feature, X_{norm} represents the value of the feature after normalization, min represents the minimum value of the feature in a range, and max represents the maximum value of the feature in a range.

4.3. EEG-Sleep Ontology. Once the EEG features are extracted, these features together with sleep-related information are used to form "EEG-Sleep ontology." This process involves the two aspects of information query mapping and information description. SPARQL language[1] is used to query the preprocessed EEG features and other sleep-related information from raw data, and then map them to "EEG-Sleep ontology." In other words, the purpose of the information query mapping is to find the classes and instances required for the original "EEG-Sleep ontology" from raw data. The information description is describing instances in the "EEG-Sleep ontology" using object property and data property. In this study, a top-down approach [40] is adopted to implement information description. A detailed description of this method is shown as Figure 2.

We can see from Figure 2, two categories (EEG and Sleep) are defined in the original "EEG-Sleep ontology," but the two categories did not contain any instances (Figure 2(1)). Information query mapping is used in both original "EEG-Sleep ontology" and raw data (Figure 2(2)) according to match keyword; the retrieved instance is then mapped to corresponding category in "EEG-Sleep ontology" (Figure 2(3)). Information description provides a clear information representation of "EEG-Sleep ontology." More specifically, the "EEG-Sleep ontology" is designed as a four-layer structure: domain layer, category layer, class layer, and instance layer. The domain layer is an indispensable component which indicates the name of ontology under modeling. The category layer contains the main categories of OBM. The detail of each category layer is shown in the class layer which contains a set of concepts (class) of OBM.

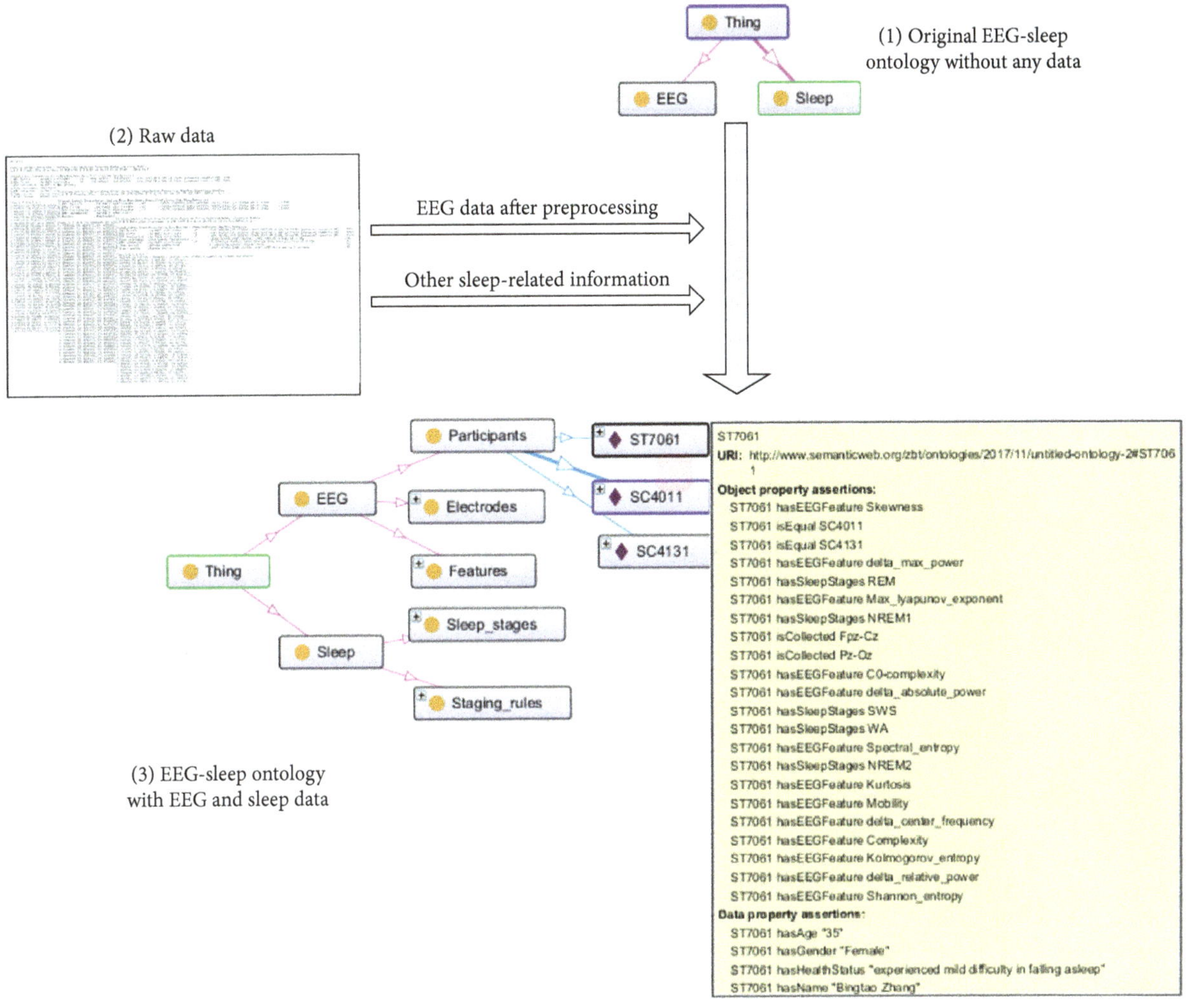

FIGURE 2: Structure diagram of EEG-sleep ontology.

At lowest layer, the instance layer defines all instances (individuals) of each class contained in the class layer. In addition, the relationship among instances is described by object properties. The relationship between instance and the basic data type (i.e., attribute) is described by data properties.

"EEG-Sleep ontology" can be described as in Figure 3. The domain name: EEG-based automatic sleep staging. This domain defines two categories, named EEG and Sleep, to represent the two main aspects of this study. Next, some concepts related to two categories are defined in the class layer. For example, EEG-related concepts (classes) include participants, electrodes, and features. Sleep-related concepts include sleep staging and staging rules. Finally, each class is materialized to create corresponding instance of the instance layer. As mentioned in the previous paragraph, there are relationships (i.e., object properties) that exist between one instance and another instance. This relationship is used to associate the entire "EEG-Sleep ontology." For example, an object property "hasEEGfeature" exists between the instance "ST7061" and the instance "Shannon entropy." Simultaneously, "ST7061" has a data property "hasAge." The data type of this property is "Integer" and the data value is "35." Several examples of object properties and data properties for the instance "ST7061" (Figure 2(3)) are listed in Tables 3 and 4.

Next, a correlation analysis is implemented for the instance information stored in the "EEG-Sleep ontology."

4.4. Correlation Analysis. An improved algorithm of Pearson correlation coefficient based on self-adaptive correlation analysis is designed (see Algorithm 1 of Section 5.2). Recent studies [41, 42] have found that there are differences in sleep structure between different genders. Therefore, the effect of gender differences on sleep staging is considered. The correlation between EEG features and sleep stages is revealed by gender contrast diagram form. As illustrated in Figure 4, there is a significant difference in the Complexity feature extracted on Fpz-Cz electrode for different gender. It can be see that for female, the Pearson correlation coefficient of Complexity feature is significantly higher than that of male. The results show that the Complexity features have a significant impact on the accuracy of female's sleep staging, but it may have less effect on the accuracy of male's sleep staging.

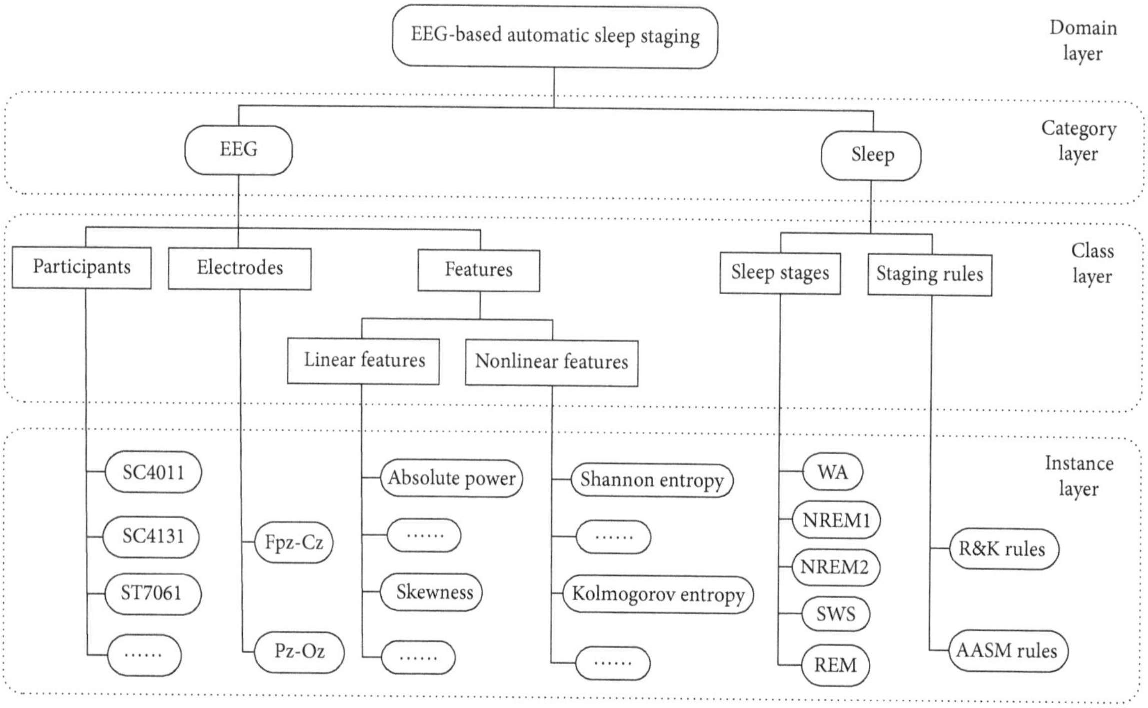

FIGURE 3: The structure of "EEG-Sleep ontology."

TABLE 3: Several objective properties of the instance "ST7061."

Instance 1	Objective properties	Instance 2
ST7061	hasEEGfeature	Shannon entropy
ST7061	hasSleepStages	REM
ST7061	isCollected	Pz-Oz

TABLE 4: Several data properties of the instance "ST7061."

Instance	Data properties	Value	Data type
ST7061	hasAge	35	Integer
ST7061	hasGender	Female	String
ST7061	hasHealthStatus	Mild difficulty in falling asleep	String

4.5. Classification and Improvement. RF algorithm [43, 44] consists of multiple independent decision trees. Building each decision tree starts at the top of tree with the entire training data. The core task during decision tree-building process is to select a most appropriate attribute (feature) at the node (root node or branch node) and then to split the training data into different subsets based on the attribute.

The criterion of node selection and splitting are done according to an information gain (IG) of the node attribute, and the attribute with the maximum information gain is selected as the splitting criterion. The IG of splitting a training set (S) into subset (S_i) can be defined as

$$\mathrm{IG}(S_i) = -\sum_{i} \frac{|S_i|}{|S|} E(S_i), \tag{12}$$

$E(S_i)$ is the information entropy of the subset S_i calculated as

$$E(S_i) = -\sum_{i=1}^{c} p_i \log_2(p_i), \tag{13}$$

where c denotes the number of sleep stages and p_i is a proportion of sleep stage i in the subset S_i.

Each node attribute is applied to calculate IG, and the attribute with the maximum IG is selected as the splitting criterion. This process is repeated recursively at each node until either every attribute is selected or this process reaches a leaf node that is a classification output.

Building RF as shown in Figure 5:

(1) Each decision tree is generated by a training data S with sample size of k, and the random vector θ_k is an independent identical distribution.

(2) RF is a set of all decision trees: $h(S, \theta_k)$. Meanwhile, each decision tree is also a model.

(3) Each decision tree model $h(S, \theta_k)$ has the right to select a result for the input variable x.

$$H(x) = \max_{Y} \sum_{i=1}^{k} I(h_i(x) = Y), \tag{14}$$

where $H(x)$ represents the classification result of RF, $h_i(x)$ is a classification result of a single decision tree, and Y is a classification object, that is, the number of sleep stages.

4.5.1. Weighting Feature Analysis and Improved Classification Algorithm. The magnitude of the feature weight reflects the relative importance of each feature in the classification assessment, which indirectly affects the

Input: $p_{\text{best}} = 0$, $AC_{\text{best}} = 0$, $S_{\text{female}} = \phi$, $S_{\text{male}} = \phi$;
% Initialize the optimal threshold value, the highest accuracy rate, the best subset of features.
(1) for $p = [0.0005: 0.0005:1]$
(2) S_1, S_2; % Two sets of feature subsets for different genders were produced based on correlation analysis.
(3) S_1, $S_2 \longrightarrow$ RF classifier; % S_1 and S_2 were forwarded to the RF classifier to classify five-state sleep stages.
(4) AC; %The classification accuracy obtained by RF.
(5) if $AC_{\text{best}} \leq AC$
(6) $AC_{\text{best}} = AC$, $p_{\text{best}} = p$, $S_{\text{female}} = S_1$, $S_{\text{male}} = S_2$;
(7) end
(8) end
Output: p_{best}, S_{female}, S_{male};

Algorithm 1: Finding the best threshold value.

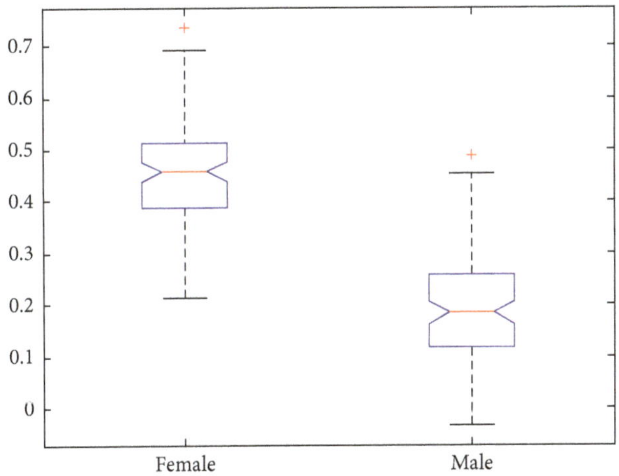

Figure 4: Mean and standard deviation of the Complexity feature for different gender.

accuracy of automatic sleep staging to a remarkable extent. Traditionally, the calculation methods of feature weight mainly include the Delphi method [45], paired comparison method [46], and so on. However, these traditional methods are generally based on the experience of experts in some fields, and the feature weight coefficient is determined by a prior hypothesis.

The classification process of RF is a partition process based on feature values. It is easy to determine the weight coefficients corresponding to each feature due to the features used for classification are determined before classification. However, except RF and C4.5, it is difficult to determine the corresponding relationship between features and weight coefficients for other classifiers. Additionally, RF consists of multiple decision trees, and its performance is obviously better than C4.5. Therefore, in order to reflect the role of weight coefficients in automatic sleep staging, an improved RF algorithm based on the weight coefficient of standard deviation is proposed. Firstly, we assume that the contribution of each attribute to the classification results is different in a dataset with multiple features. If the standard deviation of an attribute is relatively small, there is little difference in the value of records on the attribute. Meanwhile, the weight coefficient of an attribute is relatively small when the attribute value does not have much effect on the classification results. On the contrary, if the standard deviation of an attribute is relatively large, the weight coefficient of this attribute is also larger. Secondly, on the basis of the first step theory, we calculate the standard deviation after normalization of each attribute in the optimal feature subset according to the following formula:

$$s_i = \sqrt{\frac{\sum_{i=1}^{n} (x_i - \overline{x})^2}{n}}, \tag{15}$$

where $\overline{x}$ represents the mean value of each attribute after normalization.

The weight coefficient of each attribute in the optimal feature subset was calculated according to the following formula:

$$w_i = \frac{s_i}{\sum_{i=1}^{n} s_i}, \tag{16}$$

where s_i represents the standard deviation of the ith attribute.

Finally, to emphasize the role of the weight coefficient in the classification decision-making process, formula (12) is further improved.

$$\mathrm{IG_{imp}}(S_i) = w_i * \mathrm{IG}(S_i) = -\sum_i w_i * \left(\frac{|S_i|}{|S|} E(S_i)\right). \tag{17}$$

It is used as a new criterion for the selection and splitting of decision tree nodes in RF after the formula is improved. Selecting new nodes can not only greatly improve the accuracy of the classifier but also reduce the depth of the decision tree. In addition, this improvement can change the outcome of each decision tree, and then it has a positive effect on the classification results of RF.

5. Experiment and Discussions

"EEG-Sleep ontology" is used to represent and mange sleep-related information, such as EEG features, participants, and so on. So we first analyze and discuss the relationship between participants' age distribution and sleep staging.

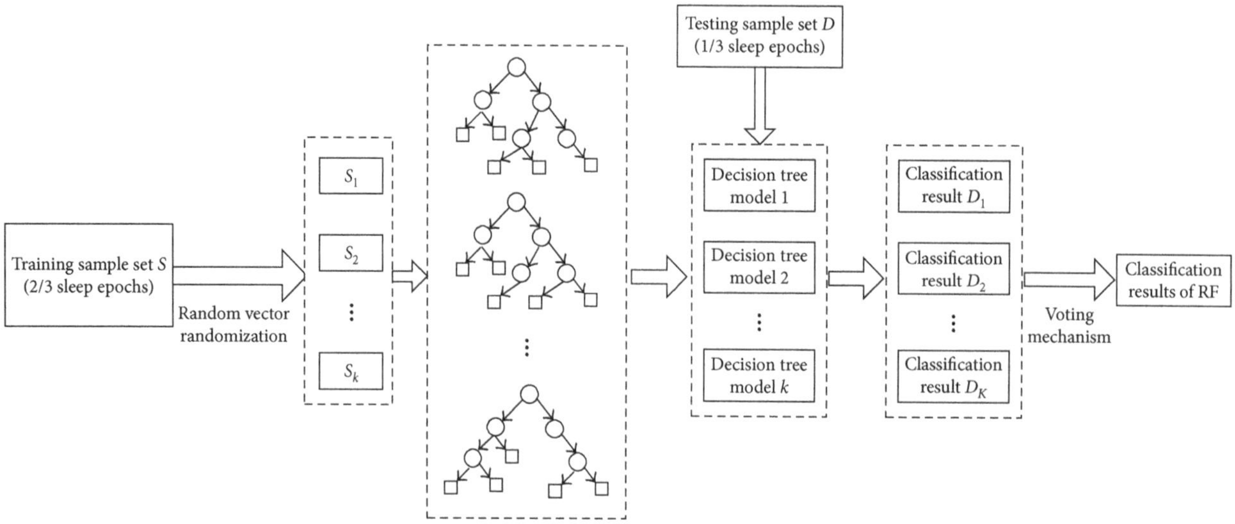

Figure 5: Building RF and classification decision.

5.1. Sample Characteristics of Participants' Age. Means and standard deviations of participants' age are shown in Table 5. Beyond that, Table 5 also shows significant differences of age in different genders. It can be seen that there is no significant difference in the female and male groups for age, $p = 0.16$. Thus, we can infer that the effect of age distribution for the sleep-staging accuracy can be neglected in this study.

Table 5: Basic information on the age of participants.

	Mean	SD	p value
Female	34.74	13.78	$p = 0.16$
Male	30.27	6.04	

$p < 0.01$: extremely significant difference. $0.01 < p < 0.05$: significant difference. $p > 0.05$: no difference.

5.2. The Result and Discussion of Correlation Analysis. The sleep-related information and EEG features derived from 61 data records were used to establish "EEG-Sleep ontology." And then, two sets of optimal feature subsets for different gender were calculated from EEG features of 1/2 which were stored in "EEG-Sleep ontology" using improved Pearson correlation coefficient. That is, half of sleep segments for female (14753 sleep segment) and male (10867 sleep segment) were used to calculate optimal feature subsets.

The results of Pearson analysis for 60 EEG features are shown in Figure 6. In this experiment, we selected the threshold of $p = 0.005$, and the calculation process of self-adaptive threshold value p are shown in Algorithm 1. The salient channels and features were obtained using improved Pearson analysis, and the result is reported in Table 6. As you can see in Table 6, the optimal feature subset of female contains 8 features, while the optimal feature subset of male contains 7 features. The most relevant features of both male and female were mainly derived from EEG linear features, and this result indirectly shows that linear features were useful for understanding the potential sleep structure.

5.3. Classification and Comparison Analysis. To find an effective classification algorithm, we have designed the following experiment: (1) We adopt optimal subset of different gender in Table 6 as the input of five classical classification algorithms to classify five-state sleep stages (WA, NREM1, NREM2, SWS, and REM). (2) Seven performance evaluation indexes were calculated for different genders, respectively. (3) Different classification algorithms were compared, the purpose of which was to observe the accuracy of the classification results obtained by the different classifiers in different evaluation indexes.

Five classical classification algorithms, with the exception RF, also include Bayesian network (BN) based on the graph model of probability, decision tree (C4.5) based on information entropy theory, multilayer perception (MLP) based on logistic regression, and support vector machine (SVM) based on statistical learning theory. Seven indicators of performance evaluation [47] involved in this experiment include the following: (1) Accuracy (AC) is defined as $TP + TN/(TP + TN + FP + FN)$ (2) The true positive (TPR) or sensitivity is defined as $TP/(TP + FN)$; it is also called the recall in some fields. (3) The true negative rate (TNR) or specificity is defined as $TN/(TN + FP)$. (4) Precision defined is defined as $TP/(TP+FP)$. (5) F-measure is defined as $2 * \text{Precision} * \text{Recall}/(\text{Precision} + \text{Recall})$. (6) ROC Area is a combination of sensitivity and specificity under different threshold. (7) Kappa statistic is defined as $2 * (TN * TP - FP * FN)/((TN + FN) * (FN + TP) + (FP + TP) * (TN + FP))$.

Weka software package [48] was used to perform all the classification tasks in the experiment. To save the running time of RF, we modified the number of decision tree from 100 to 30. The remaining parameters of RF and the parameters of other classifiers use the default values. The detailed usage of data in this experiment is shown in Table 7.

A standard experiment was implemented 10 times in Weka using the above scheme, and Figure 7 illustrates the

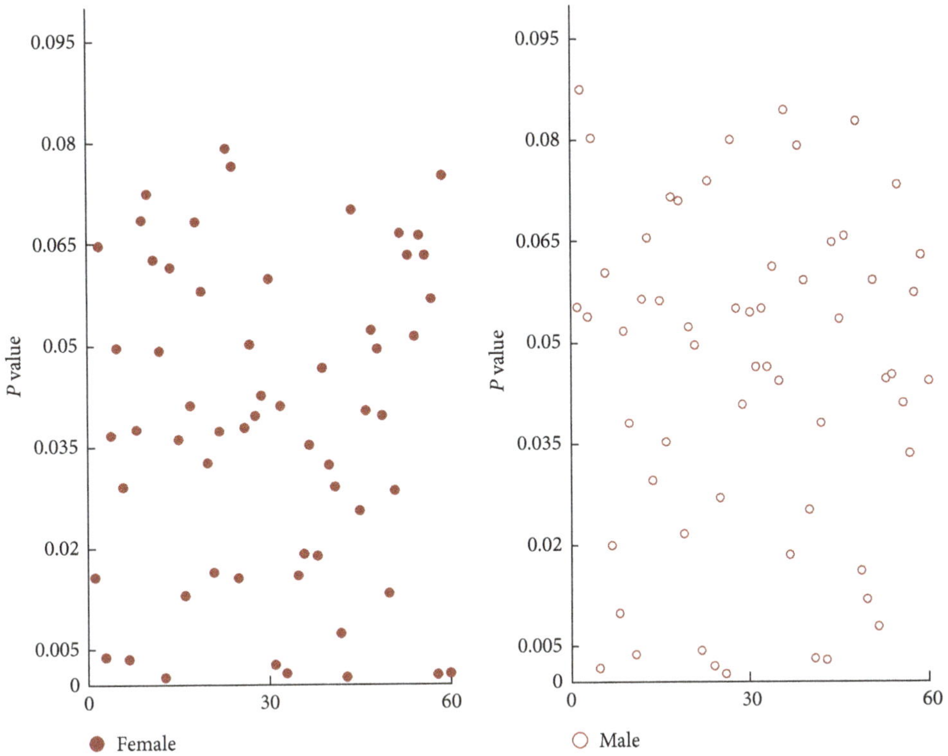

(a) (b)

FIGURE 6: Scatter plot of correlation analysis results for 60 EEG features.

TABLE 6: The significant EEG features of different gender based on Pearson correlation ($p < 0.005$).

Gender	Electrode	Features
Female	Fpz-Cz	Delta relative power, Complexity, Activity, Sawtooth relative power
	Pz-Oz	Delta relative power, Activity, Mobility, Skewness
Male	Fpz-Cz	Delta relative power, Alpha center frequency, Complexity, Skewness
	Pz-Oz	Delta relative power, Alpha center frequency, Kurtosis

TABLE 7: Data usage in this experiment (unit: epoch or segment).

	Training data	Testing data	Total
Female	19671	9835	29506
Male	14489	7245	21734
			51240

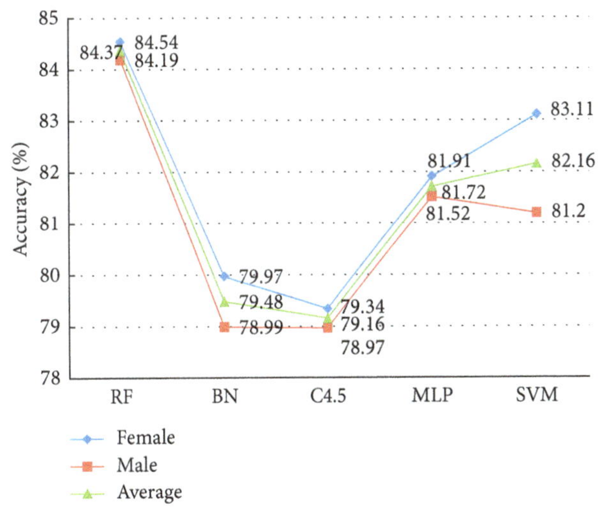

FIGURE 7: The classification accuracy of different classifiers on five-state sleep stages.

average of classification accuracy of different classifiers for five-state sleep stages. For the classification accuracy, there were some differences between the genders, but this difference has little effect on the results of sleep staging. So in the follow-up experiments, we only considered the average accuracy and ignored gender differences.

From Figure 7, it appears that the best average classification results from the RF with an accuracy of 84.37%. Why the RF classifier is superior to other classifiers? We believe that the main reason can be attributed to as follows:

(1) The randomness of RF makes the structure of each decision tree not exactly the same, and the comprehensive voting will greatly improve the classification accuracy. For example, suppose that an RF was made up of three decision trees, the error rate of each tree was 40%, and the error rate was reduced to 35.2% after comprehensive voting. Therefore, the accuracy of RF classification was obviously higher than that of C4.5.

$$1 \times 0.4^3 + 3 \times 0.4^2 \times (1 - 0.4)^1 = 0.352. \quad (18)$$

(2) Ensure that the data in each record have an equal use probability. The data for this study were not continuous and just a random integration of data from all night (Section 2). The data use strategy was consistent with the random sampling mechanism of RF [18], which enables RF to have strong generalization ability, and it can dig out the hidden rules behind the data, so as to obtain a high classification rate.

To confirm this result, we also compared and analyzed the remaining six performance evaluation indexes, and the specific experimental results are shown in Figure 8. For these six indexes, we expect higher values of TPR, Precision, Recall, F-measure, and Kappa statistic and lower value of TNR because these mean a better classification result. As we expected, the experimental results support our prediction that RF can achieve higher TPR, Precision, F-measure, ROC Area, and Kappa statistic than other four classification classifiers. However, it is remarkable that TNR of BN is a little lower than RF. But based on a comprehensive comparison of the various aspects (including accuracy and other five evaluation indicators), we finally selected the RF as classifier to identify different sleep stages.

5.4. Results of Improved Classification Algorithm. The core idea of weighting feature analysis and algorithm improvement has been previously explained in Section 4.5.1. The following was a specific process of experimental design and implementation: (1) We eliminated the 25620 sleep segments used in Section 5.2 correlation analysis and selected the remaining half of sleep segments as a sample library of weighting feature analysis and classification. This was done to prevent overfitting with the above experiments of correlation analysis. (2) The weight coefficient obtained in the first step was used as the input of formula (17) and applied to the calculation of the IG in the subsequent experiments. (3) The results of sleep staging in different cases were tested.

Based on formulas (15) and (16), the standard deviation and weight coefficient of each feature in Table 6 were calculated, and the results are shown in Tables 8 and 9.

To investigate the performance of the proposed method, the following five cases were tested. Meanwhile, to ensure the scientificity and rationality of the follow-up experimental comparison, we applied the other half of the data except Section 5.2. This is done to keep that in consistence with data used to calculate the weight coefficient, and the detailed data usage of the experiment is shown in Table 10. The Matlab2Weka toolbox was used to implement RF classification tasks. In addition to the improved part of algorithm, the number of decision trees in RF is still 30, and the other parameters were set as the default values.

Case 1. Five-state sleep stages were classified based on RF without weight coefficient.

Case 2. Five-state sleep stages were classified based on RF with weight coefficient.

Case 3. Four-state sleep stages were classified based on RF with weight coefficient.

Case 4. Three-state sleep stages were classified based on RF with weight coefficient.

Case 5. Two-state sleep stages were classified based on RF with weight coefficient.

We only list the average performance evaluation results of sleep staging in these five cases. This was done to facilitate the comparison of the subsequent experimental results. And beyond that, the experiment of Section 5.3 also showed that there was little difference in the results of sleep staging between different genders. Based on the above five cases, the five sets of standard experiment was conducted 10 times, respectively, using the data from Table 10.

The confusion matrix and average sensitivities of the first two cases are listed in Tables 11 and 12. In these two tables, each column represents sleep stage as predicted by classifier, while each row represents the actual classification of sleep stage by well-trained technicians according to the R&K manual. From the experimental results of the both tables, it is obvious to observe that the classification sensitivity is improved after adding the weight coefficients to RF. In particular, the sensitivity of REM stage was improved most obviously (79.00% up to 88.27%). The most noticeable results in these two tables were that the lowest classification sensitivity was related to the NREM1 stage, with 58.98% and 63.21%, respectively. Considering the result of NREM1, while it represents the lowest classification sensitivity, this result may be expected because of the following: (1) NREM1 and REM exhibit similar EEG wave patterns. (2) NREM1 was a transition phase of WA and NREM2 [49]. The result was consistent with the conclusion reported in literatures [50, 51] that NREM1 stage was easily mistakenly categorized as any of WA, NREM2, and REM stages. More importantly, the results of Tables 11 and 12 verify this conclusion: NREM1 stage was mainly mistakenly classified into WA, NREM2, and REM stages.

Table 13 presents the classification accuracies and kappa statistics for the five cases. The average accuracy was 84.31% in Case 1 and kappa k was 0.77, while the average accuracy was 88.98% in Case 2 and kappa k was 0.82. From comparison results of Case 2 to Case 5, it is clear that the classification accuracy increases with the decrease of sleep categories. And the classification accuracies were 88.98%, 89.73%, 92.78%, and 95.57%, respectively.

5.5. Classification Accuracy and the Number of Sleep Segments. Case 1 in Table 13 compared with Figure 7, we can observe that the accuracy of sleep staging has changed with the decrease in the number of sleep segments. To further analyze this result, the following experiment was conducted to test the accuracy of sleep staging under different numbers of

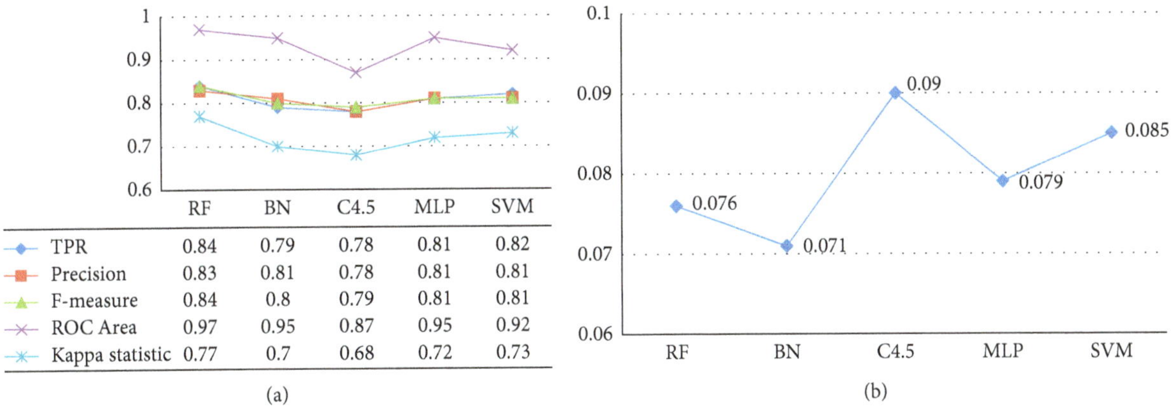

FIGURE 8: The experimental comparison results of six performance indexes on five classifiers. (a) TPR, Precision, F-measure, ROC Area, and Kappa statistic. (b) TNR.

TABLE 8: Standard deviation and weight coefficient of different features (female).

Feature	P_{rel} (Cz, δ)	Com (Cz)	Act (Cz)	P_{rel} (Cz, saw)	P_{rel} (Oz, δ)	Act (Oz)	Mob (Oz)	Skew (Oz)
Standard deviation	0.373142	0.198726	0.265727	0.305066	0.171219	0.352091	0.252303	0.144717
Weight coefficient	0.180874	0.096328	0.128807	0.147876	0.082996	0.170670	0.122300	0.070149

Note: P_{rel} (Cz, δ): the relative spectral power of delta on Fpz-Cz electrode; Com (C_z): complexity on Fpz-Cz electrode; Act (C_z): activity on Fpz-C_z electrode; P_{rel} (C_z, saw): the relative spectral power of sawtooth on Fpz-Cz electrode; P_{rel} (O_z, δ): the relative spectral power of delta on Pz-Oz electrode; Act (O_z): activity on Pz-Oz electrode; Mob (O_z): mobility on Pz-Oz electrode; Skew (O_z): skewness on Pz-Oz electrode.

TABLE 9: Standard deviation and weight coefficient of different features (male).

Feature	P_{rel} (C_z, δ)	P_{cent} (C_z, α)	Com (C_z)	Skew (C_z)	P_{rel} (O_z, δ)	P_{cent} (O_z, α)	Kurt (O_z)
Standard deviation	0.378916	0.335212	0.254561	0.183792	0.381092	0.314814	0.242165
Weight coefficient	0.181252	0.160346	0.121767	0.087915	0.182293	0.150589	0.115838

TABLE 10: Data usage in this experiment (unit: epoch or segment).

	Training data	Testing data	Total
Female	9835	4918	14753
Male	7245	3622	10867
			25620

sleep segments. The selected segments (25620, 32025, 38430, 44835, and 51240) were divided into two sets, the training (2/3) and testing (1/3) sets. Then, the weight coefficients were calculated and used as input of RF (with weight coefficient) to classify five-state sleep stages.

For the classification accuracy, Figure 9 shows a comparison of classification accuracy among different numbers of sleep segments when the improved classification algorithm was adopted. According to the obtained results, there are no significant differences in the average classification accuracy among 5 different number of sleep segments. The classification accuracy was ranged between 88.5% and 89.5%. So it was indirectly revealed that the improvement classification algorithm of our proposed method was less affected by the number of sleep segments. Weighting feature analysis used in RF achieved the highest classification accuracy. It improves the classification accuracy by 5.00% (from 84.37% to 89.37%). If the correlation analysis was not considered, the raising space of accuracy rate should be higher.

5.6. Comparison with Existing Methods. It is difficult to compare the effectiveness of various methods for automatic sleep staging because of the differences among datasets and the differences in the specific use of data. To mitigate the impact of these problems on performance comparison, our proposed method was tested on the Sleep-EDF database [34, 52] which is a subset of the Sleep-EDF database (EXPANDED) and widely used in many studies. This dataset contains 8 data recordings: sc4002e0, sc4012e0, sc4102e0, sc4112e0, st7022j0, st7052j0, st7121j0, and st7132j0. The performance of the proposed method is compared to that of the previously published studies. Table 14 provides performance comparison of various state-of-the-art methods that utilize Sleep-EDF dataset.

Based on our proposed scheme, the comparison between Tables 13 and 14 can be observed that classification accuracy from Sleep-EDF database yields better than Sleep-EDF database (EXPANDED). We think that this result may be caused by data quality. However, this conclusion is the only conjecture, and it needs further verification by sleep experts. As shown in Table 14, for four-state sleep stages

TABLE 11: Confusion matrix and average sensitivity on five-state sleep stages (without weight coefficient).

		Classifier					
		WA	NREM1	NREM2	SWS	REM	Sensitivity (%)
Experts	WA	581	51	37	7	15	84.08
	NREM1	64	404	123	5	89	58.98
	NREM2	71	137	3459	184	118	87.15
	SWS	5	3	164	1759	2	91.00
	REM	24	74	154	13	997	79.00

TABLE 12: Confusion matrix and average sensitivity on five-state sleep stages (with weight coefficient).

		Classifier					
		WA	NREM1	NREM2	SWS	REM	Sensitivity (%)
Experts	WA	594	46	33	6	12	85.96
	NREM1	59	433	109	3	81	63.21
	NREM2	64	99	3645	105	56	91.84
	SWS	4	3	111	1813	2	93.79
	REM	6	31	97	14	1114	88.27

TABLE 13: Classification accuracies and Kappa statistic based on the five cases.

Case	Case 1	Case 2	Case 3	Case 4	Case 5
Accuracy rate (%)	84.31	88.98	89.73	92.78	95.57
Kappa statistic	0.77	0.82	0.82	0.85	0.89

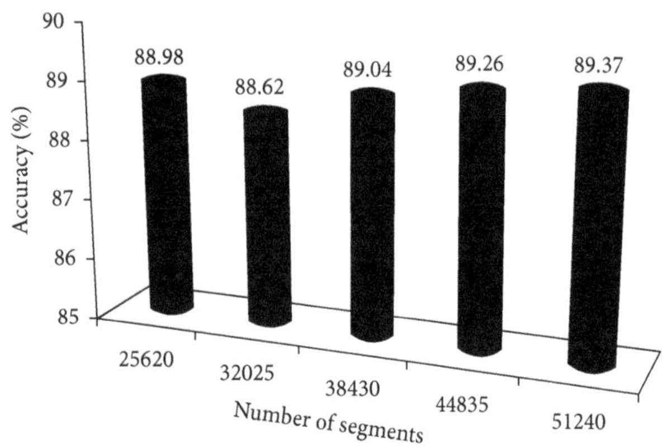

FIGURE 9: Accuracy comparison based on the number of sleep segments.

and three-state sleep stages, our proposed method outperforms others cases. For five-state sleep stages and two-state sleep stages, it gains high accuracy despite not surpassing the accuracy of some prior studies.

5.7. Discussion. In this section, we first discuss the main advantages of the proposed scheme, which makes it superior to other existing automatic sleep staging methods.

(1) Multimodal physiological signals fusion-based methods (EEG, EOG, ECG, EMG, etc.) [24, 53, 54] suffer from some limitations. For example, EOG and ECG recordings require sticky electrode placement, and this often poses restrictions on the subject's movements [6, 13]. Meanwhile, positions of subjects frequently interfere with the wire, thus diminishing the quality of EOG, ECG, and EMG signals. To overcome these limitations, we have proposed the EEG-based method.

TABLE 14: Performance comparison of various state-of-the-art methods in terms of accuracy (%).

Researchers	Five-state (%)	Four-state (%)	Three-state (%)	Two-state (%)
Sharma et al. [6]	**91.70**	92.10	93.90	**98.30**
Liang et al. [9]	83.60	—	—	—
Sharma et al. [10]	91.13	92.29	94.66	98.02
Zhu et al. [13]	88.90	89.30	92.60	97.90
Hsu et al. [17]	87.20	—	—	—
Berthomier et al. [23]	71.20	74.50	88.30	95.40
Our proposed method	90.57	**92.41**	**94.85**	97.21

Boldface indicates that the highest classification accuracy values. "—" indicates the missing cases.

(2) A major roadblock in sleep staging based on visual inspection is the identification of the sleep stages NREM1 and REM. In fact, it is also a notable disadvantage of some existing automatic sleep-staging schemes. As in [9, 13], a sensitivity of 18.75% and 15.80%, respectively, are reported for NREM1 in five-state sleep staging. Aiming at this problem, a salient feature of our proposed method is that it can effectively identify REM. As seen from the confusion matrix in Table 12, the proposed method gives a sensitivity value of 63.21%, which is much higher than those of [9, 13].

(3) "EEG-Sleep ontology" is designed and applied to manage a huge bulk of sleep EEG features and other context information. Its major advantage lies in the

realization of massive information's hierarchical management, which is convenient for human's reading and computer inquiries.

(4) A large number of EEG features (more than 50) have been used in studies [55] to achieve high classification performance. In this study, an improved method of self-adaptive correlation analysis is proposed to select the most effective EEG features. This method significantly reduces the number of selected features. For each subject, the number of selected features is less than 10.

(5) Weighting feature analysis-based improved RF not only greatly improves the classification accuracy but also avoids the overfitting in a certain extent because each tree of the forest only selects partial features.

From the above discussions, we can know that our work has made some achievements. Additionally, there are also some disadvantages. For example, under the same conditions, the dual-channel EEG sleep-staging method adopted in this paper increases the computational cost compared to the single-channel EEG. To solve this problem, on the one hand, we can use the advanced feature selection algorithm (such as an improved method of adaptive correlation analysis proposed in this study) to reduce feature dimension and computational cost. On the other hand, with the rapid development of hardware technology, we can apply advanced hardware acceleration technology to make up for the dual-channel computing cost problem. For example, the proposed algorithm can be implemented in Compute Unified Device Architecture- (CUDA-) based Graphical Processing Unit (GPU) to speed up the whole process. Although the dual-channel EEG increases the computational cost, we can effectively solve it by the above two methods. We thus conclude that our proposed scheme for automatic sleep scoring is effective and efficient. In addition, we need to explain that the single-channel ideas adopted in document [9, 10, 13] are worthy of our learning and reference. In the future, sleep stages based on single-channel EEG will become one of our research directions.

6. Conclusion and Future Work

In this paper, an automatic sleep-staging method based on EEG is proposed. The ontology-based model and the weighting feature analysis are used to represent related information of sleep staging and to improve classification algorithm, respectively. It can be seen from this paper that OBM can use a few simple terms to express and manage most of the sleep-related information, so as to solve the problem of information structured management in the process of automatic sleep staging. An improved correlation analysis algorithm based on self-adaptive correlation analysis is designed to explore the optimal subsets of different genders. And the experimental results show that linear features were useful for understanding the potential sleep structure. More importantly, the improved RF algorithm based on weighting feature analysis increases the classification accuracy by 5%. We not only verified the validity of this method by experiments but also analyzed and summarized the effects of the number of sleep segments on the classification results. Furthermore, our proposed scheme yields good NREM1 detection accuracy.

As we have shown, in the future work, our proposal scheme is feasible and potentially usable in real world for clinical assistant diagnosis of sleep disorders. Meanwhile, our classification method can be extended to other classification problems based on physiological signals, such as depressive, OSA, epilepsy, and so on. Additionally, we shall focus on exploring deep learning techniques to further improve the classification performance. To conclude, we can anticipate that our proposed method of automatic sleep staging will alleviate the burden of sleep technicians and benefit humans.

Conflicts of Interest

The authors declare that there are no conflicts of interest regarding the publication of this paper.

Acknowledgments

This work was supported by the National Basic Research Program of China (973 Program) (no. 2014CB744600), the National Natural Science Foundation of China (nos. 61632014, 61210010, 61461025, 61761027, 61871259, and 61811530325), the Program of Beijing Municipal Science and Technology Commission (no. Z171100000117005), the Program of International S&T Cooperation of MOST (no. 2013DFA11140), and the Yong Scholar Fund of Lanzhou Jiaotong University (no. 2016004).

References

[1] U. R. Acharya, S. Bhat, O. Faust et al., "Nonlinear dynamics measures for automated EEG-based sleep stage detection," *European Neurology*, vol. 74, no. 5-6, pp. 268–287, 2015.

[2] C. L. Drake, T. Roehrs, and T. Roth, "Insomnia causes, consequences, and therapeutics: an overview," *Depression and Anxiety*, vol. 18, no. 4, pp. 163–176, 2003.

[3] O. Andlauer, H. Moore, L. Jouhier, C. Drake, and P. E. Peppard, "Nocturnal rapid eye movement sleep latency for identifying patients with narcolepsy/hypocretin deficiency," *JAMA Neurology*, vol. 70, no. 7, p. 891, 2013.

[4] H. H. Jiang, B. Hu, Z. Y. Liu, L. H. Yan, and T. Y. Wang, "Investigation of different speech types and emotions for detecting depression using different classifiers," *Speech Communication*, vol. 90, no. 6, pp. 39–46, 2017.

[5] X. W. Li, Z. Jing, B. Hu, J. Zhu, and N. Zhong, "A resting-state brain functional network study in MDD based on minimum spanning tree analysis and the hierarchical clustering," *Complexity*, vol. 2017, Article ID 9514369, p. 11, 2017.

[6] M. Sharma, D. Goyal, P. V. Achuth, and U. R. Acharya, "An accurate sleep stages classification system using a new class of optimally time-frequency localized three-band wavelet filter

bank," *Computers in Biology and Medicine*, vol. 98, pp. 58–75, 2018.

[7] A. Rechtschaffen and A. Kales, *A Manual of Standardized Terminology, Techniques and Scoring Systems for Sleep Stages of Human Subjects*, U. G. P. Office, Public Health Service, Washington DC, USA, 1968.

[8] C. Iber, S. Ancoli-Israel, A. L. Chesson, and S. F. Quan, *The AASM Manual for the Scoring of Sleep and Associated Events: Rules, Terminology and Technical Specification*, American Academy of Sleep Medicine, Westchester, NY, USA, 2007.

[9] S. F. Liang, C. E. Kuo, Y. H. Hu, Y. H. Pan, and Y. H. Wang, "Automatic stage scoring of single-channel sleep EEG by using multiscale entropy and autoregressive models," *IEEE Transactions on Instrumentation and Measurement*, vol. 61, no. 6, pp. 1649–1657, 2012.

[10] R. Sharma, R. B. Pachori, and A. Upadhyay, "Automatic sleep stages classification based on iterative filtering of electroencephalogram signals," *Neural Computing and Applications*, vol. 28, no. 10, pp. 2959–2978, 2017.

[11] U. R. Acharya, E. C. Chua, K. C. Chua, L. C. Min, and T. Tamura, "Analysis and automatic identification of sleep stages using higher order spectra," *International Journal of Neural Systems*, vol. 20, no. 6, pp. 509–521, 2010.

[12] Z. H. Ge and Y. F. Sun, "Sleep stages classification using neural networks with multi-channel neural data," in *Lecture Notes in Computer Science*, Springer, Berlin, Germany, pp. 306–316, 2015.

[13] G. H. Zhu, Y. Li, and P. P. Wen, "Analysis and classification of sleep stages based on difference visibility graphs from a single-channel EEG signal," *IEEE Journal of Biomedical and Health Informatics*, vol. 18, no. 6, pp. 1813–1821, 2014.

[14] X. W. Zhang, B. Hu, X. Ma, P. Moore, and J. Chen, "Ontology driven decision support for the diagnosis of mild cognitive impairment," *Computer Methods and Programs in Biomedicine*, vol. 113, no. 3, pp. 781–791, 2014.

[15] H. T. Wu, R. Talmon, and Y. L. Lo, "Assess sleep stage by modern signal processing techniques," *IEEE Transactions on Biomedical Engineering*, vol. 62, no. 4, pp. 1159–1168, 2014.

[16] T. Lei, X. h. Jia, Y. N. Zhang, L. F. He, H. Y. Meng, and A. K. Nandi, "Significantly fast and robust fuzzy c-means clustering algorithms based on morphological reconstruction and membership filtering," *IEEE Transactions on Fuzzy Systems*, 2018.

[17] Y. L. Hsu, Y. T. Yang, J. S. Wang, and C. Y. Hsu, "Automatic sleep stage recurrent neural classifier using energy features of EEG signals," *Neurocomputing*, vol. 104, no. 2, pp. 105–114, 2013.

[18] L. Fraiwan, K. Lweesy, N. Khasawneh, H. Wenz, and H. Dickhaus, "Automated sleep stage identification system based on time-frequency analysis of a single EEG channel and random forest classifier," *Computer Methods and Programs in Biomedicine*, vol. 108, no. 1, pp. 10–19, 2012.

[19] M. Hanaoka, M. Kobayashi, and H. Yamazaki, "Automatic sleep stage scoring based on waveform recognition method and decision-tree learning," *Systems and Computers in Japan*, vol. 33, no. 11, pp. 1–13, 2002.

[20] S. T. Pan, C. E. Kuo, J.-H. Zeng, and S. F. Liang, "A transition-constrained discrete hidden Markov model for automatic sleep staging," *BioMedical Engineering OnLine*, vol. 11, no. 1, p. 52, 2012.

[21] M. Diykh, Y. Li, and W. Peng, "EEG sleep stages classification based on time domain features and structural graph similarity," *IEEE Transactions on Neural Systems and Rehabilitation Engineering*, vol. 24, no. 11, pp. 1159–1168, 2016.

[22] M. Diykh and Y. Li, "Complex networks approach for EEG signal sleep stages classification," *Expert Systems with Applications*, vol. 63, no. 11, pp. 241–248, 2016.

[23] C. Berthomier, X. Drouot, and M. H. Stoica, "Automatic analysis of single-channel sleep EEG: validation in healthy individuals," *Sleep*, vol. 30, no. 11, pp. 1587–1595, 2007.

[24] P. Piñero, P. Garcia, L. Arco, A. Alvarez, M. M. Garcia, and R. Bonal, "Sleep stage classification using fuzzy sets and machine learning techniques," *Neurocomputing*, vol. 58–60, pp. 1137–1143, 2004.

[25] B. C. Jantzen, D. G. Mayo, and L. Patton, "Ontology and methodology," *Synthese*, vol. 192, no. 11, pp. 3413–3423, 2015.

[26] X. W. Zhang, B. Hu, J. Chen, and P. Moore, "Ontology-based context modeling for emotion recognition in an intelligent web," *World Wide Web*, vol. 16, no. 4, pp. 497–513, 2013.

[27] J. Chen, B. Hu, P. Moore, X. Zhang, and X. Ma, "Electroencephalogram-based emotion assessment system using ontology and data mining techniques," *Applied Soft Computing*, vol. 30, no. 5, pp. 663–674, 2015.

[28] X. P. Hu, J. Cheng, M. C. Zhou et al., "Emotion-aware cognitive system in multi-channel cognitive radio ad hoc networks," *IEEE Communications Magazine*, vol. 56, no. 4, pp. 180–187, 2018.

[29] T. Feng, G. N. Sun, Y. H. Xia, and J. F. Ma, "Modeling and detection of ontology-based byzantine attacks," *Journal of Graduate University of Chinese Academy of Sciences*, vol. 28, no. 5, pp. 696–705, 2011.

[30] D. Soergel, B. Lauser, A. Liang, F. Fisseha, J. Keizer, and S. Katz, "Reengineering thesauri for new applications: the AGROVOC example," *Journal of Digital Information*, vol. 4, no. 4, pp. 1–23, 2004.

[31] S. K. Choi, "An ontological model to support communications of situation-aware vehicles," *Transportation Research Part C: Emerging Technologies*, vol. 53, no. 5, pp. 112–133, 2015.

[32] C. T. Bau, R. C. Chen, and C. Y. Huang, "Construction of a clinical decision support system for undergoing surgery based on domain ontology and rules reasoning," *Telemedicine and e-Health*, vol. 20, no. 5, pp. 460–472, 2015.

[33] J. Y. Wang, T. Mendori, and J. Xiong, "A language learning support system using course-centered ontology and its evaluation," *Computer and Education*, vol. 78, no. 9, pp. 278–293, 2014.

[34] B. Kemp, A. H. Zwinderman, B. Tuk, H. A. C. Kamphuisen, and J. J. L. Oberye, "Analysis of a sleep-dependent neuronal feedback loop: the slow-wave microcontinuity of the EEG," *IEEE Transactions on Biomedical Engineering*, vol. 47, no. 9, pp. 1185–1194, 2000.

[35] A. L. Goldberger, L. A. N. Amaral, L. Glass et al., "Physiobank, physiotoolkit, and physionet: components of a new research resource for complex physiologic signals," *Circulation*, vol. 101, no. 23, pp. e215–e220, 2000.

[36] B. Hu, D. Majoe, M. Ratcliffe et al., "EEG-based cognitive interfaces for ubiquitous applications: developments and challenges," *IEEE Intelligent Systems*, vol. 26, no. 5, pp. 46–53, 2011.

[37] X. W. Li, B. Hu, and S. T. Sun, "EEG-based mild depressive detection using feature selection methods and classifiers," *Computer Methods and Programs in Biomedicine*, vol. 136, no. 8, pp. 151–161, 2016.

[38] X. Li, D. Song, P. Zhang, Y. X. Hou, and B. Hu, "Deep fusion of multi-channel neurophysiological signal for emotion

recognition and monitoring," *International Journal of Data Mining and Bioinformatics*, vol. 18, no. 1, p. 1, 2017.

[39] A. Jain, K. Nandakumar, and A. Ross, "Score normalization in multimodal biometric systems," *Pattern Recognition*, vol. 38, no. 12, pp. 2270–2285, 2005.

[40] J. F. Sowa, "Top-level ontological categories," *International Journal of Human-Computer Studies*, vol. 43, no. 5-6, pp. 669–685, 1995.

[41] V. Krishnan and N. A. Collop, "Gender differences in sleep disorders," *Current Opinion in Pulmonary Medicine*, vol. 12, no. 6, pp. 383–389, 2006.

[42] B. Zhang and Y. K. Wing, "Sex differences in insomnia: a meta-analysis," *Sleep*, vol. 29, no. 1, pp. 85–93, 2006.

[43] L. Breiman, "Random forests," *Machine Learning*, vol. 45, no. 1, pp. 5–32, 2001.

[44] S. D. Río, V. López, and J. M. Benítez, "On the use of map reduce for imbalanced big data using random forest," *Inform Sciences*, vol. 285, no. 11, pp. 112–137, 2014.

[45] N. Dalkey and O. Helmer, "An experimental application of the DELPHI method to the use of experts," *Management Science*, vol. 9, no. 3, pp. 458–467, 1963.

[46] J. Bi, "Replicated similarity testing using paired comparison method," *Journal of Sensory Studies*, vol. 22, no. 2, pp. 500–507, 2007.

[47] X. W. Zhang, B. Hu, and M. Xu, "Resting-state whole-brain functional connectivity networks for MCI classification using L2-regularized logistic regression," *IEEE Transactions on NanoBioscience*, vol. 14, no. 2, pp. 237–247, 2015.

[48] Weka Packages, *Waikato Environment for Knowledge Analysis (WEKA) Package, Weka Packages*, University of Waikato, Hamilton, New Zealand, 2016, https://www.cs.waikato.ac.nz/ml/weka/.

[49] S. L. Himanen and J. Hasan, "Limitations of rechtschaffen and kales," *Sleep Medicine Reviews*, vol. 4, no. 2, pp. 149–167, 2004.

[50] M. Corsi-Cabrera, Z. Torres, Y. Portilla, and M. A. Guevara, "Power and coherent oscillations distinguish REM sleep, stage 1 and wakefulness," *International Journal of Psychophysiology*, vol. 60, no. 1, pp. 59–66, 2006.

[51] S. Khalighi, T. Sousa, G. Pires, and U. Nunes, "Automatic sleep staging: a computer assisted approach for optimal combination of features and polysomnographic channels," *Expert Systems with Applications*, vol. 40, no. 17, pp. 7046–7059, 2013.

[52] B. Kemp, *The Sleep-EDF Database Online*, 2000, http://www.physionet.org/physiobank/database/sleep-edf/.

[53] R. Agarwal and J. Gotman, "Computer-assisted sleep staging," *IEEE Transactions on Biomedical Engineering*, vol. 48, no. 12, pp. 1412–1423, 2001.

[54] J. Virkkala, J. Hasan, A. Värri, S. Himanen, and K. Müller, "Automatic sleep stage classification using two-channel electro-oculography," *Journal of Neuroscience Methods*, vol. 166, no. 1, pp. 109–115, 2007.

[55] C. S. Huang, C. L. Lin, L. W. Ko, S. Y. Liu, T. P. Su, and C. T. Lin, "A hierarchical classification system for sleep stage scoring via forehead EEG signals," in *Proceedings of IEEE Symposium on Computational Intelligence*, pp. 1–5, Singapore, April 2013.

8

Large-Deformation Image Registration of CT-TEE for Surgical Navigation of Congenital Heart Disease

Shuiping Gou,[1] Linlin Chen,[1] Yu Gu,[1] Liyu Huang,[2] Meiping Huang,[3] and Jian Zhuang[4]

[1] *Key Laboratory of Intelligent Perception and Image Understanding of Ministry of Education of China, Xidian University, Xi'an 710071, China*
[2] *Department of Biological Science and Engineering, Xidian University, Xi'an 710071, China*
[3] *Catheterization Lab, Guangdong Cardiovascular Institute, Guangdong Provincial Key Laboratory of South China Structural Heart Disease, Guangdong General Hospital, Guangdong Academy of Medical Sciences, Guangzhou 510000, China*
[4] *Department of Cardiac Surgery, Guangdong Cardiovascular Institute, Guangdong Provincial Key Laboratory of South China Structural Heart Disease, Guangdong General Hospital, Guangdong Academy of Medical Sciences, Guangzhou 510000, China*

Correspondence should be addressed to Shuiping Gou; shpgou@mail.xidian.edu.cn

Academic Editor: Po-Hsiang Tsui

The surgical treatment of congenital heart disease requires navigational assistance with transesophageal echocardiography (TEE); however, TEE images are often difficult to interpret and provide very limited anatomical information. Registering preoperative CT images to intraoperative TEE images provides surgeons with richer and more useful anatomical information. Yet, CT and TEE images differ substantially in terms of scale and geometry. In the present research, we propose a novel method for the registration of CT and TEE images for navigation during surgical repair of large defects in patients with congenital heart disease. Valve data was used for the coarse registration to determine the basic location. This was followed by the use of an enhanced probability model map to overcome gray-level differences between the two imaging modalities. Finally, the rapid optimization of mutual information was achieved by migrating parameters. This method was tested on a dataset of 240 images from 12 infant, children ($\leq$ 3 years old), and adult patients with congenital heart disease. Compared to the "bronze standard" registration, the proposed method was more accurate with an average Dice coefficient of 0.91 and an average root mean square of target registration error of 1.2655 mm.

1. Introduction

Congenital heart disease accounts for 28% of all congenital malformations. In China, the incidence of congenital heart disease is 0.4–1% among infants with 150000–200000 newborns annually. Imaging of congenital heart disease includes cardiovascular magnetic resonance (CMR), computed tomography (CT), transesophageal echocardiography (TEE), and other modalities. CMR provides comprehensive data on the mechanism of congenital heart disease without the need for radiography or invasive procedures; however, it requires subjects to remain stationary for 15–60 min, which can be challenging in infants and children. Additionally, the high cost of CMR limits its clinical application. High-resolution CT can be rapidly imaging and provide a wealth of anatomical data; however, the equipment of it is not easy to move and cannot be used in surgery. TEE allows a continuous visualization of the movement of the visceral organ without trauma and the observation of the heartbeat in real time, which is convenient to use in surgery; thus, it can be useful for navigation during the surgery. However, its images provide very limited data on cardiac anatomy, it often requires the use of another imaging modality, and there is no established method for accurately registering anatomical structures to TEE images. Each type of imaging modality has advantages and disadvantages, so image fusion is necessary.

Image registration is the premise of image fusion, which has significant value in the fields of remote sensing data analysis, computer vision, and image processing. Feature point-based registration is one of the most commonly used image registration methods. Typical feature points include corner points, line intersections, centroids of closed curves, maximum curvature points on the contour, local curvature discontinuities detected by Gabor wavelet [1], and local maximum points of wavelet transform [2]. Feature points can be extracted manually or automatically. There are many methods to automatically extract feature points. Feng et al. [3] proposed a 3D feature point extraction using a neural network. Cheng et al. [4] adapted multilevel feature point extraction into a Naïve Bayesian classification framework. Chen et al. [5] presented a novel method of 3D XYT feature point extraction. Dán et al. [6] extracted feature points by characterizing of SURF and BRISK Interest Point Distribution. Wang et al. [7] introduced a multiscale and hierarchical feature extraction method. The feature point-based registration can greatly reduce the computational complexity of the matching process and has better adaptability to the change of gray level, image deformation, and occlusion. However, when the image modalities differ greatly, the algorithm based on automatically selecting the feature points has poor registration accuracy and may even result in completely failed results. Therefore, typically, the coarse registration was implemented by selecting feature points, and then the fine registration was carried out based on the correlation matching techniques.

Similarity measures are critical for the correlation matching techniques. The sum of squared differences (SSD), sum of absolute differences (SAD) [8], normalized cross correlation (NCC) [9], and mutual information (MI) [10] are commonly used similarity measures. SSD and SAD are particularly sensitive to the gray scale transformation of the input image and are often used for monomodality image registration. The sensitivity of NCC to abnormal points affects the registration. MI was previously proposed by Viola et al. [10] and Maes et al. [11] as an insensitive method to differences in intensities between images and as a method that does not require feature extraction or segmentation that has cultivated attention recently. It is widely used for the registration of multimodal images; however, it is not a universal method. The registration function of MI can be ambiguous and can contain local extrema [12]. Therefore, it is common to combine MI with other imaging features with the aim of improving registration and having an accurate correspondence on difficult registration issues. Gong et al. [13] combined MI with SIFT to realized an automatic image registration for remote sensing images. Woo et al. [14] demonstrated a multimodal Registration via Mutual Information Incorporating Geometric and Spatial Context. Atli et al. [15] combined contrast enhancement techniques with MI-based image registration. Legg et al. [16] used the feature neighbourhood mutual information for multimodal image registration. Pradhan et al. [17] incorporated utility measure into enhanced mutual information as a weighted information to the joint histogram of the images.

Most of the aforementioned similarity measures use joint probability models of single pixels or voxels. By randomly changing the location of pixels or voxels on images and evaluating the statistical criterion, the same similarity is obtained [18]. This means that when spatial information between pixels or voxels is ignored, registration may fail because the matching criterion is not able to quantify the difference between the two images. Using spatial distributions of pixels or voxels rather than single pixels or voxels, Penney et al. [19] successfully aligned ultrasound images and MR images of the liver by registering vessel probability images. This technique used a probability map to enhance corresponding features and depress those that were noncorresponding between the images. However, the generation of a probability density functions as proposed by the authors required a large amount of training data that were different from the actual test data. This approach cannot be easily applied to cardiac CT and TEE image registration. Additionally, the probability map proposed by Penney et al. yielded evident deviations, which may not be acceptable for the registration of cardiac CT and TEE images. In this study, an enhanced probability map model is proposed to overcome the shortage of training data and the deviation of the aforementioned probability map.

The difficulties in cardiac CT-TEE registration lie in the displacement caused by cardiac motion, as well as the considerable differences in the distribution of gray levels, structural features, and edge features. Determining methods to overcome the above difficulties is the key issue in cardiac CT-TEE registration. At present, there are few studies on cardiac CT/MR-TEE registration. D. S. Cho et al. [20] and C. A. Linte et al. [21] adopted a landmark-based registration method to reduce image displacement and modal differences, both of which relied on custom-developed equipment to extract landmarks. P. Lang et al. [22] focused only on the cardiac CT-TEE registration of the aortic root. F. Li et al. [23] performed a registration using a two-stage approach. In the first step, ICP algorithm was performed on the basis of semiautomatic segmentation. In the second step, sample points were extracted by setting intensity thresholds; then, these sample points were used to perform a finer registration step based on MI. This method involves setting different intensity thresholds manually according to different images.

Most of the existing cardiac CT-TEE registrations require the assistance of hardware devices and manual intervention. Some studies only focused on the cardiac registration of certain special sections. In this paper, a registration method is proposed for all cardiac sections, which can minimize manual intervention and does not require additional assistance of hardware equipment. Considering that heart valves are clearly visible on TEE images and are also easily located on CT images, this paper introduced the spatial positions of valves as a priori information. Then, coarse registration was performed based on valve positions to narrow the position difference in CT-TEE registration. To overcome the limitation of the probability map proposed by Penney et al., an enhanced probability map model was proposed by generating the probability map based on the enhancement of regions of interest. It can better characterize the spatial distribution of gray levels and can directly generate the probability density function by using images for subsequent registration. Moreover, we used the transformation parameters from coarse registration as the initial values for the optimization algorithm in the

Input: Cardiac CT image $I_R(x)$ and TEE image $I_F(x)$
Output:Result of final registration $S_{final}(x)$
Procedure of the Proposed Method:
Step 1: Cardiac CT $I_R(x)$ and TEE $I_F(x)$ images were segmented interactively to yield the CT $G_R(x)$ and TEE $G_F(x)$ image segmentations separately.
Step 2: During interactive segmentations, the spatial positions of heart valves were used as priori information.
Step 3: Based on the valve data, CT $I_R(x)$ and TEE $I_F(x)$ images were registered to yield the transformation matrix T_{basic} and the results of the basic registration $S_{basic}(x)$.
Step 4: ROI enhancement was performed for the $I_R(x)$ and $I_F(x)$ to obtain the enhanced CT $I_{Rh}(x)$ and the enhanced TEE $I_{Fh}(x)$ images. During segmentation, gray-scale enhancement was performed to obtain the enhanced segmentations of CT $G_{Rh}(x)$ and TEE $G_{Fh}(x)$ images.
Step 5: The probability map of the CT $P_R(x)$ image was generated based on the enhanced CT$I_{Rh}(x)$ image and its enhanced segmentation $G_{Rh}(x)$. The probability map of TEE $P_F(x)$ image was generated similarly.
Step 6: Finally, the probability maps $P_R(x)$ and $P_F(x)$ were used to perform the final registration by maximizing similarity, producing the final registration transformation T_{final} and the result of final registration $S_{final}(x)$.

ALGORITHM 1: Procedure of the proposed method.

final registration to prevent the optimization algorithm from falling into a local optimum.

2. Materials and Methods

2.1. Method Overview. Registration was performed using valves as a landmark. Probability map-based MI (VPMMI) consisted of basic registration using cardiac valves as input data, while final registration was based on the probability map. This method was named VPMMI. Considering the demand for valve positions in actual application in this research context, the spatial locations of valves were introduced as a priori information to improve the effectiveness of CT and TEE image registration. With the a priori information of valve, we performed a simple and quick coarse registration. Errors in determining the initial values may cause an optimization algorithm to fall into a local optimum. To avoid this, we used the transformation parameters from coarse registration as the initial values for the optimization algorithm in the next phase of registration. Finally, in order to further improve the registration accuracy, the enhanced probability map was used to perform a finer registration step, which can overcome the shortage of training data. To sum up, our method is as follows.

Cardiac CT $I_R(x)$ was used as the reference image, while TEE $I_F(x)$ was used as the floating image, where x denoted a point on the 2 image spaces Ω_R and Ω_F. The goal of image registration was to find a suitable transformation $T:\Omega_F \longrightarrow \Omega_R$ in order to maximize the similarity $S(T; I_F; I_R)$ between $I_R(x)$ and $T(I_F(x))$. In this paper, transformation T comprised T_{basic} and T_{final}. T_{basic} was the transformation obtained from basic registration that was used as the initial parameter for the optimization algorithm for the final registration. T_{final} transformed positions from the floating to the reference images.

To achieve the aforementioned goals we performed the steps shown in Algorithm 1.

A flowchart of the proposed algorithm is shown in Figure 1.

2.2. Basic Registration

2.2.1. Interactive Segmentation. User interactions were adopted to introduce feature points and generate subsequent probability maps. We employed the grab-cut algorithm [24] for interactive segmentations, which is an image segmentation algorithm used to minimize energy function for boundary detection. In order to rapidly obtain segmentation images, the physician manually selected ROIs and marked the target and background areas on CT and TEE images. Figure 2 shows the interactive segmentation of CT and TEE images.

2.2.2. Introducing Heart Valve Locations as A Priori Information. Heart valves play an important role in cardiac circulation. Atrioventricular (between atria and ventricles) and semilunar valves (in arteries leaving the heart) help prevent the backward flow of blood. They can be clearly visualized on TEE and their approximate locations can be determined on CT scans by professionals. Therefore, it is reasonable to use valve positions as a priori information.

During interactive segmentation, three endpoints or midpoints coordinates $x_{1r} = (i_{1r}, j_{1r}), x_{2r} = (i_{2r}, j_{2r})$, $x_{3r} = (i_{3r}, j_{3r})$ of heart valves on the CT $I_R(x)$ image were determined as three feature points on CT. Similarly, three feature points $x_{1f} = (i_{1f}, j_{1f}), x_{2f} = (i_{2f}, j_{2f}), x_{3f} = (i_{3f}, j_{3f})$ were determined on the TEE image $I_F(x)$.

2.2.3. Basic Registration Based on Feature Points. Using the three CT feature points $x_{1r} = (i_{1r}, j_{1r}), x_{2r} = (i_{2r}, j_{2r}), x_{3r} = (i_{3r}, j_{3r})$ and the three TEE feature points

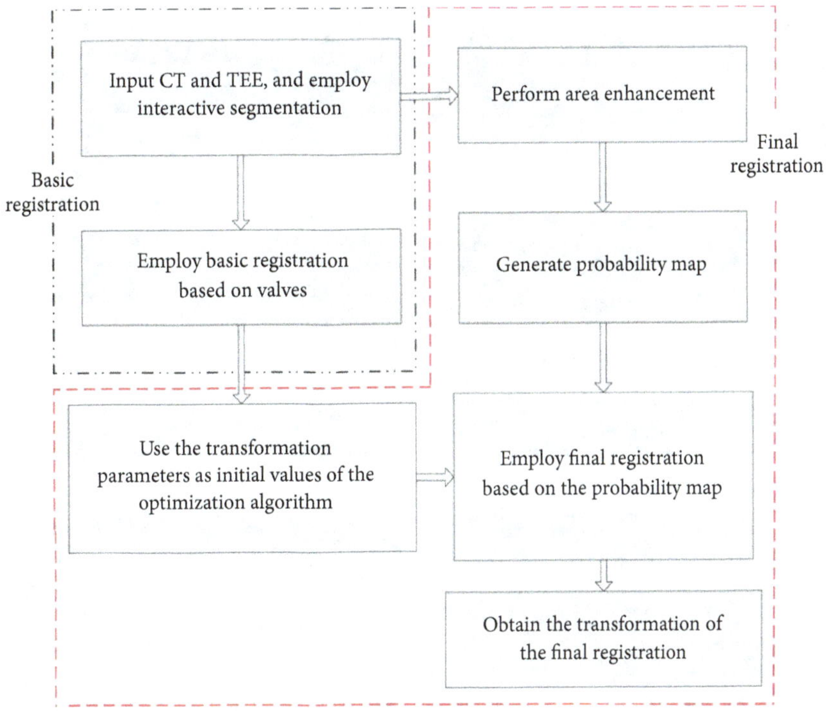

FIGURE 1: Flowchart of the proposed algorithm. CT, computed tomography; TEE, transesophageal echocardiography.

$x_{1f} = (i_{1f}, j_{1f}), x_{2f} = (i_{2f}, j_{2f}), x_{3f} = (i_{3f}, j_{3f})$, the coordinate transformation matrix T_{basic} of the basic registration below was obtained.

$$\begin{bmatrix} x_{1r} \\ x_{2r} \\ x_{3r} \end{bmatrix} = T_{basic} \times \begin{bmatrix} x_{1f} \\ x_{2f} \\ x_{3f} \end{bmatrix} \quad (1)$$

Based on T_{basic}, the results of basic registration $S_{basic}(x)$ were obtained by performing an affine transformation and a cubic interpolation to the floating image $I_F(x)$.

2.3. Final Registration

2.3.1. Enhancement of CT and TEE Images. In order to generate an improved probability map for good performance in the subsequent registration, we performed region enhancement to improve the contrasts between the foreground and the background. First, we set an enhancement matrix $V_R(x)$ for the CT image $I_R(x)$. $V_R(x)$ was constructed such that the size of $V_R(x)$ was the same as the size of the CT image $I_R(x)$ and the pixels corresponding to the coordinates of the ROI on $I_R(x)$ were set to a number N (N set to 80 in our experiment; this value was based on the average ROI intensity and empirical experiments) and 0 for the remaining pixels. The process of region enhancement is shown in Figure 3.

Subsequently, we enhanced $I_R(x)$ and its segmentation $G_R(x)$ to obtain the enhanced CT image $I_{Rh}(x)$ and enhanced segmentation $G_{Rh}(x)$:

$$I_{Rh}(x) = I_R(x) + V_R(x) \quad (2)$$

$$G_{Rh}(x) = G_R(x) + V_R(x) \quad (3)$$

Similarly, we generated a TEE image enhancement matrix $V_F(x)$. $V_F(x)$ was constructed such that the size of $V_F(x)$ was the same as the size of the TEE image $I_R(x)$ and the pixels corresponding to the coordinates of the ROI in $I_R(x)$ were set to a number M (M was 100 in our experiment; it was an empirical value based on the average ROI intensity of many experiments) and 0 for the remaining pixels. Finally, $I_F(x)$ and its segmentation $G_F(x)$ were enhanced to obtain the enhanced TEE image $I_{Fh}(x)$ and enhanced segmentation $G_{Fh}(x)$:

$$I_{Fh}(x) = I_F(x) + V_F(x) \quad (4)$$

$$G_{Fh}(x) = G_F(x) + V_F(x) \quad (5)$$

It is worth mentioning that although we strengthened the pixel intensity on this step, we did not change the spatial relationship between pixels. Therefore, the subsequent probability map model was not affected.

2.3.2. Probability Map Generation. Instead of using pixel intensity, the probability map described image information,

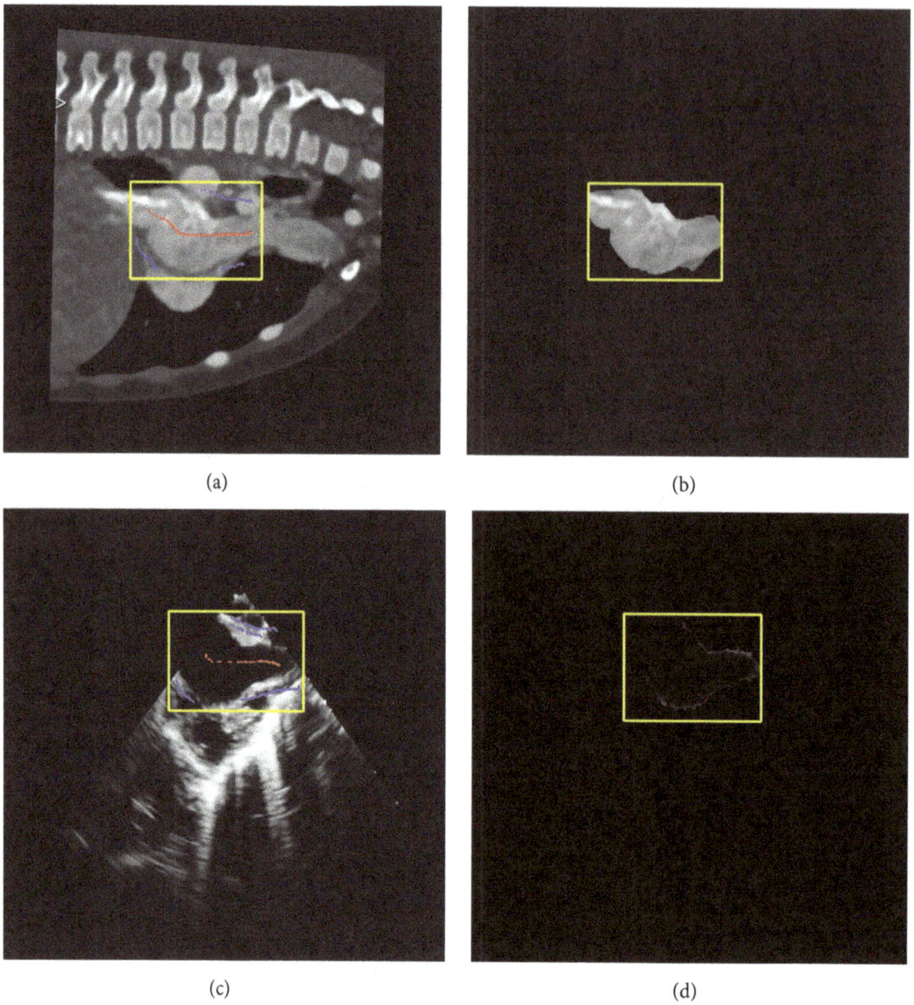

FIGURE 2: The process of interactive segmentation. (a) Interactive segmentation of a CT image. (b) The result of CT image segmentation. (c) Interactive segmentation of a TEE image. (d) The result of TEE image segmentation. Note: during interactive segmentation, the user selected a region of interest (yellow box) and marked the target (red dots) and the background areas (blue dots). CT, computed tomography; TEE, transesophageal echocardiography.

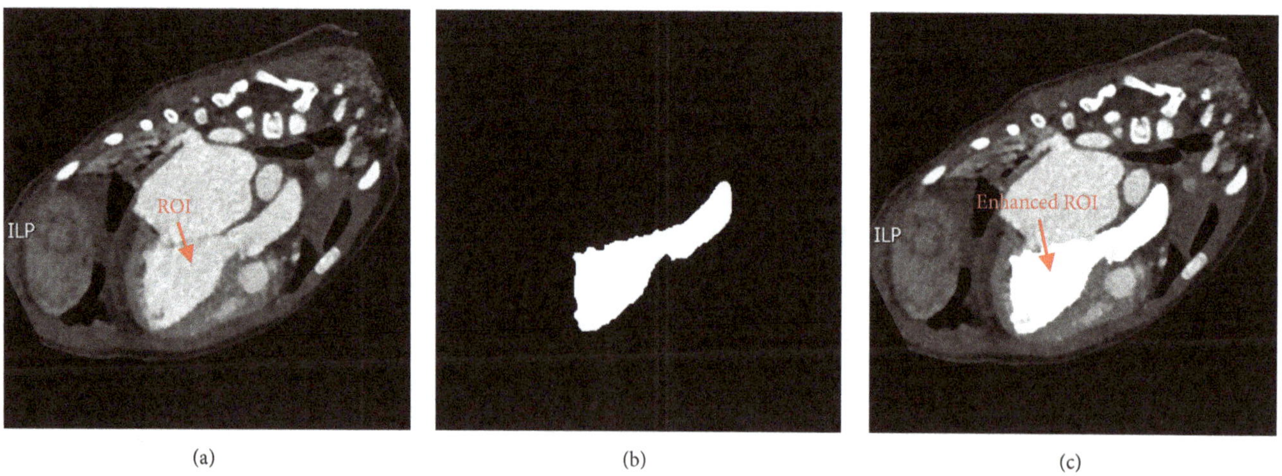

FIGURE 3: The process of region enhancement. (a) Original CT image $I_R(x)$. (b) Enhancement matrix $V_R(x)$. (c) Enhanced CT image. CT, computed tomography; ROI, region of interest.

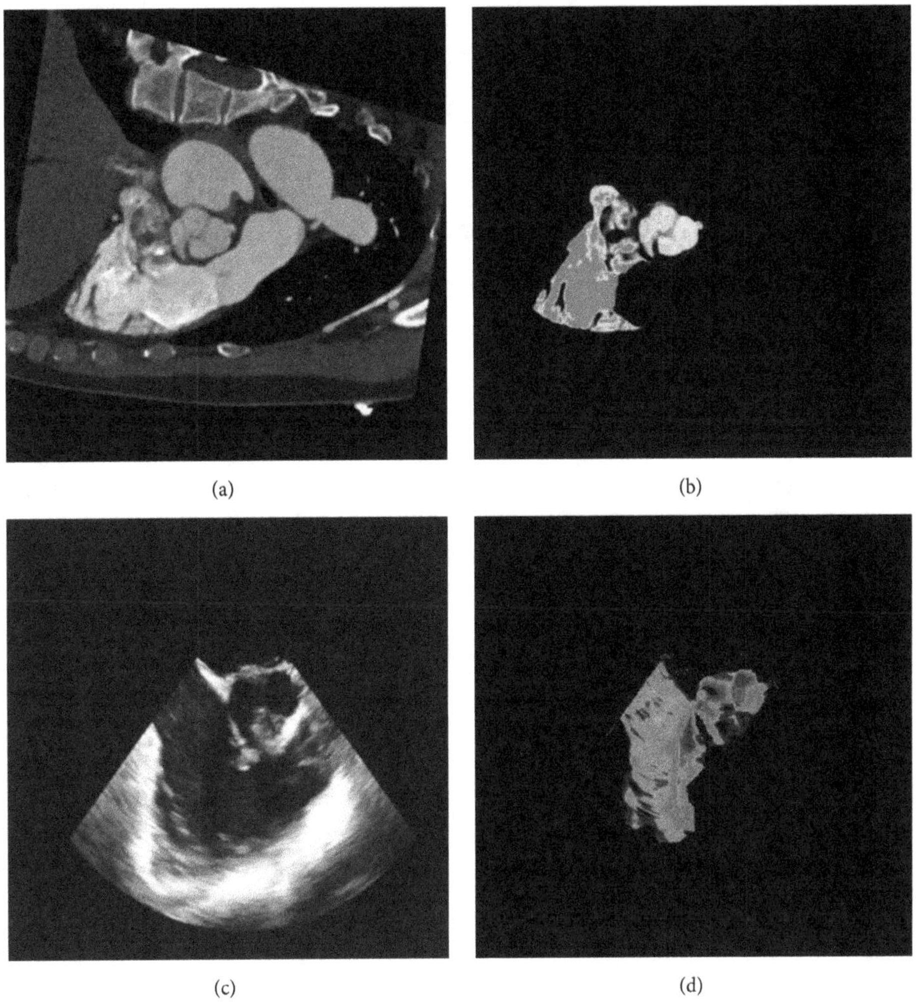

FIGURE 4: CT and TEE probability maps. (a) The original CT image. (b) The CT probability map. (c) The original TEE image. (d) The TEE probability map. CT, computed tomography; TEE, transesophageal electrocardiography.

i.e., the probability that a pixel was located inside the ROI. Probability mapping enhances relevant features and depresses irrelevant features between two images. Based on the region-enhanced CT image $I_{Rh}(x)$ and its enhanced segmentation $G_{Rh}(x)$, the probability density function of CT $f_R(i)$ was generated as follows:

$$f_R(i) = \frac{Number\ of\ G_{Rh}(x)\ pixels\ of\ intensity\ i}{Number\ of\ I_{Rh}(x)\ pixels\ of\ intensity\ i}\% \quad (6)$$

The probability density function $f_R(i)$ corresponds to a retrieval table that outputs a probability that a given pixel is located within the ROI when a pixel value is used as an input. In the same way, we derived the probability density function of TEE $f_F(i)$ as follows:

$$f_F(i) = \frac{Number\ of\ G_{Fh}(x)\ pixels\ of\ intensity\ i}{Number\ of\ I_{Fh}(x)\ pixels\ of\ intensity\ i}\% \quad (7)$$

Subsequently, we used the region-enhanced CT image $I_{Rh}(x)$ as the input of $f_R(i)$ to develop the probability map $P_R(x)$: $P_R(x) = f_R(i), i \in I_{Rh}(x)$. Similarly, using the region-enhanced TEE image $I_{Fh}(x)$ as the input of $f_F(i)$, we developed the probability map $P_F(x)$:$P_F(x) = f_F(i), i \in I_{Fh}(x)$. CT and TEE probability maps are shown in Figure 4.

2.3.3. Final Registration Based on the Maximum Mutual Information of the Probability Map. We used Powell's algorithm as an optimization algorithm with initial parameters obtained from transforming matrix T_{basic}. The transform matrix of the final registration T_{final} was yielded when the normalized mutual information (NMI) between $P_R(x)$ and $P_F(x)$ was maximized through Powell's algorithm. Afterward, the final registration result $S_{final}(x)$ was derived as follows: $S_{final}(x) = T_{final} \times I_F(x)$.

3. Results and Discussion

The proposed algorithm was tested on a dataset consisting of 240 images from 12 patients with congenital heart disease (identified as Patient 1–Patient 12). Patients 11 and 12 were adults and the remaining 10 patients were infants and young

Table 1: Comparison of several registration algorithms on the maximum TRE.

Registration algorithm	D. S. Cho et al. [23]	C. A. Linte et al. [24]	P. Lang et al. [25]	Our method
TRE(mm)	5.1	4.8	5	2.3

TRE, target registration error.

Table 2: Registration indexes.

Patient	Average Dice coefficient			Average TRE (mm)		
	VPMMI	VMI	ICPMI	VPMMI	VMI	ICPMI
Patient 1	0.88	0.45	0.80	1.7290	9.5789	3.0710
Patient 2	0.89	0.80	0.82	1.2701	3.6162	1.8111
Patient 3	0.90	0.45	0.79	0.6013	11.5463	2.7112
Patient 4	0.95	0.95	0.84	1.1176	1.6780	8.0261
Patient 5	0.89	0.00	0.60	1.1179	12.3223	9.3880
Patient 6	0.87	0.04	0.88	2.2345	19.0837	4.9850
Patient 7	0.89	0.65	0.59	1.4118	7.5647	9.1564
Patient 8	0.91	0.32	0.89	0.8496	7.3520	0.8964
Patient 9	0.91	0.42	0.92	0.7337	5.7968	0.6936
Patient 10	0.95	0.83	0.85	0.6760	3.8212	2.9338
Patient 11 (adult)	0.90	0.58	0.69	1.7191	13.4195	6.6400
Patient 12 (adult)	0.92	0.62	0.86	1.7260	13.9903	2.2362
Average	0.91	0.51	0.79	1.2655	9.1475	4.3791

TRE, target registration error; VMI and ICPMI, the comparison algorithms; VPMMI, the proposed method (valves as a benchmark and probability map-based mutual information method)

children $\leq$ 3 years of age. Each patient's data included CT images and corresponding TEE images. Due to the difficulty of data acquisition, TEE images for each patient only included 22 standard sections of the heart. The sizes of all CT images were 1024×1024 and the sizes of all TEE images were 600×800.

All tests were executed on an Intel i5-6500 (Intel Corp., Santa Clara, CA) CPU 3.2 GHz, and the code was written using MATLAB R2015b (MathWorks, Natick, MA) development environment.

3.1. Comparison Algorithms. Currently, there are very few studies on cardiac CT/MR-TEE registration, and each study focuses on different issues. D. S. Cho et al. [20] employed a landmark-based registration that was based on anatomical landmarks extracted using a custom-developed landmark extraction method. C. A. Linte et al. [21] studied the TEE-MR registration on porcine subjects, which used the mitral and aortic valve annuli as landmarks and required manual segmentation by professionals by using a custom-developed segmentation technique. P. Lang et al. [22] focused on the CT-TEE registration of the aortic root, which is only applicable to sections with complete aortic roots. None of the above algorithms can be directly implemented for the registration of our data. Therefore, we have only shown the target registration errors (TREs) of these algorithms in the original literatures and compared them with the maximum TRE of our algorithm in Table 1. F. Li et al. [23] performed CT-TEE registration according to the following steps: (1) semiautomatic segmentation was performed followed by execution of the initial registration based on ICP algorithm; (2) after the first-step alignment, the intensity threshold was used to extract the sample points, which were then applied to the registration based on the NMI. We employed this method (denoted by ICPMI) for our data as one of the comparison algorithms; the results are presented in Figures 5 and 6 and the indicators in Table 2. In addition, we used a method that optimized the NMI between the original CT and TEE images generated by the proposed basic registration as the comparison algorithm (denoted by VMI). The only difference between this comparison algorithm and the proposed algorithm is the use of the enhanced probability map; thus, our approach can be used to illustrate the performance of the enhanced probability map.

3.2. Visualization of the Results. The study was divided into two stages consisting of the basic and final registration. First, ROIs were manually segmented on both CT and TEE images using the grab-cut algorithm to locate the relevant valves. Based on those locations, we performed a simple and fast coarse registration of the original images to obtain the results of the basic registration. Afterward, the selected ROIs were enhanced and probability maps were generated. Finally, based on the transformation matrix obtained from the basic registration, we registered the enhanced probability map by optimizing the MI metric using Powell's optimizer.

Figures 5 and 6 show the process of registration for sets of sections from an infant patient (Patient 8 in Table 2) and an adult patient (Patient 12 in Table 2). Our algorithm did not produce any difference in the registration of infant and adult data. On comparing the fusion images of the basic and final registrations, we found that the basic registration was only a

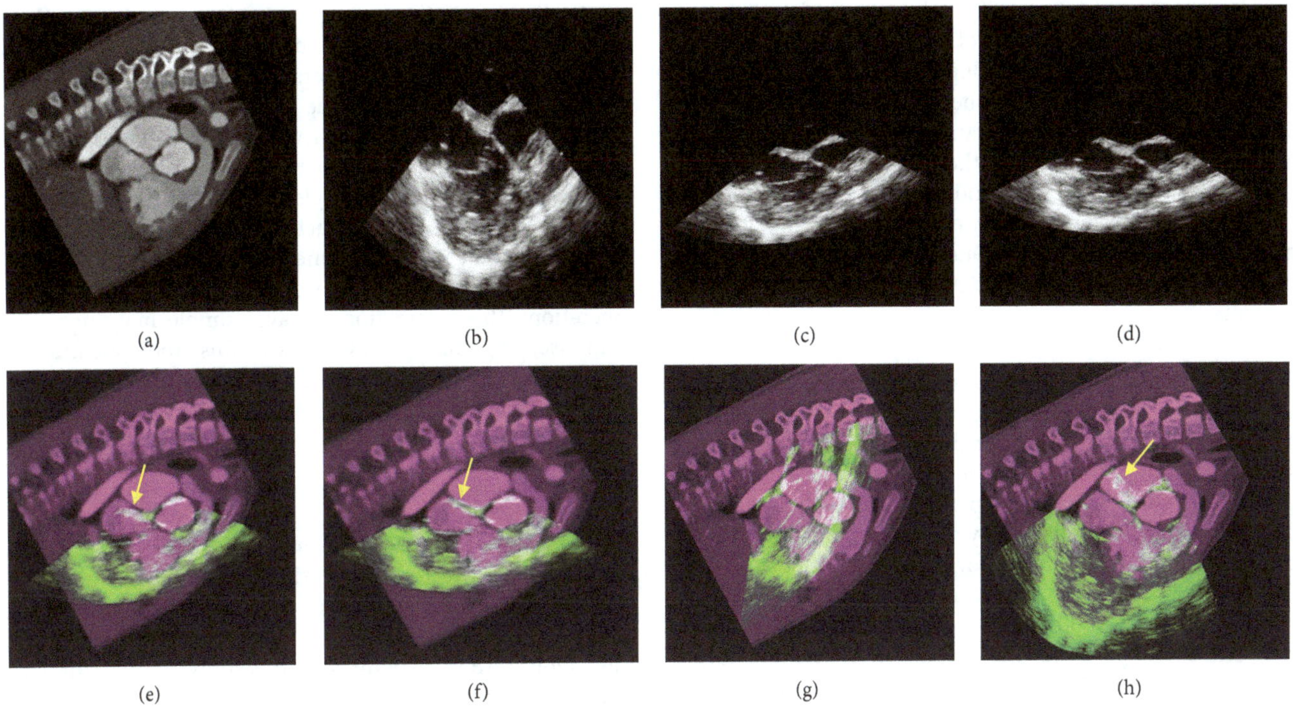

FIGURE 5: Registration process for a set of sections from an infant patient (Patient 8 in Table 1). (a) The original CT image. (b) The original TEE image. (c) The result of basic registration. (d) The result of final registration. (e) The fusion result based on the basic registration. (f) The fusion result based on the final registration. (g) The fusion result obtained from the comparison algorithm VMI. (h) The fusion result obtained from the comparison algorithm ICPMI. Note: the yellow arrows in (e), (f), and (h) indicate the alignment of the heart valves. CT, computed tomography; TEE, transesophageal electrocardiography; VMI and ICPMI, comparison algorithms.

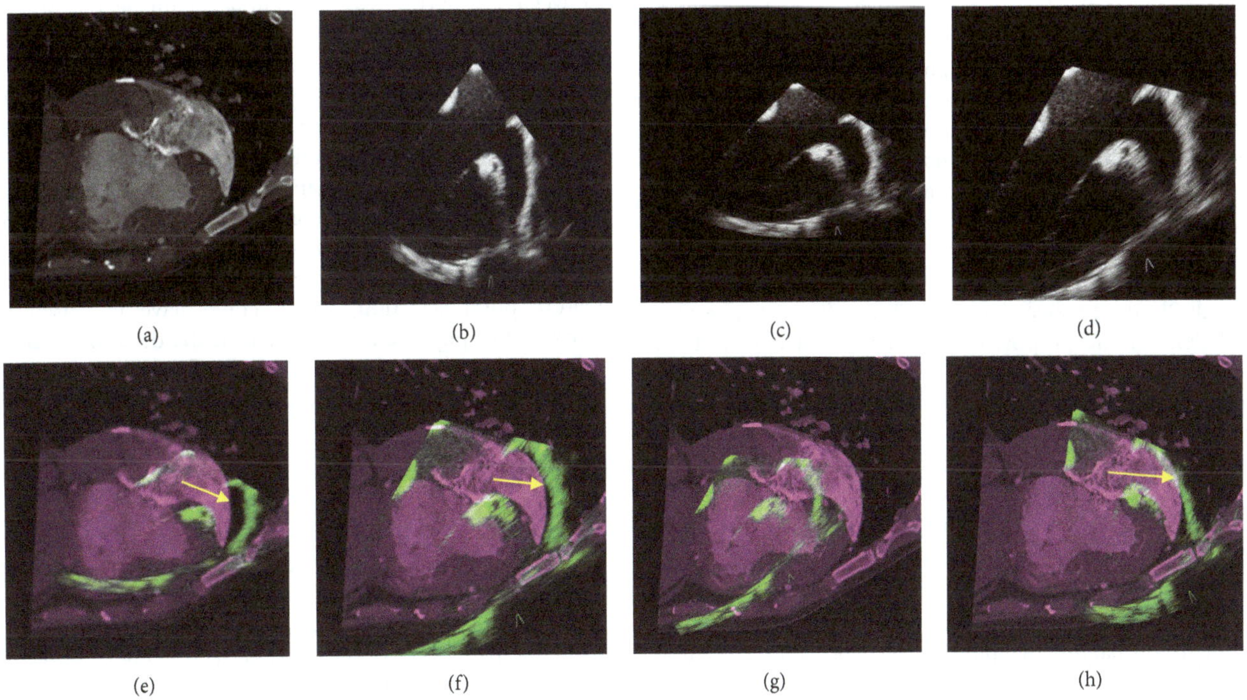

FIGURE 6: Registration process for a set of sections from an adult patient (Patient 12 in Table 1). (a) The original CT image. (b) The original TEE image. (c) The results of the basic registration. (d) The result of the final registration. (e) The fusion result based on the basic registration. (f) The fusion result based on the final registration. (g) The fusion result obtained from the comparison algorithm VMI. (h) The fusion result obtained from the comparison algorithm ICPMI. Note: the yellow arrows in (e), (f), and (h) indicate the alignment of the heart valves. CT, computed tomography; TEE, transesophageal electrocardiography; VMI and ICPMI, comparison algorithms.

rough alignment of CT and TEE, while the final registration enabled the precise alignment of the valves (yellow arrows). Fusion images of the VMI (Figures 5(g) and 6(g)) were almost a complete failure of the registration performance, while the registration performance of ICPMI (Figures 5(h) and 6(h)) is also worse than the proposed algorithm. The differences between the VMI and the proposed method only lied in the enhanced probability map, which verified its great performance in registration. Our results revealed the exact alignment of the atria, ventricles, vessels, and valves in all patients.

3.3. Index Analysis. Index analysis was performed in order to evaluate quantitatively the performance of the proposed method and comparison algorithms. Dice coefficients [25] and registration errors are shown in Table 2. Final registration parameters were referenced to rigid registration methods [26]. First, ROI was obtained by manual segmentation. Each pixel within the ROI was transformed using both parameters of the proposed registration and the "bronze standard" registration [26]. And the root mean square (RMS) distance between these positions was calculated to yield a target registration error (TRE) (Table 2). Dice coefficients were obtained as follows:

$$Dice\ coefficient = \frac{2\,|A \cap B|}{|A + B|} \tag{8}$$

with A and B representing the number of pixels in regions generated by transforming the artificial segmentation using the proposed method (or contrast algorithm) and the bronze standard registration, respectively, and $A \cap B$ represents the number of pixels in the overlapping area between them.

Table 1 shows the comparison of the maximum TREs between the existing methods and our method. Although the resolution and modality of the dataset used by each method are different, we can use this comparison to verify that our registration accuracy is within the acceptable range described in the current research on cardiac registration.

All measurements in Table 2 were based on an average over all 2D image slices for each patient. The average Dice coefficient of the proposed method reached 0.91 and the average TRE approached 1.2655 mm. These two indexes were notably better than those obtained by VMI from the 12 patients and ICPMI from the 11 patients besides Patient 9. In summary, the registration performance of our method was beyond the VMI and ICPMI in both infants and adult. The similarity of the enhanced probability maps in this paper worked better than the similarity of original images.

3.4. Performance Analysis. Our findings demonstrated the validity of probability mapping for the registration of cardiac CT and TEE images. Conventionally, infant data is more challenging to register compared to adult data due to the incomplete development of the heart and the smaller organs in infants. In our investigation, the registration performance of the proposed method was satisfying for both infant and adult data, which shows its good applicability to a variety of cardiac data.

Furthermore, the average time for optimizing MI of the probability map in our method was 7.45 s, which is substantially less than the average of 25.63 s for optimizing the original image under the same parameter conditions (Windows 7, 64-bit system, MATLAB 2015b operating environment). Therefore, the use of the probability map improved the observable accuracy of registration and reduced the registration time. Since CT interactive segmentation and feature point selection can be performed before surgery, interaction with TEE images is the only procedure necessary during the operation. This interaction involves simple markings with a mouse, on which professionals in this study spent a few seconds on each. Therefore, our method may be able to meet the requirements for real-time image guided surgery.

4. Conclusions

The imaging of congenital heart disease utilizes CMR, CT, TEE, and other technologies. Most existing image guidance surgery systems are based on traditional monomodal imaging and are therefore unable to provide comprehensive information about cardiac structures. In this study, we hypothesized that registering preoperative CT images to intraoperative TEE images can provide detailed navigation for congenital heart surgery. Compared to CT-MRI, MRI-US, and CT-PET registration of other organs in the human body, cardiac CT-TEE registration in infants is very challenging. Morphological differences caused by the deformity of heart tissues and heartbeat in addition to significant differences in gray levels between CT and TEE can complicate image registration. To address these issues, we proposed a large-deformation dynamic image registration method to meet the needs of surgical navigation in infants, young children, and adults with congenital heart disease. Our approach used valve positions to perform coarse registration, which was an enhanced probability map model for final registration and parameter transfer. These strategies were rapid and accurate. However, there were several limitations to our algorithm. First, basic registration required intervention during segmentation and feature point selection, which could have increased the workload of physicians and extended the duration of registration. Second, due to the difficulty of reconstructing TEE images, we implemented 2D-2D image registration without extending registration to a 3D space.

Data Availability

All test data in this research were obtained from Guangdong General Hospital in China. Sets of 22 standard TEE slices were collected from 12 patients with congenital heart disease using a PHILIPS S7-3t (Philips Medical Systems Technologies Ltd., Amsterdam, Netherlands) Pediatric with thermal index for soft tissue=0.1, mechanical index=0.3, frequency of the vibration=78 Hz, and 128 bpm. CT images were acquired from the same patients using a Siemens Somatom Definition (Siemens Healthcare, Forcheim, Germany) with voltage=100 kV, electric current=178 mA, multiplane reconstruction=0.8 mm, and Ctvidol=6.59. The image size of the TEE was 600 ×

800 (pixel size=0.224 mm × 0.224 mm) and that of CT was 1024 × 1024 (pixel size=0.32 mm × 0.32 mm).

Funding

This work was supported in part by the National Natural Science Foundation of China (Nos. 61472306, U1401255), Technological Application Research Plan Program for public welfare in Zhejiang Province of China (No. 2015C31167), Scientific and Technological Plan Program in Hangzhou City of China (No. 20160533B71), and Guangdong province science and technology planning project of China (No. 2014A020212228).

Acknowledgments

The authors thank the Key Laboratory of Intelligent Perception and Image Understanding of Ministry of Education of China, Xidian University and the help of Guangdong General Hospital. The authors also thank the National Natural Science Foundation of China (Nos. 61472306, U1401255), the National Basic Research Program (973 Program) of China (No. 2013CB329402) and Guangdong province science and technology planning project of China (No. 2014A020212228) for the support.

References

[1] B. S. Manjunath, C. Shekhar, and R. Chellappa, "A new approach to image feature detection with applications," *Pattern Recognition*, vol. 29, no. 4, pp. 627–640, 1996.

[2] L. M. G. Fonseca and M. H. M. Costa, "Automatic registration of satellite images," in *Proceedings of the 1997 10th Brazilian Symposium of Computer Graphic and Image Processing, SIBGRAPI 97*, pp. 219–226, Campos do Jordao, Brazil, October 1997.

[3] Y. Feng, A. Schlichting, and C. Brenner, "3D Feature point extraction from lidar data using a neural network," *International Archives of the Photogrammetry, Remote Sensing and Spatial Information Sciences - ISPRS Archives 41*, vol. XLI-B1, pp. 563–569, 2016.

[4] Y. Cheng and T. Dubois, "Object Recognition by Using Multi-level Feature Point Extraction," *Computer Vision and Pattern Recognition*, 2017.

[5] W. Y. Chen, T. F. Pan, and L. G. Chen, "A spatial first 3-D XYT feature point extraction algorithm for efficient human action recognition," in *Proceedings of the 2014 IEEE International Conference on Consumer Electronics - Taiwan (ICCE-TW)*, pp. 101-102, Taipei, Taiwan, May 2014.

[6] G. Dán, M. A. Khan, and V. Fodor, "Characterization of SURF and brisk interest point distribution for distributed feature extraction in visual sensor networks," *IEEE Transactions on Multimedia*, vol. 17, no. 5, pp. 591–602, 2015.

[7] Z. Wang, L. Zhang, T. Fang et al., "A multiscale and hierarchical feature extraction method for terrestrial laser scanning point cloud classification," *IEEE Transactions on Geoscience and Remote Sensing*, vol. 53, no. 5, pp. 2409–2425, 2015.

[8] C. K. Hoh, M. Dahlbom, G. Harris et al., "Automated iterative three-dimensional registration of positron emission tomography images," *Journal of Nuclear Medicine*, vol. 34, no. 11, pp. 2009–2018, 1993.

[9] L. Lemieux and D. G. T. Thomas, "A patient-to-computed-tomography image registration method based on digitally reconstructed radiographs," *Medical Physics*, vol. 21, no. 11, pp. 1749–1760, 1994.

[10] P. Viola and W. Wells, "Alignment by maximization of mutual information," in *Proceedings of the 5th IEEE International Conference on Computer Vision*, pp. 16–23, Cambridge, MA, USA, 1995.

[11] F. Maes, A. Collignon, D. Vandermeulen, G. Marchal, and P. Suetens, "Multimodality image registration by maximization of mutual information," *IEEE Transactions on Medical Imaging*, vol. 16, no. 2, pp. 187–198, 1997.

[12] J. P. W. Pluim, J. B. A. Maintz, and M. A. Viergever, "Image registration by maximization of combined mutual information and gradient information," *IEEE Transactions on Medical Imaging*, vol. 19, no. 8, pp. 809–814, 2000.

[13] M. Gong, S. Zhao, L. Jiao, D. Tian, and S. Wang, "A novel coarse-to-fine scheme for automatic image registration based on SIFT and mutual information," *IEEE Transactions on Geoscience and Remote Sensing*, vol. 52, no. 7, pp. 4328–4338, 2014.

[14] J. Woo, M. Stone, and J. L. Prince, "Multimodal registration via mutual information incorporating geometric and spatial context," *IEEE Transactions on Image Processing*, vol. 24, no. 2, pp. 757–759, 2015.

[15] I. Atli, A. Saracoglu, and O. S. Gedik, "The effect of contrast enhancement techniques on mutual information based image registration," in *Proceedings of the 25th Signal Processing and Communications Applications Conference, SIU 2017*, May 2017.

[16] P. A. Legg, P. L. Rosin, D. Marshall, and J. E. Morgan, "Feature Neighbourhood Mutual Information for multi-modal image registration: An application to eye fundus imaging," *Pattern Recognition*, vol. 48, no. 6, pp. 1937–1946, 2015.

[17] S. Pradhan, A. Singh, and D. Patra, "Enhanced mutual information-based multimodal brain MR image registration using phase congruency," *Progress in Intelligent Computing Techniques: Theory, Practice, and Applications*, vol. 518, pp. 197–205, 2018.

[18] D. Rueckert, M. J. Clarkson, D. L. Hill, D. J. Hawkes, and K. M. Hanson, "Non-rigid registration using higher-order mutual information," in *Proceedings of the International Society for optics and Photonics, Medical Imaging 2000*, pp. 438–447, San Diego, CA, USA, 2000.

[19] G. P. Penney, J. M. Blackall, M. S. Hamady, T. Sabharwal, A. Adam, and D. J. Hawkes, "Registration of freehand 3D ultrasound and magnetic resonance liver images," *Medical Image Analysis*, vol. 8, no. 1, pp. 81–91, 2004.

[20] D. S. Cho, C. Linte, E. C. S. Chen et al., "Predicting target vessel location on robot-assisted coronary artery bypass graft using CT to ultrasound registration," *Medical Physics*, vol. 39, no. 3, pp. 1579–1587, 2012.

[21] C. A. Linte, J. Moore, C. Wedlake et al., "Inside the beating heart: An in vivo feasibility study on fusing pre- and intra-operative imaging for minimally invasive therapy," *International Journal for Computer Assisted Radiology and Surgery*, vol. 4, no. 2, pp. 113–123, 2009.

[22] P. Lang, M. W. A. Chu, D. Bainbridge, G. M. Guiraudon, D. L. Jones, and T. M. Peters, "Surface-based CT-TEE registration of the aortic root," *IEEE Transactions on Biomedical Engineering*, vol. 60, no. 12, pp. 3382–3390, 2013.

[23] F. Li, D. R. Holmes III, K. H. Wong et al., "Towards real-time 3D US-CT registration on the beating heart for guidance of minimally invasive cardiac interventions," in *Proceedings of the Medical Imaging 2012: Image-Guided Procedures, Robotic Interventions, and Modeling. International Society for Optics and Photonics*, San Diego, Calif, USA, 2012.

[24] C. Rother, V. Kolmogorov, and A. Blake, ""Grab Cut" - interactive foreground extraction using iterated graph cuts," *Acm Transactions on Graphics*, vol. 23, no. 3, pp. 309–314, 2004.

[25] L. R. Dice, "Measures of the amount of ecologic association between species," *Ecology*, vol. 26, no. 3, pp. 297–302, 1945.

[26] J. M. Fitzpatrick, J. B. West, and C. R. Maurer Jr., "Predicting error in rigid-body point-based registration," *IEEE Transactions on Medical Imaging*, vol. 17, no. 5, pp. 694–702, 1998.

9

A Feasible Computational Fluid Dynamics Study for Relationships of Structural and Functional Alterations with Particle Depositions in Severe Asthmatic Lungs

Sanghun Choi,[1] **Shinjiro Miyawaki**,[2,3] **and Ching-Long Lin**[3]

[1] *School of Mechanical Engineering, Kyungpook National University, Daegu, Republic of Korea*
[2] *Jacobs, 1100 NE Circle Blvd., Suite 300, Corvallis, Oregon 97330, USA*
[3] *Department of Mechanical and Industrial Engineering, IIHR-Hydroscience and Engineering, University of Iowa, Iowa City, Iowa, USA*

Correspondence should be addressed to Ching-Long Lin; ching-long-lin@uiowa.edu

Academic Editor: Yu Feng

This study aims to investigate the effect of altered structures and functions in severe asthma on particle deposition by using computational fluid dynamics (CFD) models. Airway geometrical models of two healthy subjects and two severe asthmatics were reconstructed from computed tomography (CT) images. Subject-specific flow boundary conditions were obtained by image registration to account for regional functional alterations of severe asthmatics. A large eddy simulation (LES) model for transitional and turbulent flows was applied to simulate airflows, and particle transport simulations were then performed for 2.5, 5, and 10 μm particles using CFD-predicted flow fields. Compared to the healthy subjects, the severe asthmatics had a smaller air-volume change in the lower lobes and a larger air-volume change in the upper lobes. Both severe asthmatics had smaller airway circularity (*Cr*), but one of them had a significant reduction of hydraulic diameter (D_h). In severe asthmatics, the larger air-volume change in the upper lobes resulted in more particles in the upper lobes, especially for the small 2.5 μm particles. The structural alterations measured by *Cr* and D_h were associated with a higher particle deposition. D_h was found to be the most important metric which affects the specific location of particle deposition. This study demonstrates the relationship of CT-based structural and functional alterations in severe asthma with flow and particle dynamics.

1. Introduction

Asthma is pathologically characterized by combined phenotypes of airflow obstruction, bronchial hyperresponsiveness, and airway inflammation [1]. However, how structural and functional alterations of asthma affect flow structure and particle deposition is yet to be investigated. In imaging studies of asthma, ventilation defects and airway structural changes have been investigated by using magnetic resonance image (MRI), positron emission tomography (PET), and single-photon emission computed tomography (SPECT) [2–4]. In addition, via quantitative computed tomography (QCT) imaging, several studies [5–8] have demonstrated significant alterations such as reduced airway diameter as well as increased wall thickness and air trapping. Although QCT can provide structural measurements of the airways up to segmental airways (~2 mm), quantification of local functional variables is still limited.

Image registration technique has been utilized to provide functional information by matching images at different inflation levels [9]. The registration derived-variables were validated by comparing ventilation maps from different imaging modalities [10]. Furthermore, this technique has shown strengths when characterizing functional alterations of diseased lungs [11, 12]. For instance, a study used the technique to differentiate airway vs. parenchymal phenotypes in a chronic obstructive pulmonary disease (COPD) [11]. We have recently demonstrated that volume changes of severe asthmatic lungs are preferentially smaller near basal regions and the smaller volume changes are compensated with air-volume change in apical regions [12]. We also demonstrated that severe asthmatics were characterized by the reduced airway diameters and noncircular shapes, unlike healthy and nonsevere asthmatics [13].

Computational fluid dynamics (CFD) technique has been used to analyze flow characteristics and particle depositions in the human lungs [14]. With regard to CFD simulations of asthmatic lungs, a study [15] showed alterations of particle depositions with an asthmatic subject before and after asthma attack, and another study [16] correlated forced expiratory volume in one second (FEV_1) with CFD-based resistance before and after bronchodilator. However, both of them imposed parabolic velocity profiles to the trachea regions and uniform pressure boundary conditions to QCT-resolved ending branches, under the Reynolds-averaged Navier-Stokes equations (RANS) model or laminar assumption. Heenan et al. [17] and Jayaraju et al. [18] have compared CFD simulations using RANS model with experimental results, concluding that ones with RANS provided less accurate prediction in both flow structure and particle deposition. On the other hand, Longest et al. [19] and Tian et al. [20] have demonstrated that low-Reynolds number k-ω approximation with near-wall correction improves the prediction of particle deposition. With five asthmatics, Vinchurkar et al. [21] have also demonstrated a good agreement of aerosol delivery between imaging-based CFD method and in vivo experimental data. Meanwhile, Backer et al. [22] have demonstrated that lobar distributions of air-volume change between SPECT and QCT are consistent, so they have emphasized the importance of subject-specific boundary conditions in CFD simulations.

The main objective of this study is to investigate how functional and structural alterations of severe asthmatic lungs affect flow structures and particle depositions. This study further evaluates clinical applications of CFD in localizing hot spots. According to our existing population-based analysis [8, 12, 13], severe asthmatics were characterized by the shifted lobar air-volume change from lower lobes to upper lobes, decreased airway circularity (Cr), and reduced hydraulic diameter (D_h). To reproduce regional air-volume change, we employed an image registration technique for a subject-specific boundary condition. Next, to reflect subject-specific airway features such as Cr and D_h, we employed a surface fitting method from 1D skeleton to 3D CT-resolved airways [23]. We then performed computational fluid-particle dynamics simulations to obtain local particle distribution and deposition. To achieve main objective, three major components in this study are (1) reproducing regional air-volume change, (2) quantifying subject-specific airway features, and (3) quantifying local particle distribution and deposition.

2. Methods

2.1. Human Subjects. The imaging study and protocols for acquiring QCT images at both total lung capacity (TLC) and functional residual capacity (FRC) were approved by Institutional Review Board (IRB) of University of Pittsburgh as a part of multicenter Severe Asthma Research Program (SARP) [24]. Four human subjects were employed in this study, among whom two subjects were healthy and two subjects were severe asthmatic (Table 1). The CT images were taken via GE helical VCT-64 slice scanner with the slice thickness of 0.625 mm. Major criteria used to define severe asthma included treatments with oral corticosteroids and high-dose inhaled corticosteroids, besides several minor criteria [24]. At least, one major and two minor criteria were required to be classified as severe asthmatics. The CT scans were acquired in a supine position, and 3D airways, 1D skeletons, lungs, and lobes were segmented using Apollo software (VIDA Diagnostics, Coralville, Iowa).

2.2. Flow Simulation and Boundary Condition. Filtered continuity and incompressible Navier-Stokes equations read

$$\nabla \cdot \overline{\mathbf{u}} = 0, \tag{1a}$$

$$\frac{\partial \overline{\mathbf{u}}}{\partial t} + (\overline{\mathbf{u}} \cdot \nabla)\,\overline{\mathbf{u}} = -\frac{1}{\rho_f}\nabla \overline{p} + \nabla \cdot \left[(\nu_f + \nu_T)\,\nabla \overline{\mathbf{u}}\right] \tag{1b}$$

where $\overline{\mathbf{u}}$, ρ_f, $\overline{p}$, ν_f, and ν_T are filtered velocity vector, fluid density, filtered pressure, fluid kinematic viscosity, and subgrid-scale turbulent eddy viscosity [25], respectively. A large eddy simulation (LES) model was adopted to resolve laminar-transitional-turbulent flows in the central airways. The properties of ρ_f and ν_f were set to 1.12 kg/m^3 and 1.64×10^{-5} m^2/s, respectively. A characteristic Galerkin finite element method [26] was employed to discretize (1a) and (1b). A moderate steady-inspiratory flow-rate of 3.27×10^{-4} m^3/s ($\approx$20 liters/min) was imposed as the inlet condition, being equivalent to the peak flow-rate of a sinusoidal waveform with a tidal volume of 500 mL and a period of 4.8 s. Reynolds numbers (Re) in trachea range from 1300 to 1700 in the given flow-rate and individual tracheal sizes of two healthy and two severe asthmatics.

Figure 1(a) shows a flow chart from the image segmentation and registration to a realistic CFD simulation. First, an image registration technique [9] was employed to estimate local air-volume change between TLC and FRC at lung parenchyma (top row), and one-dimensional (1D) centerline (CL) tree structures obtained by a volume filling method [27] bridged three-dimensional- (3D-) resolved ending branches and lung parenchyma (middle row). Air-volume change measured at parenchyma was summed at CT-resolved ending branch to estimate regional flow-rate ratios. With the flow-rate ratios, parabolic profiles of velocity were essentially imposed at the CT ending branch. Lastly, a surface fitting method [23] together with Gmesh [28] was used to fit an initial CL-based 3D surface geometry to CT-segmented airway surface and construct a CL-CT-based airway geometrical model (bottom row). The resulting model captured subject-specific airway features such as Cr and D_h. At this step, we created a synthetic turbulence [29] right above the glottal constricted regions, because the SARP study [24] did not gather oropharynx scans. The turbulent intensity and largest eddy size were set to 0.29 and 8 mm, respectively, to mimic the turbulent flows found in the CFD simulations with oropharynx [30]. The number of elements ranges from 3.9 to 5.0 million for four cases. Time steps of 5.0×10^{-6} s (HS 1,

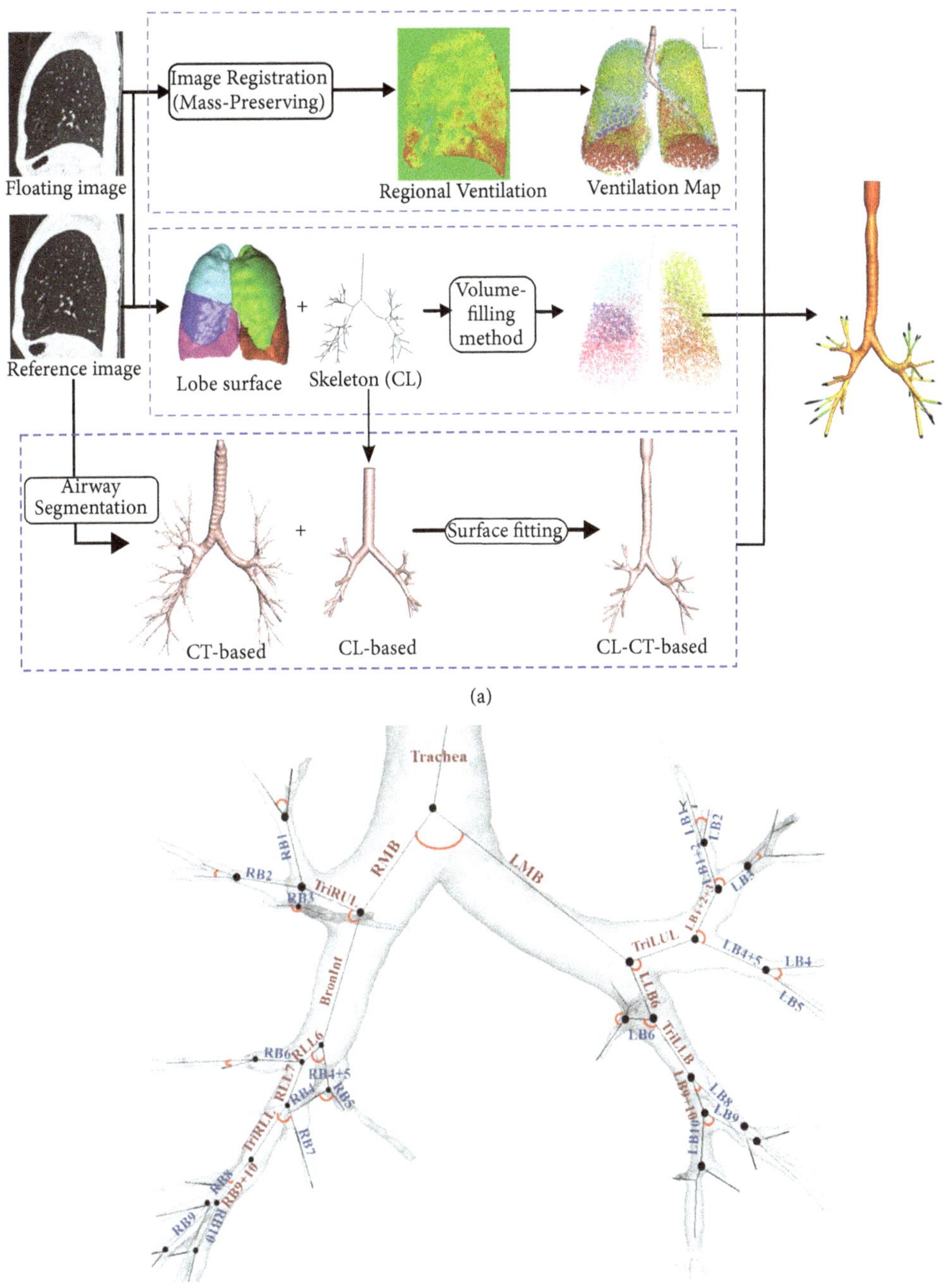

FIGURE 1: (a) Flow charts of connecting CT imaging-based structural and functional data for physiologically consistent CFD simulations. (b) Segmental names of airways: each angle of the segment represents the bifurcation angle between two daughter branches.

HS 2, and SA 1) and 3.0×10^{-6} s (SA 2) were chosen to satisfy Courant–Friedrichs–Lewy (CFL) number less than 1.

2.3. Structural Metrics Associated with Aerosol Delivery. Cr and D_h could be directly associated with airflows and particle delivery. In this study, we measured the structural metrics at 31 segmental airways (Figure 1(b)). Cr was computed to assess the degree of elliptical shape of an airway cross section as follows:

$$Cr = \frac{\text{Perimeter of an area-equivalent circle}}{\text{Perimeter of a luminal area}} = \frac{\pi D_{\text{ave}}}{P_e}, \quad (2)$$

where D_{ave} and P_e are average diameter and perimeter of an airway. D_{ave} was computed as follows: $\sqrt{4 \times A_c/\pi}$, where A_c is the cross-sectional area of the airway. Next, in order to assess a degree of airway narrowing, D_h was computed as follows:

$$D_h = \frac{4A_c}{P_e}. \quad (3)$$

Table 1: Demographic, PFT baseline (prebronchodilator), and maximal (postbronchodilator) lung functions and CT-based air-volumes of two healthy subjects and two severe asthmatics.

	Healthy subjects (HS)		Severe asthmatics (SA)	
	HS 1	HS 2	SA 1	SA 2
Demographics				
Gender (F, Female)	F	F	F	F
Age (yrs.)	59	29	61	48
BMI (kg/m^2)	23.7	22.2	32.4	23.9
Asthma duration (yrs.)	-	-	10.6	19.7
Baseline lung function (pre-bronchodilator)				
FEV$_1$ (liters) (% predicted)	2.97 (100%)	3.15 (94%)	1.19 (34%)	1.02 (40%)
FVC (liters) (% predicted)	3.82 (100%)	4.00 (101%)	1.83 (39%)	2.52 (80%)
FEV1/FVC (%)	78%	79%	65%	41%
Maximal lung function (post-bronchodilator)				
FEV$_1$ (liters) (% predicted)	3.36 (107%)	3.17 (100%)	2.05 (72%)	1.17 (46%)
FVC (liters) (% predicted)	4.04 (103%)	3.94 (102%)	2.97 (74%)	3.11 (98%)
FEV1/FVC (%)	83%	80%	69%	38%
CT-based air-volumes				
TLC (liters)	5.37	4.31	5.64	4.33
FRC (liters)	2.04	1.79	2.85	1.77

All subjects are Caucasians and nonsmokers. PFT and CT measurements were obtained in upright and supine positions, respectively. CT scans were performed after bronchodilator.

Note that Cr decreases as airway lumen becomes elliptic, and D_h decreases as airway narrowing is dominant. Cr and D_h were decreased in severe asthmatics compared with healthy subjects, in a population-based comparison [13]. In this study, Cr decreases in both severe asthmatics (SA 1 and SA 2), but D_h only decreases in a severe asthmatic subject (SA 2). In Results, the effects of Cr and D_h on particle delivery are investigated.

2.4. Particle Simulation. To compare the characteristics of particle distribution and deposition such as pharmaceutical aerosols between healthy subjects and severe asthmatics, particle transport analysis was conducted using LES-predicted airflow fields. Quasi-steady airflow fields for 2.4 s were collected, e.g., 400 datasets × 0.006 s, after initial 2.4 s airflow fields. For particle simulation, Lagrangian particle tracking algorithm [31] was adopted to track particle trajectories as follows:

$$\frac{d\mathbf{u}_p}{dt} = \frac{U}{\text{Stk} \cdot D_{\text{ave}}}\left(\bar{\mathbf{u}} - \mathbf{u}_p\right) + \frac{\rho_p - \rho_f}{\rho_p}\mathbf{g}, \tag{4}$$

where $\mathbf{u}_p$, ρ_p, and $\mathbf{g}$ are particle velocity, particle density, and gravitational acceleration, respectively. In the equation, Stokes number (Stk) is defined as follows:

$$\text{Stk} = \frac{2Q\rho_p d^2}{9\pi\mu_f D_{\text{ave}}^{\;3}} C_c \alpha^{3.7} \tag{5}$$

where Q, d, μ_f, C_c, and α are the flow-rate of the branch, the diameter of particles, fluid dynamic viscosity, the Cunningham slip correction factor, and the particle-particle interaction factor, respectively. The detailed descriptions of particle transport simulation could be found in [30]. At the beginning of particle simulation, the spherical particles were uniformly distributed in a cylinder with a radius of 10 mm and a depth of 4 mm at the trachea inlet. They were then released 9 different times to obtain an ensemble average of particle deposition. The number of particles was set to 10,000, and three different spherical particle sizes of 2.5, 5, and 10 μm were chosen in this study. The aerosol size was chosen with a general distribution of aerosols using dry powder inhaler (DPI) and soft mist inhaler (SMI) [32]. This is to investigate the relationship between aerosol size and airway structure. The particle density and mean free path were given as 1000 kg/m^3 and 68 nm, respectively. "Particle distribution", "deposition", and "advection" by lobe are defined as "particles entering each lobe", "those deposited in 3D segments of each lobe", and "those exiting 3D ending branches of each lobe without being deposited in 3D segments", respectively.

3. Results

3.1. Pulmonary Function Test (PFT). Table 1 shows demographic as well as PFT- and CT-based measurements. Bronchodilator was performed for both healthy subjects and severe asthmatics to obtain maximal lung functions, and CT scans were acquired after bronchodilator because the aim of SARP study was to assess lung function of stable asthma [33]. First, the baseline and maximal FEV_1 and FVC %predicted values in two healthy subjects were close to normal ranges (~100% predicted). In contrast, baseline FEV_1% predicted values of SA 1 (34%) and SA 2 (40%) were much smaller than normal ranges (≥80%). The maximal FEV_1% predicted value (~46%) of SA 2 was still lower even with bronchodilator. In the same subject, the maximal FVC % predicted value was very close to normal ranges, leading to a significantly reduced FEV_1/FVC (38%). Thus, the baseline FEV_1% predicted values of both SA 1 and SA 2 were small values, but the lung function of SA 1 (38% ↑) is relatively reversible than that of SA 2 (6% ↑). According to baseline and maximal PFTs, SA 2 is likely to have significant airway narrowing despite bronchodilator, whereas airways of SA 1 might have dilated with the aid of bronchodilator.

3.2. CT-Based Structural and Functional Characteristics. We compared structural quantities of *Cr* and D_h in two healthy subjects and two severe asthmatics (Figure 2). Figure 2(a) shows that *Cr* of two severe asthmatics was reduced compared to those of two healthy subjects. For example, *Cr* of Trachea, RMB, TriRUL, RLL7, RB6, RB9+10, and LMB in two severe asthmatics were deviated from those of the same airway segments in healthy subjects. Both SA 1 and SA 2 had the prominently reduced *Cr*, but only SA 2 demonstrated significantly smaller D_h compared with HS 1, HS 2, and SA 1 in CT-resolved airways (Figure 2(b)), which was also reflected in PFT measurements at postbronchodilator. The effect of reduced *Cr* in RMB and TriRUL is assessed between "HS 1, HS 2" and "SA 1, SA 2". Next, the effect of constricted branch on particle deposition is mainly assessed with SA 2. In addition, SA 1 and SA 2 were characterized by reduced air-volume change in lower lungs along with elevated air-volume change in upper lungs, as compared with the healthy subjects, being quantified by the air-volume change in upper lobes to air-volume change in middle and lower lobes, i.e., U/(M+L)|v (Figure 3).

3.3. Regional Volume Change vs. Particle Distribution. Given the regional air-volume change, particle distribution was evaluated with three different particle sizes (Figure 3). With 2.5 μm particles, the $U/(M + L)|_{dist}$ (the particle distribution ratio of upper lobes to middle and lower lobes) was close to the ratio of air-volume changes ($U/(M + L)|_v$), because small particles are likely to follow airflow streamlines. Next, $U/(M + L)|_{dist}$ decreased as particle size increases due in part to the inertial effect that particles can easily move along with the flow to the lower lobes. Overall, SA 1 and SA 2 had an increased particle distribution to upper lobes compared with HS 1 and HS 2, being consistent with the alteration of air-volume change distribution in severe asthmatics. Furthermore, the particle distribution to upper lobes in severe asthmatics became more evident with decreasing particle size.

TABLE 2: Circularity *Cr*, D_{ave}, and D_h of two healthy subjects (HS 1 and HS 2) and two severe asthmatics (SA 1 and SA 2) in RMB and TriRUL regions (see Figure 1(b)).

	Healthy subjects		Severe asthmatics	
	HS 1	HS 2	SA 1	SA 2
RMB				
Cr	0.974	0.962	0.926	0.927
D_{ave} (mm)	12.8	14.1	15.9	12.5
D_h (mm)	12.5	13.6	14.7	11.6
TriRUL				
Cr	0.953	0.959	0.883	0.820
D_{ave} (mm)	8.3	8.2	8.5	6.1
D_h (mm)	7.9	7.8	7.6	5.0

3.4. Circularity and Particle Deposition. Around RMB and TriRUL regions (Figure 1(b)), we compared particle depositions of two healthy subjects without elliptic shapes vs. two severe asthmatics with elliptic shapes, because SA 1 and SA 2 exhibit significantly decreased *Cr* in both RMB and TriRUL (Table 2). The two selected severe asthmatics (83° and 88°) have similar bifurcation angles as two healthy subjects (90° and 91°) at RMB, expecting similar particle depositions given the same range of Stk numbers. In Table 2, D_{ave} of TriRUL in SA 1 is the largest among four subjects, but D_h was slightly smaller than those of both HS 1 and HS 2 due to the reduced *Cr*. Figure 4(a) shows the increased particle deposition of SA 1 and SA 2 in TriRUL, as compared with HS 1 and HS 2 in the same Stk range of the parent branch (RMB). Note that the comparisons based on the same Stk numbers were made to control the effects of flow-rate and airway narrowing. Figure 5 also displays the distributions of particle deposition in TriRUL. Thus, an increase of particle deposition in both severe asthmatics was possibly due to the decreased airway *Cr*.

3.5. Constriction vs. Particle Deposition. Particle deposition efficiency of the branches based on Stk numbers, daughter branches of RB9+10 and LB10, were plotted in Figures 4(b) and 4(c), respectively. Figure 6(a) (around RB9+10) and Figure 6(b) (around LB10) show particle depositions of SA 2 with narrowed and elliptic airways. Compared with HS 1, HS 2, and SA 1, Stk is increased for SA 2 due to airway narrowing, because Stk is proportional to $1/D_{ave}^{\ 3}$ with a given flow-rate (see (5)). In the case of 5 μm particles, particle deposition efficiency reached 60% and 90% for RB9+10 and LB10, respectively. For 10 μm particles, particle deposition reached 100%; thus large particles could not ventilate beyond these airways. These results imply that particle deposition would be very sensitive to the reduced airway diameters due to airway constriction (see Figures 6(a) and 6(b)). We further investigated flow structure, wall shear stress, and pressure drop around LB10 (Figure 6(c)). In this airway, constriction-induced jet-flow, higher wall shear stress and large pressure drop were simultaneously observed.

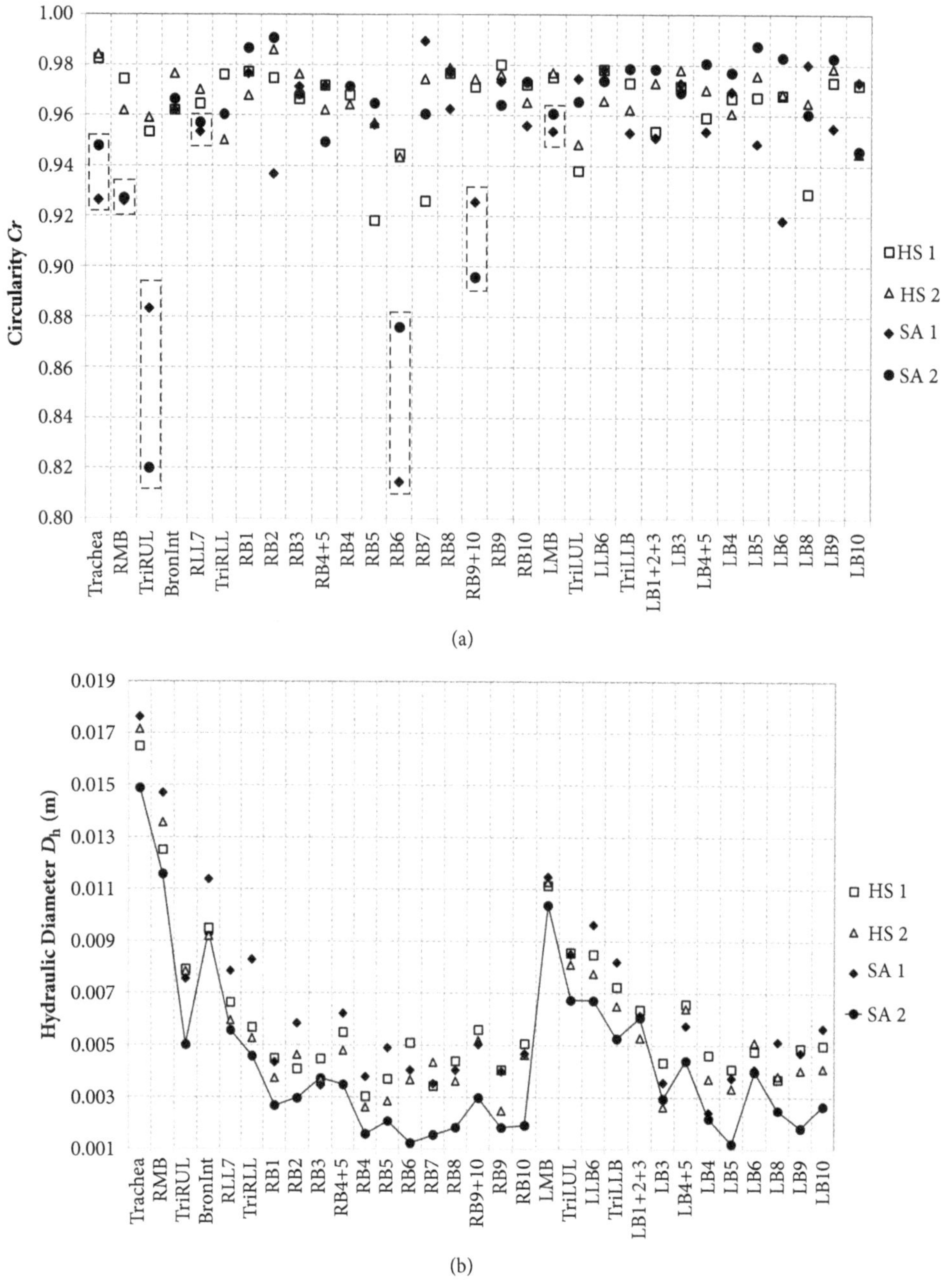

FIGURE 2: Structural variability of (a) *Cr* and (b) D_h for 31 segmental airways after 1st generation in two healthy subjects and two severe asthmatics.

4. Discussion

Our previous studies [8, 12] have demonstrated that air-volume change is smaller in lower lobes and larger in the upper lobes in severe asthmatics, as compared with healthy subjects. In addition, severe asthmatics were structurally characterized by smaller *Cr* and D_h. In this study, we focused on the effect of structural and functional alterations in severe asthmatics on particle depositions under a moderate inspiratory flow condition (~20 liters/min). The two severe asthmatics had altered lung function characterized by larger U/(M + L)|v than the two healthy subjects. Both SA 1 and SA 2 had small FEV_1% predicted values in the baseline lung function and showed smaller *Cr*, whereas only SA 2 subject had small FEV_1% predicted in the maximal lung function and showed smaller D_h in CT-resolved regions. This observation implied that bronchodilator may help recover narrowed airways, but it may not recover noncircular airway shape. This emphasizes the importance of investigating airway shape besides airway diameter in particle delivery. In this study, SA 1 and SA 2 were used to evaluate the effect of *Cr* on particle deposition, and SA 2 was used to evaluate the effect of D_h on particle deposition.

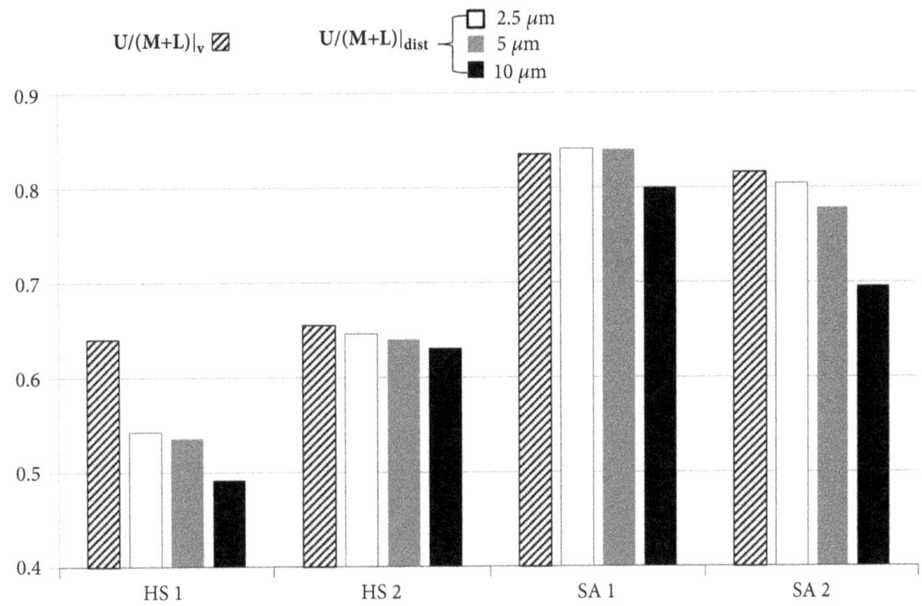

FIGURE 3: U/(M + L)|v (the ratio of air-volume change in upper lobes to middle and lower lobes) and $U/(M+L)|_{dist}$ (the distribution ratio of particles in upper lobes to middle and lower lobes) according to the particle sizes.

First, investigating characteristics of particle deposition in severe asthmatics is important for inhaled pharmaceutical drugs [21, 34], airborne bacteria, or air pollutants. Lobar distribution of small particles was consistent with air-volume distribution (Figure 3). With increasing particle size, the delivery of particles to lower lobes may increase because of an increased inertial effect of large particles. When targeting subject-specific treatments of aerosols, understanding the mechanism of particle distribution would be critical. For example, if lower lobar bronchi need be targeted for some specific subjects, relatively larger aerosol should be treated. However, one should also consider sizes of targeted branches, because larger aerosols could not be delivered to small airways with higher generation.

We previously found that in severe asthmatics *Cr* of RMB is significantly smaller [13], so we investigated the effect of smaller *Cr* on particle deposition. With the same Stk and similar bifurcation angle of RMB, particle depositions of SA 1 and SA 2 were greater in TriRUL than healthy subjects, which could be attributed to smaller *Cr*. Airway structures with smaller *Cr* might increase the sensitivity of bacterial inflammation due to particle deposition, which would potentially increase airway wall thickness. While both SA 1 and SA 2 demonstrated smaller *Cr*, only SA 2 had constricted airways, especially in lower lobes. Particle deposition efficiency is the function of $1/D_{ave}^3$ based on Stk numbers, given flow-rate. In other words, airway constriction could be the most important structural characteristic in association with increased particle deposition efficiency. Thus, if airways are chronically constricted in a subject, the subject may have an increased exposure of pollutants, so further reducing airway diameters along with airway inflammation. Mechanisms among airway constriction, wall shear stress, pressure drop, and particle deposition may be strongly coupled as described in Figure 6(c). First, constriction-induced high velocity creates high wall shear stress in association with high velocity gradient on the wall. A large pressure drop is required to overcome large wall shear stress. In addition, the constriction-induced high velocity might lead to an increase of particle deposition due to particle impaction in both constricted parent and daughter branches (Figures 6(a) and 6(b)). As a result, both high velocity and small D_{ave} may contribute to be a large Stk (see (5)), leading to an increase of particle deposition.

In this study, we only investigated two healthy subjects and two severe asthmatics, which could prevent statistical reliability, so a CFD-based population study is necessary. However, performing many CFD simulations to obtain normal distribution is computationally expensive. We have recently performed CT imaging-based clustering analysis and found clinically meaningful subgroups [35]. Since the clustering membership employed airway structure and lung function, each cluster has similar airway structure and lung function. We believe that such an approach using clustering analysis possibly reduces the number of samples by transitioning an interest from intersubject study to intercluster study. In this study, we imposed the same flow-rate on trachea inlet, due to a lack of subject-specific flow-rate information. If the subject-specific flow-rate is measured, we would further improve the comparison between healthy subjects and severe asthmatics. Furthermore, we have recently investigated airway resistance with symmetric branching angles up to the 20th generation for inspiration and expiration [36]. Then an empirical airway resistance model was introduced to estimate pressure drop due to kinetic energy and viscous dissipation. However, we have not discussed airway resistance

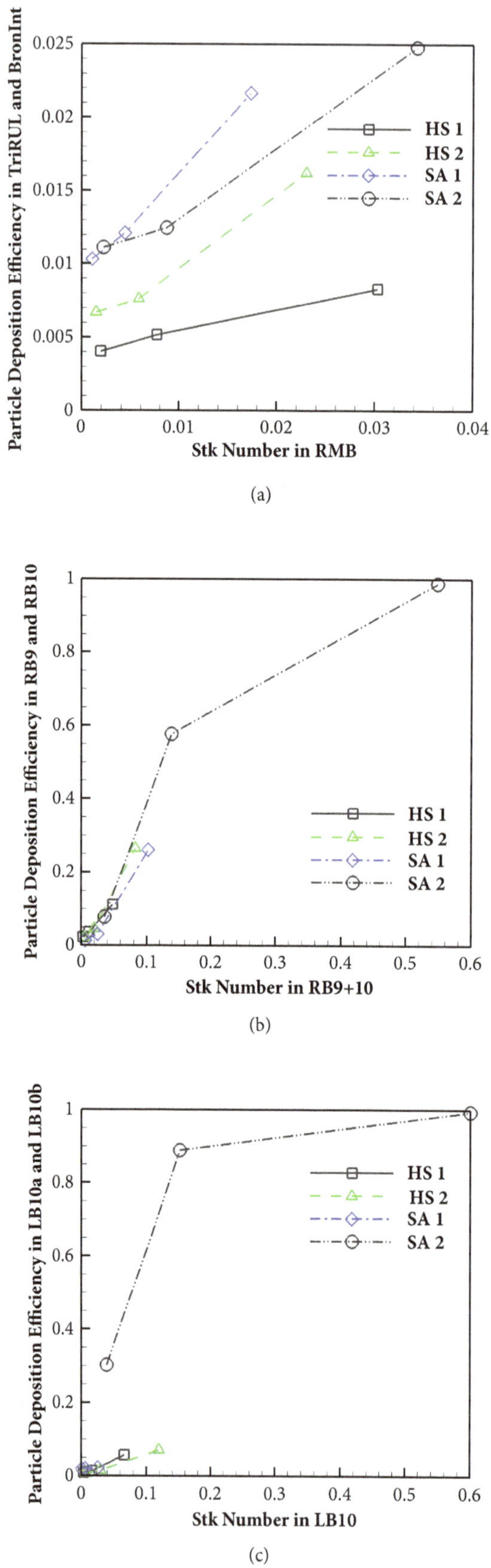

FIGURE 4: Particle deposition efficiency in (a) TriRUL based on Stk in RMB, (b) RB9 and RB10 based on Stk in RB9+10, and (c) LB10a and LB10b based on Stk in LB10.

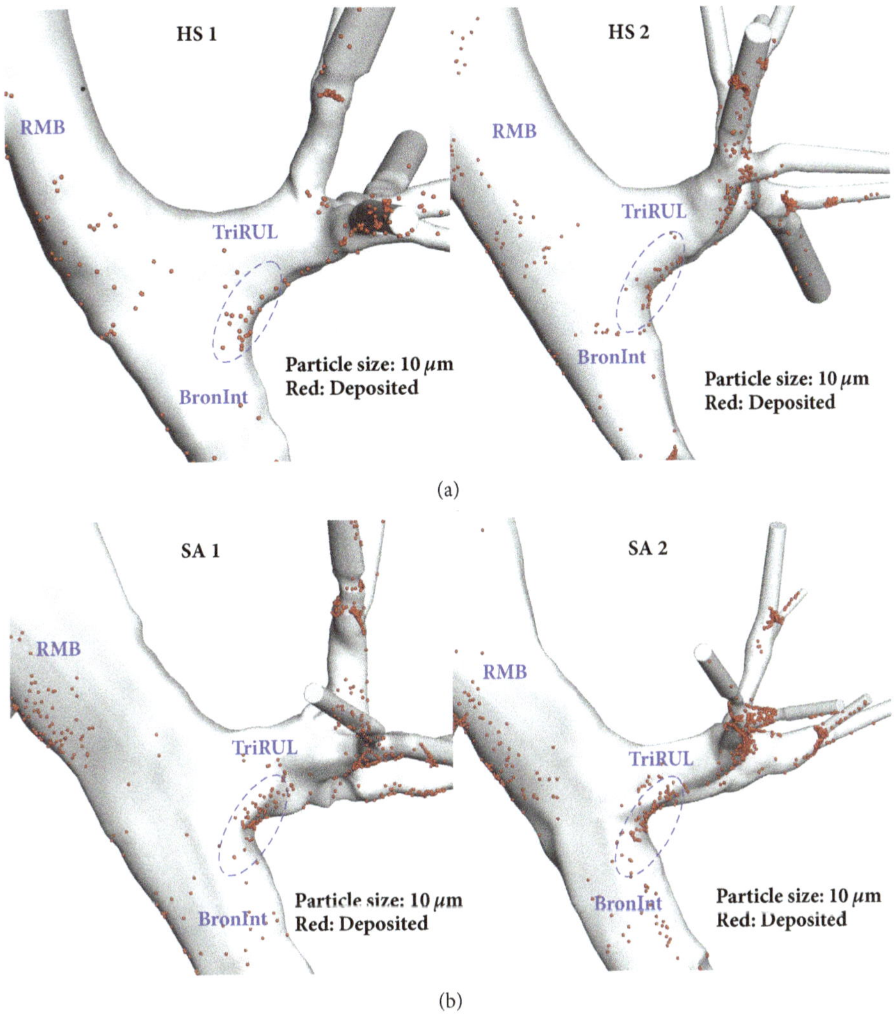

FIGURE 5: Relationships among noncircularity and particle deposition in RMB, TriRUL, and BronInt regions between (a) healthy subjects and (b) severe asthmatics. All of the images are plotted as backside-view.

in this study, because the number of airways was limited to investigate airway resistance. In the future, airway resistance study of severe asthmatics with realistic multiscale airways would be desirable.

In conclusion, we applied a high-fidelity CFD model together with CT image-based airway models to study particle deposition in both healthy and severe asthmatic lungs. With the aid of image registration technique, the subject-specific physiologically realistic-flow boundary condition is derived based on air-volume difference between two CT lung images of the same human subject. As expected from flow-rate distribution, particle distribution to upper lobes was larger in severe asthmatics, relative to healthy subjects. This phenomenon was more prominent when using smaller particles. In both healthy subjects and severe asthmatics, with increasing particle size, particles are distributed more toward lower lobar regions due to inertial effects. Alterations of airway Cr and D_h were found to be associated with particle deposition. With the same Stk, reduced Cr increases particle deposition. On the other hand, reduced D_h significantly elevates Stk, resulting in greater deposition efficiency. The constricted airways contribute to high wall shear stress, elevated pressure drop, and significantly increased particle deposition. It is concluded that $U/(M+L)|v$, Cr, and D_h shall be carefully considered to target subject-specific aerosols.

Disclosure

Preliminary results of the current manuscript were presented as an abstract in the following link: http://meetings.aps.org/Meeting/DFD14/Session/E15.1.

Conflicts of Interest

The authors declare that they have no conflicts of interest.

Acknowledgments

The authors thank Drs. Eric A Hoffman, Sally E Wenzel, and Jiwoong Choi for assisting in imaging data acquisition and analysis. Computational time was supported by XSEDE

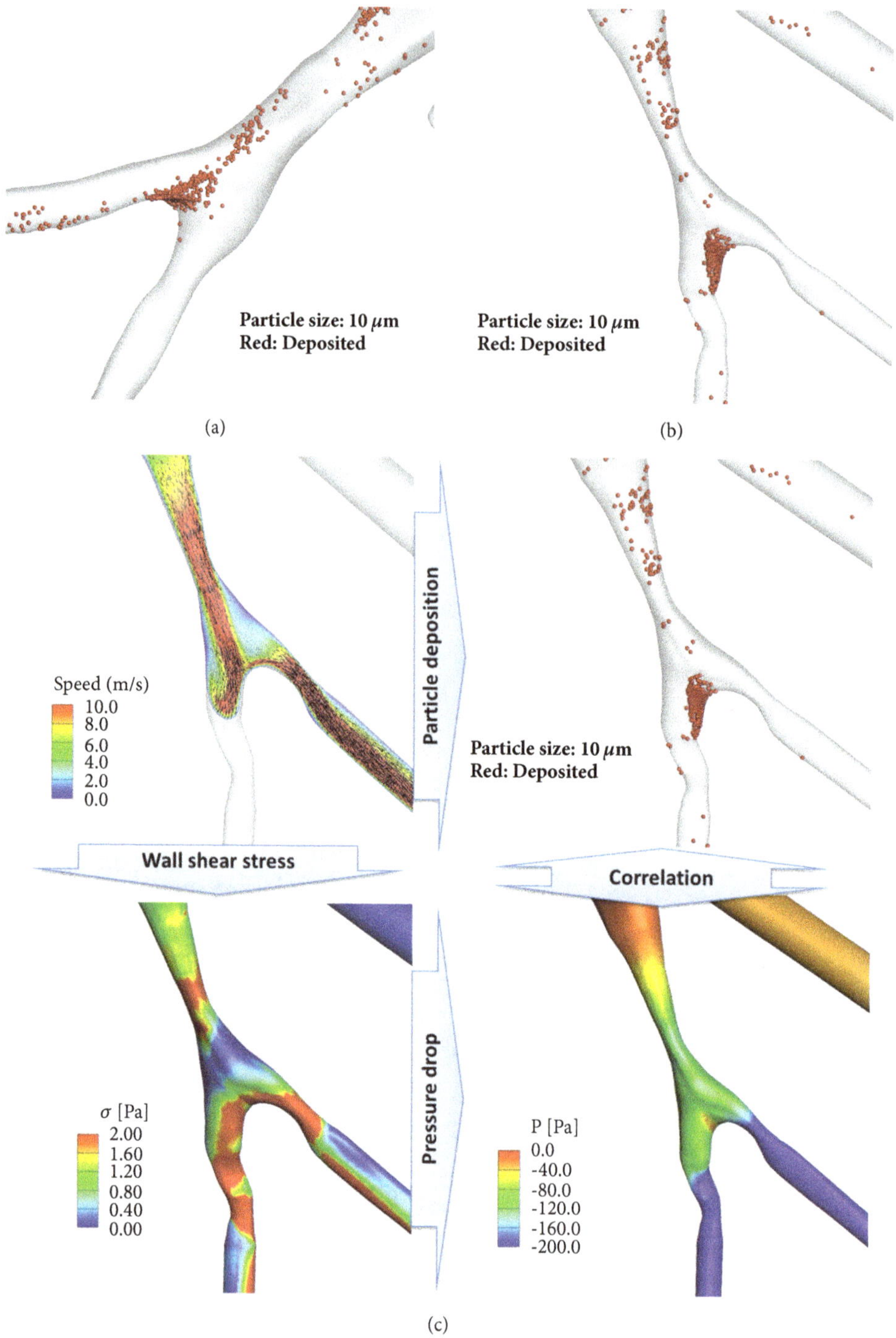

FIGURE 6: Two representative constricted regions of (a) RB9+10 and (b) LB10 in lower lobes in SA 2 subject and (c) correlations among airway constriction, wall shear stress, pressure drop, and particle deposition.

and by the National Institute of Supercomputing and Network/Korea Institute of Science and Technology Information with supercomputing resources including technical support (KSC-2017-S1-0021). This study was supported in part by NIH Grants U01 HL114494, R01 HL094315, and S10 RR022421 and in part by Basic Science Research Program through the National Research Foundation of Korea (NRF) funded by the Ministry of Education (NRF-2017R1D1A1B03034157).

References

[1] W. W. Busse and R. F. Lemanske Jr., "Asthma," *The New England Journal of Medicine*, vol. 344, no. 5, pp. 350–362, 2001.

[2] R. S. Harris, H. Fujii-Rios, T. Winkler, G. Musch, M. F. Vidal Melo, and J. G. Venegas, "Ventilation Defect Formation in Healthy and Asthma Subjects Is Determined by Lung Inflation," *PLoS ONE*, vol. 7, no. 12, Article ID e53216, 2012.

[3] G. G. King, B. Harris, and S. Mahadev, "V/Q SPECT: Utility for investigation of pulmonary physiology," *Seminars in Nuclear Medicine*, vol. 40, no. 6, pp. 467–473, 2010.

[4] J. G. Venegas, T. Schroeder, S. Harris, R. T. Winkler, and M. F. Vidal Melo, "The distribution of ventilation during bronchoconstriction is patchy and bimodal: A PET imaging study," *Respiratory Physiology & Neurobiology*, vol. 148, no. 1-2, pp. 57–64, 2005.

[5] M. Montaudon, M. Lederlin, S. Reich et al., "Bronchial measurements in patients with asthma: Comparison of quantitative thin-section CT findings with those in healthy subjects and correlation with pathologic findings," *Radiology*, vol. 253, no. 3, pp. 844–853, 2009.

[6] A. Busacker, J. D. Newell Jr., T. Keefe et al., "A multivariate analysis of risk factors for the air-trapping asthmatic phenotype as measured by quantitative CT analysis," *CHEST*, vol. 135, no. 1, pp. 48–56, 2009.

[7] R. S. Aysola, E. A. Hoffman, D. Gierada et al., "Airway remodeling measured by multidetector CT is increased in severe asthma and correlates with pathology," *CHEST*, vol. 134, no. 6, pp. 1183–1191, 2008.

[8] S. Choi, E. A. Hoffman, S. E. Wenzel, M. Castro, and C.-L. Lin, "Improved CT-based estimate of pulmonary gas trapping accounting for scanner and lung-volume variations in a multicenter asthmatic study," *Journal of Applied Physiology*, vol. 117, no. 6, pp. 593–603, 2014.

[9] Y. Yin, E. A. Hoffman, and C.-L. Lin, "Mass preserving nonrigid registration of CT lung images using cubic B-spline," *Medical Physics*, vol. 36, no. 9, pp. 4213–4222, 2009.

[10] J. M. Reinhardt, K. Ding, K. Cao, G. E. Christensen, E. A. Hoffman, and S. V. Bodas, "Registration-based estimates of local lung tissue expansion compared to xenon CT measures of specific ventilation," *Medical Image Analysis*, vol. 12, no. 6, pp. 752–763, 2008.

[11] C. J. Galbán, M. K. Han, J. L. Boes et al., "Computed tomography-based biomarker provides unique signature for diagnosis of COPD phenotypes and disease progression," *Nature Medicine*, vol. 18, no. 11, pp. 1711–1715, 2012.

[12] S. Choi, E. A. Hoffman, S. E. Wenzel et al., "Registration-based assessment of regional lung function via volumetric CT images of normal subjects vs. severe asthmatics," *Journal of Applied Physiology*, vol. 115, no. 5, pp. 730–742, 2013.

[13] S. Choi, E. A. Hoffman, S. E. Wenzel et al., "Quantitative assessment of multiscale structural and functional alterations in asthmatic populations," *Journal of Applied Physiology*, vol. 118, no. 10, pp. 1286–1298, 2015.

[14] H. Zhang and G. Papadakis, "Computational analysis of flow structure and particle deposition in a single asthmatic human airway bifurcation," *Journal of Biomechanics*, vol. 43, no. 13, pp. 2453–2459, 2010.

[15] K. Inthavong, J. Tu, Y. Ye, S. Ding, A. Subic, and F. Thien, "Effects of airway obstruction induced by asthma attack on particle deposition," *Journal of Aerosol Science*, vol. 41, no. 6, pp. 587–601, 2010.

[16] J. W. De Backer, W. G. Vos, A. Devolder et al., "Computational fluid dynamics can detect changes in airway resistance in asthmatics after acute bronchodilation," *Journal of Biomechanics*, vol. 41, no. 1, pp. 106–113, 2008.

[17] A. F. Heenan, E. Matida, A. Pollard, and W. H. Finlay, "Experimental measurements and computational modeling of the flow field in an idealized human oropharynx," *Experiments in Fluids*, vol. 35, no. 1, pp. 70–84, 2003.

[18] S. T. Jayaraju, M. Brouns, C. Lacor, B. Belkassem, and S. Verbanck, "Large eddy and detached eddy simulations of fluid flow and particle deposition in a human mouth-throat," *Journal of Aerosol Science*, vol. 39, no. 10, pp. 862–875, 2008.

[19] P. W. Longest, M. Hindle, S. D. Choudhuri, and P. R. Byron, "Numerical simulations of capillary aerosol generation: CFD model development and comparisons with experimental data," *Aerosol Science and Technology*, vol. 41, no. 10, pp. 952–973, 2007.

[20] G. Tian, P. W. Longest, G. Su, R. L. Walenga, and M. Hindle, "Development of a stochastic individual path (SIP) model for predicting the tracheobronchial deposition of pharmaceutical aerosols: Effects of transient inhalation and sampling the airways," *Journal of Aerosol Science*, vol. 42, no. 11, pp. 781–799, 2011.

[21] S. Vinchurkar, L. De Backer, W. Vos, C. Van Holsbeke, J. De Backer, and W. De Backer, "A case series on lung deposition analysis of inhaled medication using functional imaging based computational fluid dynamics in asthmatic patients: Effect of upper airway morphology and comparison with in vivo data," *Inhalation Toxicology*, vol. 24, no. 2, pp. 81–88, 2012.

[22] J. W. De Backer, W. G. Vos, S. C. Vinchurkar et al., "Validation of computational fluid dynamics in CT-based airway models with SPECT/CT," *Radiology*, vol. 257, no. 3, pp. 854–862, 2010.

[23] S. Miyawaki, M. H. Tawhai, E. A. Hoffman, S. E. Wenzel, and C.-L. Lin, "Automatic construction of subject-specific human airway geometry including trifurcations based on a CT-segmented airway skeleton and surface," *Biomechanics and Modeling in Mechanobiology*, vol. 16, no. 2, pp. 583–596, 2017.

[24] S. E. Wenzel and W. W. Busse, "Severe asthma: Lessons from the Severe Asthma Research Program," *The Journal of Allergy and Clinical Immunology*, vol. 119, no. 1, pp. 14–21, 2007.

[25] A. W. Vreman, "An eddy-viscosity subgrid-scale model for turbulent shear flow: algebraic theory and applications," *Physics of Fluids*, vol. 16, no. 10, pp. 3670–3681, 2004.

[26] C.-L. Lin, H. Lee, T. Lee, and L. J. Weber, "A level set characteristic Galerkin finite element method for free surface flows," *International Journal for Numerical Methods in Fluids*, vol. 49, no. 5, pp. 521–547, 2005.

[27] M. Howatson Tawhai, A. J. Pullan, and P. J. Hunter, "Generation of an Anatomically Based Three-Dimensional Model of the Conducting Airways," *Annals of Biomedical Engineering*, vol. 28, no. 7, pp. 793–802, 2000.

[28] C. Geuzaine and J. F. Remacle, "Gmsh: A 3-D finite element mesh generator with built-in pre- and post-processing facilities," *International Journal for Numerical Methods in Engineering*, vol. 79, no. 11, pp. 1309–1331, 2009.

[29] S. Miyawaki, E. A. Hoffman, and C.-L. Lin, "Numerical simulations of aerosol delivery to the human lung with an idealized laryngeal model, image-based airway model, and automatic meshing algorithm," *Computers & Fluids*, vol. 148, pp. 1–9, 2017.

[30] S. Miyawaki, M. H. Tawhai, E. A. Hoffman, and C.-L. Lin, "Effect of carrier gas properties on aerosol distribution in a CT-based human airway numerical model," *Annals of Biomedical Engineering*, vol. 40, no. 7, pp. 1495–1507, 2012.

[31] M. R. Maxey and J. J. Riley, "Equation of motion for a small rigid sphere in a nonuniform flow," *Physics of Fluids*, vol. 26, no. 4, pp. 883–889, 1983.

[32] P. W. Longest, G. Tian, R. L. Walenga, and M. Hindle, "Comparing MDI and DPI aerosol deposition using in vitro experiments and a new stochastic individual path (SIP) model of the conducting airways," *Pharmaceutical Research*, vol. 29, no. 6, pp. 1670–1688, 2012.

[33] R. L. Sorkness, E. R. Bleecker, W. W. Busse et al., "Lung function in adults with stable but severe asthma: Air trapping and incomplete reversal of obstruction with bronchodilation," *Journal of Applied Physiology*, vol. 104, no. 2, pp. 394–403, 2008.

[34] R. L. Walenga and P. W. Longest, "Current Inhalers Deliver Very Small Doses to the Lower Tracheobronchial Airways: Assessment of Healthy and Constricted Lungs," *Journal of Pharmaceutical Sciences*, vol. 105, no. 1, pp. 147–159, 2016.

[35] S. Choi, E. A. Hoffman, S. E. Wenzel et al., "Quantitative computed tomographic imaging-based clustering differentiates asthmatic subgroups with distinctive clinical phenotypes," *The Journal of Allergy and Clinical Immunology*, vol. 140, no. 3, pp. 690–700.e8, 2017.

[36] S. Choi, J. Choi, and C.-L. Lin, "Contributions of Kinetic Energy and Viscous Dissipation to Airway Resistance in Pulmonary Inspiratory and Expiratory Airflows in Successive Symmetric Airway Models with Various Bifurcation Angles," *Journal of Biomechanical Engineering*, vol. 140, no. 1, 2018.

10

Qualitative and Sensitivity Analysis of the Effect of Electronic Cigarettes on Smoking Cessation

Jae Hun Jung, Anna Park, and Il Hyo Jung

Department of Mathematics, Pusan National University, Busan 46241, Republic of Korea

Correspondence should be addressed to Il Hyo Jung; ilhjung@pusan.ac.kr

Academic Editor: Ruisheng Wang

Recently, the role of the electronic cigarettes (e-cigarettes) in a way to reduce smoking is increasing. E-cigarettes are a device that delivers only the nicotine, and its use is considered less harmful to health compared with tobacco cigarettes. Smokers frequently make use of e-cigarettes as one of the nonsmoking aid devices. In this work, we propose a mathematical model to analyze the effect of e-cigarettes on smoking cessation. The stability and the bifurcation of the model have been discussed. The parameter estimations from the observed data are drawn, and using the parameters, a reasonable smoking model has been designed. Moreover, by considering the sensitivity results depending on the basic reproduction number R_0, the effective strategies that reduce the smokers are investigated. Numerical simulations of the model show that e-cigarettes may somewhat diminish the numbers of smokers, but it does not reduce the number of quitters ultimately.

1. Introduction

Smoking is well known as one of the most serious global public health problems. According to the report by the World Health Organization (WHO) [1], smoking is the legalization of a drug that kills many of its users. In other words, smoking leads to disease and disability and harms almost every organ of the body by both active and passive (second-hand) smoking. Moreover, smoking also induces an addiction, so that smokers who want to stop smoking cannot do it. Each year, active smoking is responsible for the death of about 5 million people in the world. In terms of casualties, it is more lethal than tuberculosis, human immunodeficiency virus/acquired immunodeficiency syndrome (HIV/AIDS), and malaria combined. Overall, 600,000 people are estimated to die annually due to the effects of second-hand smoke. Therefore, smoking is a serious health risk.

For that reason, the government encourages enrollment in smoking cessation programs, as well as the use of nicotine patches and nicotine gums so as to reduce the number of smokers. In addition, it establishes no smoking areas. However, since smoking is very difficult to quit, some smokers tend to use electronic cigarettes as a substitute for tobacco cigarettes. E-cigarettes are also considered to be less harmful than tobacco cigarettes [2], as, unlike cigarettes whose smoke contains thousands of harmful substances, such as tar and carbon monoxide, the e-cigarettes contain only nicotine. In fact, from the medical point of view, e-cigarettes have been studied extensively [3, 4]. For example, in [5] and [6], e-cigarettes are described as a valuable product for smoking cessation. However, in [7], the authors suggested that advertising for e-cigarettes must be banned until scientific evidence appears.

On the one hand, from the mathematical modelling point of view, the smoking cessation models have been studied by using mathematical modelling [8–11]. For instance, Sharma and Misra have studied a mathematical model of smoking cessation with media campaigns and bifurcation analysis [12]. Furthermore, Pang et al. have proposed a mathematical model with a saturated incidence rate to explore the effect of controlling smoking [13]. Similarly, Zaman has proposed several mathematical models of an SIR type [14–16].

Although there have been many studies on smoking, to the best of our knowledge, none of previous studies have analyzed the effect of e-cigarettes on smoking cessation using a smoking model. It may be difficult to judge whether e-cigarettes actually help smoking cessation. Therefore, it is worthwhile to study a mathematical model that can identify the characteristics of nonsmokers and quitters, unlike statistical methods and experiments that confirm only the characteristics of simple smokers.

In the present study, we investigate such a smoking model based on the real data from the US Department of Health and Human Services [17]. The aim of this paper is to demonstrate both the addictive nature of smoking and the efficacy of e-cigarettes as an aid in smoking cessation using a mathematical model. Numerical simulations of the model confirm the dynamics of two aspects of the effect of e-cigarettes: the numbers of smokers and nonsmokers.

The rest of the article is composed as follows: In Section 2, a mathematical model to assess the effect of e-cigarettes on smoking cessation is proposed; furthermore, the dynamics of the model, the basic reproduction number, and the stability of equilibria are investigated. In particular, the condition for occurring bifurcation situation is presented. In Section 3, the parameter estimations from real data are provided. The sensitivity analysis and numerical simulation confirm the results that are obtained analytically. Finally, the conclusions are briefly summarized in Section 4.

2. Mathematical Model and Analysis

2.1. Model. The traditional epidemiologic model of agent, host, vector, and environment is useful for studying the interplay of various influences on patterns of tobacco use in populations. Since the dynamics of smoking cause disease and addiction that is defined when occurring craving, tolerance, withdrawal symptoms, and loss of control, it is very similar to that of an epidemic [18]. For example, a nonsmoker comes into contact with smokers and starts smoking from his influence. Based on epidemic models, therefore, we propose a nonlinear mathematical model to assess the effect of e-cigarettes on smoking cessation.

First, we set up a region with the total population N at time t. The total population is divided into four subpopulations [19]: potential smokers (P) who do not smoke yet but might become smokers in the future; smokers (S) who have smoked more than 100 cigarettes in their lifetime and reported smoking "everyday" or "some days" at the time; e-smokers (E) who currently use e-cigarettes; and quitters (Q) who had smoked more than 100 cigarettes in their lifetime and reported smoking "not at all" at the time. We assume that all of the natural death rates equal to the birth rates is μ, and the mortality rate due to the specific diseases that caused by smoking is not considered.

We have considered the effective contact rate, that is, the average number of visits to social gatherings of potential smokers for influential contact of smokers per unit time and the probability of the proportion of becoming smokers after a casual smoking. The constant β_1 is the effective contact rate that potential smokers can become smokers due to peer influence and come in the smokers class at rate $\beta_1 SP/N$. The constant β_2 is the effective contact rate, but this case is a relapse to smoking by peer influence. Then quitters can become smokers due to peer influence at rate $\beta_2 SQ/N$. In this part, we have considered the effects of peer influence in dynamics of smoking and presented the nonlinear term. Furthermore, we assume that the start smoking rate β_1 is greater than the relapse rate β_2 because the influence of the curiosity of the people who has not smoked is more than people who quit smoking.

To quit smoking, however, we consider two kinds of groups: smokers who may quit smoking by their own will at rate γ_1 or smokers who use e-cigarettes at rate g. In the latter case, e-smokers who used e-cigarettes are likely to revert to smokers at rate α. The constant γ_2 represents the per capita quit rate that stops smoking by use of e-cigarettes. The model parameters are summarized in Table 1. The proposed model is shown in Figure 1 and is described as follows:

$$\begin{aligned}
\frac{dP}{dt} &= \mu N - \beta_1 \frac{PS}{N} - \mu P, \\
\frac{dS}{dt} &= \beta_1 \frac{PS}{N} + \beta_2 \frac{SQ}{N} + \alpha E - \gamma_1 S - gS - \mu S, \\
\frac{dE}{dt} &= gS - \alpha E - \gamma_2 E - \mu E, \\
\frac{dQ}{dt} &= \gamma_1 S + \gamma_2 E - \beta_2 \frac{SQ}{N} - \mu Q,
\end{aligned} \tag{1}$$

where the initial conditions are given by

$$\begin{aligned}
P(0) &= P_0 \geq 0, \\
S(0) &= S_0 \geq 0, \\
E(0) &= E_0 \geq 0, \\
Q(0) &= Q_0 \geq 0.
\end{aligned} \tag{2}$$

In system (1), adding all of the equations gives $dN/dt = 0$ and so the total population is constant, which is equal to N. Since system (1) monitors the human population, all the state variables and parameters are nonnegative for all $t \geq 0$. System (1) shows that the region

$$\Lambda = \left\{(P, S, E, Q) \in \mathbb{R}_+^4 \ : \ P + S + E + Q = N\right\} \tag{3}$$

is positively invariant, which is the region of attraction for the system.

2.2. Equilibria. In this section, to find the equilibrium of system (1), it is useful to consider the fractions $P/N = p$, $S/N = s$, $E/N = e$, and $Q/N = q$, with $p + s + e + q = 1$ for convenience of calculation with the reduced form. Then, we can get

$$\begin{aligned}
\frac{dp}{dt} &= \mu - \beta_1 ps - \mu p, \\
\frac{ds}{dt} &= \beta_1 ps + \beta_2 sq + \alpha e - \gamma_1 s - gs - \mu s, \\
\frac{de}{dt} &= gs - \alpha e - \gamma_2 e - \mu e, \\
\frac{dq}{dt} &= \gamma_1 s + \gamma_2 e - \beta_2 sq - \mu q.
\end{aligned} \tag{4}$$

TABLE 1: Parameter description.

Par.	Description	Value ($year^{-1}$)	Reference
μ	Mortality rate	1/79.8	[20]
β_1	Rate at which someone is transitioned to smoking	0.1961	Estimated in [16]
β_2	Rate of relapse	0.0101	Estimated in [16]
γ_1	Treatment rate of people quitting to smoke by their own will	0.0772	Estimated in [16]
γ_2	Treatment rate of people who used e-cigarettes	0.1008	Estimated
α	Rates at which people stop using e-cigarettes	0.0822	Estimated
g	Rates at which people begin using e-cigarettes	0.1245	Estimated

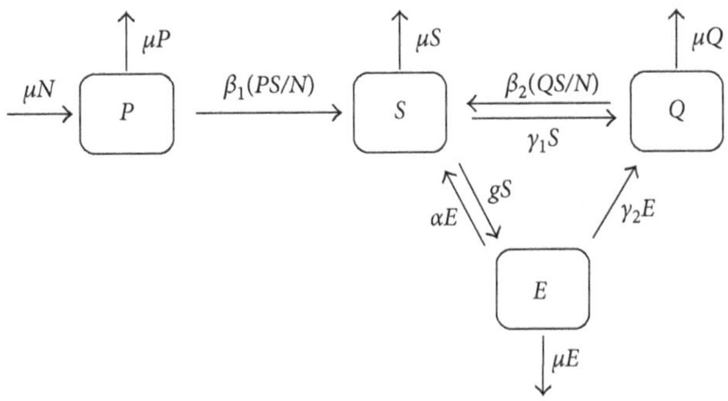

FIGURE 1: Schematic diagram of system (1).

Substituting $p = 1 - s - e - q$, system (4) can be reduced as follows:

$$\begin{aligned} s' &= \beta_1 s(1-s-e-q) + \beta_2 sq + \alpha e - \gamma_1 s - gs - \mu s, \\ e' &= gs - \alpha e - \gamma_2 e - \mu e, \\ q' &= \gamma_1 s + \gamma_2 e - \beta_2 sq - \mu q. \end{aligned} \tag{5}$$

Here, we investigate the basic reproduction number, R_0, which is defined as the average number of secondary smoking cases produced by one case of smoking that is completely free from e-cigarettes over its smoking period in an entirely potential smoker. If $R_0 < 1$, then a few smokers will completely replace potential smokers, and the smoking will not spread. If $R_0 > 1$, then the number of smokers will increase with each generation, and the smoking will spread. To obtain R_0, we have calculated the spectral radius of the next generation matrix used in [21]. Let $x = (s, e, q)^T$, and thus system (5) can be written as

$$\frac{dx_i}{dt} = \mathcal{F} - \mathcal{V}, \quad i = 1, 2, 3, \tag{6}$$

where

$$\begin{aligned} \mathcal{F} &= \begin{pmatrix} \mathcal{F}_1(x) \\ \mathcal{F}_2(x) \\ \mathcal{F}_3(x) \end{pmatrix} = \begin{pmatrix} \beta_1 s(1-s-e-q) + \beta_2 q \\ 0 \\ 0 \end{pmatrix}, \\ \mathcal{V} &= \begin{pmatrix} \mathcal{V}_1(x) \\ \mathcal{V}_2(x) \\ \mathcal{V}_3(x) \end{pmatrix} = \begin{pmatrix} -\alpha e + (\gamma_1 + g + \mu)s \\ -gs + (\alpha + \gamma_2 + \mu)e \\ -\gamma_1 s - \gamma_2 e + \beta_2 sq + \mu q \end{pmatrix}. \end{aligned} \tag{7}$$

By calculating the Jacobian matrices at $s = e = q = 0$, we find that the derivatives of $\mathcal{F}$ and $\mathcal{V}$ lead to the following expressions for F and V:

$$\begin{aligned} F &= \begin{pmatrix} \beta_1 & 0 & 0 \\ 0 & 0 & 0 \\ 0 & 0 & 0 \end{pmatrix}, \\ V &= \begin{pmatrix} \gamma_1 + g + \mu & -\alpha & 0 \\ -g & \alpha + \gamma_2 + \mu & 0 \\ -\gamma_1 & -\gamma_2 & \mu \end{pmatrix}. \end{aligned} \tag{8}$$

Hence, we have simply calculated

$$\begin{aligned} V^{-1} &= \frac{1}{(g + \gamma_1 + \mu)(\alpha + \gamma_2 + \mu) - \alpha g} \\ &\cdot \begin{pmatrix} \alpha + \gamma_2 + \mu & \alpha \\ g & g + \gamma_1 + \mu \end{pmatrix}. \end{aligned} \tag{9}$$

Therefore, we get

$$R_0 = \rho(FV^{-1}) = \frac{\beta_1(\alpha + \gamma_2 + \mu)}{(g + \gamma_1 + \mu)(\alpha + \gamma_2 + \mu) - \alpha g}, \tag{10}$$

where FV^{-1} denotes the next generation matrix and $\rho(A)$ denotes the spectral radius of matrix A.

Obviously, system (5) always has a smoking-free equilibrium $e_0 = (0, 0, 0)$ which indicates that smoking does not exist. In addition, system (5) has a smoking-present equilibrium $e_* = (s^*, e^*, q^*)$. Letting the left side equations of system (5) be equal to zero, the smoking-present equilibrium is calculated as follows:

$$e^* = \frac{gs^*}{\alpha + \gamma_2 + \mu},$$
$$q^* = \frac{\gamma_1 s^* + \gamma_2 e^*}{\beta_2 s^* + \mu}. \quad (11)$$

Substituting the first equation of system (5) into (11), we get the following quadratic equation:

$$a(s^*)^2 + b(s^*) + c = 0, \quad (12)$$

where

$$\begin{aligned} a &= \beta_1\beta_2(\alpha + \gamma_2 + g + \mu), \\ b &= (\beta_1 - \beta_2)\{(\alpha + \mu)\gamma_1 + (\gamma_1 + g)\gamma_2\} + \beta_1\mu(\alpha + \gamma_2 + \mu + g) \\ &\quad + \beta_2\{(\alpha + \gamma_2 + \mu)(g + \gamma_1 + \mu) - \alpha g\}(1 - R_0), \\ c &= \mu\{(\alpha + \gamma_2 + \mu)(g + \gamma_1 + \mu) - \alpha g\}(1 - R_0). \end{aligned} \quad (13)$$

From (12), we get

$$s_+^* = \frac{-b + \sqrt{b^2 - 4ac}}{2a},$$
$$s_-^* = \frac{-b - \sqrt{b^2 - 4ac}}{2a}. \quad (14)$$

Here, the coefficient a is always positive for all parameter values. But the sign of b is changed by parameters, and the sign of c depends on R_0. We can consider two cases: (i) if $R_0 > 1$, then $c < 0$. In this case, each has one negative and one positive solution. It follows that (12) has a unique positive equilibrium. (ii) If $R_0 < 1$, then $c > 0$. When $b < 0$, (12) has two positive solutions and the number of endemic equilibria may be two.

2.3. Stability Analysis. Now, the stability of the equilibria is explored and the following results are obtained.

Theorem 1. *System (5) has the smoking-free equilibrium $e_0 = (0, 0, 0)$ with $p + s + e + q = 1$. If $R_0 < 1$, then e_0 is locally asymptotically stable. Otherwise, if $R_0 > 1$, then e_0 is unstable.*

Proof. The Jacobian matrix at e_0 of system (5) is

$$J(e_0) = \begin{pmatrix} \beta_1 - (g + \gamma_1 + \mu) & \alpha & 0 \\ g & -(\alpha + \gamma_2 + \mu) & 0 \\ \gamma_1 & \gamma_2 & -\mu \end{pmatrix}. \quad (15)$$

The eigenvalue from (15) is given by $\lambda_1 = -\mu < 0$, and the other eigenvalues are determined by the sub-Jacobian matrix:

$$J_s = \begin{pmatrix} \beta_1 - (g + \gamma_1 + \mu) & \alpha \\ g & -(\alpha + \gamma_2 + \mu) \end{pmatrix}. \quad (16)$$

Then, the characteristic equation of J_s is as follows:

$$\lambda^2 + A_1\lambda + A_2 = 0, \quad (17)$$

where

$$\begin{aligned} A_1 &= (\alpha + \gamma_2 + \mu) + (g + \gamma_1 + \mu), \\ A_2 &= \{(\alpha + \gamma_2 + \mu)(g + \gamma_1 + \mu) - \alpha g\}(1 - R_0). \end{aligned} \quad (18)$$

It follows that $A_1 > 0$ and $A_2 > 0$ (i.e., $R_0 < 1$), and then e_0 is locally asymptotically stable if $R_0 < 1$, otherwise unstable if $R_0 > 1$.

Theorem 2. *System (5) has the smoking-present equilibrium $e_* = (s^*, e^*, q^*)$, and it is locally asymptotically stable if $\gamma_1 > \beta_2 q^*$ and $\beta_1 > (\alpha/s^*)$.*

Proof. By setting $ds/dt = 0$, $de/dt = 0$, and $dq/dt = 0$ and dividing the first equation of (5) by s^*, the following can be obtained:

$$\beta_1(1 - s^* - e^* - q^*) + \beta_2 q^* + \alpha\frac{e^*}{s^*} - (\gamma_1 + g + \mu) = 0. \quad (19)$$

From the above equation and system (5), the Jacobian matrix at e_* of system (5) is

$$J(e_*) = \begin{pmatrix} -\beta_1 s^* - \alpha\frac{e^*}{s^*} & \beta_1 + \frac{\alpha}{s^*} & -\beta_1 + \beta_2 \\ g & -(\alpha + \gamma_2 + \mu) & 0 \\ \gamma_1 - \beta_2 q^* & \gamma_2 & -(\beta_2 s^* + \mu) \end{pmatrix}. \quad (20)$$

The characteristic equation of $J(e_*)$ is

$$\lambda^3 + A_1\lambda^2 + A_2\lambda + A_3 = 0, \quad (21)$$

where the coefficients are given by

$$\begin{aligned} A_1 &= \beta_1 s^* + \beta_2 s^* + \frac{\alpha e^*}{s^*} + \alpha + \gamma_2 + 2\mu, \\ A_2 &= (\beta_2 s^* + \mu)\left(\beta_1 s^* + \frac{\alpha e^*}{s^*} + \alpha + \gamma_2 + \mu\right) + \left(\beta_1 s^* + \frac{\alpha e^*}{s^*}\right) \\ &\quad \cdot (\alpha + \gamma_2 + \mu) + g\left(\beta_1 - \frac{\alpha}{s^*}\right) + (\beta_1 - \beta_2)(\gamma_1 - \beta_2 q^*), \\ A_3 &= (\beta_1 - \beta_2)\{g\gamma_2 + (\gamma_1 - \beta_2 q^*)(\alpha + \gamma_2 + \mu)\} + (\beta_2 s^* + \mu) \\ &\quad \cdot \left\{\left(\beta_1 s^* + \frac{\alpha e^*}{s^*}\right)(\alpha + \gamma_2 + \mu) + g\left(\beta_1 - \frac{\alpha}{s^*}\right)\right\}. \end{aligned} \quad (22)$$

And we have

$$A_1A_2 - A_3 = \left\{(\alpha+\gamma_2+\mu)(\beta_2 s^* + \mu) + \left(\beta_1 s^* + \frac{\alpha e^*}{s^*}\right)\right\} \cdot \left(\beta_1 s^* + \beta_2 s^* + \frac{\alpha e^*}{s^*} + \alpha + \gamma_2 + 2\mu\right) + g\left(\alpha+\gamma_2+\mu+\beta_1 s^* + \frac{\alpha e^*}{s^*}\right)\left(\beta_1 - \frac{\alpha}{s^*}\right) + (\beta_2 s^* + \mu)\left(\beta_1 s^* + \frac{\alpha e^*}{s^*}\right) + \left(\beta_2 s^* + \beta_1 s^* + \frac{\alpha e^*}{s^*} + \mu\right) + (\beta_1 - \beta_2) \cdot \left\{(\gamma_1 - \beta_2 q^*)\left(\beta_2 s^* + \mu + \beta_1 s^* + \frac{\alpha e^*}{s^*}\right) - g\gamma_2\right\}. \quad (23)$$

If $\gamma_1 > \beta_2 q^*$ and $\beta_1 > \alpha/s^*$, then it is apparent that $A_1 > 0$, $A_2 > 0$, $A_3 > 0$, and $A_1A_2 - A_3 > 0$. Hence, by Routh–Hurwitz criterion, the smoking-present equilibrium e_* is locally asymptotically stable when the stated condition holds.

2.4. Bifurcation. In this section, we provide the dynamics of system (5) at $R_0 = 1$, and this transition is called a transcritical (its direction is forward) bifurcation. This means that there is a change in equilibrium behavior at $R_0 = 1$, but the equilibrium of smoking population depends on R_0. Besides, we perform a qualitative analysis to investigate the occurrence of backward bifurcations that play an appropriate role in disease control and eradication. In fact, it is well known that in disease transmission modelling, a necessary condition for disease eradication is that the basic reproduction number R_0 must be less than unity [22]. If forward bifurcation occurs when $R_0 < 1$, then it is a necessary and sufficient condition for smoking extinction. However, if backward bifurcation occurs, an smoking-present equilibrium may also exist for $R_0 < 1$. In system (5), we calculate a quadratic equation substituting the result of endemic equilibrium e_* as follows.

Now, applying Theorem 4.1 in [23] to system (5), we obtain a rigorous proof for the forward bifurcation of system (5). We focus on the smoking-free equilibrium e_0 and investigate the occurrence of the transcritical bifurcation with $R_0 = 1$ corresponding to

$$\beta_1 = \beta_1^* = \frac{\alpha(\gamma_1+\mu) + (\gamma_2+\mu)(g+\gamma_1+\mu)}{\alpha+\gamma_2+\mu}. \quad (24)$$

First of all, observe the eigenvalues of the matrix:

$$J(e_0;\beta_1^*) = \begin{pmatrix} \frac{-\alpha g}{\alpha+\gamma_2+\mu} & \alpha & 0 \\ g & -(\alpha+\gamma_2+\mu) & 0 \\ \gamma_1 & \gamma_2 & -\mu \end{pmatrix}, \quad (25)$$

which are the roots of the characteristic equation given by

$$\lambda_1 = 0,$$
$$\lambda_2 = -\mu, \quad (26)$$
$$\lambda_3 = -(\alpha+\gamma_2+\mu) - \frac{\alpha g}{\alpha+\gamma_2+\mu}.$$

Since $J(e_0;\beta_1^*)$ has simple eigenvalues with the zero real part and all eigenvalues have negative real part, system (5) with $R_0 = 1$ has a nonhyperbolic equilibrium. Now, a right eigenvalue corresponding to zero eigenvalue, λ_1, denoted by $\mathbf{w} = (w_1, w_2, w_3)^T$ is obtained as follows:

$$\begin{pmatrix} \frac{-\alpha g}{\alpha+\gamma_2+\mu} & \alpha & 0 \\ g & -(\alpha+\gamma_2+\mu) & 0 \\ \gamma_1 & \gamma_2 & -\mu \end{pmatrix} \begin{pmatrix} w_1 \\ w_2 \\ w_3 \end{pmatrix} = \begin{pmatrix} 0 \\ 0 \\ 0 \end{pmatrix}, \quad (27)$$

$$-\frac{\alpha g}{\alpha+\gamma_2+\mu} w_1 + \alpha w_2 = 0,$$
$$g w_1 - (\alpha+\gamma_2+\mu) w_2 = 0, \quad (28)$$
$$\gamma_1 w_1 + \gamma_2 w_3 - \mu w_3 = 0.$$

Thus, we get

$$\mathbf{w} = \left[1, \frac{g}{\alpha+\gamma_2+\mu}, \frac{\gamma_1(\alpha+\gamma_2+\mu) + \gamma_2 g}{\mu(\alpha+\gamma_2+\mu)}\right]^T. \quad (29)$$

Note that all components of $\mathbf{w}$ are positive. Furthermore, all components of the left eigenvector $\mathbf{v} = (v_1, v_2, v_3)$ corresponding to λ_1 which satisfies $\mathbf{v} \cdot \mathbf{w} = 1$ given by

$$(v_1\ v_2\ v_3) \begin{pmatrix} \frac{-\alpha g}{\alpha+\gamma_2+\mu} & \alpha & 0 \\ g & -(\alpha+\gamma_2+\mu) & 0 \\ \gamma_1 & \gamma_2 & -\mu \end{pmatrix} = \begin{pmatrix} 0 \\ 0 \\ 0 \end{pmatrix}, \quad (30)$$

are positive, and we get the following:

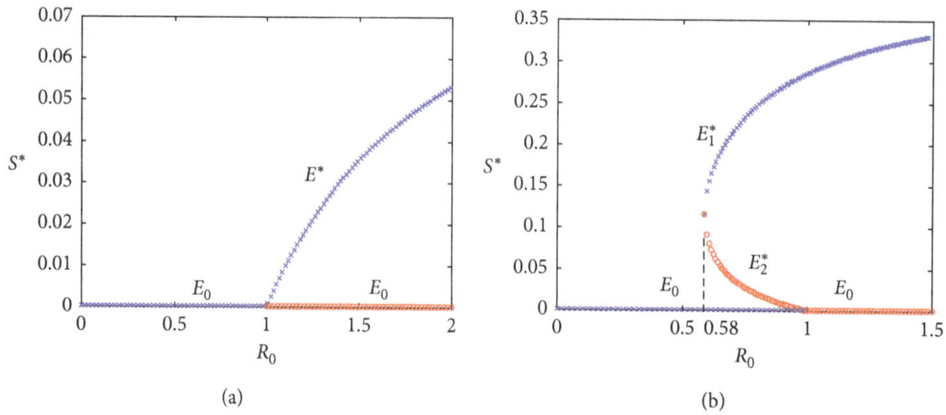

Figure 2: (a) The bifurcation diagram for forward direction ($\beta_1 = 0.1961, \beta_2 = 0.0101$); (b) the bifurcation diagram for backward direction ($\beta_1 = 0.1961, \beta_2 = 0.4$); cross mark denotes stable equilibrium and dot mark denotes unstable equilibrium.

$$
\begin{aligned}
-\frac{\alpha g}{\alpha + \gamma_2 + \mu} v_1 + g v_2 + \gamma_1 v_3 &= 0, \\
\alpha v_1 - (\alpha + \gamma_2 + \mu) v_2 + \gamma_2 v_3 &= 0, \\
-\mu v_3 &= 0.
\end{aligned} \tag{31}
$$

The left eigenvector $\mathbf{v}$ is given by

$$
\mathbf{v} = \left[1, \frac{\alpha}{\alpha + \gamma_2 + \mu}, 0 \right]. \tag{32}
$$

The coefficients $\bar{a}$ and $\bar{b}$ defined in Theorem 4.1 in [23] are as follows:

$$
\begin{aligned}
\bar{a} &= \sum_{k,i,j=1}^{3} v_k w_i w_j \frac{\partial^2 f_k}{\partial x_i \partial x_j} (e_0, \beta_1^*), \\
\bar{b} &= \sum_{k,i=1}^{3} v_k w_i \frac{\partial^2 f_k}{\partial x_i \partial \beta_1} (e_0, \beta_1^*).
\end{aligned} \tag{33}
$$

Computing them yields the following:

$$
\begin{aligned}
\bar{a} &= \frac{2\beta_2 \{(\gamma_1 - g)\alpha + \gamma_1 (\gamma_2 + \mu)\} - \{\alpha(\gamma_1 - g) + \mu(\alpha + g) + (\gamma_2 + \mu)(\gamma_1 + \mu)\}}{\mu(\alpha + \gamma_2 + \mu)}, \\
\bar{b} &= \frac{(\alpha + \gamma_2 + \mu)(2\alpha + \gamma_2 + \mu)}{\alpha g + (\alpha + \gamma_2 + \mu)^2}.
\end{aligned} \tag{34}
$$

Therefore, we have the result shown in Figure 2(a).

Theorem 3. *System (5) exhibits transcritical bifurcation at $R_0 = 1$ and its direction is forward if*

$$
\beta_2 \{(\gamma_1 - g)\alpha + \gamma_1 (\gamma_2 + \mu)\} < \alpha(\gamma_1 - g) + \mu(\alpha + g) + (\gamma_2 + \mu)(\gamma_1 + \mu). \tag{35}
$$

On the contrary, we have also checked the occurrence of backward bifurcation in Figure 2(b) because it has a significant impact on prevention strategies and control policies. Indeed, if $\gamma_1 > g$ and $\beta_2 > \beta_1$, then the condition $\bar{a}$ is positive by Theorem 4.1 [23]. Biologically, this means that if a quitting policy is enforced without any preventing measures, then the threshold value that would lead to a state of smoking elimination has to be kept in mind. If the government does not have the policy to reduce smokers, such as smoking cessation campaigns, subsidies for smoking cessation, prevention education, and so on, then the number of smokers will increase because they are not aware of seriousness.

3. Numerical Results

3.1. Parameter Estimation. In this section, we simulate the parameter estimation for system (1), which allows support for modelling as well as reality. The estimation method is to find the best-fitting curve using the least-squares method in MATLAB. That is, the best-fit sense minimizes the sum of squared residuals that being the difference between the observed value and the fitted value provided by system (1). Here, the mortality rate, μ, is estimated by the inverse of life expectancy at birth about total population in the United States [20].

The real data used in the present study were obtained from the smoking state for 50 years at National Health Interview Survey (NHIS) [17], which reflected the trends in prevalence of current smoking among adults, 18 years

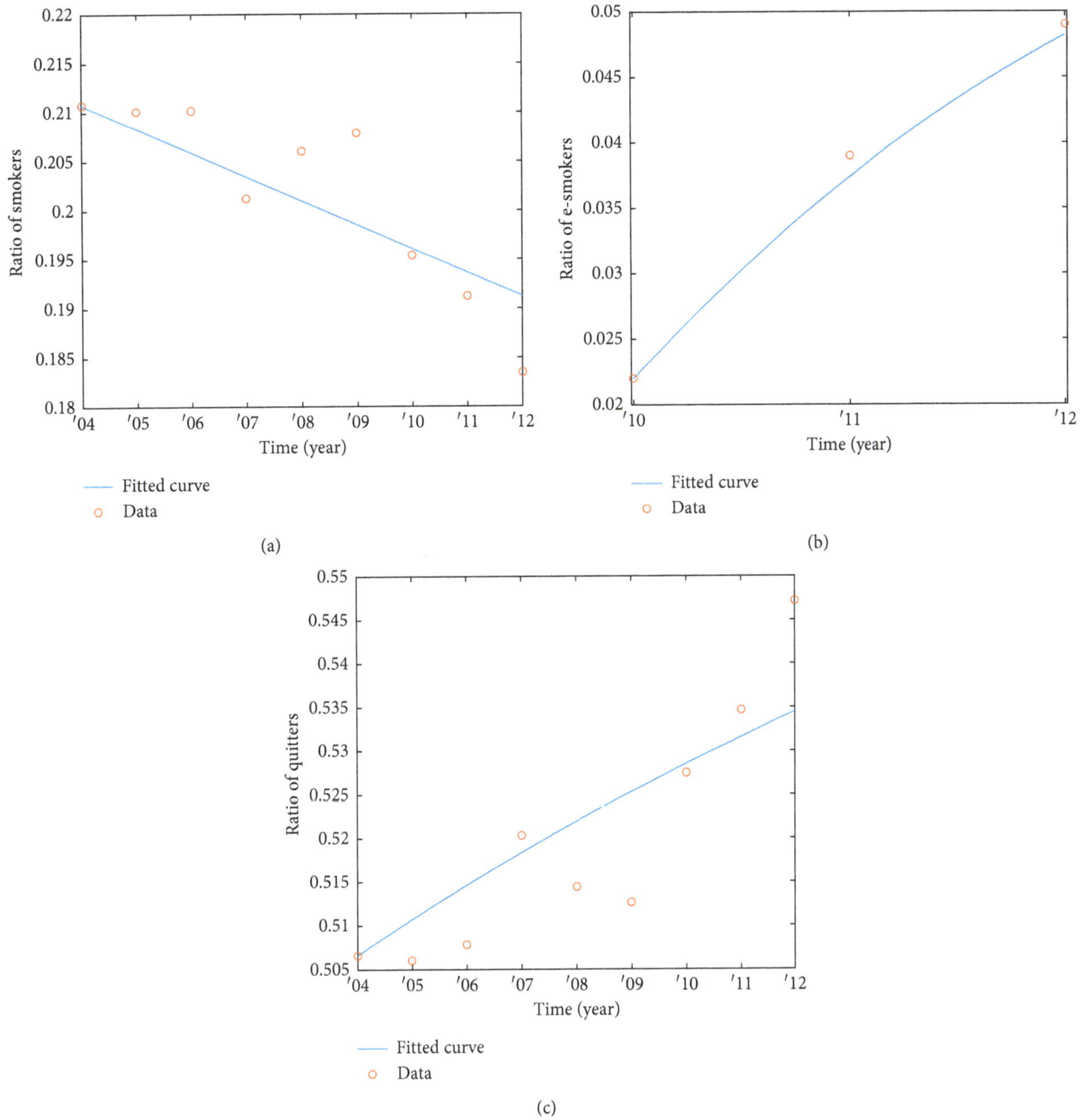

FIGURE 3: Parameters estimation: (a) the ratio of smokers, (b) the ratio of e-cigarette smokers, and (c) the ratio of quitters.

of age, and older. First, we had to estimate with data in the absence of e-cigarettes from 2004 to 2012. It is necessary to obtain the transition rate, the recovery rate, and the relapse rate in the original smoking model to verify the efficacy of e-cigarettes later. Therefore, we have obtained the parameter values of β_1, γ_1, and β_2. On the other hand, we could refer to the relationship to cravings [24]. Figures 3(a) and 3(c) show both the number of smokers and quitters referred by NHIS [17], and the data on the number of e-smokers are shown in Figure 3(b), because e-cigarettes were introduced in 2007, but it has been used by a lot of people from 2010 [19]. From these data, we estimate the parameters that are needed to system (1). From the above e-smokers data, we can obtain a ratio of people using e-cigarettes, g, and a ratio of people smoking again using e-cigarettes, α.

3.2. Sensitivity Analysis. In this section, we will analyze sensitivity which decides most effective parameters in progressively reducing the spread of smoking. To do this, we will investigate appropriately all of the possible parameters about the effect of various parameters to reduce basic reproduction number. R_0, close to one. The normalized sensitivity index $\mathcal{S}_x$ is the ratio of the corresponding normalized changes for parameters and is defined as follows:

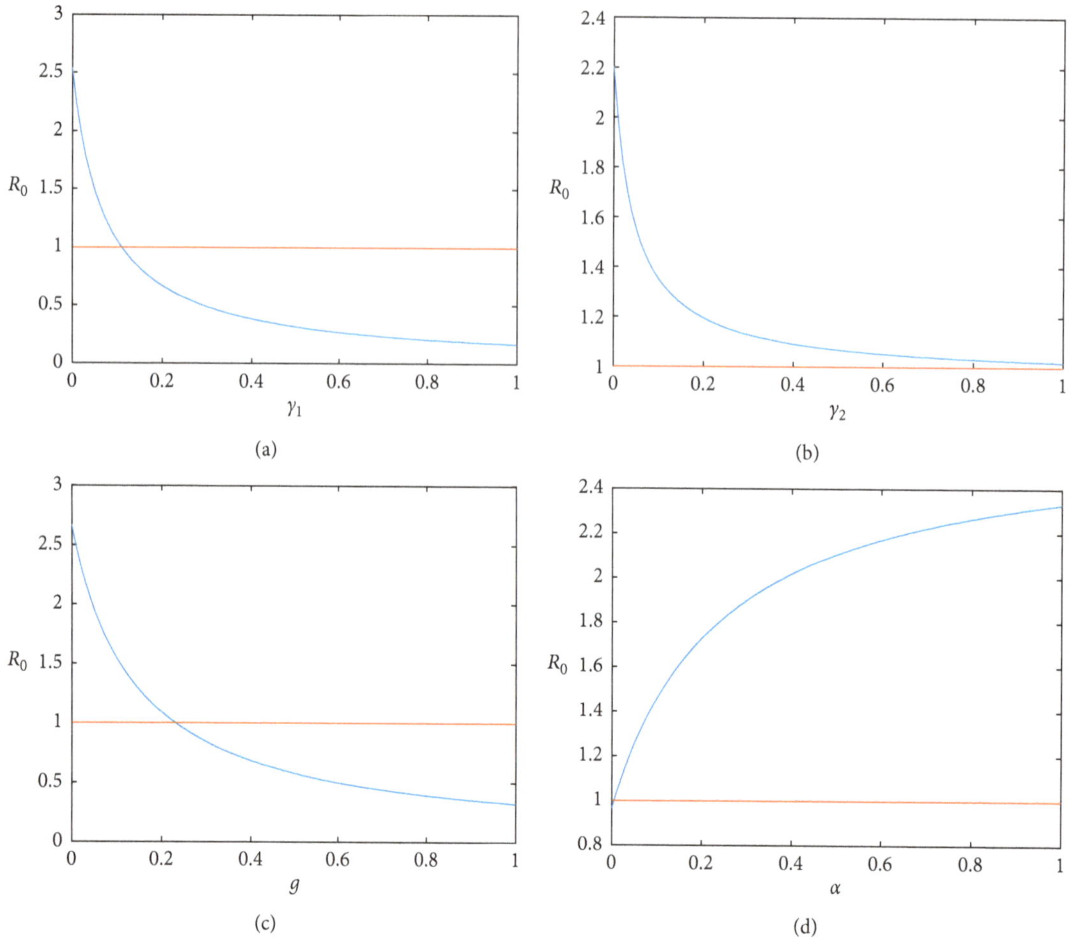

FIGURE 4: The value of sensitivity indices from the change of each configured parameter value: (a) the rate of quitting smoking by their own will, (b) the rate of quitting smoking by using e-cigarettes, (c) the rate of stopping the use of e-cigarettes, and (d) the rate of using e-cigarettes.

$$\mathcal{S}_x = \frac{\partial R_0/R_0}{\partial x/x} = \frac{\partial R_0}{\partial x}\frac{x}{R_0}. \tag{36}$$

Then, we calculate the following:

$$\begin{aligned}
\mathcal{S}_\alpha &= \frac{\alpha g(\gamma_2+\mu)}{M(\alpha+\gamma_2+\mu)} > 0,\\
\mathcal{S}_{\gamma_2} &= -\frac{\gamma_2\alpha g}{M(\alpha+\gamma_2+\mu)} < 0,\\
\mathcal{S}_{\gamma_1} &= -\frac{\gamma_1(\alpha+\gamma_2+\mu)}{M} < 0,\\
\mathcal{S}_g &= -\frac{g(\gamma_2+\mu)}{M} < 0,
\end{aligned} \tag{37}$$

where

$$M = (\alpha+\gamma_2+\mu)(g+\gamma_1+\mu) - \alpha g. \tag{38}$$

We show that the sign of each sensitivity index determines the relative change of the value of R_0. For instance, if the sensitivity index is positive, then the value of R_0 is an increasing function of the sensitivity index value. On the contrary, the form of the function is decreasing. In Figure 4(d), the sensitivity index of $\mathcal{S}_\alpha$ is positive; therefore, as the value of α increases, the value of R_0 increases as well. This means that if the government adopts a policy that restricts the use of e-cigarettes, then e-smokers would revert to smoking.

By contrast, in Figures 4(a)–4(c), the sensitivity indices of $\mathcal{S}_{\gamma_1}$, $\mathcal{S}_{\gamma_2}$, and $\mathcal{S}_g$ are negative, and these parameters are factors that reduce the number of smokers. As the values of γ_1 and g increase, the value of R_0 rapidly decreases. And, these two parameters have a similar tendency to reduce R_0. Biologically, this means that if smokers quit smoking on their own will or switch to e-cigarettes, then the number of smokers will drop substantially. Therefore, these two methods represent the most significant effect on reducing smoking. On the contrary, even if the value of γ_2 increases, the value of R_0 does not fall below one. The impact of γ_2 is weaker than that of the other variables. That is, γ_2 does not have a significant effect on R_0. In addition, the impact of α can cause the side effect of e-cigarettes with increasing R_0.Thus, it can be difficult to reduce smokers when using e-cigarettes.

In order to understand how to stop smoking from the perspective of each of the smokers and the quitters, we show the difference of the sensitivity index of R_0 for each

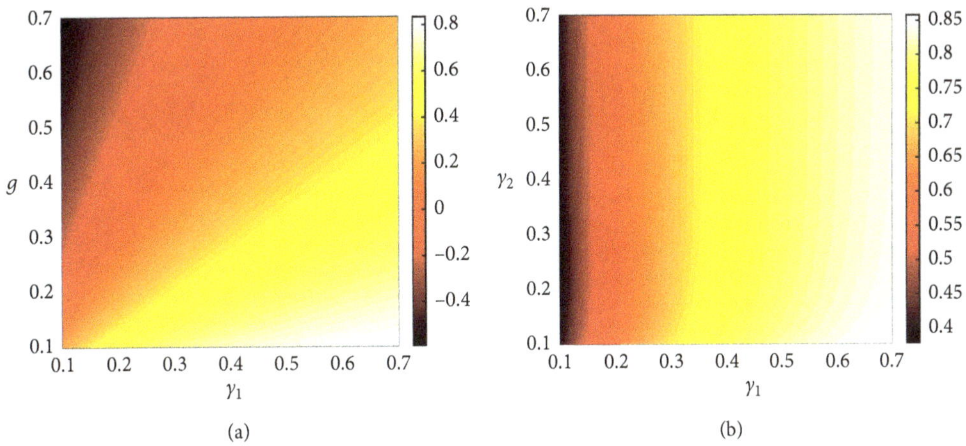

FIGURE 5: (a) A comparison value of the sensitivity index of R_0 if smoking is stopped by their own will (γ_1) and using e-cigarettes (g), that is, $\mathcal{S}_{\gamma_1} - \mathcal{S}_g$; (b) a comparison value of the sensitivity index of R_0 when smoking is stopped by their own will (γ_1) and when smoking is stopped by using e-cigarettes (γ_2), that is, $\mathcal{S}_{\gamma_1} - \mathcal{S}_{\gamma_2}$.

parameter regarding the use of e-cigarettes and recovery rate. In Figure 5(a), the value of the sensitivity index of R_0 has both positive and negative signs in relation to two parameters that show a similar pattern. It is effective to both quitting smoking by their own will or using e-cigarettes to reduce smoking. In Figure 5(b), when comparing γ_1 and γ_2, as the value of sensitivity index of R_0 has only positive values, the influence of γ_1 turns out to be stronger. Quitting smoking by using e-cigarettes is less effective than quitting by their own will.

Actually, other parameter sets which used the values in Table 1 and Figure 6 are shown to confirm the trend as the rate of using e-cigarettes (g) changes. When use of e-cigarettes increases, we can show that the ratio of smokers gradually decreases. In this part, we will consider only the tendency of smokers and determine how e-cigarettes affect smokers.

The effect of e-cigarettes on the ratio of potential smokers, quitters, and nonsmokers can be seen in Figure 7 when used according to the recovery rate of e-cigarettes. Figure 7(a) shows the high and low recovery rate γ_2 of e-cigarettes; the ratio of potential smokers increased more rapidly when the recovery rate was higher than when the recovery rate was lower. On the contrary, in Figure 7(b), according to the rate of use of e-cigarettes, the ratio of quitters decreased when the recovery rate was higher than when the recovery rate was lower. That is, the ratio of potential smokers and quitters may have contradicting results depending on the recovery rate.

Biologically, we can think of the rate of recovery as information for e-cigarette efficacy because there can be a variety of results in places where there is an advertisement by the government, publicity of e-cigarette companies, information by a medical institution, and so on. Therefore, when the effect is large, the usage is increased, that is, the ratio of people using e-cigarettes is increased and the ratio of people who smoke is decreased; smokers are less encouraged smoking to potential smokers, which increases potential smokers. However, although e-cigarettes are used

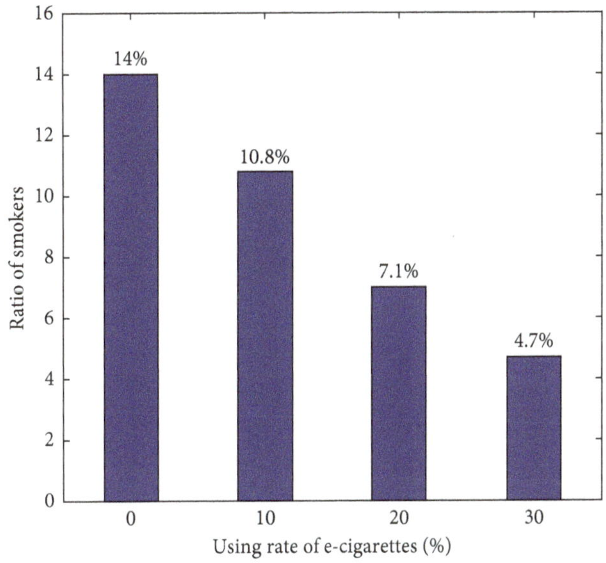

FIGURE 6: As the using e-cigarettes increases, the ratio of smokers decreases.

to stop smoking, the number of cases returning to smoking without being able to overcome the temptation to smoke is increased. That is why quitters are increased. Therefore, when considering two aspects, the number of smokers decreases and the number of the combined potential smokers and quitters increases. Using e-cigarettes may help smokers quit smoking, but a lot of e-cigarettes do not help to quit smoking easily, and quitting e-cigarettes may itself be difficult.

Figure 7(c) represents the sum of Figures 7(a) and 7(b) as a ratio of nonsmokers. From this point of view, we can see how the ratio of nonsmokers varies depending on the rate of use of e-cigarettes. It can be seen that the ratio of nonsmokers decreases evenly as the ratio of using e-cigarettes increases when the recovery rate is remarkably lower. However, when the recovery rate is higher than a certain

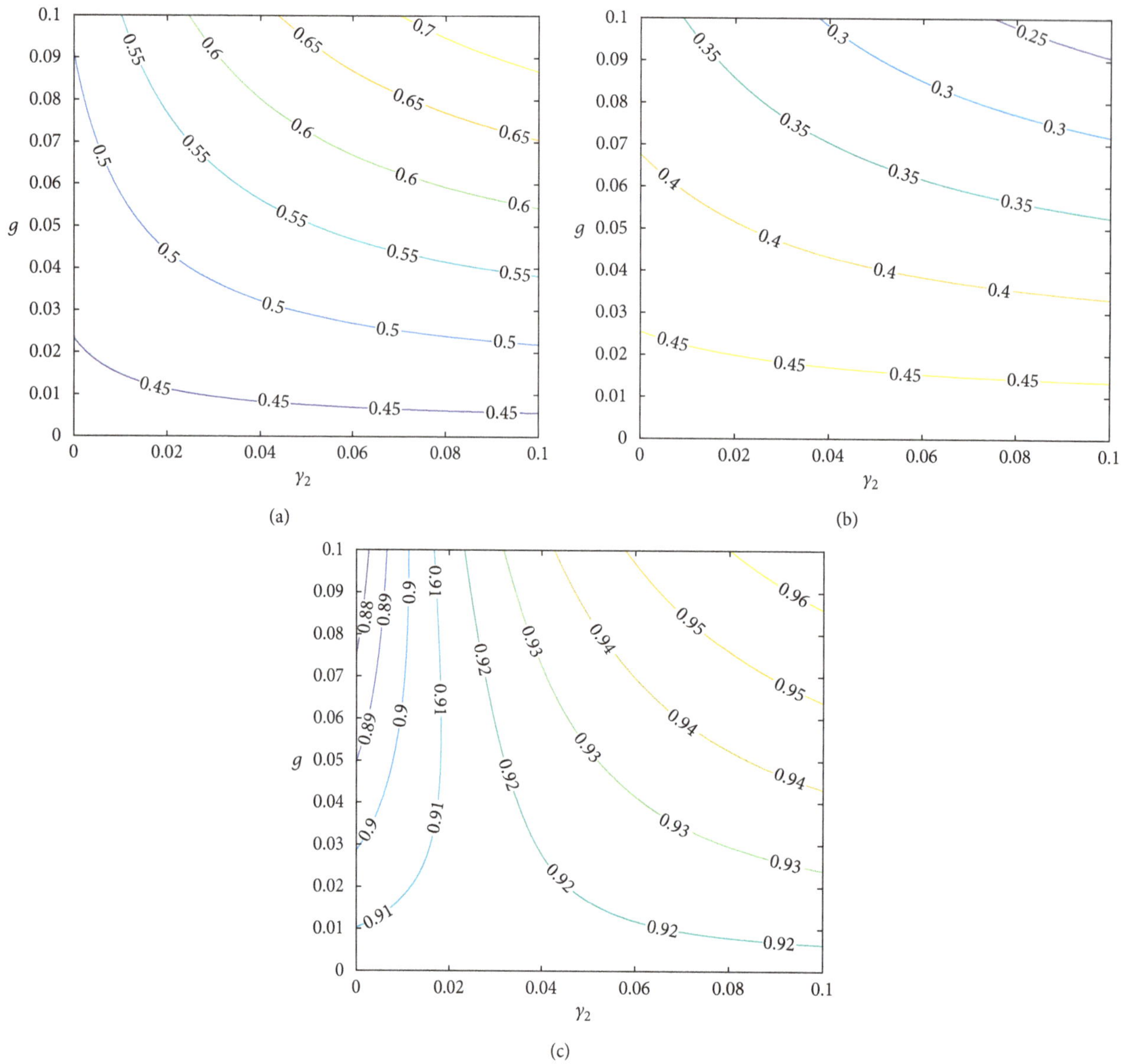

FIGURE 7: The impact of e-cigarettes on people who do not smoke: (a) the ratio of potential smokers, (b) the ratio of quitters, and (c) the ratio of nonsmokers, that is, potential smokers + quitters.

level, the ratio of nonsmokers increases as the use of e-cigarettes increases.

4. Conclusion

In this paper, our main interest was to analyze the effect of e-cigarettes on smoking cessation. We established a mathematical model for smoking cessation with e-cigarettes and presented the theoretical study for the effect of e-cigarettes. The observed real data of the ratio of potential smokers, smokers, the use of e-cigarettes, and quitters are obtained, and the parameters are estimated by the least-square method.

As our results demonstrate, the proposed model can support the reason for modeling because it follows current trends and there are few errors in this estimation. In epidemiology, one of the most important parts, basic reproduction number, R_0 is obtained by the next generation method. From this, we observed that when $R_0 < 1$, the smoking-free equilibrium e_0 is locally asymptotically stable, whereas when $R_0 > 1$, the smoking-present equilibrium e_* is locally asymptotically stable. This means that, if we control the smoking for public health, then the number of smokers in a region can persist if the value of R_0 is greater than unity, or else its users diminish.

The bifurcation analysis presents forward bifurcation at $R_0 = 1$, namely, smoking control may be carried out by reducing β_1 below the transcritical value β_1^* in order to make $R_0 < 1$. On the contrary, we should check the backward bifurcation scenario in other conditions where $\gamma_1 > g$ and $\beta_2 > \beta_1$, because smoking has two biggest factors, which are the risk of peer influence and relapse. If smokers stop smoking on their own, rather than using e-cigarettes, there may be a possibility that they will again start smoking if they are at places where there are a lot of smokers. But in the opposite case, if they are in a place where there are not many smokers, they may possibly stop smoking gradually without using e-cigarettes. For example, if smokers go back to smoking when they stop smoking on their own will, they think that the smokers will be able to quit smoking again by themselves. The sensitivity analysis about each parameter can show the effect for R_0, which checks the sign of the sensitivity index. In order to show the effect of e-cigarettes, especially, we consider the ratio of smokers using e-cigarettes and the ratio of non-smokers by the parameters γ_2 and g. From the perspective of the reducing number of smokers, using e-cigarettes can reduce the ratio of smokers, regardless of the recovery rate γ_2. On the contrary, in regard to the increasing number of nonsmokers, potential smokers and quitters tend to be the opposite when the use of e-cigarettes increases.

In conclusion, we verified the efficacy of e-cigarettes from a theoretical standpoint by constructing a mathematical model for various scenarios that could not be captured by statistical or experimental methods. Our results revealed that meaningful differences in the level of use and recovery rate of e-cigarettes change within a group, and we could explain its utility. We consider that this fact may be due to addiction to e-cigarettes. Therefore, the proper use of e-cigarettes can reduce the craving for tobacco cigarettes and may be an effective way to postpone smoking. However, smokers and e-cigarette smokers should never to be overlooked for a possibility of addiction and relapse. Using e-cigarettes may somewhat diminish the numbers of smokers, but it does not increase the number of quitters significantly. Therefore, people who know the seriousness of smoking should prevent the people around from trying to smoke and raise awareness about the adverse impacts of smoking on health, which are pivotal for prevention of smoking.

Conflicts of Interest

The authors declare that they have no conflicts of interest.

Acknowledgments

This research was supported by the Basic Science Research Program through the National Research Foundation of Korea (NRF) funded by the Ministry of Education (NRF-2015R1D1A1A02062131).

References

[1] World Health Organization, "Who global report on mortality attributable to tobacco," in *WHO Global Report on Mortality Attributable to Tobacco*, WHO, Geneva, Switzerland, 2012.

[2] D. J Nutt, L. D Phillips, D. Balfour et al., "E-cigarettes are less harmful than smoking," *The Lancet*, vol. 387, no. 10024, pp. 1160–1162, 2016.

[3] S. Feirman, E. Donaldson, J. Pearson et al., "Mathematical modelling in tobacco control research: protocol for a systematic review," *BMJ Open*, vol. 5, no. 4, article e007269, 2015.

[4] S. C. Walley and B. P. Jenssen, "Electronic nicotine delivery systems," *Pediatrics*, vol. 136, no. 5, pp. 1018–1026, 2015.

[5] M. B. Siegel, K. L. Tanwar, and K. S Wood, "Electronic cigarettes as a smoking-cessation tool: results from an online survey," *American Journal of Preventive Medicine*, vol. 40, no. 4, pp. 472–475, 2011.

[6] C. Bullen, C. Howe, M. Laugesen et al., "Electronic cigarettes for smoking cessation: a randomised controlled trial," *The Lancet*, vol. 382, no. 9905, pp. 1629–1637, 2013.

[7] R. A. Grana, L. Popova, and P. M. Ling, "A longitudinal analysis of electronic cigarette use and smoking cessation," *JAMA Internal Medicine*, vol. 174, no. 5, pp. 812-813, 2014.

[8] O. Sharomi and A. B. Gumel, "Curtailing smoking dynamics: a mathematical modeling approach," *Applied Mathematics and Computation*, vol. 195, no. 2, pp. 475–499, 2008.

[9] Z. Alkhudhari, S. Al-Sheikh, and S. Al-Tuwairqi, "Stability analysis of a giving up smoking model," *International Journal of Applied Mathematical Research*, vol. 3, no. 2, pp. 168–177, 2014.

[10] Z. Alkhudhari, S. Al-Sheikh, and S. Al-Tuwairqi, "Global dynamics of a mathematical model on smoking," *ISRN Applied Mathematics*, vol. 2014, Article ID 847075, 7 pages, 2014.

[11] A. Zeb, G. Zaman, and S. Momani, "Square-root dynamics of a giving up smoking model," *Applied Mathematical Modelling*, vol. 37, no. 7, pp. 5326–5334, 2013.

[12] A. Sharma and A. K. Misra, "Backward bifurcation in a smoking cessation model with media campaigns," *Applied Mathematical Modelling*, vol. 39, no. 3, pp. 1087–1098, 2015.

[13] L. Pang, Z. Zhao, S. Liu, and X. Zhang, "A mathematical model approach for tobacco control in China," *Applied Mathematics and Computation*, vol. 259, pp. 497–509, 2015.

[14] G. Zaman, "Qualitative behavior of giving up smoking models," *Bulletin of the Malaysian Mathematical Sciences Society*, vol. 34, no. 2, pp. 403–415, 2011.

[15] G. Zaman, Y. Han Kang, and I. H. Jung, "Stability analysis and optimal vaccination of an SIR epidemic model," *Biosystems*, vol. 93, no. 3, pp. 240–249, 2008.

[16] C. Castillo-Garsow, G. Jordan-Salivia, A. Rodriguez-Herrera et al., "Mathematical models for the dynamics of tobacco use, recovery, and relapse," *Public Health*, vol. 84, no. 4, pp. 543–547, 1997.

[17] US Department of Health and Human Services, Surgeon General, *The Health Consequences of Smoking 50 Years of Progress: A Report of the Surgeon General*, US Department of Health and Human Services, Citeseer, Washington, DC, USA, 2014.

[18] G. A Giovino, "Epidemiology of tobacco use in the united states," *Oncogene*, vol. 21, no. 48, pp. 7326–7340, 2002.

[19] B. A. King, R. Patel, K. H. Nguyen, and S. R. Dube, "Trends in awareness and use of electronic cigarettes among us adults, 2010–2013," *Nicotine and Tobacco Research*, vol. 17, no. 2, pp. 219–227, 2014.

[20] National Foreign Assessment Center (US) and United States. Central Intelligence Agency, *The World Factbook*, Central Intelligence Agency, Langley, VA, USA, 1982.

[21] P. Van den Driessche and J. Watmough, "Further notes on the basic reproduction number," in *Mathematical Epidemiology*, pp. 159–178, Springer, Berlin, Germany, 2008.

[22] B. Buonomo and D. Lacitignola, "On the backward bifurcation of a vaccination model with nonlinear incidence," *Nonlinear Analysis: Modelling and Control*, vol. 16, no. 1, pp. 30–46, 2011.

[23] C. Castillo-Chavez and B. Song, "Dynamical models of tuberculosis and their applications," *Mathematical Biosciences and Engineering*, vol. 1, no. 2, pp. 361–404, 2004.

[24] M.-H. Kim and S. Min, "Craving and treatment in smoking," *Journal of the Korean Medical Association*, vol. 54, no. 4, p. 401, 2011.

11

Pulmonary Nodule Recognition Based on Multiple Kernel Learning Support Vector Machine-PSO

Yang Li,[1,2] Zhichuan Zhu,[1,3] Alin Hou,[2] Qingdong Zhao,[2] Liwei Liu,[2] and Lijuan Zhang[2]

[1]*School of Mathematics and Statistics, Northeast Normal University, Changchun, Jilin 130024, China*
[2]*School of Computer Science and Engineering, Changchun University of Technology, Jilin 130012, China*
[3]*Faculty of Statistics, Jilin University of Finance and Economics, Changchun, Jilin 130117, China*

Correspondence should be addressed to Alin Hou; alinhou@163.com

Academic Editor: Nadia A. Chuzhanova

Pulmonary nodule recognition is the core module of lung CAD. The Support Vector Machine (SVM) algorithm has been widely used in pulmonary nodule recognition, and the algorithm of Multiple Kernel Learning Support Vector Machine (MKL-SVM) has achieved good results therein. Based on grid search, however, the MKL-SVM algorithm needs long optimization time in course of parameter optimization; also its identification accuracy depends on the fineness of grid. In the paper, swarm intelligence is introduced and the Particle Swarm Optimization (PSO) is combined with MKL-SVM algorithm to be MKL-SVM-PSO algorithm so as to realize global optimization of parameters rapidly. In order to obtain the global optimal solution, different inertia weights such as constant inertia weight, linear inertia weight, and nonlinear inertia weight are applied to pulmonary nodules recognition. The experimental results show that the model training time of the proposed MKL-SVM-PSO algorithm is only 1/7 of the training time of the MKL-SVM grid search algorithm, achieving better recognition effect. Moreover, Euclidean norm of normalized error vector is proposed to measure the proximity between the average fitness curve and the optimal fitness curve after convergence. Through statistical analysis of the average of 20 times operation results with different inertial weights, it can be seen that the dynamic inertial weight is superior to the constant inertia weight in the MKL-SVM-PSO algorithm. In the dynamic inertial weight algorithm, the parameter optimization time of nonlinear inertia weight is shorter; the average fitness value after convergence is much closer to the optimal fitness value, which is better than the linear inertial weight. Besides, a better nonlinear inertial weight is verified.

1. Introduction

The number of deaths from lung cancer is as high as 137 million annually around the world, accounting for 18% of cancer mortality ratio. Early surgical treatment is the most effective treatment for lung cancer, but most patients are diagnosed in the late stage of the disease. In 2015, the European Academy of Imaging and the European Respiratory Society published the latest white paper on lung cancer screening in European Respiratory Journal (ERJ) to guide clinical lung cancer screening for early detection and early treatment of lung cancer.

As early representation form of lung cancer in the lung CT image, a pulmonary nodule is defined as a nearly spherical opacity with a diameter smaller than 3 cm. Computed Tomography (CT) technology is an important means of early detection of pulmonary nodules disease. According to the CT characterization, pulmonary nodules can be divided into solid nodules (such as solitary pulmonary nodules, pulmonary wall adhesion nodules, and vascular adhesion nodules), ground glass nodules, and cavitary nodules.

Computer-Aided Detection (CAD) system of lung is one of the applications of machine vision; it can reduce overload visual fatigue of the radiologist and decrease the possibility of the resulting miscarriage or omission and also provide auxiliary diagnosis results for the doctor as "third party." Usually, the lung CAD system includes the following modules: acquisition of the lung CT image data, preprocessing of CT image, lung parenchyma segmentation, detection of VOI (Volume of Interest) or ROI (Region of

Interest) in candidate nodules images (mainly refers to the extraction or segmentation), calculation and selection of ROI or VOI features, and recognition of pulmonary nodules, where pulmonary nodules recognition is the core module of the CAD system. The algorithm of Support Vector Machine (SVM) has been widely used in the detection and recognition of pulmonary nodules (see, e.g., [1–15]). Among them, Li et al. [1] applied mixed kernel SVM algorithm to distinguish benign and malignant lung nodules, making the recognition accuracy (ACC) reach 92% and the sensitivity index reach 92.59%; Wang et al. [2] detected lung lesions by use of three-dimensional SVM with Latent Variable algorithm. Furthermore, Demir and Çamurcu [16] and Chang et al. [17] imported the algorithm of Particle Swarm Optimization (PSO) into SVM and selected the optimal parameter group by PSO and then used SVM for identification. In addition, Ma et al. [18] adopted the method of multiple classifiers fusion for lung disease identification.

The Multiple Kernel Learning Support Vector Machine (MKL-SVM) algorithm has achieved good recognition accuracy not just in recognition of lung nodules in [1] but also in other application fields (see [19, 20]). In [21], the Multiple Kernel Learning (MKL) method was elaborated and the latest research progresses were presented in this field. However, the MKL-SVM algorithm involves a large number of parameters, and the selection of parameters will have an important impact on the actual results. In [1], the selection of the optimal parameters is obtained by the grid search algorithm. The advantage of the grid search algorithm is the easiness to get the global optimal solution in the case of dense mesh division, but the disadvantage of the method is that it has a large amount of computation, a long time to search, and a poor real-time performance, which is not easy to form online identification algorithm. The selection of the relevant parameters is an urgent problem to be solved in the MKL-SVM, and the Particle Swarm Optimization (PSO) algorithm based on swarm intelligence algorithm provides an idea to solve the problem.

In this paper, the PSO algorithm and MKL-SVM algorithm are combined to realize the parameter optimization of the MKL-SVM. On this basis, the PSO algorithm with different inertia weights was compared and analyzed in order to obtain the parametric array similar or superior to that of the grid search algorithm aiming at quickly searching the optimal parametric array and the reasonable inertia weight and then precise identification of the pulmonary nodules.

2. Multiple Kernel Learning Support Vector Machine (MKL-SVM)

2.1. Support Vector Machine. SVM is a learning method using small amount of samples, which can be applied to predict or classify unknown samples by structural risk minimization. The training sample is represented as follows:

$$T = \{(x_i, y_i)\}, \quad i = 1, 2, \ldots, l, \tag{1}$$

where l is the number of training samples, x_i denotes the input vector of SVM, corresponding to the feature of the above N-dimension region of interest (ROI), $x_i \in R^N$; $y_i \in \{-1, +1\}$ indicates category label; here, $y_i = 1$ corresponds to nodules and $y_i = -1$ corresponds to nonnodules.

When SVM is used in the two classification problems, the original model can be written as the following nonlinear optimization problem:

$$\begin{aligned} \min_{w,b,\xi} \quad & \frac{1}{2}\|w\|^2 + C\sum_{i=1}^{l}\xi_i \\ \text{s.t.} \quad & y_i\left((w \bullet \Phi(x_i)) + b\right) \geq 1 - \xi_i, \quad i = 1, 2, \ldots, l \\ & \xi_i \geq 0, \; i = 1, 2, \ldots, l, \end{aligned} \tag{2}$$

where w is the weight vector and b is the threshold, and the aim of SVM is to maximize the classification interval $2/\|w\|$, that is, minimization of $\|w\|^2$. C is the regularization coefficient or penalty parameter, which describes the degree of penalty for misclassification samples. The greater C is, the more obvious the penalty for misclassification is. When the data cannot be completely separated, the maximum interval will be negative, thus introducing slack variables ξ_i which can measure the distance between the actual output y_i and the Support Vector Machine output.

In the feature space, SVM is used to map the input data (X_i) into a high-dimensional feature space Z by nonlinear transformation $\Phi(X)$, and then the optimal classification hyperplane is constructed in the high-dimensional feature space Z to realize the SVM. In the process of constructing the hyperplane in the feature space Z, the training algorithm uses the dot product and the kernel function $K(x_i, x_j)$ to represent the inner products $\Phi(x_i)$ and $\Phi(x_j)$; that is, a function K can be found to form the next formula:

$$K\left(x_i, x_j\right) = \Phi\left(x_i\right) \cdot \Phi\left(x_j\right). \tag{3}$$

Thus, it is not necessary to construct and solve the convex quadratic programming problem for a given training sample, and the problem is transformed into the following optimization problem by using Lagrange multiplier:

$$\begin{aligned} \min_{\alpha} \quad & \frac{1}{2}\sum_{i-1}^{l}\sum_{j=1}^{l} y_i y_j K\left(x_i, x_j\right)\alpha_i\alpha_j - \sum_{j=1}^{l}\alpha_j \\ \text{s.t.} \quad & \sum_{i=1}^{l} y_i\alpha_i = 0, \\ & 0 \leq \alpha_i \leq C, \; i = 1, 2, \ldots, l. \end{aligned} \tag{4}$$

The offset b in (2) can be solved by means of the following equation:

$$b = y_j - \sum_{i=1}^{l} y_i\alpha_i K\left(x_i, x_j\right). \tag{5}$$

The decision function is constructed as follows:

$$f(x) = \operatorname{sgn}\left(g(x)\right), \tag{6}$$

where

$$g(x) = \sum_{i=1}^{l} \alpha_i y_i K\left(x_i, x_j\right) + b. \tag{7}$$

2.2. Multiple Kernel Learning SVM (MKL-SVM). Various kernel functions have diverse advantages. One of the keys to improve the performance of SVM is to design an appropriate kernel function for a given problem. The common basic kernel functions are polynomial kernel function and radial basis function (RBF), which are presented, respectively, as follows:

$$K_{\text{poly}}(x,y) = (x^t y + 1)^d$$
$$K_{\text{rbf}}(x,y) = \exp\left(-\frac{\|x-y\|^2}{2g^2}\right), \tag{8}$$

where the parameter d represents the order of the polynomial kernel, the parameter g denotes the width of the RBF kernel, and d and g need to be given in advance.

Proposition 1. *The convex combination form of the kernel function is still a kernel function:*

$$K_{\text{mix}}(x,x') = \sum_{p=1}^{U} m_p K_p(x,x'), \tag{9}$$

where

$$\sum_{p=1}^{U} m_p = 1, \quad 0 < m_p < 0, \; p = 1,\ldots,U, \tag{10}$$

and K_p is the pth species of basic kernel function and m_p corresponds to the weights of the pth species of basic kernel function in the total multiple kernel function. U species basic kernel functions are used in the multiple kernel function, and the weight sum of various basic kernel functions is one so as to limit the weight proportion of various basic kernel functions in the multiple kernel functions in proportion.

Proof. Let $\{x_1, x_2, \ldots, x_l\}$ be a set of l points in any given R^N; we just need to prove that the Gram matrix in (9) is positive semidefinite matrix.

Let $K_1, K_2, \ldots, K_P$ be the Gram matrix of $K_1(x,x')$, $K_2(x,x'), \ldots, K_P(x,x')$ for $\{x_1, x_2, \ldots, x_l\}$; for any $\alpha \in R^l$, we obtain

$$\begin{aligned}\alpha^T K_{\text{mix}}(x,x')\alpha &= \alpha^T\left(\sum_{p=1}^{U} m_p K_p(x,x')\right)\alpha \\ &= \alpha^T\left(m_1 K_1(x,x') + m_2 K_2(x,x') + \cdots \right. \\ &\quad \left. + m_U K_U(x,x')\right)\alpha = m_1 \alpha^T K_1(x,x')\alpha \\ &\quad + m_2 \alpha^T K_2(x,x')\alpha + \cdots + m_U \alpha^T K_U(x,x')\alpha \ge 0.\end{aligned} \tag{11}$$

So $K_{\text{mix}}(x,x') = \sum_{p=1}^{U} m_p K_p(x,x')$ is positive semidefinite matrix; that is, $K_{\text{mix}}(x,x')$ is a kernel function, and the evidence is proven. □

It is proven that the kernel function expressed by (9) satisfies the Mercer condition and can be used for the training and classification of SVM. By using the above MKL-SVM, we can use nonlinear transformation of the sample points to get the corresponding kernel matrix so as to obtain the classification results during the training of SVM classifier.

RBF kernel has a strong ability to learn, and polynomial kernel has strong generalization ability; thus the combination of the two can take into account the ability of both learning and generalization. If we use only two kinds of basic kernel functions of both RBF kernel and polynomial kernel, that is, $U = 2$, $K_1 = K_{\text{poly}}$, and $K_2 = K_{\text{rbf}}$, the multiple kernel function of (12) is able to be formed. Compared with the single kernel function, we need to estimate a set of kernel parameters and a weight coefficient m. The weight coefficient m can regulate freely the proportion of different kernel functions mixed in multiple kernels, adjust flexibly the ability of learning and generalization, and make the results unbiased towards the promotion of a particular target.

$$K(x,x') = mK_{\text{poly}}(x,x') + (1-m)K_{\text{rbf}}(x,x'). \tag{12}$$

In [1], the grid search algorithm in the sense of CV is used to find the optimal regularization coefficient C, the order d of the polynomial kernel, the kernel width g of the RBF kernel, and the weight coefficient m of the multiple kernels. The optimal parameter group can be obtained by the grid search algorithm during the CV process corresponding to the highest classification accuracy. Lots of parameters and short step length of mesh induce large amount of calculation and long running time. The global optimal solution could be found by the heuristic algorithm, not needing to traverse all the parameter points in the grid.

2.3. MKL-SVM Based on Modified Particle Swarm Optimization Algorithm. Particle Swarm Optimization (PSO) is a typical heuristic algorithm on the basis of swarm intelligence optimization theory. In 1955, PSO was first proposed by Kennedy and Eberhart in [22], whose basic principle was originated from the research on the predation behavior of artificial life and birds. When birds prey upon food, the simplest and most effective method of finding food is to search the current area around the food nearest to birds. Compared with Generic Algorithm (GA), PSO searches the optimal particles by tracking the particles in the solution space without selection, crossover, and mutation.

It is assumed that the population $X = (X_1, X_2, \ldots, X_n)$ consists of n particles in a D-dimensional search space, where X_i represents the position of the ith particle in D-dimensional search space and also is a candidate solution of problem denoted by a vector of D dimensions as $X_i = (X_{i1}, X_{i2}, \ldots, X_{iD})^T$. According to the objective function, we can calculate the fitness value of each particle position X_i. The speed of the ith particle is $V_i = (V_{i1}, V_{i2}, \ldots, V_{iD})^T$, and its individual extreme value and group extreme value are $P_i = (P_{i1}, P_{i2}, \ldots, P_{iD})^T$ and $P_g = (P_{g1}, P_{g2}, \ldots, P_{gD})^T$, respectively. During each iteration, the particle updates its velocity and position by the individual extrema and the group extrema, which are given, respectively, as follows:

$$V_{id}^{k+1} = \omega V_{id}^{k} + c_1 r_1 \left(P_{id}^{k} - X_{id}^{k}\right) + c_2 r_2 \left(P_{gd}^{k} - X_{id}^{k}\right),$$
$$X_{id}^{k+1} = X_{id}^{k} + V_{id}^{k+1}, \tag{13}$$

where ω is an inertia weight; $d = 1, 2, \ldots, D$; D represents the number of parameters to be searched; k is the number of the present iterations; V_{id} is the velocity of particles, c_1 and c_2 are acceleration factors, which are nonnegative constants, and r_1 and r_2 are random numbers distributed within the interval $[0, 1]$. In order to prevent the blind search of particles, the position and velocity are usually limited to the range of $[-X_{\max}, X_{\max}]$ and $[-V_{\max}, V_{\max}]$.

The PSO algorithm is applied into MKL-SVM algorithm of (12). Because the corresponding order of polynomial kernel is defined as positive integer for $d \geqslant 2$ and with the increase of D, generalization ability of polynomial kernel decreases gradually, so only the two values $d = 2$ and $d = 3$ were calculated, and there is no need to search other parameters. Here the dimension of the search space of the particle is set to $D = 3$; $X_i = (X_{i1}, X_{i2}, X_{i3})^T$ represents the solution of the ith particles, where X_{i1}, X_{i2}, and X_{i3} of each dimension are corresponding to the regularization coefficient C, the kernel width g of RBF, and the multiple kernel weight m to be searched, respectively.

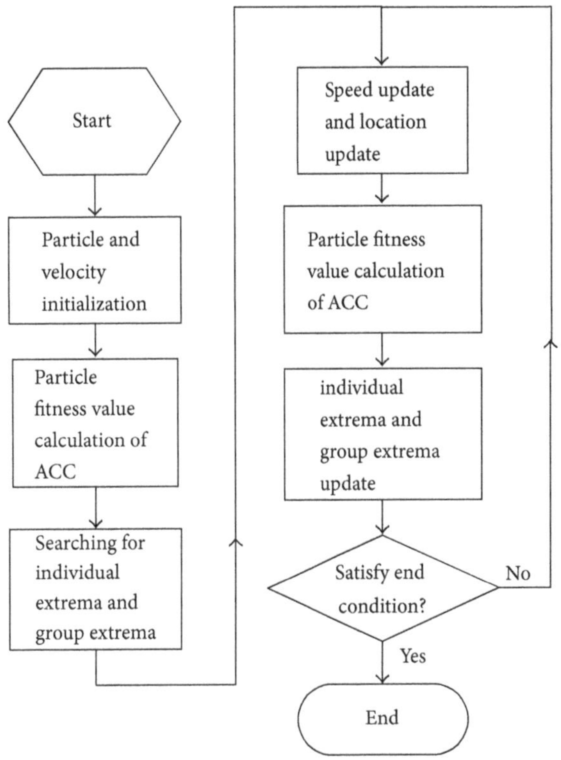

Figure 1: The diagram of MKL-SVM-PSO algorithm.

3. Application of MKL-SVM-PSO Algorithm in Pulmonary Nodule Recognition

After introducing the classic PSO algorithm, the recognition accuracy rate (ACC) of pulmonary nodules in the sense of CV is regarded as the final target and determined as the fitness function value of PSO, and then ACC is defined as follows:

$$\text{ACC} = \frac{(\text{TP} + \text{TN})}{\text{TP} + \text{TN} + \text{FP} + \text{FN}}, \tag{14}$$

where TP denotes the detected true positive nodule; FP is the detected false positive nodule; FN indicates the undetected false negative nodule; TN is the detected true negative nodule, that is, nonnodule. ACC measures total recognition accuracy to measure the actual detection rate of pulmonary nodules; the SEN is defined as follows:

$$\text{SEN} = \frac{\text{TP}}{(\text{TP} + \text{FN})}. \tag{15}$$

The parameter optimization algorithm of MKL-SVM-PSO algorithm is described in Figure 1.

4. Experimental Results and Analysis

4.1. The Experimental Data and the Results Analysis of MKL-SVM-PSO Algorithm. The experimental data were collected from 20 groups from a third-grade class A hospital in Jilin province with a total of about 700 images, and each group was diagnosed with the diagnostic criteria of doctor. The size of each CT image was 512 × 512, and the slice thickness is 5.0 mm. 270 ROI are extracted: 80 pulmonary nodules and 190 false positive nodules. After feature selection, the data samples were randomly divided into two groups: the training group including 170 samples and the testing group including 100 samples.

The simulation experiments are carried out using the platform MATLAB with libsvm toolbox. In the process of model parameter optimization, 5-fold cross-validation is used to obtain the optimal parameter set corresponding to the highest ACC. Let the number n of cluster particles be 20, that is, $n = 20$ in the MKL-SVM-PSO algorithm, and the dimension D of each particle be 3, that is, $D = 3$, and then the maximum number of iterations maxgen of the algorithm is set to be 200, and the inertia weight is constant, that is, $\omega = 1$. The simulation experiment will show the convergence performance through the optimal fitness curve and the average fitness curve as in Figure 2.

The optimal individual fitness curve of MKL-SVM-PSO algorithm is obtained as shown in Figure 2. The most optimal individual fitness value, namely, the nodule recognition accuracy, is 94.1176% through 5-fold cross-validation in training set; and the correspondent optimal particle position is $C = 29.7267$, $g = 19.0653$, and $d = 2$, $m = 0.8$, respectively. In this case, the running time of the proposed MKL-SVM-PSO algorithm is 363.4640 s, less than 2815.0786 s compared with 3178.5426 s of the grid search algorithm [1], and it accounts for just 11.43% of the grid search algorithm running time. Applying the proposed method to the test set, the test results of ACC reached 91% and those of SEN reached 88.89%.

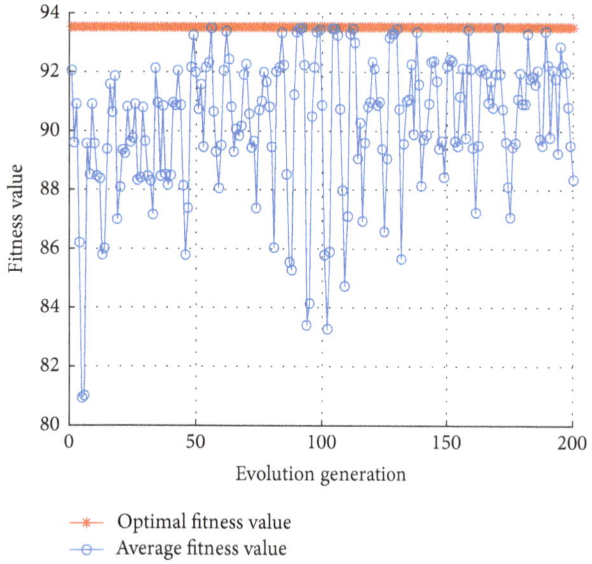

FIGURE 2: The fitness curve of the MKL-SVM-PSO algorithm with constant inertia weight.

4.2. *The Influence and Comparison of Different Inertia Weight on Lung Nodule Recognition.*

Compared with the grid search algorithm, the computational time of MKL-SVM-PSO algorithm requires shorter time, but as the iteration times gradually increase, the shock amplitude of average fitness value in each generation is more severe, and a certain gap exists with the optimal fitness value, which can be found in Figure 2.

In [23], the inertia weight ω was first introduced into PSO algorithm, and the larger inertia weight value that was conducive to the global search was pointed out, and the smaller inertia weight that was more conducive to the local search was also presented, and, moreover, the inertia weight that can reflect the ability of the particle to inherit the previous speed was also discussed. In order to balance the global search ability and local search ability of the algorithm, the linear decreasing inertia weight (LDIW) method proposed by Shi and Eberhart [23] will be used to reassign ω as follows:

$$\omega(k) = \frac{\omega_{\text{start}}(\omega_{\text{start}} - \omega_{\text{end}})(T_{\max} - k)}{T_{\max}}, \quad (16)$$

where ω_{start} is the initial inertia weight; ω_{end} is the inertia weight of the iteration to the maximum number of times; k is the number of the current iteration generations; $T_{\max}$ is the maximal iteration number. In order to ensure that the above iteration algorithm not only has better global search ability in initial phase but also has strong local search ability to obtain the optimal solution in later iterations, commonly let $\omega_{\text{start}} = 0.9$ and $\omega_{\text{end}} = 0.4$; the inertia weight decreases linearly from initial 0.9 to 0.4, and this is also an empirical approach. The fitness curve of the optimal parameters obtained by the MKL-SVM-PSO algorithm is shown in Figure 3.

It can be seen from Figure 3 that the shock amplitude is reduced by using the inertia weight of (16) rather than using constant inertia weight, close to the optimal solution in the early stage. Besides, commonly used linear inertia weight is shown in the following equation:

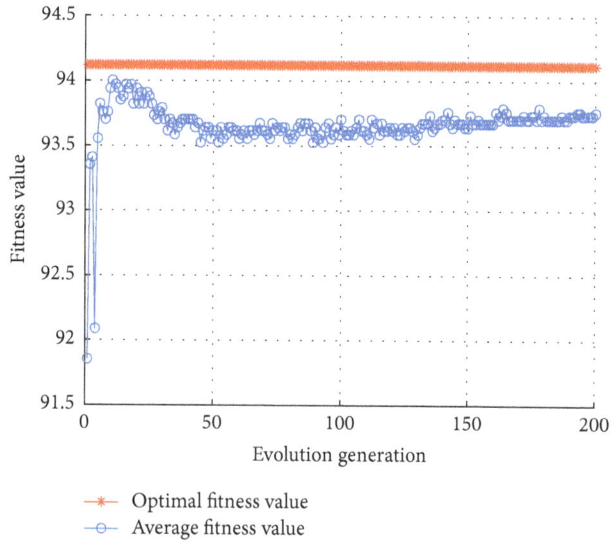

FIGURE 3: The fitness curve of MKL-SVM-PSO algorithm of (16).

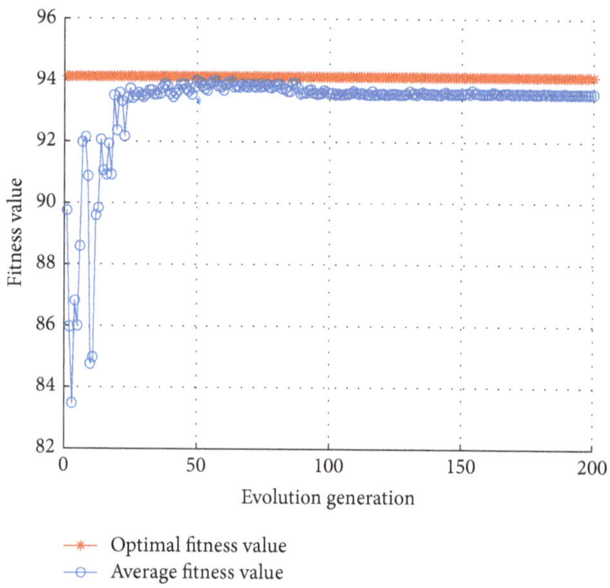

FIGURE 4: The fitness curve using the MKL-SVM-PSO algorithm of (17).

$$\omega(k) = \omega_{\text{end}} + (\omega_{\text{start}} - \omega_{\text{end}})\left(\frac{T_{\max} - k}{T_{\max}}\right). \quad (17)$$

Figure 4 shows the fitness curve corresponding to (17). In Figure 4, the shock amplitude induced by inertia weight of (17) is slightly larger, but it quickly converges to the optimal individual fitness value. The linear inertia weight represented by (16) and (17) can make the average fitness curve smooth, though it is easy to fall into local optimum in the early stage.

In order to ensure obtaining the global optimal solution, the following three kinds of nonlinear inertia weight are adopted to control convergence precision and convergence

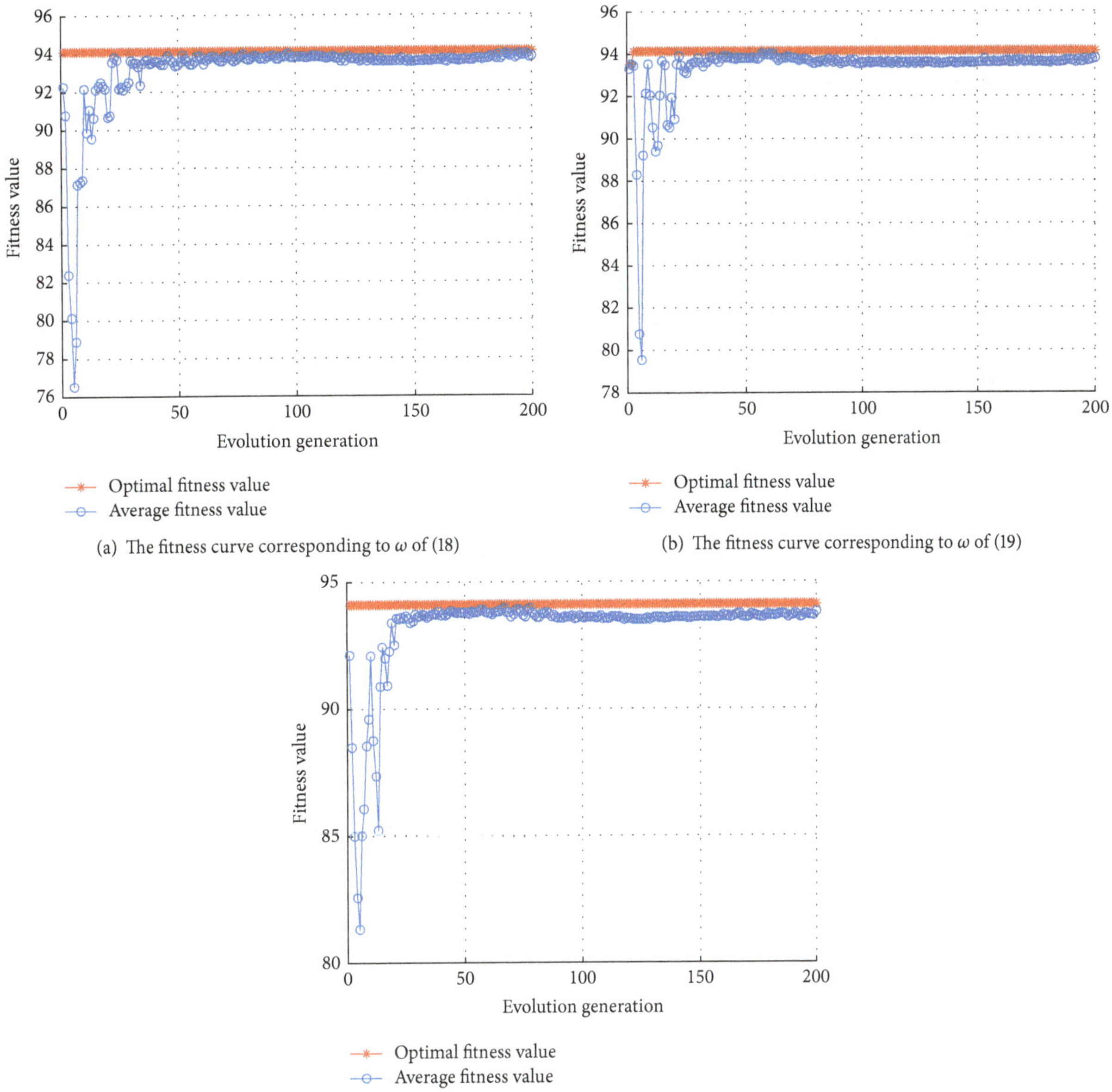

(a) The fitness curve corresponding to ω of (18)

(b) The fitness curve corresponding to ω of (19)

(c) The fitness curve corresponding to ω of (20)

FIGURE 5: The fitness curve of optimal parameters group searching by MKL-SVM-PSO algorithm.

speed, so that the average fitness values reach the best fitness value index quickly and smoothly.

$$\omega(k) = \omega_{\text{start}} - (\omega_{\text{start}} - \omega_{\text{end}}) \left(\frac{k}{T_{\max}} \right)^2 \quad (18)$$

$$\omega(k) = \omega_{\text{start}} - (\omega_{\text{start}} - \omega_{\text{end}}) \left(\frac{2k}{T_{\max}} - \left(\frac{k}{T_{\max}} \right)^2 \right) \quad (19)$$

$$\omega(k) = \omega_{\text{end}} \left(\frac{\omega_{\text{start}}}{\omega_{\text{end}}} \right)^{1/(1+ck/T_{\max})}. \quad (20)$$

Using the nonlinear parameter of the above three methods, respectively, the fitness function curve represents total recognition accuracy of pulmonary nodules as shown in Figure 5.

In order to compare the influence of different kinds of inertia weight on the parameter optimization, Figure 6 describes the curve of five kinds of dynamic weight corresponding to (16)–(20) changing along with the number of iterations. In the early stage of iteration, the larger inertia weight can make the algorithm maintain a strong global search ability, and the small inertia weight can make the algorithm search precisely in the late stage of iteration. As we know from the variation curve of several dynamic weights, the dynamic weight of (18) changes slowly in the early stage, and the value is larger so as to maintain the global search ability of the proposed algorithm; moreover, the dynamic weights change rapidly in the late stage and improve greatly the local searching ability of the algorithm; furthermore, the parameter optimization has also got a good result with the

TABLE 1: Comparison of various indexes in parameter optimization stage of different algorithms.

	Different inertial weight algorithm	Average parameter optimization time (s)	Average optimal fitness value	Average ACC value obtained from test set	Average SEN value obtained from test set
1	The constant is 1	370.7950	94.1176%	90.45%	86.85%
2	(16)	462.4134	94.1176%	91%	88.89%
3	(17)	457.0022	94.1176%	91%	88.89%
4	(18)	416.0204	94.1176%	91%	88.89%
5	(19)	448.1536	94.1176%	91%	88.89%
6	(20)	450.4456	94.1176%	91%	88.89%
7	Grid search algorithm	3096.1427	94.1176%	91%	88.89%

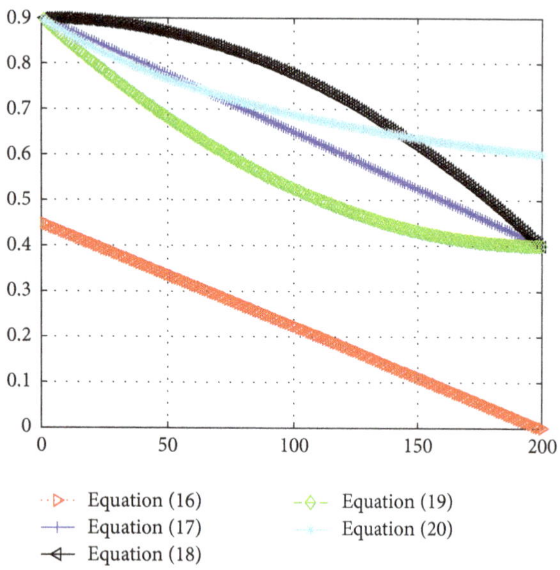

FIGURE 6: The variation curves of several dynamic inertia weights.

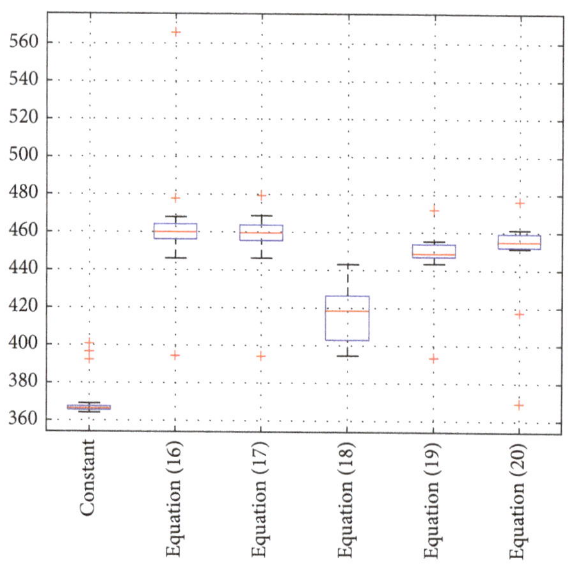

FIGURE 7: The statistical boxplot of parameter optimization time.

corresponding fitness curve, which is the optimal way of dynamic weights.

In summary, the Particle Swarm Optimization algorithm with constant weight has a fast convergence speed, but in the later stages it is easy to fall into local optimal solution with little accuracy. The linear inertia weights of (16) and (17) are easy to fall into local optimum. If we adopt several dynamic nonlinear inertia weight methods from (18) to (20), the algorithm converges slowly in initial stage, but in the later period local search ability is enhanced, which makes the algorithm jump out of local optimum and get the global optimal solution, so as to improve the accuracy of the algorithm. The form in (18) is the optimal nonlinear inertia weight.

In order to compare several kinds of Particle Swarm Optimization algorithms with different inertial weights and the parameters optimization time and recognition results of grid search algorithm, each algorithm is operated 20 times, and the average results of 20 times are listed in Table 1.

From the experimental results in Table 1, it can be seen that the parameter optimization time of the MKL-SVM grid search algorithm is the longest, which is almost 7 times of that of the MKL-SVM-PSO algorithm, and the running speed is much slower than that of the MKL-SVM-PSO grid search algorithm. In the MKL-SVM-PSO algorithm, the inertia weight is, respectively, set to a constant value, a linear dynamic weight, and a nonlinear dynamic weight, and the parameter optimization time has a little difference as well. When the inertia weight is set to a constant value, the average running time is the shortest, and the optimum fitness is 94.1176% in the training stage, but its generalization ability is not as good as the experimental result obtained by dynamically searching the inertia weight. The average ACC value obtained from the test set is 90.45% and the average SEN is only 86.85%. The optimal fitness value is always 94.1176% during each operation process in the training stage using dynamic inertia weight from (16) to (20). The same results of every test are obtained on the test set. The ACC value is 91% and the SEN is 88.89%; only the training times are different. Among them, the average running time of (18) is the shortest, that is, 416.0204 s, which is the best among all the algorithms. The nonlinear inertial weight algorithm is faster than the linear inertial weight algorithm, which should be due to the fast convergence of the dynamic inertial weight algorithm.

Figure 7 shows the 20 times statistical results corresponding to parameter optimization times. The statistical values of the boxplot are listed in Table 2.

TABLE 2: Statistical indexes corresponding to Figure 7.

	Different inertial weight algorithm	Upper adjacent(s)	Lower adjacent(s)	Median value(s)	Number of outliers
1	The constant is 1	369.049	364.151	366.343	3
2	(16)	468.018	446.098	459.983	3
3	(17)	468.646	446.225	459.6345	2
4	(18)	443.129	394.638	148.49	0
5	(19)	455.38	443.456	448.798	2
6	(20)	461.342	451.524	455.122	3

From the data in Figure 7 and Table 2, it can be seen that when the inertia weight is constant, the box is located at the bottom, and the three outliers are also very low, and also the training time is the shortest. Combined with Figure 2, the convergence rate of the algorithm is much slow, and no obvious generation number of convergence can be found. From the test results of test set in Table 1, the generalization ability is not as good as that of the dynamic inertial weight algorithm, so it is not the optimal one. When dynamic inertial weight algorithm is adopted, the boxplot of the model training time corresponding to the three nonlinear inertial weight algorithms is located under the boxplot corresponding to the linear inertial weight. It means that the training time (i.e., parameter optimization time) of nonlinear inertial weight is less than that of linear inertial weight, and the parameter optimization time obtained by (18) is the shortest and there are no outliers in it. The upper adjacent was 443.129 s, the lower adjacent was 394.638 s, and the median value was 418.49 s, respectively. The lower adjacent optimization time of the inertia weight obtained by (19) is 443.456 s, which is close to the upper adjacent value of the box obtained by (18). Furthermore, the training time of the former method is second only to that of (18). The position of box obtained by (20) is superior to that of linear inertial weight, but it is not good at nonlinear weight. The positions of the two boxes obtained by the data of both (16) and (17) are very close. The upper adjacent and lower adjacent of the box corresponding to (17) are all higher than those of (16), but the median value of (17) box is lower than that of (16) box. Therefore, the parameter optimization time of (18) is optimal.

Since the value of each inertial weight is different, the maximum number of convergence generations is different in various algorithms. Therefore, it is not reasonable to compare the Euclidean norm error between the optimal fitness value and the average fitness value after convergence, because it is difficult to express the merits and demerits of each algorithm. In order to reasonably express the Euclidean distance between the optimal fitness curve and the average fitness curve after convergence, we define the normalized Euclidean norm error as follows:

$$F_{gy} = \left\| V_{\text{best}}(t:\max\text{gen}) - V_{\text{Average}}(t:\max\text{gen}) \right\|_2 = \frac{\sqrt{(V_{\text{best}}(t) - V_{\text{Ave}}(t))^2 + (V_{\text{best}}(t+1) - V_{\text{Ave}}(t+1))^2 + \cdots + (V_{\text{best}}(\max\text{gen}) - V_{\text{Ave}}(\max\text{gen}))^2}}{\max\text{gen} - t + 1}, \tag{21}$$

where the average fitness curve converges from the tth generation. V_{best} and V_{ave} are both maxgen dimensional vectors and represent the optimal and average fitness value, respectively. Here the normalized Euclidean norm error is defined as F_{gy} and used to express the Euclidean distance between the average fitness value and the optimal fitness value of each generation after convergence.

The average values of the indexes obtained after 20 operation times are shown in Table 3. We compared and analyzed several key parameters such as the convergence generation number of the average fitness curve, the normalized Euclidean distance between the average fitness value and the optimal fitness value after reaching the convergence generation number, the median value, the maximum, the mean value, and total Euclidean norm of error vector with 200 generations in the average fitness curve. Each index is the statistical average of 20 operation results.

From above experimental results, it can be seen that when the inertia weight is constant, it is very difficult to find which generation of curves converges obviously in Table 3, so it is impossible to calculate the Euclidean norm of normalized error vector after reaching maximum number of convergence generations. The median and mean value of the average fitness curve are also the lowest among all the algorithms, which further indicates that the effect of the constant inertia weight is not as good as that of dynamic inertia weight. In the dynamic inertial weight algorithm, the maximum, mean, and median values of the average fitness curve are the highest in all the algorithms, and the Euclidean distance between the mean fitness curve and the optimal fitness curve is the smallest; that is, the least squares norm of the error vector is the smallest. The convergence is earlier, and the normalized Euclidean norm of the error vector is the smallest. However, there exists the particle premature phenomenon, so it is easy

Table 3: Comparison of various indexes under different inertia weights.

Different inertial weight algorithm	Maximum of V_{best}	Mean value of V_{ave}	Median value of V_{ave}	The Euclidean norm of the global error vector	Convergence generation number	The Euclidean norm of the normalized error vector after reaching the convergent generation number
The constant is 1	93.6044%	90.1465%	63.3698%	90.6721	—	—
(16)	94.0279%	93.6873%	93.7169%	8.4488	9	0.0294
(17)	93.9941%	93.2348%	93.6419%	28.2732	30	0.0348
(18)	94.0290%	93.2050%	93.6719%	31.5911	36	0.0320
(19)	94.0044%	93.2996%	93.6618%	27.2316	22	0.0324
(20)	93.9926%	93.2350%	93.6206%	28.1948	27	0.0341

to fall into the local optimal solution. It is further proven that the global search performance in the early stage is limited. The mean value of average fitness corresponding to nonlinear inertial weight of (18) is 93.2050%, and the maximum is 94.0290%; the median value is 93.6719%. The mean value is low, and also the convergence generation number is large, and the Euclidean norm of global error vector is large because of the sharp oscillation in the initial iteration. It also reflects that the global search ability of the algorithm in the initial iteration prevents fully the particle premature convergence. Thus, the three indexes, the maximum, the median value, and the Euclidean norm of the normalized error vector after reaching convergence, are able to reflect better convergence performance after convergence. The indexes obtained by (19) and (20) are basically superior to that of (17); that is, the nonlinear inertial weight algorithm is superior to the linear inertial weight algorithm generally.

In summary, the Particle Swarm Optimization algorithm with dynamic inertia weight is better than the one with constant inertia weight, and the algorithm using nonlinear inertia weight is better than that one using linear inertia weight. The MKL-SVM-PSO algorithm has gained good results by use of dynamic nonlinear inertial weight of (18) in this paper. The algorithm has the ability of global searching at the beginning of iteration. After reaching the convergence generation number, the average fitness value can approach the optimal fitness value more smoothly and quickly, which makes it easier to find the global optimal solution.

5. Conclusion

In this paper, a MKL-SVM-PSO algorithm with nonlinear inertial weight is proposed to search the optimal parameter set of hybrid kernel Support Vector Machine quickly and accurately and achieved better effects in pulmonary nodule recognition. The main innovative work goes as follows:

(1) The PSO algorithm is introduced into the mixture Kernels SVM algorithm and used for the discrimination of benign and malignant pulmonary nodules.

(2) On the basis of changing dynamic weights, the similarities and differences between linear weights and nonlinear weights are discussed, and the optimal dynamic nonlinear weights are obtained. The average fitness value of the algorithm is close to the optimal fitness value quickly and smoothly, so that the global optimal solution is easy to be obtained.

(3) The Euclidean norm index of normalized error vector is proposed to measure the difference between the optimal fitness curve and the average fitness curve after convergence with different inertial weights. The index solves the problem that different convergence generations of different algorithms result in different dimensions of error vectors in various algorithms, and it is difficult to compare the discrepancy. The validity of dynamic inertial weight algorithm is verified from the point of view of statistics.

The experimental results show that the model training time of MKL-SVM-PSO algorithm is only 1/7 of the training time of MKL-SVM grid search algorithm with better recognition effect. It can be seen that the dynamic inertia weight is better than constant inertia weight in the MKL-SVM-PSO algorithm from Table 3. Compared with the linear inertial weight algorithm, the parameter optimization speed of nonlinear inertial weight algorithm is rapid, and the average fitness value after convergence is much closer to the optimal fitness value. The dynamic inertial weight corresponding to (18) is the optimal method in this paper.

Although ACC, as a fitness value, has obtained good experimental results in this method, medical attention is often paid to the SEN index to prevent missed detection. Our next job is to extend the proposed MKL-SVM-PSO algorithm to multitarget search in order to achieve accurate identification and nonmissed detection of pulmonary nodules.

Conflicts of Interest

The authors declare that there are no conflicts of interest regarding the publication of this paper.

Acknowledgments

The authors gratefully acknowledge the support of the China Postdoctoral Science Foundation (no. 2016M591468), Jilin Province Postdoctoral Science Foundation (111900358), the Jilin Provincial Science and Technology Development Plan funded project (20170520050JH), the Education Department of Jilin Province (no. JJKH20170575KJ and no. JJKH20181041KJ), Jilin Province Postdoctoral Science Foundation, and the Key Program of the National Natural Science Foundation of China (nos. NSFC11690012 and NSFC11631003).

References

[1] Y. Li, D. Wen, K. Wang, and A. Hou, "Mixed kernel function SVM for pulmonary nodule recognition," *Lecture Notes in Computer Science*, vol. 8157, no. 2, pp. 449–458, 2013.

[2] Q. Wang, W. Kang, C. Wu, and B. Wang, "Computer-aided detection of lung nodules by SVM based on 3D matrix patterns," *Clinical Imaging*, vol. 37, no. 1, pp. 62–69, 2013.

[3] M. Javaid, M. Javid, M. Z. U. Rehman, and S. I. A. Shah, "A novel approach to CAD system for the detection of lung nodules in CT images," *Computer Methods and Programs in Biomedicine*, vol. 135, pp. 125–139, 2016.

[4] S. Shen, A. A. T. Bui, J. Cong, and W. Hsu, "An automated lung segmentation approach using bidirectional chain codes to improve nodule detection accuracy," *Computers in Biology and Medicine*, vol. 57, pp. 139–149, 2015.

[5] O. H. Madero, O. O. Vergara Villegas, V. G. Cruz Sanchez, D. H. J. Ochoa, and A. M. J. Nandayapa, "Automated system for lung nodules classification based on wavelet feature descriptor and support vector machine," *BioMedical Engineering OnLine*, vol. 14, pp. 1–20, 2015.

[6] M. Firmino, G. Angelo, H. Morais, M. R. Dantas, and R. Valentim, "Computer-aided detection (CADe) and diagnosis (CADx) system for lung cancer with likelihood of malignancy," *Biomedical Engineering Online*, vol. 15, no. 1, article no. 2, pp. 1–17, 2016.

[7] Y. Sui, Y. Wei, and D. Zhao, "Computer-aided lung nodule recognition by SVM classifier based on combination of random undersampling and SMOTE," *Computational and Mathematical Methods in Medicine*, vol. 12, Article ID 368674, pp. 1–13, 2015.

[8] Z. Zhao, J. Chen, X. Yin, H. Song, X. Wang, and J. Wang, "Establishing assistant diagnosis models of solitary pulmonary nodules based on intelligent algorithms," *Cellular Physiology and Biochemistry*, vol. 35, no. 6, pp. 2463–2471, 2015.

[9] A. Hussain, "Intensity-based statistical features for classification of lungs CT scan nodules using artificial intelligence techniques," *Journal of Experimental and Theoretical Artificial Intelligence*, vol. 27, pp. 1–15, 2015.

[10] T. Manikandan and N. Bharathi, "Lung cancer detection using fuzzy auto-seed cluster means morphological segmentation and SVM classifier," *Journal of Medical Systems*, vol. 40, no. 7, article no. 181, 2016.

[11] B. R. Froz, A. O. D. C. Filho, A. C. Silva, A. C. D. Paiva, R. A. Nunes, and M. Gattass, "Lung nodule classification using artificial crawlers, directional texture and support vector machine," *Expert Systems with Applications*, vol. 69, pp. 176–188, 2017.

[12] S. A. Khan, K. Kenza, M. Nazir, and M. Usman, "Proficient lungs nodule detection and classification using machine learning techniques," *Journal of Intelligent & Fuzzy Systems: Applications in Engineering and Technology*, vol. 28, no. 2, pp. 905–917, 2015.

[13] T. Zhou, H. Lu, J. Zhang, and H. Shi, "Pulmonary nodule detection model based on SVM and CT image feature-level fusion with rough sets," *BioMed Research International*, vol. 2016, Article ID 8052436, pp. 1–13, 2016.

[14] M. Bergtholdt, R. Wiemker, and T. Klinder, "Pulmonary nodule detection using a cascaded SVM classifier," *Spie Medical Imaging*, vol. 9785, 2016.

[15] A. Sheeraz, J. M. Younus, A. M. Usman, Q. Usman, and H. Ali, "Pulmonary nodules detection and classification using hybrid features from computerized tomographic images," *Journal of Medical Imaging and Health Informatics*, vol. 6, no. 1, pp. 252–259, 2016.

[16] Ö. Demir and A. Y. Çamurcu, "Computer-aided detection of lung nodules using outer surface features," *Bio-Medical Materials and Engineering*, vol. 26, pp. S1213–S1222, 2015.

[17] B. M. Chang, H. H. Tsai, and C. Y. Yen, "SVM-PSO based rotation-invariant image texture classification in SVD and DWT domains," *Journal of medical imaging Health Informatics*, vol. 52, pp. 96–107, 2016.

[18] L. Ma, X. Liu, L. Song, C. Zhou, X. Zhao, and Y. Zhao, "A new classifier fusion method based on historical and on-line classification reliability for recognizing common CT imaging signs of lung diseases," *Computerized Medical Imaging and Graphics*, vol. 40, pp. 39–48, 2015.

[19] M. Li, X. Lu, X. Wang, S. Lu, and N. Zhong, "Biomedical classification application and parameters optimization of mixed kernel SVM based on the information entropy particle swarm optimization," *Computer Assisted Surgery*, vol. 21, pp. 132–141, 2016.

[20] S. Fei-Yan, T. Yu-Bo, and R. Zuo-Lin, "Modeling the resonant frequency of compact microstrip antenna by the PSO-based SVM with the hybrid kernel function," *International Journal of Numerical Modelling: Electronic Networks, Devices and Fields*, vol. 29, no. 6, pp. 1129–1139, 2016.

[21] S. S. Bucak, R. Jin, and A. K. Jain, "Multiple kernel learning for visual object recognition: a review," *IEEE Transactions on Pattern Analysis and Machine Intelligence*, vol. 36, no. 7, pp. 1354–1369, 2014.

[22] J. Kennedy and R. Eberhart, "Particle swarm optimization," in *Proceedings of the IEEE International Conference on Neural Networks*, pp. 1942–1948, Perth, Australia, December 1995.

[23] Y. Shi and R. C. Eberhart, "Empirical study of particle swarm optimization," *Congress on Evolutionary Computation*, vol. 3, pp. 1945–1950, 1999.

12

Maximin Efficiencies under Treatment-Dependent Costs and Outcome Variances for Parallel, AA/BB, and AB/BA Designs

Math J. J. M. Candel

Department Methodology and Statistics, CAPHRI Care and Public Health Research Institute, Maastricht University, Maastricht, Netherlands

Correspondence should be addressed to Math J. J. M. Candel; math.candel@maastrichtuniversity.nl

Academic Editor: Reinoud Maex

If there are no carryover effects, AB/BA crossover designs are more efficient than parallel (A/B) and extended parallel (AA/BB) group designs. This study extends these results in that (a) optimal instead of equal treatment allocation is examined, (b) allowance for treatment-dependent outcome variances is made, and (c) next to treatment effects, also treatment by period interaction effects are examined. Starting from a linear mixed model analysis, the optimal allocation requires knowledge on intraclass correlations in A and B, which typically is rather vague. To solve this, maximin versions of the designs are derived, which guarantee a power level across plausible ranges of the intraclass correlations at the lowest research costs. For the *treatment effect*, an extensive numerical evaluation shows that if the treatment costs of A and B are equal, or if the sum of the costs of one treatment and measurement per person is less than the remaining subject-specific costs (e.g., recruitment costs), the maximin crossover design is most efficient for ranges of intraclass correlations starting at 0.15 or higher. For other cost scenarios, the maximin parallel or extended parallel design can also become most efficient. For the *treatment by period interaction*, the maximin AA/BB design can be proven to be the most efficient. A simulation study supports these asymptotic results for small samples.

1. Introduction

The standard design of a randomized clinical trial is the parallel group design: subjects are randomly assigned to one of two treatments, say A or B. An alternative, well-known design is the AB/BA crossover trial in which subjects receive both treatments, A and B, but the sequencing of the treatments is opposite for two randomly allocated groups [1, 2]. An AB/BA crossover trial is considered most suited when examining treatments for chronic or ongoing diseases, such as rheumatism, chronic obstructive pulmonary disease, or (frequent) heartburn. In these cases, there is no real possibility that the disease gets cured, and the aim is to moderate the effects of the disease [2]. A third design that we will consider involves treatment sequences AA and BB. This design extends the parallel design across two treatment periods, allows for testing treatment by time interaction effects, and is a realistic alternative for the AB/BA design in case the treatment regime should continue.

If the outcome variable is continuous and (approximately) normally distributed, the data can be analyzed by mixed effects regression [3]. Of primary interest is testing the treatment effect of, for instance, a new medication for chronic obstructive pulmonary disease. A relevant issue then is which design is the most efficient in estimating the treatment effect, thereby yielding maximum power for testing this effect. Such optimality has already been examined when comparing crossover and parallel designs [2] and when comparing all three designs introduced before [4, 5]. If there are no carryover effects and no dropouts, the sample sizes are equal and equally allocated to the treatments, an AB/BA design yields more efficient estimates of the treatment effect than a parallel and extended parallel design and consequently, will yield more power to test this effect.

The present study extends results on the relative efficiencies of these designs in that (a) optimal instead of equal treatment allocation is examined, (b) allowance is made for treatment-dependent outcome variances, and (c) next to treatment effects, also treatment by period interaction effects are examined.

Outcome variances may differ between treatments [6, 7]. This also is to be expected if treatments differ in terms of their effectiveness. Furthermore, since research costs and outcome variances may differ between treatments, equal allocation to treatments may not be the most efficient. The issue then is how to allocate subjects to treatments such that a design's efficiency is optimized, and how different designs relate in terms of efficiency under such optimal allocation. Optimal allocation requires a priori knowledge on parameters of the analysis model, that is, intraclass correlations for the mixed effects model that we consider. Since this knowledge typically is rather vague, optimal allocations and corresponding efficiencies for maximin versions of the (extended) parallel design and crossover design will be derived. These maximin designs guarantee a power level across plausible ranges of the intraclass correlations at the lowest research costs.

In designs where treatments are successively given to the same group of subjects, carryover may occur. For the AB/BA trial, it may be that, in the AB sequence treatment, A still has an effect on the outcome, when B has been given and the second measurement is done. When in the BA sequence, the effect of B is present, once A has been administered and this effect differs from the carryover effect for the AB sequence, differential carryover occurs. The present study assumes that differential carryover can be safely excluded or is negligible and that this effect does not need to be estimated in analyzing the data.

The paper is structured as follows. Section 2 will present the linear mixed model for analyzing data from each of the three designs. Section 3 will introduce the efficiency criterion and will provide asymptotic expressions for this criterion in the case of maximum likelihood estimation of the treatment effect. Starting from a flexible cost function, optimal allocations to treatments will be derived as well as resulting design efficiencies. Since the efficiencies depend on the intraclass correlations and knowledge on these parameters is often limited, in Section 4, we will derive maximin designs. Section 5 will show to what extent the asymptotic efficiencies translate into desired power levels for small sample sizes. Section 6 will give an application of the results, and Section 7 will discuss some issues for further research.

2. Linear Mixed Effects Models

In the case of a parallel design, an extended parallel design, and a crossover design, the subjects are randomly allocated to one of the two arms. In a parallel design to treatment A or treatment B, in an extended parallel design, they are allocated to treatment sequence AA or BB, and in a crossover trial to treatment sequence AB or BA. We consider a quantitative outcome variable, denoted as y_{ij} for person j ($j = 1, \ldots, N$) at measurement occasion i, and assume y_{ij} is (approximately) normally distributed.

For a parallel design and outcome variances that differ between treatments A and B, simple linear regression with heterogeneous variances may be an adequate tool for data analysis:

$$y_{ij} = \beta_0 + \beta_1 \text{treat}_{ij} + \delta_{ij}^{\text{A}}\left(1 - \text{treat}_{ij}\right) + \delta_{ij}^{\text{B}}\text{treat}_{ij}, \qquad (1)$$

where treatment is coded 0 for persons having treatment A and coded 1 for persons having treatment B, and δ_{ij}^{A} and δ_{ij}^{B} are normally distributed, with mean 0 and variances σ_{A}^2 and σ_{B}^2, respectively. The random terms δ_{ij}^{A} and δ_{ij}^{B} can be thought of as consisting of a random person (between-subject) effect, u_{0j}, and a treatment-dependent random error (within-subject) effect, $\varepsilon_{ij}^{\text{A}}$ and $\varepsilon_{ij}^{\text{B}}$. In formula, $\delta_{ij}^{\text{A}} = u_{0j} + \varepsilon_{ij}^{\text{A}}$, and $\delta_{ij}^{\text{B}} = u_{0j} + \varepsilon_{ij}^{\text{B}}$. These two sources of random variation cannot be separated in a single-period parallel trial.

For a crossover AB/BA design and an extended, two-period, parallel AA/BB design, however, the variances of u_{0j} and of $\varepsilon_{ij}^{\text{A}}$ and $\varepsilon_{ij}^{\text{B}}$ can be identified. The linear regression model can then be extended with a random intercept as well as a fixed effect of time, yielding the following mixed effects model:

$$y_{ij} = \beta_0 + \beta_1 \text{treat}_{ij} + \beta_2 \text{time}_{ij} + u_{oj} + \varepsilon_{ij}^{\text{A}}\left(1 - \text{treat}_{ij}\right) + \varepsilon_{ij}^{\text{B}}\text{treat}_{ij}. \qquad (2)$$

In (2), time is coded 0 for observations at the first measurement and coded 1 for observations at the second measurement. The random terms u_{0j}, $\varepsilon_{ij}^{\text{A}}$, and $\varepsilon_{ij}^{\text{B}}$ are independently normally distributed, with mean 0 and variances σ_0^2, $\sigma_{\varepsilon\text{A}}^2$, and $\sigma_{\varepsilon\text{B}}^2$, respectively. Their relation with the variances in (1) is $\sigma_{\text{A}}^2 = \sigma_0^2 + \sigma_{\varepsilon\text{A}}^2$ for treatment A and $\sigma_{\text{B}}^2 = \sigma_0^2 + \sigma_{\varepsilon\text{B}}^2$ for treatment B.

In the case we want to examine whether there is an interaction between treatment and period, the model in (2) is extended as follows:

$$y_{ij} = \beta_0 + \beta_1 \text{treat}_{ij} + \beta_2 \text{time}_{ij} + \beta_3 \text{treat}_{ij} \times \text{time}_{ij} + u_{oj} + \varepsilon_{ij}^{\text{A}}\left(1 - \text{treat}_{ij}\right) + \varepsilon_{ij}^{\text{B}}\text{treat}_{ij}, \qquad (3)$$

where β_3 represents the treatment by period interaction effect. The parameters in (1)–(3) can be estimated through maximum likelihood (ML). In what follows, we are interested in optimally estimating β_1 in (1) and (2), which will be denoted as β_{treat}, and in optimally estimating β_3 in (3), which will be denoted as $\beta_{\text{treat}\times\text{time}}$. A relevant concept is the intraclass correlation, which is between-subject variation on the outcome as compared to the total outcome variation. For the models in (2) and (3),this can be expressed as $\rho_{\text{A}} = \sigma_0^2/(\sigma_0^2 + \sigma_{\varepsilon\text{A}}^2)$ and $\rho_{\text{B}} = \sigma_0^2/(\sigma_0^2 + \sigma_{\varepsilon\text{B}}^2)$ for treatments A and B, respectively. The larger the person (between-subject) variance as compared to the error (within-subject) variance, the larger the intraclass correlations. Note that we assume a common between-subject variance, but allow for treatment-dependent within-subject variances, leading to treatment-dependent within-subject correlations. We also define a variance ratio $\phi = (\sigma_0^2 + \sigma_{\varepsilon\text{A}}^2)/(\sigma_0^2 + \sigma_{\varepsilon\text{B}}^2) = \sigma_{\text{A}}^2/\sigma_{\text{B}}^2$, which can be expressed as a ratio of the intraclass correlations, $\phi = \rho_{\text{B}}/\rho_{\text{A}}$.

3. Optimal Allocations and Corresponding Design Efficiencies

Let $\text{Var}(\hat{\beta}_x \mid \xi)$ denote the variance of the estimator of the treatment effect β_1 in (1) or (2) or the variance of the treatment by the period interaction effect β_3 in (3), given a design ξ. The efficiency of an estimator of β_x is defined as the inverse of its variance, that is, $(\mathrm{Var}(\widehat{\beta}_x | \xi))^{-1}$. In the sequel, we will consider the efficiency of one design, ξ_1, versus another design, ξ_2, which is defined as $\mathrm{Var}(\widehat{\beta}_x | \xi_2)/\mathrm{Var}(\widehat{\beta}_x | \xi_1)$ and denoted as the relative efficiency. Since no closed-form expressions are available for the variances of the maximum likelihood (ML) estimator, asymptotic variances of the ML estimator were derived (Appendices A.1 and A.2).

The optimal allocation to treatments minimizes the variance of the estimator of β_{treat} in (1) or (2) and of $\beta_{\text{treat}\times\text{time}}$ in (3), given a fixed research budget. Note that changing the coding of the treatment factor or the time factor in (1)–(3), for instance into 1 versus −1 instead of 1 versus 0, will not affect the optimal allocation. Such a change of coding leads to a linear transformation of β_{treat} or $\beta_{\text{treat}\times\text{time}}$, and this will change the variance of their estimators only by a multiplicative constant. This implies that allocations that minimize the variance of the estimators do not depend on the coding of treatment and time.

To derive the optimal allocations under a budget restriction, we need to define a budget function. Let the costs involved with each subject in the parallel design be c_{sp} euros, in an extended parallel design be c_{sep} euros, and in a crossover design be c_{sc} euros. These costs may represent financial rewards given to subjects for participating in the trial but also the (average) costs of recruiting a subject. Furthermore, for treatments A and B there are, for each subject, costs c_{A} and c_{B}, respectively, and each measurement may involve c_{t} euros. Finally, attached to each treatment sequence, there may be administration costs c_{ts}.

In the case of allocation proportions p_{A} for treatment A and $p_{\text{B}} = 1 - p_{\text{A}}$ for treatment B in a parallel design having n_{p} subjects, the following budget C^* is required:

$$C^* = 2c_{\text{ts}} + n_{\text{p}}\left(c_{\text{sp}} + c_{\text{t}}\right) + p_{\text{A}}n_{\text{p}}c_{\text{A}} + \left(1 - p_{\text{A}}\right)n_{\text{p}}c_{\text{B}}. \quad (4)$$

For the designs that we consider, this budget function can be reparametrized such that it is the same as the cost function given by Yuan and Zhou [8], thereby generalizing the cost function proposed by Brown [9] and Berger and Wong [4].

For an AB/BA crossover design, involving n_{c} subjects and allocation proportions p_{AB} for treatment sequence AB and $p_{\text{BA}} = 1 - p_{\text{AB}}$ for treatment sequence BA, noting that each subject receives both treatment A and B and is measured twice, the following budget is required:

$$C^* = 2c_{\text{ts}} + n_{\text{c}}c_{\text{sc}} + n_{\text{c}}c_{\text{A}} + n_{\text{c}}c_{\text{B}} + 2n_{\text{c}}c_{\text{t}}. \quad (5)$$

Finally, the required budget for an AA/BB design, involving n_{ep} subjects, with allocation proportions p_{AA} and $p_{\text{BB}} = 1 - p_{\text{AA}}$ for the treatment sequences AA and BB, respectively, is as follows:

$$C^* = 2c_{\text{ts}} + n_{\text{ep}}c_{\text{sep}} + 2n_{\text{ep}}p_{\text{AA}}c_{\text{A}} + 2n_{\text{ep}}\left(1 - p_{\text{AA}}\right)c_{\text{B}} + 2n_{\text{ep}}c_{\text{t}}. \quad (6)$$

Note that, for the functions in (4)–(6), the budget may simply be the total number of observations involved in a study, by setting $c_{\text{t}} = 1$ and the other costs to 0. It can also represent the total number of subjects involved, by setting $c_{\text{sp}} = c_{\text{sep}} = c_{\text{sc}} = 1$ and the remaining costs to 0.

In what follows, we will assume that the subject-specific costs of the two-period designs are the same; that is, $c_{\text{sep}} = c_{\text{sc}} = c_{\text{s\_2p}}$. Since subjects in these designs receive two treatments and a washout period may be involved, these costs are very likely larger than those of a parallel design. We also assume that the subject-specific costs for the two-period designs will not exceed 2 times the subject-specific costs for the parallel design, so that $c_{\text{sp}} \le c_{\text{s\_2p}} \le 2c_{\text{sp}}$. Finally, since each design involves two treatment sequences, the administration costs are the same for each of the three designs considered, and thus, the budgets that are available for remaining costs are identical; that is, the budget $C = C^* - 2c_{\text{st}}$ is the same for each design.

3.1. Treatment Effect. For treatment effect estimation, the optimal allocations to the treatment sequences are derived in Appendix B. The optimal allocations and corresponding (asymptotic) variances of the treatment effect estimators are shown in the second and third column of Table 1, respectively. The optimal allocation ratios of the parallel and the extended parallel design depend on the costs and intraclass correlations: the more the expensive treatment A (or the cheaper treatment B) and the larger the intraclass correlation in treatment A (or the smaller the intraclass correlation in treatment B), the more the subjects have to be assigned to treatment B. The optimal allocation ratio for a crossover design is 1, which may be expected, since both groups receive both treatment A and B.

3.2. Treatment by Period Interaction Effect. In the case the treatment by period interaction effect is of primary interest, the optimal allocations can be derived along lines similar to the derivations for the treatment effect (Appendix B). The allocations and corresponding optimal variances are displayed in Table 1. Note that, similar to treatment effect estimation, the allocation ratio for a crossover design is 1, whereas the allocation ratio for an extended parallel design depends on the treatment costs and intraclass correlations, such that more persons are allocated to treatment sequence AA if the intraclass correlation of A decreases, the intraclass correlation of B increases, the costs of treatment A decrease, or the costs of treatment B increase.

4. Maximin Designs

Choosing the optimal allocation requires knowledge on the intraclass correlations ρ_{A} and ρ_{B} (remember that the variance ratio ϕ is fixed if ρ_{A} and ρ_{B} are given). Commonly, there is only limited knowledge on these parameters. A possible solution is the maximin strategy [4], consisting of 2 steps: (1) for each design determine the minimum efficiency of the effect estimator across the plausible ranges for the intraclass correlations ρ_{A} and ρ_{B} and (2) choose that design which maximizes this minimum efficiency. Such a design optimizes a worst case scenario and is called a *maximin design*. The maximin strategy implies choosing the design

TABLE 1: Optimal allocation ratios and corresponding variances of the ML estimator of the treatment effect and the treatment by period interaction for different designs under heterogeneity of outcome variances and costs.

Design	Allocation ratio (n_1/n_2)	$\mathrm{Var}(\widehat{\beta}_{\mathrm{treat}})$
	Treatment effect	
Crossover design	1	$(c_A + c_B + 2c_t + c_{s\_2p})(1 - ((2\rho_A\rho_B)/(\rho_A + \rho_B)))\sigma_y^2/C$
Parallel design	$\sqrt{(\rho_B/\rho_A)((c_B + c_t + c_{sp})/(c_A + c_t + c_{sp}))}$	$(\sqrt{(c_A + c_t + c_{sp})(\rho_B/(\rho_A + \rho_B))} + \sqrt{(c_B + c_t + c_{sp})(\rho_A/(\rho_A + \rho_B))})^2 (\sigma_y^2/C)$
Extended parallel design	$\sqrt{(\rho_B(1+\rho_A)/(\rho_A(1+\rho_B))) \times ((2(c_B + c_t) + c_{s\_2p})/(2(c_A + c_t) + c_{s\_2p}))}$	$(\sqrt{(2(c_A + c_t) + c_{s\_2p})((\rho_B(1+\rho_A))/(\rho_A + \rho_B))} + \sqrt{(2(c_B + c_t) + c_{s\_2p})((\rho_A(1+\rho_B))/(\rho_A + \rho_B))})^2 \sigma_y^2/(2C)$
	Treatment by period interaction	
Design	Allocation ratio (n_1/n_2)	$\mathrm{Var}(\widehat{\beta}_{\mathrm{treat\times time}})$
Crossover design	1	$4(c_A + c_B + 2c_t + c_{s\_2p})(1 + 2\rho_A\rho_B/(\rho_A + \rho_B))(\sigma_y^2/C)$
Extended parallel design	$\sqrt{(\rho_B(1-\rho_A)/(\rho_A(1-\rho_B))) \times (2(c_B + c_t) + c_{s\_2p})/(2(c_A + c_t) + c_{s\_2p})}$	$2(\sqrt{(2(c_A + c_t) + c_{s\_2p})((\rho_B(1-\rho_A))/(\rho_A + \rho_B))} + \sqrt{(2(c_B + c_t) + c_{s\_2p})((\rho_A(1-\rho_B))/(\rho_A + \rho_B))})^2 (\sigma_y^2/C)$

Note. n_1: sample size for A (parallel design), AB (crossover design), or AA (extended parallel design) sequence; n_2: sample size for B (parallel design), BA (crossover design), or BB (extended parallel design) sequence; $\sigma_y^2 = \sigma_A^2 + \sigma_B^2$.

that minimizes the maximum variance of the estimator of the effect of interest. In determining sample sizes, choosing values for the intraclass correlations ρ_A and ρ_B within their plausible ranges (and thus a variance ratio ϕ within its plausible range) for which the variance is maximum will guarantee the desired power level also for all other values of these parameters. Moreover, the maximin design guarantees this power level at the lowest research costs. In what follows, we will refer to ranges of ρ_A and ρ_B that have lower bounds ρ_A^L and ρ_B^L and upper bounds ρ_A^U and ρ_B^U, respectively.

4.1. Treatment Effect. From the asymptotic variances in Table 1, one can derive for which values of ρ_A and ρ_B (and thus for which value of the variance ratio ϕ), the variance of the treatment effect estimator is maximized. These derivations are given in Appendix C. The maximin parameter values and corresponding variances for the treatment effect estimator under optimal allocation to the treatments are shown in Table 2. The corresponding optimal allocations for the maximin designs are obtained by substituting the maximin parameter values of Table 2 into the allocation ratios as given in Table 1.

If for a parallel design the maximin value for the variance ratio $\phi^* = (c_A + c_t + c_{sp})/(c_B + c_t + c_{sp})$ is within the plausible range for ϕ, that is, $[\rho_B^L/\rho_A^U, \rho_B^U/\rho_A^L]$ and $c_{s\_2p} \leq 2c_{sp}$, then a parallel design is always less efficient than a maximin crossover design. If for an extended parallel design the maximin value for one of the intraclass correlations is within the plausible range for the corresponding intraclass correlation, then also this design is less efficient than a maximin crossover design. For other scenarios, the relations between the maximin designs are more complicated, depending on the ranges for ρ_A and ρ_B, the costs of treatments, subject recruitment, and measurement.

A systematic numerical evaluation was done to examine under what conditions the crossover design is the best choice in terms of efficiency. For ρ_A and ρ_B, we consider ranges of width 0.10 (small), 0.30 (medium), and 0.60 (large). The lower bounds were $\{0.01, 0.05, 0.10, 0.15, 0.20, \ldots\}$, where the largest possible lower bound was determined by the width of the range under consideration. For instance, if the range is 0.30 (medium), the largest lower bound for the intraclass correlation is 0.70. All combinations of small, medium, and large ranges for ρ_A and ρ_B were considered. The values of the variance ratio ϕ thus considered vary from 1/100 to 100. Since in most crossover trials, the intraclass correlation exceeds 0.30 [1–3, 10, 11], ranges with lower bounds of 0.30 or higher are empirically most relevant. The empirical evidence on the costs c_A, c_B, c_t, c_{sp}, and c_{s_2p} is scarce, and we thus choose costs covering a wide range of scenarios. Let $\mathrm{CR_A} = (c_A + c_t)/c_{sp}$, $\mathrm{CR_B} = (c_B + c_t)/c_{sp}$, and $\mathrm{CR_p} = c_{s\_2p}/c_{sp}$ (note that the relative efficiencies of the maximin designs depend only on these cost ratios). $\mathrm{CR_A}$ and $\mathrm{CR_B}$ take on the values 100, 20, 10, 1, 0.1, 0.05, and 0.01. For $\mathrm{CR_p}$, we consider 1 and 2.

If the costs of treatments are identical between the treatment arms, that is, $\mathrm{CR_A} = \mathrm{CR_B}$, for most scenarios examined, the crossover maximin design turns out to be most efficient. For $\mathrm{CR_p} = 1$ and $\mathrm{CR_A} = \mathrm{CR_B} \leq 1$, the crossover design always is the most efficient. For $\mathrm{CR_p} = 1$ and $\mathrm{CR_A} = \mathrm{CR_B} > 1$, or $\mathrm{CR_p} = 2$, only if the lower bound of one of the intraclass correlations is 0.05 or lower and the ranges of the intraclass correlations do not overlap, the parallel design can become most efficient. Since in most empirical studies the intraclass correlations will exceed 0.05, this implies that, for equal costs of treatments, the crossover maximin design will almost always be the most efficient design.

In the case the treatment costs differ and $\mathrm{CR_A} \leq 1$ and $\mathrm{CR_B} \leq 1$, only in the case the lower bound of one of the intraclass correlations is 0.10 or smaller, the parallel or the extended parallel maximin design can become most efficient. The extended parallel design can only become most efficient if $\mathrm{CR_p} = 1$. Hence, in all scenarios with unequal treatment costs and $\mathrm{CR_A} \leq 1$ and $\mathrm{CR_B} \leq 1$, for intraclass correlations of 0.15 or higher, the maximin crossover design is most efficient.

TABLE 2: Values for the parameters and variances of the treatment effect estimator for each of the maximin designs.

Design	Maximin parameter values	$\mathrm{Var}(\widehat{\beta}_{treat})$ for the maximin design
Crossover design	ρ_A^L, ρ_B^L	$(c_A + c_B + 2c_t + c_{s\_2p})(1 - 2\rho_A^L\rho_B^L/(\rho_A^L + \rho_B^L))(\sigma_y^2/C)$
Parallel design	If $(\rho_B^L/\rho_A^U) \le (c_A + c_t + c_{sp})/(c_B + c_t + c_{sp}) \le (\rho_B^U/\rho_A^L)$, choose $(\rho_B/\rho_A) = (c_A + c_t + c_{sp})/(c_B + c_t + c_{sp})$	$(c_A + c_B + 2(c_t + c_{sp}))(\sigma_y^2/C)$
	If $(c_A + c_t + c_{sp})/(c_B + c_t + c_{sp}) > (\rho_B^U/\rho_A^L)$, then ρ_A^L and ρ_B^U	$(\sqrt{(c_A + c_t + c_{sp})(\rho_B^U/(\rho_A^L + \rho_B^U))} + \sqrt{(c_B + c_t + c_{sp})(\rho_A^L/(\rho_A^L + \rho_B^U))})^2(\sigma_y^2/C)$
	If $(c_A + c_t + c_{sp})/(c_B + c_t + c_{sp}) < (\rho_B^L/\rho_A^U)$, then ρ_A^U and ρ_B^L	$(\sqrt{(c_A + c_t + c_{sp})(\rho_B^L/(\rho_A^U + \rho_B^L))} + \sqrt{(c_B + c_t + c_{sp})(\rho_A^U/(\rho_A^U + \rho_B^L))})^2(\sigma_y^2/C)$
Extended parallel design	If $(1/\rho_A^U) + 1 \le ((\lambda(1-\rho_B^U)^2)/((1+\rho_B^U)\rho_B^U)) \le (1/\rho_A^L) + 1$, then $\rho_A^* = (\rho_B^U(1+\rho_B^U))/(\lambda(1-\rho_B^U)^2 - \rho_B^U(1+\rho_B^U)), \rho_B^U$	$(c_A + c_B((1+\rho_B^U)/(1-\rho_B^U)) + ((2c_t + c_{s\_2p})/(1-\rho_B^U)))(\sigma_y^2/C)$
	else if $(1/\rho_B^U) + 1 \le (((1-\rho_A^U)^2)/(\lambda(1+\rho_A^U)\rho_A^U)) \le (1/\rho_B^L) + 1$, then $\rho_A^U, \rho_B^* = (\lambda\rho_A^U(1+\rho_A^U))/((1-\rho_A^U)^2 - \lambda\rho_A^U(1+\rho_A^U))$,	$(c_A((1+\rho_A^U)/(1-\rho_A^U)) + c_B + ((2c_t + c_{s\_2p})/(1-\rho_A^U)))(\sigma_y^2/C)$
	else if $(\lambda(1-\rho_B^U)^2)/((1+\rho_B^U)\rho_B^U) < (1/\rho_A^U) + 1$, and $(1-\rho_A^U)^2/(\lambda(1+\rho_A^U)\rho_A^U) < (1/\rho_B^U) + 1$, then ρ_A^U, ρ_B^U	$(\sqrt{(2(c_A + c_t) + c_{s\_2p})(\rho_B^U(1+\rho_A^U)/(\rho_A^U + \rho_B^U))} + \sqrt{(2(c_B + c_t) + c_{s\_2p})(\rho_A^U(1+\rho_B^U)/(\rho_A^U + \rho_B^U))})^2(\sigma_y^2/(2C))$
	else if $((\lambda(1-\rho_B^U)^2)/((1+\rho_B^U)\rho_B^U)) < (1/\rho_A^U) + 1$, then ρ_A^U, ρ_B^L	$(\sqrt{(2(c_A + c_t) + c_{s\_2p})(\rho_B^L(1+\rho_A^U)/(\rho_A^U + \rho_B^L))} + \sqrt{(2(c_B + c_t) + c_{s\_2p})(\rho_A^U(1+\rho_B^L)/(\rho_A^U + \rho_B^L))})^2(\sigma_y^2/(2C))$
	else ρ_A^L, ρ_B^U	$(\sqrt{(2(c_A + c_t) + c_{s\_2p})(\rho_B^U(1+\rho_A^L)/(\rho_A^L + \rho_B^U))} + \sqrt{(2(c_B + c_t) + c_{s\_2p})(\rho_A^L(1+\rho_B^U)/(\rho_A^L + \rho_B^U))})^2(\sigma_y^2/(2C))$

Note. $\sigma_y^2 = \sigma_A^2 + \sigma_B^2$; $\lambda = (2c_A + 2c_t + c_{s\_2p})/(2c_B + 2c_t + c_{s\_2p})$.

In the case the treatment costs differ and $CR_A > 1$ or $CR_B > 1$, the maximin crossover design is less often most efficient. For these cost scenarios, also for ranges of intraclass correlations exceeding 0.15, the maximin parallel and extended parallel design may become more efficient. This especially occurs if the costs of treatment A and lower bound of the range of ρ_A are both larger (or smaller) than the costs of treatment B and lower bound of ρ_B, respectively. The efficiency improvement is large if treatment A is much more expensive than treatment B and if the costs of treatments and measurements are large compared to the subject-related costs. This is illustrated in Figure 1. The top row shows that if the costs of treatment A are larger than the costs of treatment B and the lower bound of ρ_A is larger than the lower bound of ρ_B, a parallel design is most efficient, even up to an upper bound 1 of ρ_B if $CR_A = 100$. As can also be seen, the upper bound of ρ_A is not very relevant in terms of the relative efficiencies. The left plot of the middle row of Figure 1 shows that if the lower bounds of ρ_A and ρ_B are equal, then for almost all upper bounds of ρ_B, the crossover design is most efficient. Again, as can be seen in the rightmost plot of the middle row, if the lower bound of ρ_A is higher than the lower bound of ρ_B, then for higher upper bounds of ρ_B, the parallel design is most efficient but to a lesser extent as compared to a smaller lower bound of ρ_B. As is evident from the four subplots in the top and middle row, when increasing the ratio CR_A/CR_B, the crossover design becomes less efficient as compared to the other two designs. The subplots of the bottom row furthermore show that the crossover design also becomes less efficient compared to the other designs if CR_A and CR_B increase while the ratio CR_A/CR_B remains constant. This illustrates that the efficiency of the other designs relative to the crossover design becomes larger if the costs of treatments and measurements are large compared to the subject-related costs. However, to summarize, if the treatment costs differ and $CR_A > 1$ or $CR_B > 1$, no simple rules of the thumb emerge and the most solid way to choose the most efficient design is just to calculate the maximin variances as given in Table 2.

Finally, if $CR_p = 2$, the maximin parallel design is consistently more efficient than the maximin extended parallel design (as is illustrated in Figure 1). If $CR_p = 1$, the maximin extended parallel design can also become more efficient than the maximin parallel design.

4.2. Treatment by Period Interaction Effect. The maximin parameter values and corresponding variances of the estimator of the treatment by period interaction effect are shown in Table 3. The derivations of these results can be done along lines similar to the derivations for the treatment effect estimator (Appendix C). The optimal allocation for the extended parallel design is obtained by substituting the maximin parameter values in the expression for the allocation ratio in Table 1. For a crossover design, the allocation ratio is 1. The maximin efficiency of an extended parallel design is always higher than that of a crossover design if the maximin value ρ_A^* is within the plausible range for ρ_A. This follows from

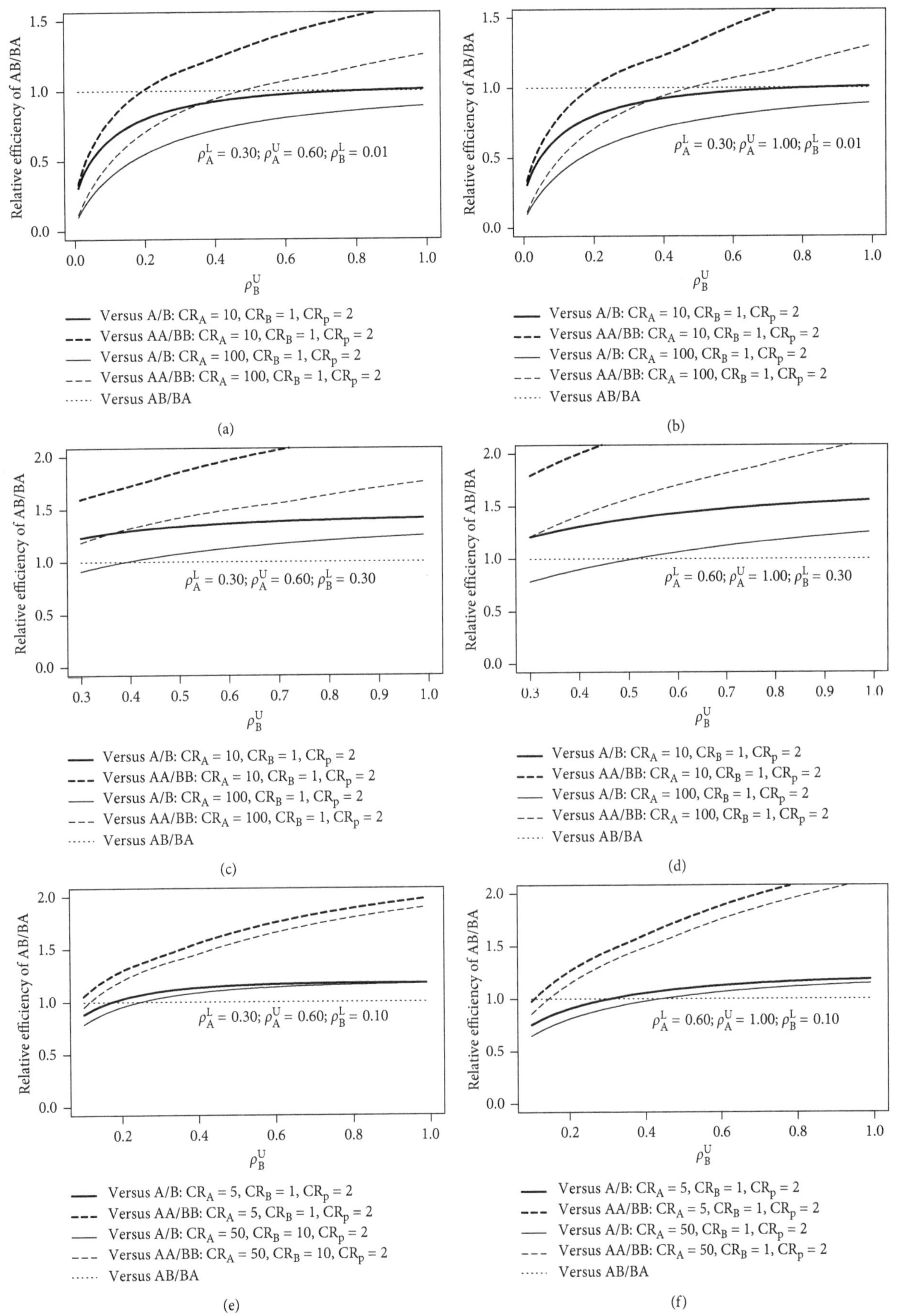

FIGURE 1: Relative efficiency for the treatment estimator of the AB/BA design versus the A/B and AA/BB designs as a function of the ranges for the intraclass correlations in treatment A (lower bound = ρ_A^L, upper bound = ρ_A^U) and B (lower bound = ρ_B^L, upper bound = ρ_B^U), in case of minimizing the research costs.

TABLE 3: Values for the parameters and variances of the treatment by period interaction effect estimator for each of the maximin designs.

Design	Maximin parameter values	$\mathrm{Var}(\widehat{\beta}_{\text{treat}\times\text{time}})$
Crossover design	ρ_A^U, ρ_B^U	$(4(c_A + c_B + 2c_t + c_{s\_2p})(1 + ((2\rho_A^U\rho_B^U)/(\rho_A^U + \rho_B^U))))(\sigma_y^2/C)$
Extended parallel design	If $(1/\rho_A^U) - 1 \le ((\lambda(1+\rho_B^L)^2)/((1-\rho_B^L)\rho_B^L)) \le (1/\rho_A^L) - 1$, then $\rho_A^* = (\rho_B^L(1-\rho_B^L))/(\lambda(1+\rho_B^L)^2 + \rho_B^L(1-\rho_B^L)), \rho_B^L$,	$((4(c_A(1+\rho_B^L) + c_B(1-\rho_B^L) + 2c_t + c_{s\_2p}))/(1+\rho_B^L))(\sigma_y^2/C)$
	else if $(1/\rho_B^U) - 1 \le (((1+\rho_A^L)^2)/(\lambda(1-\rho_A^L)\rho_A^L)) \le (1/\rho_B^L) - 1$, then $\rho_A^L, \rho_B^* = (\lambda\rho_A^L(1-\rho_A^L))/((1+\rho_A^L)^2 + \lambda\rho_A^L(1-\rho_A^L))$,	$((4(c_A(1-\rho_A^L) + c_B(1+\rho_A^L) + 2c_t + c_{s\_2p}))/(1+\rho_A^L))(\sigma_y^2/C)$
	else if $(\lambda(1+\rho_B^L)^2)/((1-\rho_B^L)\rho_B^L) > (1/\rho_A^L) - 1$ and $(1+\rho_A^L)^2/(\lambda(1-\rho_A^L)\rho_A^L) > (1/\rho_B^L) - 1$, then ρ_A^L, ρ_B^L	$2(\sqrt{((2(c_A + c_t) + c_{s\_2p})((\rho_B^L(1-\rho_A^L))/(\rho_A^L + \rho_B^L)))} + \sqrt{((2(c_B + c_t) + c_{s\_2p})(\rho_A^L(1-\rho_B^L)/(\rho_A^L + \rho_B^L)))})^2(\sigma_y^2/C)$
	else if $((\lambda(1+\rho_B^L)^2)/((1-\rho_B^L)\rho_B^L)) < (1/\rho_A^U) - 1$, then ρ_A^U, ρ_B^L	$2(\sqrt{((2(c_A + c_t) + c_{s\_2p})((\rho_B^L(1-\rho_A^U))/(\rho_A^U + \rho_B^L)))} + \sqrt{((2(c_B + c_t) + c_{s\_2p})(\rho_A^U(1-\rho_B^L)/(\rho_A^U + \rho_B^L)))})^2(\sigma_y^2/C)$
	else ρ_A^L, ρ_B^U	$2(\sqrt{((2(c_A + c_t) + c_{s\_2p})((\rho_B^U(1-\rho_A^L))/(\rho_A^L + \rho_B^U)))} + \sqrt{((2(c_B + c_t) + c_{s\_2p})(\rho_A^L(1-\rho_B^U)/(\rho_A^L + \rho_B^U)))})^2(\sigma_y^2/C)$

Note. $\sigma_y^2 = \sigma_A^2 + \sigma_B^2$; $\lambda = (2c_A + 2c_t + c_{s\_2p})/(2c_B + 2c_t + c_{s\_2p})$.

$$\frac{4\left(c_A(1+\rho_B^L) + c_B(1-\rho_B^L) + 2c_t + c_{s\_2p}\right)}{(1+\rho_B^L)}\left(\frac{\sigma_y^2}{C}\right) \le 4\left(c_A + c_B + 2c_t + c_{s\_2p}\right)\left(\frac{\sigma_y^2}{C}\right), \tag{7}$$

where the right-hand side of the inequality in turn is smaller than the variance of a maximin crossover design (Table 3). The higher maximin efficiency of the extended parallel design can also be shown to hold if the maximin value ρ_B^* is within the plausible range for ρ_B. Furthermore, if the variance maximizing values ρ_A^* and ρ_B^* are outside the plausible ranges for ρ_A and ρ_B, respectively, then values for ρ_A and ρ_B that coincide with one of the borders of their corresponding ranges should be chosen as values that maximize the variance. But in that case, even smaller variances result for the extended parallel design.

4.3. Maximin Designs That Minimize the Number of Subjects and Number of Measurements. As noted in Section 3, by setting $c_{sp} = c_{sep} = c_{sc} = 1$ and the remaining costs to 0 in (4)–(6), the budget is simply the total number of subjects involved in a study, and by setting $c_t = 1$ and the other costs to 0, the budget reduces to the total number of measurements involved. When the budget is the total sample size and interest is in estimating the treatment effect, it can be proven, based on the formulas in Table 2, that a maximin crossover design requires less subjects than a maximin parallel design. From an extensive numerical evaluation analogous to the one of Section 4.1, a maximin crossover design also appears to require less subjects than a maximin extended parallel design.

When minimizing the number of measurements, the numerical evaluation shows again that the maximin crossover design is the best choice provided the lower bounds of both intraclass correlations are 0.10 or higher. In other cases also a maximin parallel design may minimize the total number of measurements. Since in most crossover trials the intraclass correlation exceeds 0.30 [1–3, 10, 11], in practice, this implies that the maximin crossover trial also is the best choice when minimizing the number of measurements.

In the case, interest is in the treatment by period interaction, Section 4.2 showed a maximin extended parallel design to be more efficient and thus also to require less budget than a maximin crossover trial. In the special case where the number of subjects or the total number of measurements are minimized, the maximin extended parallel design will therefore also outperform the maximin crossover design.

5. Monte Carlo Evaluation of the Power of Maximin Designs

The efficiencies as derived for the maximin designs are based on the asymptotic variance of the ML estimator, $\mathrm{Var}(\widehat{\beta}_x \mid \xi)$. For sufficiently large numbers of subjects, the relation between the asymptotic variance of the ML estimator and the power level $1-\gamma$ to detect a treatment effect in a two-tailed test with type I error rate α can be approximated as follows:

$$\mathrm{Var}\left(\widehat{\beta}_x \mid \xi\right) = \left(\frac{\beta_x}{z_{1-\alpha/2} + z_{1-\gamma}}\right)^2, \tag{8}$$

where $z_{1-\alpha/2}$ and $z_{1-\gamma}$ are the 100 $(1-\alpha/2)$ and 100 $(1-\gamma)$ percentiles of the standard normal distribution. For small sample sizes calculated by (8), corrections are needed [12, 13]. For each of the three designs, these corrections will be applied. We will examine to what extent the differences between designs in asymptotic efficiencies translate into corresponding differences in power levels for small samples. Also, when planning sample sizes based on the asymptotic variances, we can check whether the commonly used power levels of 80% or 90% are realized in the case of small samples.

For the treatment effect estimator, the following expression for the required number of subjects results for a crossover design with optimal allocation:

$$n_c = \left(z_{1-\alpha/2} + z_{1-\gamma}\right)^2\left(1 - \frac{2\rho_A\rho_B}{\rho_A+\rho_B}\right)\left(\frac{\sigma_A^2+\sigma_B^2}{\beta_{\text{treat}}^2}\right). \quad (9)$$

If we let $\text{ES} = \beta_{\text{treat}}/\sqrt{0.5(\sigma_A^2+\sigma_B^2)}$ be the effect size based on the outcome variances in the treatment and control arm (cf. [14]), then (9) can be rewritten as follows:

$$n_c = 2\left(\frac{z_{1-\alpha/2} + z_{1-\gamma}}{\text{ES}}\right)^2\left(1 - \frac{2\rho_A\rho_B}{\rho_A+\rho_B}\right). \quad (10)$$

Note that, in the case of a maximin design, the expression is the same as (10), however, with ρ_A^L and ρ_B^L being substituted for ρ_A and ρ_B, respectively. Similar rewritings of the sample sizes in terms of the effect size are possible for the parallel and extended parallel design, respectively:

$$n_p = 2\left(\frac{z_{1-\alpha/2} + z_{1-\gamma}}{\text{ES}}\right)^2 \cdot \left(1 + \left(\frac{c_A + c_B + 2c_t + 2c_{sp}}{\sqrt{\left(c_A + c_t + c_{sp}\right)\left(c_B + c_t + c_{sp}\right)}}\right) \cdot \left(\sqrt{\frac{\rho_B}{\rho_A+\rho_B}}\sqrt{\frac{\rho_A}{\rho_A+\rho_B}}\right)\right),$$

$$n_{ep} = \left(\frac{z_{1-\alpha/2} + z_{1-\gamma}}{\text{ES}}\right)^2\left(1 + \frac{2\rho_A\rho_B}{\rho_A+\rho_B} + \left(\frac{2c_A + 2c_B + 4c_t + 2c_{s\_2p}}{\sqrt{\left(2c_A + 2c_t + c_{s\_2p}\right)\left(2c_B + 2c_t + c_{s\_2p}\right)}}\right) \cdot \left(\sqrt{\frac{\rho_B(1+\rho_A)}{\rho_A+\rho_B}}\sqrt{\frac{\rho_A(1+\rho_B)}{\rho_A+\rho_B}}\right)\right). \quad (11)$$

The choices to be made for ρ_A and ρ_B in the case of maximin versions of the parallel and extended parallel design are determined by the conditions as formulated in Table 2.

In the case of the treatment by period interaction effect, the following expression for the required number of subjects can be derived for a crossover design with optimal allocation:

$$n_c = 8\left(\frac{z_{1-\alpha/2} + z_{1-\gamma}}{\text{ES}}\right)^2\left(1 + \frac{2\rho_A\rho_B}{\rho_A+\rho_B}\right), \quad (12)$$

where $\text{ES} = \beta_{\text{treat}\times\text{time}}/\sqrt{0.5(\sigma_A^2+\sigma_B^2)}$. In the case of a maximin design, the expression is the same as (12), however, with ρ_A^U and ρ_B^U being substituted for ρ_A and ρ_B, respectively. The expression for the sample size of an extended parallel design, when allocating optimally, can be written as follows:

$$n_{ep} = 4\left(\frac{z_{1-\alpha/2} + z_{1-\gamma}}{\text{ES}}\right)^2\left(1 - \frac{2\rho_A\rho_B}{\rho_A+\rho_B} + \left(\frac{2c_A + 2c_B + 4c_t + 2c_{s\_2p}}{\sqrt{\left(2c_A + 2c_t + c_{s\_2p}\right)\left(2c_B + 2c_t + c_{s\_2p}\right)}}\right) \cdot \left(\sqrt{\frac{\rho_B(1-\rho_A)}{\rho_A+\rho_B}}\sqrt{\frac{\rho_A(1-\rho_B)}{\rho_A+\rho_B}}\right)\right). \quad (13)$$

The choices to be made for ρ_A and ρ_B in the case of a maximin extended parallel design are determined by the conditions formulated in Table 3.

Since maximin designs only require information on plausible ranges of model parameters, they are more practical than optimal designs. In what follows, we will therefore examine through a Monte Carlo simulation the power for maximin designs in the case of small sample sizes. First, we will discuss the factors that are varied and motivate the choices made for these factors in determining the simulation scenarios.

5.1. Choice of Ranges/Values for Relevant Factors

5.1.1. Effect Sizes. The effect size, ES, is commonly categorized into small (0.2), medium (0.5), and large (0.8) [14]. Being primarily interested in the small sample performance of (9)–(13), we only will consider $\text{ES} = 0.8$, leading to the smallest sample sizes.

5.1.2. Costs. The empirical evidence on costs is rather scarce, but we will choose the costs such that they imply minimizing the sample size of a study (i.e., $c_A = c_B = c_t = 0$ and $c_{sp} = c_{s\_2p} = 1$).

5.1.3. Intraclass Correlations. The ranges for ρ_A and ρ_B are identical to the ranges of the numerical evaluation of Section 4.1. Since we are interested in the small sample performance, for each design, we consider that pair of ranges across all combinations of ranges for the intraclass correlations (i.e., small-small, medium-medium, large-large, small-medium, small-large, and medium-large) that lead to the smallest sample sizes. Since this each time turns out to be a pair from the small-small category, the same was done for all pairs of medium and large ranges, which will be used more often in practice. For each design, the two resulting pairs of ranges of intraclass correlation are displayed in the two leftmost columns of Table 4 and the Table in Appendix D.

5.1.4. Power Level and Type I Error Rate. In sample size planning commonly used power levels are 80% and 90% in a two-tailed test with either a 5% or a 1% type I error rate. Focusing on the small sample performance, we will consider 80% power in a two-tailed test with a 5% type I error rate. For small sample sizes derived from the standard normal

Table 4: Powers from the Monte Carlo simulations for maximin designs in the case of the treatment effect. For each pair of ranges of the intraclass correlations, the asymptotic efficiency of each design versus the most efficient design is given within brackets.

			Treatment effect		
		N	Crossover design	Parallel design	Extended parallel design
Type I error rate = 0.05					
$[\rho_A^L, \rho_A^U]$	$[\rho_B^L, \rho_B^U]$				
[0.01,0.10]	[0.90,1.00]	30	0.857 (1)	0.648 (0.61)	**0.836** (0.96)
[0.01,0.10]	[0.90,1.00]	44	0.959 (1)	**0.822** (0.61)	0.950 (0.96)
[0.01,0.30]	[0.01,0.30]	36	0.916 (1)	0.805 (0.50)	**0.823** (0.76)
[0.01,0.30]	[0.70,1.00]	52	0.982 (1)	**0.819** (0.51)	0.922 (0.70)
[0.70,1.00]	[0.70,1.00]	10	**0.813** (1)	0.180 (0.15)	0.182 (0.15)
[0.90,1.00]	[0.90,1.00]	6	**0.902** (1)	0.088 (0.05)	0.087 (0.05)
Type I error rate = 0.01					
$[\rho_A^L, \rho_A^U]$	$[\rho_B^L, \rho_B^U]$				
[0.01,0.10]	[0.90,1.00]	45	0.854 (1)	0.618 (0.61)	**0.848** (0.96)
[0.01,0.10]	[0.90,1.00]	66	0.974 (1)	**0.829** (0.61)	0.967 (0.96)
[0.01,0.30]	[0.01,0.30]	55	0.931 (1)	0.817 (0.50)	**0.838** (0.76)
[0.01,0.30]	[0.70,1.00]	77	0.989 (1)	**0.822** (0.51)	0.931 (0.70)
[0.70,1.00]	[0.70,1.00]	16	**0.856** (1)	0.112 (0.15)	0.119 (0.15)
[0.90,1.00]	[0.90,1.00]	8	**0.849** (1)	0.029 (0.05)	0.030 (0.05)

Note. The power printed in bold indicates the design for which the sample calculation should yield a power of 80%. N: total sample size.

distribution (as in (9)–(13)), corrections are needed that turn out to depend on the type I error rate [12, 13]. For this reason, we will also study a 1% type I error rate.

5.2. Simulation Procedure and Testing Methods. For each of the 20 simulation scenarios, 25,000 data sets were generated. To distinguish chance deviations of the simulated power from systematic deviations, a 95% predictive interval was calculated. For a nominal power π, this is defined as $[\pi + z_{0.025}\sqrt{\pi(1-\pi)/N\text{sim}},\ \pi + z_{0.975}\sqrt{\pi(1-\pi)/N\text{sim}}]$, where $z_{0.025}$ and $z_{0.975}$ are the 2.5 and 97.5 percentiles of the standard normal distribution and Nsim is the number of simulations. Since the nominal power $\pi = 0.80$ and $N\text{sim} = 25{,}000$, the 95% predictive interval is [0.795, 0.805].

For a test of the treatment effect, the data generated for the crossover design were analyzed with a two-sample t-test on the difference scores obtained by subtracting the two measurements for each subject. The model in (2) implies homogeneity of variances for these difference scores, so that a pooled variance t-test was applied. The data generated for the parallel design were simply analyzed by a two-sample t-test on the original scores, whereas the data for the extended parallel design were analyzed with a two-sample t-test on the scores averaged across both measurements. For these parallel designs, (1) and (2) imply that the analyzed scores may have variances differing between groups, so that an unpooled variance t-test was applied. For the treatment by period interaction effect, the data generated for the crossover design were analyzed with a two-sample (pooled variance) t-test on the scores averaged for each subject across both measurements, whereas the data generated for the extended parallel design were analyzed with a two-sample (unpooled variance) t-test on the differences between the two measurements (3). These different t-tests follow for each of the designs (involving equal numbers of measurements per subject) from the analysis models in (1)–(3) and do not require asymptotic assumptions.

In calculating the required sample size, (9)–(11) were used, when interest is in testing the treatment effect, and (12) and (13) were used, when interest is in testing the treatment by period interaction. The optimal allocations for each design are given in Table 1, taking the maximin values for ρ_A and ρ_B as determined from Tables 2 and 3 for the treatment main and treatment by period interaction effect, respectively. Since the outcome variances are unknown, the calculated sample sizes were subsequently corrected. For an unpooled variance t-test, if the number of persons per arm is 8 or more, sufficient corrections for two-tailed tests are 2 extra persons per arm if $\alpha = 0.05$ and 4 extra persons per arm if $\alpha = 0.01$. With less than 8 persons in one or both arms, sufficient corrections are 3 extra persons per arm if $\alpha = 0.05$ and 4 extra persons per arm if $\alpha = 0.01$ ([13], p. 568). For a pooled variance t-test, we only need to add 1 person per arm if $\alpha = 0.05$ (two-tailed) and 2 persons per arm if $\alpha = 0.01$ (two-tailed) ([12], p. 1216-1217). These are sufficient corrections for planned powers of 80% and 90%. The simulations, statistical tests, and power calculations were done in R, version 3.1.3 [15].

5.3. Results. As can be seen in Table 4 and Appendix D, for all sample size-design combinations that should yield 80% power, the simulated powers were either within or above the 95% predictive intervals. This indicates that the asymptotic results, supplemented with simple correction rules for using the standard normal distribution, yield sample sizes that guarantee the desired level of power. The realized power levels generally are higher than 80%, since the small sample corrections are sufficient and in some cases smaller corrections would have been appropriate [12, 13].

The power differences between the designs can become rather large and are in line with the asymptotic relative efficiencies. For the examples of Table 4, the crossover design always is most efficient and in the simulation also has the highest power. Additional simulations show that similar conclusions can be drawn for ranges of intraclass correlations for which the crossover design is not most efficient. As expected, when testing the treatment by period interaction, the extended parallel design has more power than the crossover design (Appendix D).

6. Application in Planning a Trial

Suppose one would like to perform a randomized trial on the effectiveness of indacaterol versus tiotropium, among subjects suffering from chronic obstructive pulmonary disease, similar to Donohue et al. [16]. After 12 weeks of treatment, one plans to evaluate the effect of 18 μg of tiotropium versus 300 μg of indacaterol on the bronchodilator efficacy of 24 h postdose forced expiratory volume in 1 s (FEV1 in mL). The variance of FEV1 in the indacaterol and the tiotropium conditions in the study by Donohue et al. [16] differed significantly from each other, their ratio being 1.58. No information was available on the intraclass correlations and the research costs. Suppose reasonable guesses on the intraclass correlations are $\rho_A \in [0.10, 0.70]$ for indacaterol and $\rho_B \in [0.30, 0.90]$ for tiotropium. If one aims at minimizing the total sample size (so we set $c_A = c_B = c_t = 0$ and $c_{sp} = c_{s\_2p} = 1$), the maximin crossover design is more efficient than the AA/BB and A/B designs requiring only 48% and 43%, respectively, of the number of subjects of these designs. To be able to detect a medium effect (ES = 0.5, see below (9)) with 80% power in a two-tailed test with a 5% type I error rate, taking as maximin parameter values $\rho_A^* = 0.10$ and $\rho_B^* = 0.30$ in (10), 54 subjects are needed. Since the sample size calculation in (10) is based on the standard normal, whereas the test statistic follows a t-distribution, we add 1 subject to each treatment sequence [12], yielding a total sample size of 56 subjects with 28 subjects being allocated to each of the two treatment sequences of the crossover design.

7. Conclusion and Discussion

We examined the asymptotic efficiency of the ML estimator of the treatment and the treatment by period interaction effect for three two-treatment designs: a parallel, an extended parallel, and a crossover design. For a flexible cost function, the optimal allocations to the treatment sequences and corresponding optimal efficiencies were derived. Since commonly the intraclass correlations for each of the treatments and the ratio of treatment-dependent variances are not precisely known, also maximin designs were derived, which guarantee a power level across plausible ranges of values for the intraclass correlations at the lowest costs.

When interested in testing the main effects of the treatments, the relations between the efficiencies of the maximin versions of the A/B, AB/BA, and AA/BB designs depend on assumed ranges of the intraclass correlations, on the costs of the treatments and the costs of recruiting and measuring subjects. A numerical investigation shows that if A and B are equally expensive or the sum of the costs of one treatment and measurement per person are less than the remaining subject-specific costs (such as recruitment costs), then the crossover design is most efficient for ranges of intraclass correlations starting at 0.15 or higher. In other cost scenarios, also for ranges of the intraclass correlations above 0.15, the parallel design or its extended version may become most efficient. Then, the efficiency relations are complicated, and the most efficient design is best determined by the results in Table 2. For the treatment by period interaction, however, the maximin AA/BB trial is proven to be more efficient than the maximin AB/BA design.

Since the efficiency comparisons of the maximin designs were based on asymptotic variances, a Monte Carlo simulation study was done for small samples. After applying correction factors in sample size planning based on the standard normal distribution, it was shown that (a) the asymptotic relative efficiencies translate into corresponding relative power levels and (b) power levels targeted in sample size planning are realized. This illustrates the practical utility of these results for sample size calculation.

If prerandomization measurements of the outcome variable are available, these could be included as covariates in the analysis [1]. Adding covariates in a randomized trial will not change the treatment effect of interest but will lead to a reduction of the intercept variance and thus of the intraclass correlations [3]. Provided that the costs of prerandomization measurements are the same for all designs (and there are no missing values on these prerandomization covariates), the results of the present study also apply.

The present study did not consider carry over in deriving optimal and maximin designs. If there is *self carry over*, that is, carry over from a treatment onto itself, this implies that steady-state did not yet occur in the first period, and then the total treatment effect would be the relevant effect, that is, the direct effect in the first period plus the carry over effect in the second period [5]. If there is self carry over, one-period designs and the AB/BA design are not suitable, as they do not allow for estimating the total treatment effect, leaving only the AA/BB design as a suitable option. There may also be *steady-state carry over*, which can only occur if there is a switch of treatments [17]. Such carry over would affect the efficiency of the crossover design. Although one commonly tries to avoid carry over, examining to what extent steady-state carry over affects the relative efficiency of the maximin crossover design would be an interesting issue for further research.

Appendix

A.1. Asymptotic Variance for the ML Estimator of the Treatment Effect

Let the vector of observed scores on the dependent variable for person $j(= 1, \ldots, N)$ in a two-period design be denoted as $\mathbf{y}_j = \begin{bmatrix} y_{1j} \\ y_{2j} \end{bmatrix}$. The linear mixed model for the scores $\mathbf{y}_j$ can be expressed as follows:

$$\mathbf{y}_j = \mathbf{X}_j\boldsymbol{\beta} + \mathbf{1}u_{0j} + \mathbf{Q}\boldsymbol{\varepsilon}_j^{A} + (\mathbf{I} - \mathbf{Q})\boldsymbol{\varepsilon}_j^{B}, \qquad (A.1)$$

where $\mathbf{X}_j$ is the design matrix for subject j, $\boldsymbol{\beta}$ is the vector of regression coefficients, u_{0j} is the random person effect,

$\boldsymbol{\varepsilon}_j^{A} = \begin{bmatrix} \varepsilon_{1j}^{A} \\ \varepsilon_{2j}^{A} \end{bmatrix}$ is the vector of residual scores under treatment A, and $\boldsymbol{\varepsilon}_j^{B} = \begin{bmatrix} \varepsilon_{1j}^{B} \\ \varepsilon_{2j}^{B} \end{bmatrix}$ is the vector of residual scores under treatment B. In (A.1), $\mathbf{1} = \begin{bmatrix} 1 \\ 1 \end{bmatrix}$, $\mathbf{I}$ is the identity matrix, and $\mathbf{Q}$ is a matrix which indicates whether treatment A has been given or not in a particular period, so for a person j with treatment sequence AB, we have $\mathbf{Q} = \begin{bmatrix} 1 & 0 \\ 0 & 0 \end{bmatrix}$, indicating that A has been given in period 1 and treatment B in period 2.

Let $\mathbf{J}$ be a matrix with only ones (of order 2 by 2). The variance-covariance matrix of $\mathbf{y}_j$ can be derived as follows:

$$\mathbf{V}_j = \mathbf{Q}^T\mathbf{Q}\sigma_{\varepsilon A}^2 + (\mathbf{I}-\mathbf{Q})^T(\mathbf{I}-\mathbf{Q})\sigma_{\varepsilon B}^2 + \sigma_0^2\mathbf{J}. \quad (A.2)$$

For the AA sequence, we have $\mathbf{Q}=\mathbf{I}$, and (A.2) can be rewritten as follows:

$$\mathbf{V}_j = \sigma_{\varepsilon A}^2\mathbf{I} + \sigma_0^2\mathbf{J}. \quad (A.3)$$

By applying the result that, for an $n\times n$ matrix, $\mathbf{X} = a\mathbf{I} + b\mathbf{J}$, with $a \neq 0$ and $a \neq -\text{nb}$, the inverse is $\mathbf{X}^{-1} = (1/a)(\mathbf{I}-\mathbf{J}(b/(\text{nb}+a)))$ ([18], p. 443), the inverse of the matrix in (A.3) can be written as follows:

$$\mathbf{V}_j^{-1} = \frac{1}{\sigma_{\varepsilon A}^2}\left(\mathbf{I}-\mathbf{J}\left(\frac{\sigma_0^2}{2\sigma_0^2+\sigma_{\varepsilon A}^2}\right)\right) = \frac{1}{\sigma_{\varepsilon A}^2}\times\frac{1}{2\sigma_0^2+\sigma_{\varepsilon A}^2}\times\begin{pmatrix} \sigma_0^2+\sigma_{\varepsilon A}^2 & -\sigma_0^2 \\ -\sigma_0^2 & \sigma_0^2+\sigma_{\varepsilon A}^2 \end{pmatrix}. \quad (A.4)$$

For the *BB* sequence, we have $\mathbf{I}-\mathbf{Q}=\mathbf{I}$, and in a similar way, we obtain

$$\mathbf{V}_j^{-1} = \frac{1}{\sigma_{\varepsilon B}^2}\left(\mathbf{I}-\mathbf{J}\left(\frac{\sigma_0^2}{2\sigma_0^2+\sigma_{\varepsilon B}^2}\right)\right) = \frac{1}{\sigma_{\varepsilon B}^2}\times\frac{1}{2\sigma_0^2+\sigma_{\varepsilon A}^2}\times\begin{pmatrix} \sigma_0^2+\sigma_{\varepsilon B}^2 & -\sigma_0^2 \\ -\sigma_0^2 & \sigma_0^2+\sigma_{\varepsilon B}^2 \end{pmatrix}. \quad (A.5)$$

Finally, for the AB sequence and BA sequence, we can derive the following equations, respectively:

$$\mathbf{V}_j^{-1} = \frac{1}{(\sigma_0^2+\sigma_{\varepsilon A}^2)(\sigma_0^2+\sigma_{\varepsilon B}^2)-\sigma_0^4}\times\begin{pmatrix} \sigma_0^2+\sigma_{\varepsilon B}^2 & -\sigma_0^2 \\ -\sigma_0^2 & \sigma_0^2+\sigma_{\varepsilon A}^2 \end{pmatrix},$$

$$\mathbf{V}_j^{-1} = \frac{1}{(\sigma_0^2+\sigma_{\varepsilon A}^2)(\sigma_0^2+\sigma_{\varepsilon B}^2)-\sigma_0^4}\times\begin{pmatrix} \sigma_0^2+\sigma_{\varepsilon A}^2 & -\sigma_0^2 \\ -\sigma_0^2 & \sigma_0^2+\sigma_{\varepsilon B}^2 \end{pmatrix}. \quad (A.6)$$

The information matrix of the ML estimator of $\boldsymbol{\beta}$ can be written as

$$\text{Inf}\,\hat{\boldsymbol{\beta}} = \sum_{j=1}^{N}\mathbf{X}_j^T\mathbf{V}_j^{-1}\mathbf{X}_j. \quad (A.7)$$

Taking the inverse of the information matrix yields the asymptotic variance-covariance matrix of the ML estimators of $\boldsymbol{\beta}$.

Different treatment sequences not only lead to different matrices $\mathbf{V}_j$ but also to different $\mathbf{X}_j$ matrices. As an example, we consider the crossover design. In total, we have N subjects, and p is the proportion of persons being allocated to the AB sequence. Let $\boldsymbol{\beta}^T = (\beta_0, \beta_1, \beta_2)$, where the regression coefficients represent the intercept, the treatment effect, and the time effect, respectively. For persons allocated to the AB sequence, we have $\mathbf{X}_j = \begin{bmatrix} 1 & 0 & 0 \\ 1 & 1 & 1 \end{bmatrix}$, where $j = 1, \ldots, \text{Np}$. For persons allocated to the BA sequence, we have $\mathbf{X}_j = \begin{bmatrix} 1 & 1 & 0 \\ 1 & 0 & 1 \end{bmatrix}$, where $j = \text{Np}+1, \ldots, N$. We can now elaborate the matrix formulation in (A.7). For persons in the AB sequence, we have

$$\mathbf{X}_j^T\mathbf{V}_j^{-1}\mathbf{X}_j = \frac{1}{(\sigma_0^2+\sigma_{\varepsilon A}^2)\sigma_{\varepsilon B}^2+\sigma_0^2\sigma_{\varepsilon A}^2} \cdot \begin{bmatrix} \sigma_{\varepsilon A}^2+\sigma_{\varepsilon B}^2 & \sigma_{\varepsilon A}^2 & \sigma_{\varepsilon A}^2 \\ \sigma_{\varepsilon A}^2 & \sigma_0^2+\sigma_{\varepsilon A}^2 & \sigma_0^2+\sigma_{\varepsilon A}^2 \\ \sigma_{\varepsilon A}^2 & \sigma_0^2+\sigma_{\varepsilon A}^2 & \sigma_0^2+\sigma_{\varepsilon A}^2 \end{bmatrix}. \quad (A.8)$$

and for persons in the BA sequence, we obtain

$$\mathbf{X}_j^T\mathbf{V}_j^{-1}\mathbf{X}_j = \frac{1}{(\sigma_0^2+\sigma_{\varepsilon A}^2)\sigma_{\varepsilon B}^2+\sigma_0^2\sigma_{\varepsilon A}^2} \cdot \begin{bmatrix} \sigma_{\varepsilon A}^2+\sigma_{\varepsilon B}^2 & \sigma_{\varepsilon A}^2 & \sigma_{\varepsilon B}^2 \\ \sigma_{\varepsilon A}^2 & \sigma_0^2+\sigma_{\varepsilon A}^2 & -\sigma_0^2 \\ \sigma_{\varepsilon B}^2 & -\sigma_0^2 & \sigma_0^2+\sigma_{\varepsilon B}^2 \end{bmatrix}. \quad (A.9)$$

The information matrix in (A.7) can be obtained by summating the expressions in (A.8) and (A.9) across all persons in each of the treatment sequences, and the asymptotic variance-covariance matrix of the ML estimators then results by taking the inverse of this matrix. We are interested in the variance of the treatment effect estimator, which is the entry in row 2 and column 2 of the resulting variance-covariance matrix:

$$\text{Var}\left(\hat{\beta}_1 \mid \xi_{\text{crossover}}\right) = \frac{(\sigma_{\varepsilon B}^2-\sigma_{\varepsilon A}^2)^2 p(1-p) + (\sigma_0^2+\sigma_{\varepsilon A}^2)\sigma_{\varepsilon B}^2 + \sigma_0^2\cdot\sigma_{\varepsilon A}^2}{(4\sigma_0^2+\sigma_{\varepsilon A}^2+\sigma_{\varepsilon B}^2)p(1-p)N}, \quad (A.10)$$

which, noting that $\rho_A = \sigma_0^2/(\sigma_0^2+\sigma_{\varepsilon A}^2)$, $\rho_B = \sigma_0^2/(\sigma_0^2+\sigma_{\varepsilon B}^2)$, and $\sigma_y^2 = 2\sigma_0^2+\sigma_{\varepsilon A}^2+\sigma_{\varepsilon B}^2$, can be written as the expression given in Table 5. Along similar lines, the variance of the treatment effect estimator for a parallel and an extended parallel design, as shown in Table 5, can be derived.

TABLE 5: Asymptotic variances of the ML estimator of the treatment effect and treatment by period interaction effect for different designs under heterogeneity of outcome variances.

Design	$\mathrm{Var}(\widehat{\beta}_{\text{treat}})$	$\mathrm{Var}(\widehat{\beta}_{\text{time}\times\text{treat}})$
Crossover design	$((\rho_A-\rho_B)^2p(1-p)+\rho_A\rho_B(1-\rho_A\rho_B))/((\rho_A+\rho_B)(\rho_A+\rho_B+2\rho_A\rho_B))(\sigma_y^2/((1-p)pN))$	$(1+2\rho_A\rho_B/(\rho_A+\rho_B))(\sigma_y^2/((1-p)pN))$
Parallel design	$((\rho_B/(\rho_A+\rho_B))(1/p)+(\rho_A/(\rho_A+\rho_B))1/(1-p))(\sigma_y^2/N)$	na
Extended parallel design	$((\rho_B/(\rho_A+\rho_B))(1/p)+(\rho_A/(\rho_A+\rho_B))(1/(1-p))+((\rho_A\rho_B)/(\rho_A+\rho_B))(1/((1-p)p)))(\sigma_y^2/(2N))$	$(((\rho_B(1-\rho_A))/(\rho_A+\rho_B))(1/p)+((\rho_A(1-\rho_B))/(\rho_A+\rho_B))(1/(1-p)))(2\sigma_y^2/N)$

Note. $\sigma_y^2=\sigma_A^2+\sigma_B^2$; na: not applicable.

A.2. Asymptotic Variance for the ML Estimator of the Treatment by Period Interaction Effect

The derivations of the variance of treatment by period interaction estimator are also similar, but start from another model and corresponding design matrices for the fixed regression coefficients. Now $\boldsymbol{\beta}^T=(\beta_0,\beta_1,\beta_2,\beta_3)$, where the regression coefficients represent the intercept, the treatment effect, the time effect, and the treatment by period interaction effect, respectively. When considering again a crossover trial as an example, for persons assigned to the AB sequence, we have $\mathbf{X}_j=\begin{bmatrix}1&0&0&0\\1&1&1&1\end{bmatrix}$, where $j=1,\ldots,\mathrm{Np}$. For persons assigned to the BA sequence, we have $\mathbf{X}_j=\begin{bmatrix}1&1&0&0\\1&0&1&0\end{bmatrix}$, where $j=\mathrm{Np}+1,\ldots,N$. For persons in the AB sequence, we obtain

$$\mathbf{X}_j^T\mathbf{V}_j^{-1}\mathbf{X}_j=\frac{1}{(\sigma_0^2+\sigma_{\varepsilon A}^2)\sigma_{\varepsilon B}^2+\sigma_0^2\sigma_{\varepsilon A}^2}\cdot\begin{bmatrix}\sigma_{\varepsilon A}^2+\sigma_{\varepsilon B}^2 & \sigma_{\varepsilon A}^2 & \sigma_{\varepsilon A}^2 & \sigma_{\varepsilon A}^2\\ \sigma_{\varepsilon A}^2 & \sigma_0^2+\sigma_{\varepsilon A}^2 & \sigma_0^2+\sigma_{\varepsilon A}^2 & \sigma_0^2+\sigma_{\varepsilon A}^2\\ \sigma_{\varepsilon A}^2 & \sigma_0^2+\sigma_{\varepsilon A}^2 & \sigma_0^2+\sigma_{\varepsilon A}^2 & \sigma_0^2+\sigma_{\varepsilon A}^2\\ \sigma_{\varepsilon A}^2 & \sigma_0^2+\sigma_{\varepsilon A}^2 & \sigma_0^2+\sigma_{\varepsilon A}^2 & \sigma_0^2+\sigma_{\varepsilon A}^2\end{bmatrix}, \tag{A.11}$$

and for persons in the BA sequence, we obtain

$$\mathbf{X}_j^T\mathbf{V}_j^{-1}\mathbf{X}_j=\frac{1}{(\sigma_0^2+\sigma_{\varepsilon A}^2)\sigma_{\varepsilon B}^2+\sigma_0^2\sigma_{\varepsilon A}^2}\cdot\begin{bmatrix}\sigma_{\varepsilon A}^2+\sigma_{\varepsilon B}^2 & \sigma_{\varepsilon A}^2 & \sigma_{\varepsilon B}^2 & 0\\ \sigma_{\varepsilon A}^2 & \sigma_0^2+\sigma_{\varepsilon A}^2 & -\sigma_0^2 & 0\\ \sigma_{\varepsilon B}^2 & -\sigma_0^2 & \sigma_0^2+\sigma_{\varepsilon B}^2 & 0\\ 0 & 0 & 0 & 0\end{bmatrix}. \tag{A.12}$$

The information matrix in (A.7) is obtained by summating the expressions in (A.11) and (A.12) across all persons in both treatment sequences, and the asymptotic variance-covariance matrix of the ML estimators then results by taking the inverse of this matrix. We are interested in the variance of the treatment by period interaction effect estimator, which is the entry in row 4 and column 4 of the variance-covariance matrix:

$$\mathrm{Var}\left(\widehat{\beta}_3\mid\xi_{\text{cross-over}}\right)=\frac{\left(4\sigma_0^2+\sigma_{\varepsilon A}^2+\sigma_{\varepsilon B}^2\right)}{p(1-p)N}, \tag{A.13}$$

which can be rewritten in terms of ρ_A, ρ_B, and σ_y^2, as the expression in Table 5. Along similar lines, the variance of the treatment by period interaction effect estimator for an extended parallel design, as shown in Table 5, can be derived.

B. Derivation of Optimal Allocations under a Budget Constraint

The asymptotic variance of the treatment effect estimator is minimized as a function of p, the allocation proportion, given a fixed budget C. For a crossover, parallel, and extended parallel design, p is the proportion allocated to the sequence AB, A, and AA, respectively. The variance for a crossover design as given in Table 5 can be rewritten in terms of the research costs and budget C, employing the cost function in (5) and noting that $c_{sc}=c_{s\_2p}$:

$$\begin{aligned}\mathrm{Var}\left(\widehat{\beta}_1\mid\xi_{\text{crossover}}\right)&=\left(\frac{(\rho_A-\rho_B)^2p(1-p)+\rho_A\rho_B(1-\rho_A\rho_B)}{(\rho_A+\rho_B)(\rho_A+\rho_B+2\rho_A\rho_B)}\right)\cdot\left(\frac{\sigma_y^2}{(1-p)pC}\right)\left(c_A+c_B+2c_t+c_{s\_2p}\right)\\&=\left((\rho_A-\rho_B)^2+\frac{\rho_A\rho_B(1-\rho_A\rho_B)}{(1-p)p}\right)\cdot\left(\frac{\sigma_y^2\left(c_A+c_B+2c_t+c_{s\_2p}\right)}{(\rho_A+\rho_B)(\rho_A+\rho_B+2\rho_A\rho_B)C}\right).\end{aligned} \tag{B.1}$$

It is easy to see that this expression is minimized if $p(1-p)$ is maximized, which is the case if $p=0.5$.

For a parallel design, the variance of the treatment effect estimator as given in Table 5 can be rewritten in terms of the cost function in (4) as follows:

$$\mathrm{Var}\left(\widehat{\beta}_1\mid\xi_{\text{parallel}}\right)=\left(\left(\frac{\rho_B}{\rho_A+\rho_B}\right)\frac{1}{p}+\left(\frac{\rho_A}{\rho_A+\rho_B}\right)\frac{1}{(1-p)}\right)\frac{\sigma_y^2}{C}\cdot\left(c_{sp}+c_t+pc_A+(1-p)c_B\right). \tag{B.2}$$

Taking the derivative of (B.2) with respect to p yields the following expression:

$$\left(\left(\left(2\left(c_{sp}+c_t\right)+c_A+c_B\right)\left(\frac{\rho_A}{\rho_A+\rho_B}\right)-\left(c_{sp}+c_t+c_B\right)\right)p^2\right.$$
$$\left.+2\left(c_{sp}+c_t+c_B\right)\left(\frac{\rho_B}{\rho_A+\rho_B}\right)p-\left(c_{sp}+c_t+c_B\right)\left(\frac{\rho_B}{\rho_A+\rho_B}\right)\right)$$
$$\cdot\frac{\sigma_y^2}{Cp^2(1-p)^2}. \tag{B.3}$$

Solving the expression for p gives two solutions, one of which turns out to give a minimum (the second derivative of the variance as a function of p is positive for this particular value):

$$p=\frac{\sqrt{\left(c_{sp}+c_t+c_B\right)\rho_B}}{\sqrt{\left(c_{sp}+c_t+c_B\right)\rho_B}+\sqrt{\left(c_{sp}+c_t+c_A\right)\rho_A}}. \tag{B.4}$$

Finally, for an extended parallel design, the variance of the treatment effect estimator as given in Table 5 can be rewritten in terms of the cost function in (6) noting that $c_{sep}=c_{s\_2p}$:

$$\mathrm{Var}\left(\hat{\beta}_1\mid\xi_{\text{extended parallel}}\right)=\left(\left(\frac{\rho_B}{\rho_A+\rho_B}\right)\frac{1}{p}+\left(\frac{\rho_A}{\rho_A+\rho_B}\right)\frac{1}{(1-p)}\right.$$
$$\left.+\left(\frac{\rho_A\rho_B}{\rho_A+\rho_B}\right)\frac{1}{(1-p)p}\right)$$
$$\cdot\frac{\sigma_y^2\left(c_{s\_2p}+2pc_A+2(1-p)c_B+2c_t\right)}{2C}. \tag{B.5}$$

Taking the derivative with respect to p yields the following expression:

$$\left(\left(\left(c_{s\_2p}+2c_t\right)\left(\frac{\rho_A-\rho_B}{\rho_A+\rho_B}\right)-2c_B\left(\frac{(1+\rho_A)\rho_B}{\rho_A+\rho_B}\right)\right.\right.$$
$$\left.\left.+2c_A\left(\frac{(1+\rho_B)\rho_A}{\rho_A+\rho_B}\right)\right)p^2+2\left(c_{s\_2p}+2c_t+2c_B\right)\right.$$
$$\left.\cdot\left(\frac{\rho_B(1+\rho_A)}{\rho_A+\rho_B}\right)p-\left(c_{s\_2p}+2c_t+2c_B\right)\left(\frac{\rho_B(1+\rho_A)}{\rho_A+\rho_B}\right)\right)$$
$$\times\frac{\sigma_y^2}{2Cp^2(1-p)^2}. \tag{B.6}$$

Solving (B.6) for p gives two solutions, one of which turns out to give a minimum (the second derivative of the variance as a function of p is positive for his particular value):

$$p=\frac{\sqrt{\left(c_{s\_2p}+2\left(c_t+c_B\right)\right)(1+\rho_A)\rho_B}}{\sqrt{\left(c_{s\_2p}+2\left(c_t+c_B\right)\right)(1+\rho_A)\rho_B}+\sqrt{\left(c_{s\_2p}+2\left(c_t+c_A\right)\right)(1+\rho_B)\rho_A}}. \tag{B.7}$$

Substituting the optimal allocation $p=0.5$, and the allocations in (B.4) and (B.7) into the corresponding expressions for the variance of the treatment effect estimator yields the optimal variances as given in Table 1 of the main text. Derivations along similar lines can be done if interest is in minimizing the variance of the treatment by period effect estimator. These result in the allocations and variances, as also presented in Table 1 of the main text.

C. Derivation of Maximin Designs and Variances of the Effect Estimators

Maximin designs optimize the allocation to the treatment sequences under the worst case, that is, the maximum variance of the treatment effect estimator across plausible ranges of the model parameters, here the intraclass correlations ρ_A and ρ_B.

For crossover designs, the variance of the treatment effect estimator under optimal allocation to the treatments is (see Table 1 of the main text)

$$\mathrm{Var}\left(\hat{\beta}_1\mid\xi_{\text{crossover}}\right)=\left(c_A+c_B+2c_t+c_{s\_2p}\right)\left(1-\frac{2\rho_A\rho_B}{\rho_A+\rho_B}\right)\frac{\sigma_y^2}{C}. \tag{C.1}$$

By taking the derivatives with respect to ρ_A and ρ_B, it can be shown that this expression decreases as a function of these parameters. This implies that the worst case occurs for the lower bounds of the ranges for ρ_A and ρ_B. This yields the expression for the variance of the treatment effect estimator in the case of a maximin crossover design in Table 2 of the main text.

Similarly, for a parallel design, we have to examine for which values of ρ_A and ρ_B the following expression is maximized:

$$\mathrm{Var}\left(\hat{\beta}_1\mid\xi_{\text{parallel}}\right)=\left(\sqrt{\left(c_A+c_t+c_{sp}\right)\left(\frac{\rho_B}{\rho_A+\rho_B}\right)}\right.$$
$$\left.+\sqrt{\left(c_B+c_t+c_{sp}\right)\left(\frac{\rho_A}{\rho_A+\rho_B}\right)}\right)^2\frac{\sigma_y^2}{C}. \tag{C.2}$$

For derivations purposes, it is more convenient to rewrite (C.2) in terms of $\phi=\rho_B/\rho_A$:

$$\mathrm{Var}\left(\hat{\beta}_1\mid\xi_{\text{parallel}}\right)=\left(\sqrt{\left(c_A+c_t+c_{sp}\right)\left(\frac{\phi}{1+\phi}\right)}\right.$$
$$\left.+\sqrt{\left(c_B+c_t+c_{sp}\right)\left(\frac{1}{1+\phi}\right)}\right)^2\frac{\sigma_y^2}{C}. \tag{C.3}$$

Taking the derivative of (C.3) with respect to ϕ, we find that the variance increases as a function of ϕ as long as $\phi\le(c_A+c_t+c_{sp})/(c_B+c_t+c_{sp})$ and decreases as a function of ϕ if $\phi>(c_A+c_t+c_{sp})/(c_B+c_t+c_{sp})$. So, if we can choose ρ_A and ρ_B from their plausible ranges such that $\rho_B/\rho_A=(c_A+c_t+c_{sp})/(c_B+c_t+c_{sp})$, this maximizes (C.2). On the other hand, if even the lower bound of ρ_A, ρ_A^L, and the upper bound of ρ_B, ρ_B^U, yield $\rho_B^L/\rho_A^U<(c_A+c_t+c_{sp})/(c_B+c_t+c_{sp})$, then choose ρ_A^L and ρ_B^U. Furthermore, if

$(c_A + c_t + c_{sp})/(c_B + c_t + c_{sp}) < \rho_B^L/\rho_A^U$, then choose ρ_A^U and ρ_B^L. Substituting the maximin values of ρ_A and ρ_B into (C.2) will result in the variances of the treatment effect estimator as displayed in Table 2 of the main text.

For the extended parallel design, the variance for optimal allocation to the treatment sequences is given in Table 1 of the main text:

$$\mathrm{Var}(\hat{\beta}_1 \mid \xi_{\text{extended parallel}}) = \left(\sqrt{(2(c_A + c_t) + c_{s\_2p})\left(\frac{\rho_B(1+\rho_A)}{\rho_A+\rho_B}\right)} + \sqrt{(2(c_B + c_t) + c_{s\_2p})\left(\frac{\rho_A(1+\rho_B)}{\rho_A+\rho_B}\right)}\right)^2 \cdot \frac{\sigma_y^2}{2C}. \tag{C.4}$$

Taking the derivative of (C.4) with respect to ρ_A shows that the variance of the treatment effect estimator increases as a function of ρ_A as long as $\lambda(1-\rho_B)^2/(\rho_B(1+\rho_B)) \le (1+\rho_A)/\rho_A$ where $\lambda = (2(c_A + c_t) + c_{s\_2p})/(2(c_B + c_t) + c_{s\_2p})$ and decreases if $\lambda(1-\rho_B)^2/(\rho_B(1+\rho_B)) > (1+\rho_A)/\rho_A$. Taking into account a feasible range for ρ_A, the value for ρ_A maximizing the variance in (C.4), ρ_A^*, therefore is

$$\rho_A^* = \frac{\rho_B(1+\rho_B)}{\lambda(1-\rho_B)^2 - \rho_B(1+\rho_B)} \quad \text{if} \quad \rho_A^L \le \rho_A^* \le \rho_A^U$$
$$\text{or} \quad \frac{1}{\rho_A^U} + 1 \le \frac{\lambda(1-\rho_B)^2}{(1+\rho_B)\rho_B} \le \frac{1}{\rho_A^L} + 1, \tag{C.5}$$

$$\rho_A^* = \rho_A^U \quad \text{if} \quad \rho_A^U \le \rho_A^* \quad \text{or} \quad \frac{\lambda(1-\rho_B)^2}{(1+\rho_B)\rho_B} < \frac{1}{\rho_A^U} + 1, \tag{C.6}$$

$$\rho_A^* = \rho_A^L \quad \text{if} \quad \rho_A^* < \rho_A^L \quad \text{or} \quad \frac{1}{\rho_A^L} + 1 < \frac{\lambda(1-\rho_B)^2}{(1+\rho_B)\rho_B}. \tag{C.7}$$

Taking the derivative of (C.4) with respect to ρ_B shows that the variance increases as a function of ρ_B as long as $(1-\rho_A)^2/(\lambda\rho_A(1+\rho_A)) \le (1+\rho_B)/\rho_B$ and decreases if $(1-\rho_A)^2/(\lambda\rho_A(1+\rho_A)) > (1+\rho_B)/\rho_B$. So, also taking into account a feasible range for ρ_B, we have as maximin value for ρ_B:

$$\rho_B^* = \frac{\lambda\rho_A(1+\rho_A)}{(1-\rho_A)^2 - \lambda\rho_A(1+\rho_A)} \quad \text{if} \quad \rho_B^L \le \rho_B^* \le \rho_B^U$$
$$\text{or} \quad \frac{1}{\rho_B^U} + 1 \le \frac{(1-\rho_A)^2}{\lambda(1+\rho_A)\rho_A} \le \frac{1}{\rho_B^L} + 1, \tag{C.8}$$

$$\rho_B^* = \rho_B^U \quad \text{if} \quad \rho_B^U < \rho_B^* \quad \text{or} \quad \frac{(1-\rho_A)^2}{\lambda(1+\rho_A)\rho_A} < \frac{1}{\rho_B^U} + 1, \tag{C.9}$$

$$\rho_B^* = \rho_B^L \quad \text{if} \quad \rho_B^* < \rho_B^L \quad \text{or} \quad \frac{1}{\rho_B^L} + 1 < \frac{(1-\rho_A)^2}{\lambda(1+\rho_A)\rho_A}. \tag{C.10}$$

For each of the intraclass correlations, ρ_A and ρ_B, there are three possible values that maximize the variance of the treatment effect estimator. Not all values can cooccur. (C.5) and (C.8) imply that $(1+\rho_B^*)^2(1+\rho_A^*)^2 = (1-\rho_B^*)^2(1-\rho_A^*)^2$, which in turn is true only if $\rho_A^* = \rho_B^* = 0$, an unrealistic condition that we will not consider. Furthermore, if $\rho_B^* = \rho_B^L$ then $(1/\rho_B^L) + 1 < ((1-\rho_A)^2)/(\lambda(1+\rho_A)\rho_A)$(C.10), which implies that $\lambda < (((1-\rho_A)^2)/((1+\rho_A)\rho_A)) \times (\rho_B^L/(\rho_B^L+1))$. If at the same time $\rho_A^* = \rho_B(1+\rho_B)/(\lambda(1-\rho_B)^2 - \rho_B(1+\rho_B))$, then also $\lambda = ((1+\rho_A)/\rho_A) \times (\rho_B^L(\rho_B^L+1)/(1-\rho_B^L)^2)$. The latter two conditions for λ imply that $(1+\rho_A)^2(1+\rho_B^L)^2 < (1-\rho_A)^2(1-\rho_B^L)^2$, which is never true. On thc other hand, if $\rho_A^* = \rho_B(1+\rho_B)/(\lambda(1-\rho_B)^2 - \rho_B(1+\rho_B))$, then it is possible that $\rho_B^* = \rho_B^U$. It can also be shown that if $\rho_B^* = \lambda\rho_A(1+\rho_A)/((1-\rho_A)^2 - \lambda\rho_A(1+\rho_A))$, then it is only possible that $\rho_A^* = \rho_A^U$. So, all this implies that we can rewrite the "if" conditions in (C.5)–(C.7), with ρ_B replaced by ρ_B^U, and the "if" conditions in (C.8)–(C.10), with ρ_A replaced by ρ_A^U.

Finally, we can show that the combination $\rho_A^* = \rho_A^L$ and $\rho_B^* = \rho_B^L$ cannot occur. This implies the inequalities in (C.7) and (C.10), which in turn imply that $(1/\rho_A^L) + 1 < (\lambda(1-\rho_B^U)^2)/((1+\rho_B^U)\rho_B^U)$ and $(1/\rho_B^L) + 1 < ((1-\rho_A^U)^2)/(\lambda(1+\rho_A^U)\rho_A^U)$, respectively, which imply $\lambda > ((1+\rho_A^L)/\rho_A^L) \times ((1+\rho_B^U)\rho_B^U/(1-\rho_B^U)^2)$ and $\lambda < ((1-\rho_A^U)^2/((1+\rho_A^U)\rho_A^U)) \times (\rho_B^L/(\rho_B^L+1))$, respectively, which are incompatible. All other combinations of the maximin values ρ_A^* and ρ_B^* can be shown to be possible. These results lead to the procedure for determining the maximin design as delineated in Table 2 of the main text for an extended parallel design.

Similar derivations can be given for the maximin versions of a crossover and extended parallel design in the case interest is in the treatment by period interaction effect. These derivations are, upon request, available from the author.

D. Powers from the Monte Carlo Simulations for Maximin Designs in the Case of the Treatment by Period Interaction

Table 6 provides the simulated powers for maximin designs in the case of the treatment by period interaction. For each

TABLE 6

Treatment by period interaction				
		N	Crossover design	Extended parallel design
Type I error rate = 0.05				
$[\rho_A^L, \rho_A^U]$	$[\rho_B^L, \rho_B^U]$			
[0.01, 0.10]	[0.01, 0.10]	110	**0.799** (0.90)	0.840 (1)
[0.01, 0.30]	[0.01, 0.30]	130	**0.801** (0.76)	0.899 (1)
[0.70, 1.00]	[0.70, 1.00]	34	0.206 (0.15)	**0.830** (1)
[0.90, 1.00]	[0.90, 1.00]	15	0.111 (0.05)	**0.883** (1)
Type I error rate = 0.01				
$[\rho_A^L, \rho_A^U]$	$[\rho_B^L, \rho_B^U]$			
[0.01, 0.10]	[0.01, 0.10]	166	**0.808** (0.90)	0.848 (1)
[0.01, 0.30]	[0.01, 0.30]	194	**0.800** (0.76)	0.912 (1)
[0.70, 1.00]	[0.70, 1.00]	51	0.117 (0.15)	**0.840** (1)
[0.90, 1.00]	[0.90, 1.00]	23	0.048 (0.05)	**0.911** (1)

Note. The power printed in bold indicates the design for which the sample calculation should yield a power of 80%. *N*: total sample size.

pair of ranges of the intraclass correlations, the asymptotic efficiency of each design versus the most efficient design is given within brackets.

Conflicts of Interest

The author declares that there are no conflicts of interest regarding the publication of this paper.

References

[1] B. Jones and M. G. Kenward, *Design and Analysis of Cross-Over Trials*, Chapman & Hall, London, UK, 1989.

[2] S. Senn, *Cross-Over Trials in Clinical Research*, Wiley, Chichester, UK, 2002.

[3] H. Brown and R. Prescott, *Applied Mixed Models in Medicine*, Wiley, Chichester, UK, 2006.

[4] M. P. F. Berger and W. K. Wong, *An Introduction to Optimal Designs for Social and Biomedical Research*, Wiley, Chichester, UK, 2009.

[5] M. J. J. M. Candel, "Parallel, AA/BB, AB/BA and Balaam's design: efficient and maximin choices when testing the treatment effect through mixed effects linear regression," *Pharmaceutical Statistics*, vol. 11, no. 2, pp. 97–106, 2012.

[6] R. J. Grissom, "Heterogeneity of variance in clinical data," *Journal of Consulting and Clinical Psychology*, vol. 68, no. 1, pp. 155–165, 2000.

[7] L.-C. Huang, M.-J. Wen, and S. H. Cheung, "Noninferiority studies with multiple new treatments and heterogeneous variances," *Journal of Biopharmaceutical Statistics*, vol. 25, no. 5, pp. 958–971, 2015.

[8] Y. Yuan and J. Zhou, "Cost-efficient higher-order crossover designs for two-treatment clinical trials," *Pharmaceutical Statistics*, vol. 4, no. 4, pp. 245–252, 2005.

[9] B. W. Brown Jr., "The crossover experiment for clinical trials," *Biometrics*, vol. 36, no. 1, pp. 69–79, 1980.

[10] R. Garcia, M. Benet, C. Arnau, and E. Cobo, "Efficiency of the cross-over design: an empirical evaluation," *Statistics in Medicine*, vol. 23, no. 24, pp. 3773–3780, 2004.

[11] M. G. Kenward and J. H. Roger, "The use of baseline covariates in crossover studies," *Biostatistics*, vol. 11, no. 1, pp. 1–17, 2010.

[12] G. J. P. Van Breukelen and M. J. J. M. Candel, "Calculating sample sizes for cluster randomized trials: we can keep it simple and efficient!," *Journal of Clinical Epidemiology*, vol. 65, no. 11, pp. 1212–1218, 2012.

[13] M. J. J. M. Candel and G. J. P. Van Breukelen, "Sample size calculation for treatment effects in randomized trials with fixed cluster sizes and heterogeneous intraclass correlations and variances," *Statistical Methods in Medical Research*, vol. 24, no. 5, pp. 557–573, 2015.

[14] J. Cohen, "A power primer," *Psychological Bulletin*, vol. 112, no. 1, pp. 155–159, 1992.

[15] R Core Team, *R: A Language and Environment for Statistical Computing*, R Foundation for Statistical Computing, Vienna, Austria, 2013, http://www.R-project.org/.

[16] J. F. Donohue, C. Fogarty, J. Lötvall, D. A. Mahler, and H. Worth, "Once-daily bronchodilators for chronic obstructive pulmonary disease: indacaterol versus tiotropium," *American Journal of Respiratory and Critical Care Medicine*, vol. 182, no. 2, pp. 155–162, 2010.

[17] S. Senn and D. Lambrou, "Robust and realistic approaches to carry-over," *Statistics in Medicine*, vol. 17, no. 24, pp. 2849–2864, 1998.

[18] S. R. Searle, G. Casella, and C. E. McCulloch, *Variance Components*, Wiley, New York, NY, USA, 1992.

13

A Review on the Extraction of Quantitative Retinal Microvascular Image Feature

Kuryati Kipli,[1] Mohammed Enamul Hoque,[1] Lik Thai Lim,[2] Muhammad Hamdi Mahmood,[3] Siti Kudnie Sahari,[1] Rohana Sapawi,[1] Nordiana Rajaee,[1] and Annie Joseph[1]

[1]*Department of Electrical and Electronics Engineering, University Malaysia Sarawak (UNIMAS), 94300 Kota Samarahan, Kuching, Malaysia*
[2]*Department of Ophthalmology, Faculty of Medicine and Health Sciences (FMHS), University Malaysia Sarawak (UNIMAS), 94300 Kota Samarahan, Sarawak, Malaysia*
[3]*Department of Para-Clinical Sciences, Faculty of Medicine and Health Sciences (FMHS), University Malaysia Sarawak (UNIMAS), 94300 Kota Samarahan, Sarawak, Malaysia*

Correspondence should be addressed to Kuryati Kipli; kkuryati@unimas.my

Academic Editor: Andrzej Kloczkowski

Digital image processing is one of the most widely used computer vision technologies in biomedical engineering. In the present modern ophthalmological practice, biomarkers analysis through digital fundus image processing analysis greatly contributes to vision science. This further facilitates developments in medical imaging, enabling this robust technology to attain extensive scopes in biomedical engineering platform. Various diagnostic techniques are used to analyze retinal microvasculature image to enable geometric features measurements such as vessel tortuosity, branching angles, branching coefficient, vessel diameter, and fractal dimension. These extracted markers or characterized fundus digital image features provide insights and relates quantitative retinal vascular topography abnormalities to various pathologies such as diabetic retinopathy, macular degeneration, hypertensive retinopathy, transient ischemic attack, neovascular glaucoma, and cardiovascular diseases. Apart from that, this noninvasive research tool is automated, allowing it to be used in large-scale screening programs, and all are described in this present review paper. This paper will also review recent research on the image processing-based extraction techniques of the quantitative retinal microvascular feature. It mainly focuses on features associated with the early symptom of transient ischemic attack or sharp stroke.

1. Introduction

One of the most important subfields of biomedical engineering is the analysis of fundus retinal images. Analysis of the human fundus eye images has become the key point for diagnosing the various pathologies of retinal vasculature. Furthermore, image analysis provides a simple and noninvasive visualization of the retinal blood vessels in those high risk ophthalmologic medical conditions [1–3].

The fundus retinal images are directly captured from human eye that includes some other landmarks like microcirculation system of the retina, macula, optic disc, fovea, microaneurysm, and exudates [4]. This cost-effective, simple image acquisition system can be used in the large-scale screening programs and retinal image analysis developing mathematical and computational techniques. In addition, physicians can benefit from this technique as to objectively assess abnormal symptoms such as vessel tortuosity, vessel width, bifurcation angles, branching angles, and vessel caliber. All these features are useful for early detection of hypertensive and diabetic retinopathy, macular degeneration, acute stroke, neovascular glaucoma, and some other cardiovascular diseases [1, 3, 5–9].

Some distinct changes in the retinal microvasculature are recognized as the preindicator of subsequent vascular incidents like ischemic stroke or acute stroke [10]. It was found in many types of research that there is a clear relationship between the ocular funduscopic abnormalities and acute stroke even though the blood pressure and other vascular risk factors are in control [11]. According to the study of [12] with a multiethnic cohort, retinal arteriolar narrowing and retinopathy of diabetic free people have an association with increased risk of acute stroke. But the Cardiovascular Health Study stated that there is no association between retinal arteriolar caliber (diameter) and stroke but rather there is a close association between stroke and the larger venular caliber (diameter) [13]. Rotterdam cohort study also came into a decision after a long-term observation that the retinal venular diameter is associated with any stroke or ischemic stroke [14].

The authors of [15] examined the association of hypertensive retinopathy with the risk of stroke in their population base study. Retinal microvascular abnormalities like microaneurysm, arteriovenous nicking, haemorrhages, and vessel caliber are considered as associative to the stroke and indicative of death from stroke and IHD (Ischemic Heart Diseases) [1]. A population-based study collaborating with Beaver Dam Eye Study (BDES) of [1] revealed that increased diameter ratio was associated with increased stroke mortality (P = 0.02 unadjusted). The paper [16] searched MEDLINE and EMBASE to find out the relation between microvascular changes of retinal microvasculature and prevalence or incident of stroke. In the study of [16] 20659 patients were involved, 1178 patients of them had stroke, and they found the Odd Ratio (OR) of stroke and retinal arteriolar narrowing and AV nicking was 1.42 and 1.91, respectively, which indicated that these abnormalities are slightly associated with stroke. Microvascular lesions like microaneurysm and haemorrhages which were found as highly associated with stroke as their OR were 3.83 and 3.21, respectively, and the OR between stroke and arteriole narrowing and venular network are 2.28 and 1.80 indicated the association of these abnormalities with stroke [16]. A cohort study of Asian Malay persons consisting of 3189 patients, free from prevalent stroke at baseline, revealed that 51 (1.93%) participants had an incident stroke event that could be predicted by analyzing the microvascular changes of retinal vasculature [17]. Prevalent and incident stroke have the association with retinopathy and venular widening [18]. Retinal vessel widening is also the predictor of hypertensive retinopathy which also has an association with stroke [19].

Many techniques have been developed using image processing principle to measure the vessel diameter of retinal microvasculature. Suppose some of the available vessel diameter measurement techniques are based on Linear Discriminant Analysis (LDA), image gradient segmentation technique (ARG) for vessel edge detection, using active contour [20–22], mask creation [20], graph-theoretic method [23, 24] Multistep Regression Method (Higher order Gaussian modeling) [25], Adaptive Higuchi's Dimension [26], and so on.

There are several datasets for fundus retinal images that are publicly available for the researchers from all over the world. High-Resolution Fundus Image Database (HRFID) and Digital Retinal Images for Vessel Extraction (DRIVE) are two kinds of publicly open datasets that can be accessed by the researchers [6, 27]. Researchers from digital image processing discipline can use images from these datasets as the raw materials for their experiment. REVIEW (Retinal Vessel Image set for Estimation of Width) is another publicly available standard dataset that includes various normal and pathological retinal images for the evaluation of the algorithm for retinal vessel diameter measurement. REVIEW dataset includes 16 images with 193 vessel segments that demonstrate the different types of pathologies and vessel types and this dataset consists of four subsets which are HRIS (High-Resolution Image Set), VDIS (Vascular Disease Image), CLRIS (Central Light Reflex Image Set), and KPIS (Kick Point Image) [28].

This review is focused mainly on the retinal vessel diameter measurement algorithms, applications of image processing technique, to provide an overview of the recent advancement in measuring the diameter of the retinal blood vessel. In biomedical engineering, the existing retinal image analysis methods are still dependent on a bulk of human observers. The blood vessel diameter measurement methods are also not fully automated. There is limited published work on automation of blood vessel diameter measurement. For medical diagnosing systems, maximum precision is a must for detecting the diseases as it is one of the most sensitive issues for proper diagnosing. This review is a prestep of our future work of shaping a novel, automated, and computer-aided algorithm to measure the diameter of the retinal blood vessel with maximum accuracy.

In the following sections of this paper, we briefly discussed the abnormalities of retinal microvascular features, responsible for stroke and the circumstantial scenario of the existing methods for analyzing these abnormalities. Here we emphasized to manifest especially the retinal blood vessel diameter measurement techniques as it is one of the most important markers of prevalent and incident stroke. The basic principle of digital image processing and a general digital image processing procedure for extracting the features of the retinal image containing image acquisition, grey-scale image, image enhancement, restoration, segmentation, registration and vessel extraction subsection are also described in this paper in a cabalistic manner. The arising challenges of present retinal image processing techniques for measuring the blood vessel diameter and the future scopes of this field are also explained in this paper.

2. Retinal Vascularization

The retina is a highly vascular tissue, which receives supply from two sources, namely, central and choroidal circulatory system (Figure 1). The central blood vessel supplies the inner retina which made ~30% of the retinal blood flow [40]. Central retinal artery (CRA) runs through the optic disc and enters the inner layer of the retina. CRA branches into superior- and inferior- arteries with diameters of 150μm [41]. Ultimately it forms a network of capillaries with diameters of 5μm [42]. On the other hand, blood from the retina drains into the central retinal vein (CRV). Approximately 70% of the retinal blood flow is supplied by the choroidal blood vessel which nourishes the outer retina and the photoreceptor-retinal pigment epithelium (RPE) complex located adjacent

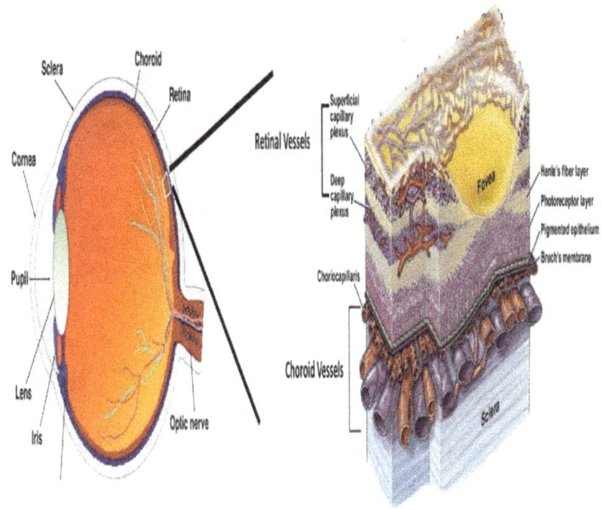

FIGURE 1: Schematic diagram of the retinal vasculature adapted from [29].

to Bruch's membrane [40]. Apart from nourishing the region, the choroidal circulatory system serves as a heat sink which removes generated metabolic heat due to light photons strike on the photopigments and the melanin of the RPE [43]. Figure 1 shows the schematic diagram of the retinal vasculature.

2.1. Pathology of Retinal Vasculature. There is a range of eye diseases particularly affecting retinal blood vessel. Among determinant factors of retinal vascular disorders are a physiological hardening of the artery walls in aging and existing vascular diseases such as high blood pressure and atherosclerosis. Most common retinal vascular disorders are hypertensive retinopathy, Retinal Vein Occlusion (RVO), Central Retinal Artery Occlusion (CRAO), and Diabetic Retinopathy. Monitoring of the retinal vasculature through retinal imaging and vascular caliber monitoring permit a direct assessment of the retinal vascular disorders.

2.2. Hypertensive Retinopathy. Hypertensive retinopathy occurs when arterioles and venules of the retina are damaged, which eventually leads to blindness. Thus, several classification systems have been developed to facilitate early identification [44]. Hypertensive retinopathy initially affects the retinal blood vessels at all tributary levels, especially the arterial vessels. This process is known as sclerosis (thickening and stiffening of the artery), which can alter the angular course of the artery and may also affect the tributary angles. Furthermore, the artery and vein junction share a common adventitial sheath. Any sclerotic changes to the artery in this enclosed common "compartment" space can further compress the weaker venular vessels, which can result in further changes to the angular course of the veins as well [45]. Such angular changes can be further studied to determine its significance in the prediction and assessment of disease process and progression.

2.3. Retinal Vein Occlusion (RVO). Retinal Vein Occlusion (RVO) disorder is commonly associated with various underlying systemic disorders including arterial hypertension, diabetes mellitus, dyslipidemia, and systemic vasculitis. RVO may be found in varying blood vessel caliber. This includes central retinal vein occlusion (CRVO); media such as hemicentral retinal vein occlusion and small-caliber veins such as branch retinal vein occlusion (BRVO) [46].

Retinal vein occlusion can be divided into central and branch. The branch retinal vein occlusion can happen at the various tributary levels, especially the first and second tributaries. However, prior to a full-blown vein occlusion, a temporary phase known as impending retinal vein occlusion may occur. In this phase, the affected veins will become more engorged, dilated, and tortuous, and the course and tributary angle of the affected veins will alter with this pathological process [50]. These angular changes can be further studied to determine its significance in the prediction and assessment of disease progression.

2.4. Central Retinal Artery Occlusion (CRAO). Central Retinal Artery Occlusion (CRAO) is an ophthalmic emergency analogous to the acute stroke of the eye. Unfortunately, presentation of such cases to the appropriate medical attention passed the stage of reversibility. The clinical presentation with the central artery occluded is invariably a pale retina with a central "cherry red spot," as the retina is devoid of oxygenated blood. It signifies end-organ ischemia and often the underlying atherosclerotic disease [55].

2.5. Diabetic Retinopathy. One-third of people with diabetes would have diabetic retinopathy (DR). This condition is ranked as the fifth most common cause of blindness worldwide [56]. DR is strongly associated with a prolonged duration of diabetes, hyperglycaemia, and hypertension. Severe stages of DR include proliferative DR, caused by vascular leakage and new retinal blood vessels growth amplified by chemokine secretion such as vasculoendothelial growth factor (VEGF) and diabetic macular oedema, in which there are exudation and oedema in the central part of the

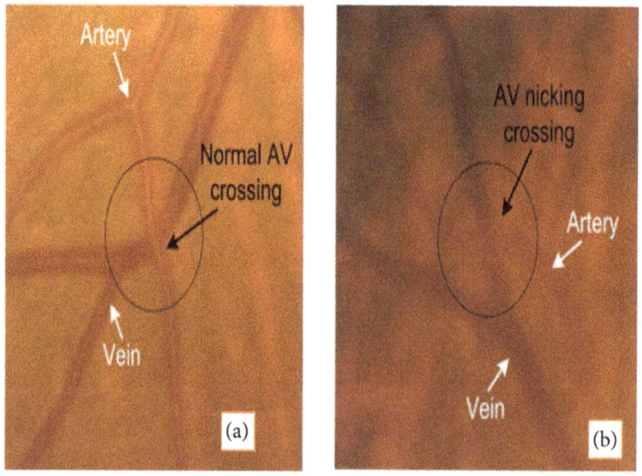

(a) (b)

FIGURE 2: Artery-vein crossing phenomenon. (a) Common AV crossing and (b) affected AV crossing (nicking) [30].

retina [57, 58]. At present, intraocular treatment modalities for diabetic eye disease include laser photocoagulation, intravitreous injections of anti-VEGF and steroid agents, and vitreoretinal surgery [58].

The pathophysiology of diabetic retinopathy initially affects the retinal blood vessels. Due to poor glycaemic control, the advanced glycation end products (AGEs) will invariably damage the pericytes which are the cellular supporting structure of the retinal capillaries. This leads to weaker and hence leaky capillaries and promoting microhypertension of the retinal blood vessels. This, in turn, will result in various retinal vascular changes including dilated, tortuous vessels with altering course and tributary angles [59]. This angular change can be further studied to determine its significance in the prediction and assessment of disease progression.

3. The Image Processing Techniques for Analyzing and Detecting the Abnormal Features of the Retinal Vasculature in Stroke

The microvascular structure of retinal blood vessel plays a significant role in diagnosing the possibility of causing stroke. Researchers found from population-based studies that diabetic retinopathy signs like microaneurysm and hard exudates and the hypertensive retinopathy signs like arteriovenous nicking and focal retinal arteriolar narrowing were also associated with the acute stroke and stroke mortality even if the people were free from other stroke risk factors [19, 48].

A cohort study in [49] investigated that the most retinal microvascular features that predicted incident stroke and ischemic stroke and that adjusted relative risk with any retinopathies are microaneurysm, soft exudates, bolt haemorrhages, flame-shaped haemorrhages, arteriovenous (AV) nicking, and decreasing Artery to Vein Ratio (AVR) [19, 49, 51, 60, 61].

3.1. AV Nicking Observation. AV nicking or AV nipping is abnormality phenomenon observed in the retinal microvascular structure during ophthalmological imaging which shows the characteristics view of crossing a small artery over a vein. Due to this arteriovenous crossing, the vein inflates either side of artery cross-section point. In several types of research related to retinal imaging, AV nicking has been found as the early indicator of eye-related diseases such as BRVO, diabetes, hypertension, and acute stroke [30, 60, 61]. Figure 2 shows the normal and affected AV crossing scenario.

The study of [30] proposed a computer-based algorithm to calculate the AV nicking severity applying the image processing techniques like image acquisition, segmentation, artery-vein classification method, and method for measuring vessel width. They validated this experiment on 47 high-resolution retinal images that were obtained from two different population-based studies. The AV nicking values obtained from this research showed a great correlation with expert grading and the accuracy of this finding is 88-89% which clarifies that severe and moderate AV nicking can be detected more precisely [30].

3.2. Microaneurysm Detection. Generally, microaneurysm is a specific small area that is seen in the microvasculature of fundus eye image which looks like a swelled balloon. Microaneurysm is recognized as the biomarker of diabetic retinopathy as well as the presign of ischemic stroke or acute stroke that is generally observed in the fundus retinal images after analyzing the eye images [19, 48]. Figure 3 shows retinal fundus image with microaneurysm (inside the white square).

An automated microaneurysm detection method was investigated in [31] which was applied on the 50 training cases from Retinopathy Online Challenge (ROC) dataset and they came out with 45% sensitivity with their proposed method when false positive rate per image was 27. Authors of [62] used MA-Tracker, a semiautomated method, to observe the relation between the state of diabetic retinopathy and

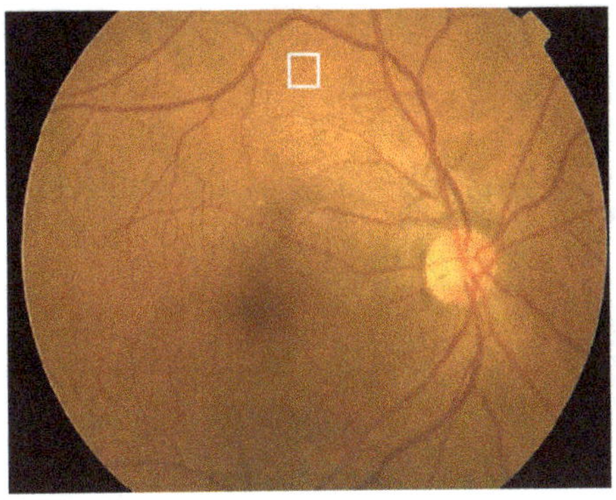

FIGURE 3: Retinal fundus image with microaneurysm (inside the white square) [31].

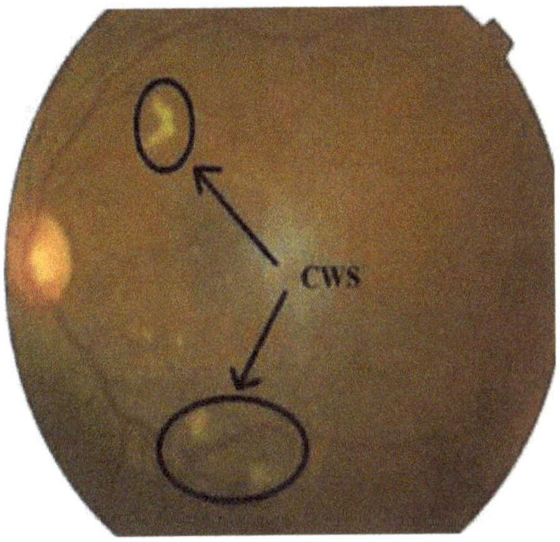

FIGURE 4: Cotton wool spot in retinal fundus image (in black circle) [32].

microaneurysm and they experienced that the formation rate of microaneurysm is a better bioindicator of the development of diabetic retinopathy to clinically significant macular edema (CSME) among the patients of type 2 diabetes with Nonproliferative Diabetic Retinopathy. Authors of [63] developed an automated algorithm that can count the microaneurysm and detect the turnover of microaneurysm from a digital Fluorescein angiogram repeatedly which dealt with 64 images of diabetes patient and they observed that this algorithm showed 82% sensitivity along with 2.0 false positive rate per image.

3.3. Cotton Wool Spot Detection. In the fundoscopic test of the retina of human eyes, Cotton Wool Spots (CWS) are found as the abnormality which looks like a white fluffy patch. They are also considered as the retinal infarcts which mean the damage of tissues as a result of insufficient oxygen supply because of the blockage of blood supply to the tissues and this mostly happens in the retina of the diabetic patient that leads them to be in risk of acute stroke [32]. The configuration of CWS is found similar in the fundus eye images of both diabetes and hypertension patient ([64]; Irshad & Akram, 2015;). Figure 4 displays the CWS in fundus retinal image inside the black circle.

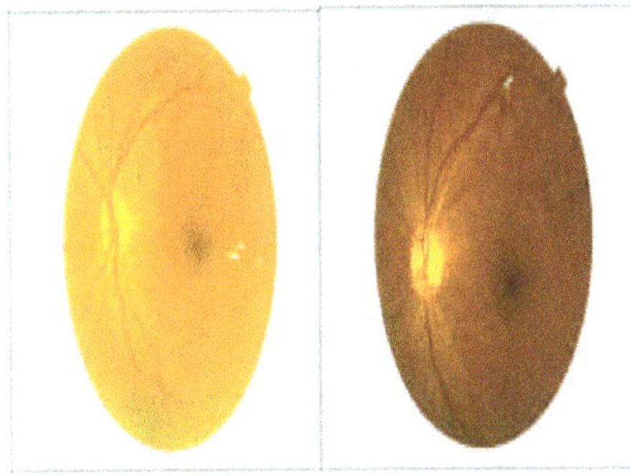

FIGURE 5: Hard exudates in fundus retinal image [33].

Researchers of [65] developed an automated system based on machine learning to detect the CWS and differentiate this from drusen existing in color images that were collected from diabetic patients and the researchers were able to obtain the sensitivity pairs of 0.95/0.86 to detect CWS.

3.4. Hard Exudates Detection. Hard exudates found yellow or white flecks in the outer layer of the retinal vasculature of those humans who are affected by diabetes. More generally, hard exudates are the microaneurysm that is found as white dots without blood in the lumen. Sometimes these exudates are deposited along with the vein of the retina. Hard exudates are bright interretinal protein deposition and considered as the hallmark of diabetes [66]. Figure 5 shows the hard exudate in the fundus retinal image.

An automatic image processing technique was developed by authors of [67] following Fisher's Linear Discriminant Analysis with a dataset containing 58 fundus retinal images concentrating on the variables like quality, color, and brightness which gained 88% sensitivity with the false positive mean number 4.83±4.64 per image. They used performance evaluation criterion based on the lesion and obtained 100% accuracy of image-based classification [67]. The study of [68] proposed an automated image segmentation algorithm to segment the exudates in fundus retinal images from DRIVE and STARE dataset using their green component and preprocessing steps such as contrast adjustment, average filtering, thresholding, and finally they experienced the 96.7% sensitivity [68]. The authors of [33] also performed an experiment that showed the improvement of accuracy in detecting hard exudates in retinal images and they achieved 100% and 74% specificity.

The authors of [69] detect hard exudates using the mixture model and thresholding technique to isolate the exudates from the background and then applied edge detection to differentiate exudates from CWS. This formula showed 90.2% sensitivity and 96.8% positive predictive value. The accuracy of the image-based classification was satisfactory as the 100% sensitivity and 90% specificity were obtained [69].

3.5. Focal Arteriolar Narrowing Measurement. Focal arteriolar narrowing is also a symptom for the early detection of hypertension and acute stroke. Focal arteriolar narrowing generally occurs with the increasing arteriolar blood pressure from 130mm Hg to 160mm Hg that affects the arteriolar wall [70]. The study of [71] investigated the relationship between the arteriolar narrowing and hypertension and found that the hypertension increasing rate is higher among the people with arteriolar narrowing. Figure 6 shows the focal arteriolar narrowing in fundus retinal image.

3.6. Vessel Width Measuring Method. As the retinal microvascular sign for predicting the stroke possibility, artery and vein diameter were widely examined to obtain more accurate measurement system and, at the very beginning of this study, AVR or arteriolar Length to Diameter Ratio (LDR) has been normalized to compute artery and vein diameter. But Atherosclerosis Risk in Communities (ARIC) and Beaver Dam Eye Study (BDES) did not find any association with AVR or LDR and stroke. However, when the measurements of retinal vessel diameter have been normalized to optic nerve head diameter association with stroke was found. Central Retinal Artery Equivalent and Central Retinal Vein Equivalent were explored to meet the challenge raised in the study of retinal vascular dimension when retinal microvascular anatomy has been found different in different individuals [72].

Retinal vessel widening is a preindication of hypertensive retinopathy and researchers found there is an association with retinal vessel diameter and acute stroke [19]. Table 1 shows the association between retinal vessel diameter and stroke.

3.7. Haemorrhages Detection. Haemorrhage is an abnormality that appeared in the retinal blood vessels of human eye due to the bleeding in the light-sensitive tissues on the back wall of the eye. Haemorrhages are generally seen in the retina of people with hypertension. Haemorrhages are observed in different shapes with red color and this shape can be correlated with the depth in the retina [35, 73]. The intensity of blood vessel and haemorrhages are similar and the only way to identify the haemorrhage is to eliminate the blood vessels from the blood vessels with haemorrhages which can be done by using ball-shaped Structuring Element (SE) of size 6 and size 25 simultaneously [74]. Figure 7 shows (a) large superficial haemorrhages, (b) fundus image with haemorrhages, (c) detail dot haemorrhages, and (d) bolt haemorrhages.

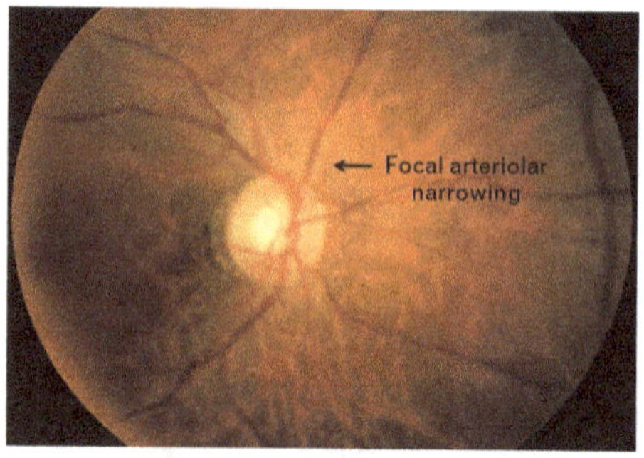

Figure 6: Focal arteriolar narrowing indicated by black arrow [34].

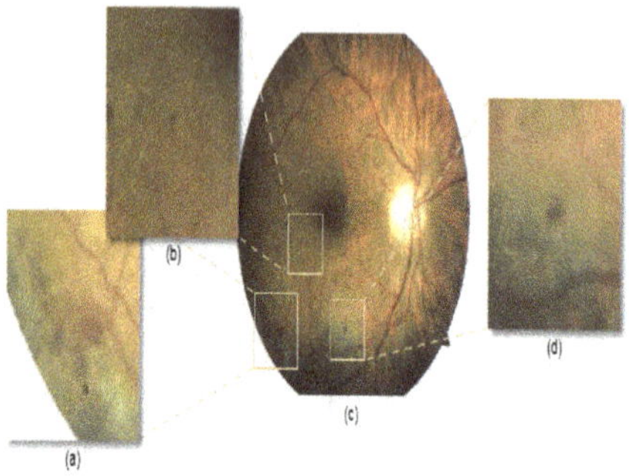

Figure 7: Different types of haemorrhages in fundus retinal image [35].

4. Generic Feature Extraction Process of Retinal Vasculature

Image processing is the method of applying mathematical operations in signal processing systems where image or video is fed as input and the output also is either image or a group of features or parameters that are related to the image [75]. Digital imaging accomplishes functions on a digital image. It is being used in image enhancement, data compression, and machine vision and deals with difficulties from edge detection to pattern recognition and reconstruction [76, 77]. In biomedical engineering, digital image processing is being applied in many researches and diagnosing the diseases, planning and supervising treatment for that disease, and monitoring the state of diseases simultaneously [78]. Digital image processing is playing an important role in medical sector to reduce the involvement of observers in avoiding unexpected errors and getting a more precise result [79].

Many life-threatening cardiovascular diseases like diabetes, hypertension, and stroke are related to the early change of the caliber of retinal microvasculature. It is the boon of modern acquisition technology that high-resolution retinal images can be captured easily for analyzing the risk of stroke among the individuals that are at risk of cardiovascular diseases and stroke. It is also responsible for enlarging the database of retinal images. The development of automated and computer-aided quantitative measurement techniques based on image processing for monitoring the changes of this microvascular caliber of the human retina has become the situation demand to ensure the maximum accuracy in detecting these destructive diseases and avoid the bulkiness of current diagnostic systems [80].

There are three major operations performed in image processing to process an image for extracting special features of the image. The first one is the enhancement of input image like contrast improvement, the second one is image restoration which means deblurring image, and the third one is the segmentation of the image, which means separating the certain portion of the image that is considered as the area of interest [81]. The related image processing techniques for extracting different features of fundus retinal image are described in Figure 8.

4.1. Acquisition of Fundus Retinal Image. The term image acquisition can be defined generally as obtaining an image from any hardware based source. Image acquisition is the

TABLE 1: Association between retinal vessel diameter and stroke [19].

Outcome	Study	Total Sample size	Methodology	Association with Retinal Vessel Diameter
Prevalent stroke	CHS [6]	2050	Clinical	0
	CHS [40]	1717	MRI	AVR
	ARIC [28]	1684	MRI	AVR*
Prevalent WML	ARIC [47]	1684	MRI	0
	CHS [40]	1717	MRI	AVR
	Rotterdam [48]	490	MRI	Venular diameter
Progression WML				
3.3-year	Rotterdam [48]	490	MRI	Venular diameter
5-year	CHS [40]	1717	MRI	AVR
Incident stroke				
3.3-year	Rotterdam [48]	490	MRI	0‡
3.5-year	ARIC [5]	10 358	Clinical	0
5-year	CHS [40]	1717	MRI	AVR
5-year	CHS [49]	1992	Clinical	Venular diameter
7-year	BMES [8]	3654	CT or MRI	0#
8.5-year	Rotterdam [50]	5540	CT or MRI	Venular diameter
10–12 years	Pooled BDES BMES [51]	7494	Clinical	0
Stroke mortality			ANDI	
7-year	BMES [8]	3654	Death certificate	0
10-year	BDES [24]	413 cases		AVR
		1251 controls	Death certificate	
10–12	Pooled BDES BMES [51]	7494	or ANDI	0

FIGURE 8: Conventional feature extraction process.

initial and most significant step of image processing. The performance of processing image for any intended job extensively depends on the performance of image acquisition. Retinal fundus imaging can be explained as the process of obtaining a two-dimensional representation of three-dimensional semitransparent tissues of retinal microvasculature projecting on to the imaging plane and then reflecting the required amount of light [82].

A technique was developed in [83] for generating seamless, high-quality, and wide field montage which could be good for real-time photo documentation of the disc and macular abnormalities. The technique used alignment with high accuracy and blending of partially overlapped slit lamp biomicroscopic fundus image [83]. The study of [84] developed an algorithm to obtain fundus intensity image using Scanning Laser Ophthalmoscope (SLO) quality from the main spectra that were measured with spectral-domain Optical Coherence Tomography (OCT) and this algorithm provides fundus and OCT images together that can avoid the complexity of registering the fundus feature of any cross-sectional OCT image. Afterwards, this algorithm was extended to generate high contrast shadow-grams of the retinal blood vessel to facilitate the OCT data registration to the further imaging systems [84].

The study of [85] demonstrated a technique for high-speed ultrahigh-resolution 3-dimensional OCT retinal imaging with and retinal imaging protocols using Fourier domain detection. This technique used a dense raster scan pattern to obtain the three-dimensional OCT data of macula and optic disc. Retinal and interretinal layer and nerve layer thickness can also be mapped using this system [85]. The study of [86] also developed a real-time imaging approach using a single pixel camera for capturing a perfect fundus retinal image.

In modern ophthalmology, there are a lot of image acquisition techniques already developed even using the smartphone. The built-in camera technology, exciting development of cloud storage, and technology for accessing electromedical history using smartphone have encouraged the physicians to use the smartphone in ophthalmic imaging and lead to a dream to have a teleophthalmology system fully based on a smartphone. Welch Allyn first developed an imaging adaptor based on smartphone attaching an iPhone with Welch Allyn Panoptic Ophthalmoscope that can take pictures of retina including the iExaminer App [87]. The noble design of a compact, slim, smooth, and 3D printed attachment that allows high-quality fundus image coupling smartphone to indirect ophthalmoscopy condensing lenses was reported in [87].

4.2. Grey-Scale Retinal Image Processing. The grey-scale image can be represented as data matrix and the value of this matrix illustrates the shades of grey. If the elements of a grey-scale image are of class "unit8," the range of their integer values will be [0, 255] and if the elements are of class "unit16" then the range will be [0,65535]. When the class of a grey-scale image is "single" or "double," normally its value can be scaled in the range [0,1] [88].

Converting an RGB color image into the grey-scale image is the first step of many image analysis workflows because in the grey-scale image the amount of information is simplified and maintained only the information related to the features of the image like edges, regions, blobs, and junctions that are needed to be analyzed [89]. The following transformation equation is used to convert the RGB color image into grey-scale image:

$$\mathrm{I_{gray}}(\mathrm{n,m}) = \alpha \mathrm{I_{color}}(\mathrm{n,m,r}) + \beta \mathrm{I_{color}}(\mathrm{n,m,g}) + \gamma \mathrm{I_{color}}(\mathrm{n,m,b})\ [89]. \tag{1}$$

Here $\mathrm{I_{gray}}$ is grey-scale image, $\mathrm{I_{color}}$ is color image, $\alpha = 0.2989$, $\beta = 0.5870$, $\gamma = 0.1140$, (n,m) is pixel location with grey-scale image, and (n,m,c) is a channel at pixel location (n,m) in the color image for channel c in red r, blue b, and green image channel [89].

There is the possibility of losing the important features of an image like structure, sharpness, contrast, and shadow during the conversion of the color image into grey-scale image. The study of [90] proposed an algorithm for converting the RGB image into the grey-scale image which can do RGB approximation, reduction, and addition of luminance and chrominance and sustain the structure, sharpness, contrast, and shadow of the original RGB color image in the resultant image. The study of [91] presented a formula based on the Singular Value Decomposition (SVD) to measure the graphical and scaler values of distortion caused by the noise sources. As a step of fundus retinal image processing, the raw RGB color image is converted into grey-scale using the green channel because the blood vessels of retinal microvasculature appear more contrasted through green channel [36]. Figure 9 shows the original image (a) and grey-scale image (b) after conversion.

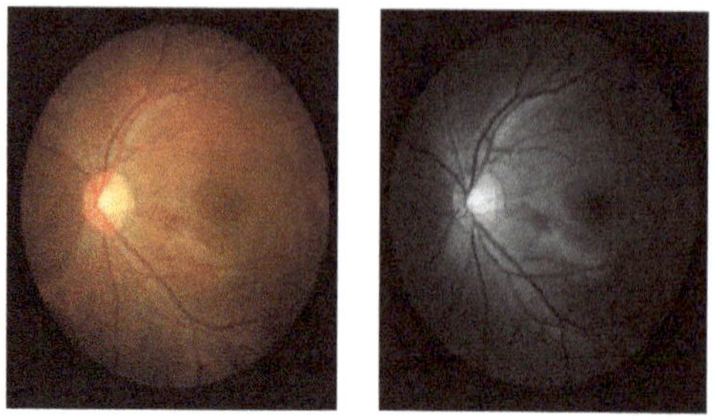

(a) (b)

FIGURE 9: Original image (a) and grey-scale image (b) of the retina [36].

4.3. Enhancement of Fundus Retinal Image. Image enhancement is the procedure to synthesize the digital images for making the output more acceptable to display or for further analysis of that image. Image enhancement is also performed to adjust the contrast and normalize the images. Most often the fundus retinal images do not illuminate uniformly and show local luminosity and irregularity in contrast [92]. The fundus retinal images of human have different outlook due to the skin pigmentation of the subject. To gain a good sensitivity and specificity in detecting lesion of fundus retinal images following any automated analytical method, color normalization of that image is a must [93].

Some of the popular image enhancement techniques are removing the noise from the image, sharpening the edges of the image, blurring the image, and so on which can be accomplished by spatial domain filtering because spatial domain filtering can directly act on the images and change the pixel values of the image through some specific procedure [89].

The authors of [93] demonstrated a system of color normalization performing intraimage shade-correction interimage histogram normalization and they detected the microaneurysm in fundus retinal images applying the effect of their technique. The authors of [92] also proposed a method for normalizing the contrast and luminosity in both intra- and interimages of the human retina and the researchers came out with the average luminosity irregularity reduction of 19% covering maximum 45% and an average improvement of contrast of 34% covering maximum 85%.

4.3.1. Retinal Image Restoration. Image restoration can be defined as recovering or reconstructing a degraded image utilizing the earlier concept of image degradation phenomenon. The techniques of restoring an image are designed towards modeling the degradation and using the reverse procedures to regain the real image.

In fundus retinal image analysis restoration process plays a vital role to avoid the blurring and uneven illumination due to the image acquisition process. The authors of [47] proposed a technique for the restoration of the color retinal

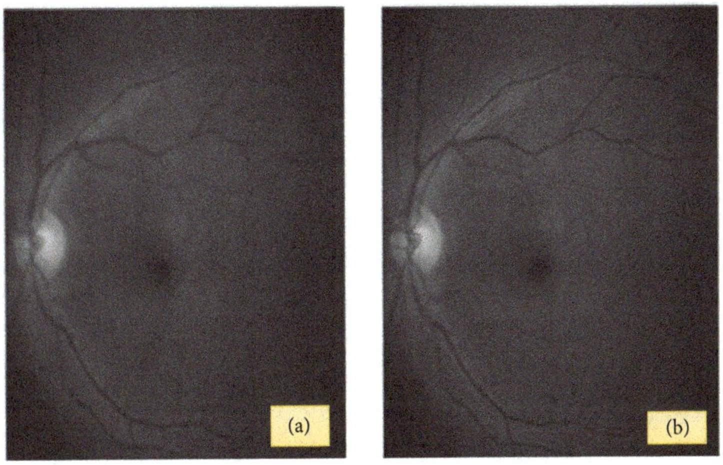

(a) (b)

FIGURE 10: (a) Original and (b) restored image using Space-Variant Point-Spread Function [37].

image through multichannel blind deconvolution. Basically, this technique is designed by composing image registration, uneven illumination compensation, and segmentation that had been validated by applying on both synthetic and original retinal image. It was experienced that this method is able to restore the degraded retinal image and can detect and picturize the structural changes of the retinal image too [47].

Fundus eye image can be degraded with blur due to the inappropriate acquisition or congenital optical shedding in the eye. The restoration process is either space-invariant or space-variant and maximum existing algorithm for image deblurring can deal with the space-invariant blur but cannot work with space-variant (SV) blur image [37]. The authors of [37, 77] presented a retinal image restoring algorithm that can deal with both unknown and space-variant blur and formulate this algorithm and they shaped the blur, interpreting linear operation for convolution with Point-Spread Function (PSF) that changes with the position in the image.

The study of [94] also presented a method to obtain a true estimation of Point-Spread Function (PSF) for restoring retinal image through space-invariant or space-invariant blind deconvolution based on the decomposition in Zernike coefficient of the estimated PFSs to determine the actual PSFs. Figure 10 shows the original and restored fundus retinal image.

The study of [95] designed a method to recover the spectral retinal image of the common RGB image using fuzzy c-means clustering for quantizing the data of image and radial function network for learning the mapping from RGB representation to the spectral space. And authors of [95] used a spectral quality metric to evaluate their experiment result compared with a set of retinal images having both spectral and RGB image and found the accuracy of their output relatively high.

4.4. Segmentation of Fundus Retinal Image. Image segmentation means the process of dividing an image into its specific constituent Regions of Object of Interest (ROI) to make that image more significant and smooth to analyze. Segmentation of an image can be done for isolating wide ROI and once the ROI is achieved segmentation process can be stopped.

Extraction of different properties of fundus retinal image is either qualitatively or quantitatively. The extracted features have become the key fact for diagnosing many severe cardiovascular diseases. As the changes of different features of retinal microvasculature such as vessel tortuosity, branching coefficient, branching angle, vessel widening or narrowing, arteriovenous nicking, existence of hard exudates, CWS, microaneurysm, and haemorrhages have been detected as early signs of many cardiovascular diseases leading to stroke [1, 3, 5–9, 19, 49, 60, 61], segmentation of fundus retinal image is a must to detect and measure the abnormalities of retinal microvasculature. There are a lot of segmentation procedures based on image modalities, automation or semiautomation, application domain, and some other different factors. Image segmentation algorithms and techniques can be categorized into six main sections such as rigid based vessel segmentation, parallel multiscale feature extraction, and region growing, artificial intelligence based methods, hybrid filtering, miscellaneous tube-like object detection methods, and neural network (NN) based methods [96]. Figure 11 shows the original fundus retinal image of a left eye with normal blood vessel network and segmented blood vessel image.

4.5. Edge Detection for Segmentation. Edge detection is one of the particular procedures of image segmentation to extract boundaries, discontinuities in the intensity of an image that can be used for analyzing that image because edges convey important clues for finding out the key fact of interest. Some structural information like the boundary of the object, illumination, geometry, and reflectance can be obtained analyzing the detected edge [97].

There are several edge detectors based on the principle of function edge that performs better in images with less noise. The available edge detectors based on the function edge are Sobel Edge Detector, Prewitt Edge Detector, Roberts Edge Detector, Laplacian of Gaussian (LoG) Edge Detector,

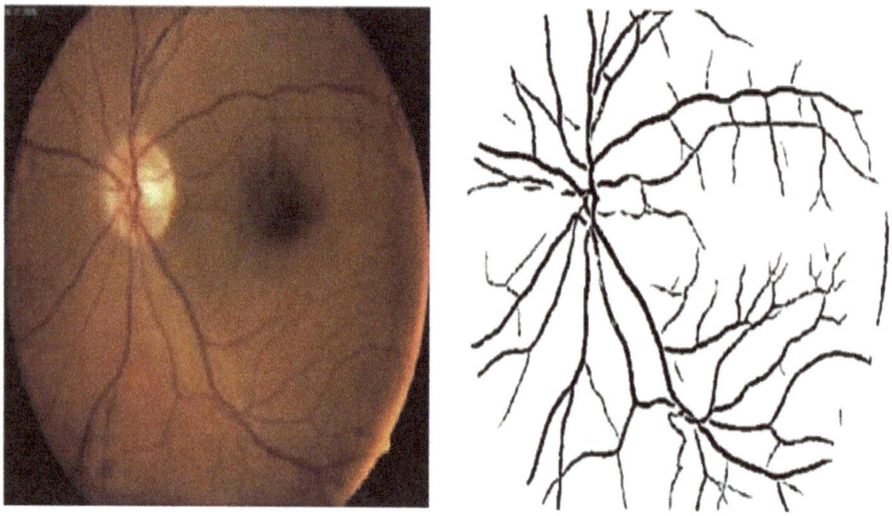

(a) (b)

FIGURE 11: (a) Fundus retinal image, (b) original and segmented [38].

Zero Crossing Edge Detector, and Canny Edge Detector (Gonzalez, Woods, & Eddins, 2009).

Edge detection is an important way to extract the key features of retinal microvasculature. A variety of edge detecting algorithms are being developed in fundus retinal image processing. The authors of [98] developed a fundus coordinate system combining region growing and edge detection to detect the exudates, an important feature of the retinal image, and modified active shape model to detect the disc boundary of fundus retinal image. The success rate of their algorithm in terms of detecting optic disc boundary is 94% and the sensitivity and specificity of detecting exudates are 100% and 71%, respectively [98].

A template-based retinal image segmentation was presented in [99] in which morphological and edge detection technique employing the circular Hough transform for approximating the circular optic disc boundary was applied.

4.6. Image Thresholding. Image thresholding technique is basically an image segmentation technique that is applied to alter a grey-scale image into the binary image to differentiate the objective point of interest. Thresholding is most suitable for analyzing the images with high contrast.

Thresholding is an important technique that is widely used for the segmentation of fundus retinal images to analyze the microvasculature. Thresholding is used in retinal image processing only to highlight the features of interest and avoid the features that are not important for analyzing the image. The authors of [100] used entropy-based thresholding as subtechnique to keep the spatial structure of vascular tree segments in their retinal image analysis algorithm to detect and extract the blood vessels. A knowledge-guided adaptive local thresholding algorithm based on the verification-based multithresholding probing scheme was proposed in [101] to detect the blood vessel in fundus retinal image.

4.7. Image Registration. Image registration is an image processing technique designed for aligning multiple images of the same configuration. Generally, image registration is the procedure to transform multiple datasets, photographs, and times depth into one coordinate system to accomplish several tasks like image rotating, scaling, and skewing. The image configuration of the same object can vary due to the acquisition at different times, different acquisition devices or acquisition inefficiency caused by the variation of camera angle, movement of the objects, orientation, sensor resolution, distance, including other crucial facts and the image registration process; organized to perform the alignment of the geometrical aberrated images, based on a standard image [88].

Image registration plays a supreme act in analyzing the fundus retinal image because it is the way to protect and provide the accurate information that is critically important for diagnosing the related diseases. The study of [102] proposed an algorithm applying a nonlinear registration method based on correlation tracking to enhance the retinal imaging with a high spatial resolution for clear and precise detection of retinal abnormalities. The local correlation was also analyzed in [102] experiment to observe the actual movement of an image in a different time period like the variation in optical flows that can be the feature of interest of diseases diagnosing.

A hybrid retinal image registration technique for Early Treatment Diabetic Retinopathy Studies (ETDRS) was designed in [103]. This technique was able to extract retinal microvascular structure applying local entropy-based thresholding and maximized the mutual information of binary image pair to estimate the zeroth order translation. Image quality was assessed regarding the definition of ETDS based on the translation model and then finally affine/quadratic model estimation had been applied after image pair was accepted [103]. The necessity of temporal image registration is to observe the different steps of disease and the detection of lesions in fundus retinal image can be improved by multimodal image registration. The study of [39] presented a temporal and multimodal retinal image registration technique based on point correspondence. This method first detects the vascular tree and labeled the bifurcation point. After that, it matches the probability for matching two points computing an angle-based invariant and uses a Bayesian Hough transform to distinguish the respective similarities and finally it computes a fine estimation to choose the best similar transformation for registration [39]. Figure 12 shows (a) original fluorescein image; (b) fluorescein image two years later of temporal registration; (c) final result of the registration [39].

An advanced sequential processing method for the retinal image was proposed by authors of [104] where they used cross-correlation followed by a fine registration employing parabolic interpolation on the peak of the cross-correlation, maximum-likelihood estimation for precise registration, and a combination of peak tracking and Procrustes transformation to measure angle rotation of the fundus retinal image.

4.8. Vessel Extraction of Fundus Retinal Image. Extracting the vessel tree of retinal microvasculature is an important step to analyze the microcirculation for the detection of retinal diseases. The study of [105] reported a retinal blood vessel segmentation algorithm following the scale-space analysis of the first and second derivative of the intensity image. A parallel multiscale feature extraction and region growing algorithm based on ITK (Insight Segmentation and Registration Toolkit) was developed for retinal blood vessel segmentation and showed that this method is effective for high-resolution retinal image analysis [80]. The study of [106] designed an automated segmentation and reconstruction method for 3-dimensional retinal vessel tree extraction, vessel detection, and vessel calibers estimation assembling the near-infrared reflectance retinography information with OCT section.

A postprocessing model for extracting features of fundus retinal image was introduced in [107–109] that segment the retinal microvasculature to extract the blood vessel applying Krisch Edge Detector and identify the true vessel using Graph Tracer. An automated segmentation method was developed in [110] for segmenting the image of retinal microvascular structure to identify the true vessel using pixel's feature vectors that are the combination of pixel's intensity and continuous two-dimensional Morelet Wavelet transform response adopted at multiple scales.

The study of [111] proposed a vessel segmentation technique using nonlinear diffusion filter for the smoothening vessel to their principal direction, compound vessel enhancement filter combining eigenvalues of the Hessian matrix, matched filter response, and edge constraints of multiple scales for vessel enhancement and then multiple thresholding was applied for centerline tracking. The study of [112] introduced an automatic system to enhance and segment true blood vessel of fundus retinal image using 2-dimensional Gabor wavelet and multilayered thresholding, respectively.

The authors of [113, 114] combined several techniques for the segmentation of fundus retinal image that can detect the vessel centerline and slice the morphological bit plane to extract the vessel tree of the human retina. The study of [38] presented two different methods for retinal blood vessel segmentation where the first method is the procedure for region growing using the hysteresis thresholding which later applied to the response vector similarities of adjacent pixels within the fundus image and the second method was developed based on region growing and directional response vector similarities. For the latter method response vector was calculated through template matching with general Gabor function [38].

The study of [115] proposed an algorithm for vessel segmentation and vascular network extraction based on the multiscale line-tracking [116] procedure where map quantization of the multiscale confident matrix was applied to generate the initial vessel network and then disconnected vessel line was restored and noisy line eliminated applying the median filtering to the generated vascular network and finally directional attributes of vessel and morphological reconstruction was applied as postprocessing for avoiding the faulty areas. An automatic blood vessel extraction technique was presented in [117] in which curvelet-based contrast enhancement, match filtering, curvelet-based edge extraction, and length filtering were used to extract the blood vessel.

The study of [118] proposed a vessel segmentation method to detect the blood vessel in the retinal image based on a probabilistic tracking method. The study of [119] presented a multiconcavity-based segmentation model that deals with the lesion in retinal microvasculature. There were four different techniques demonstrated in [120] for blood vessel segmentation based on Edge Enhancement Edge Detection, Image-Line Cross-Section, Continuation Algorithm, and Modified Matched Filtering that can work with abnormal retinal images having exudates, drusen, and low vessel contrast. A supervised method for detecting blood vessel in fundus retinal images was designed in [121] based on a blueprint of NN for pixel classification and a 7D vector computation that is composed of moment-invariant based features for pixel representation and grey-level. The study of [98] examined a novel method for feature extraction of color retinal image applying Principal Component Analysis, Active Shape Model, fundus coordinates system, and a combined region growing and edge detection technique to locate optic disc, detect the shape of an optic disc, describe the features, and detect the exudates in color retinal image, respectively.

The blood vessels in a fundus retinal image that appear in a higher illumination variance area are found missing if the background removal method for segmentation is applied because the background and intensity values of the blood vessel are almost the same. The study of [122] presented a robust method for blood vessel segmentation to change the illumination intensity applying background estimation that was calculated by a weighted surface fitting method with a higher degree polynomial.

4.8.1. Automatic Identification of Vessels. Identification of retinal artery and vein in fundus retinal images is one of the most important stages for detecting the changes of either qualitative or quantitative features in retinal microvasculature and further extraction of the respective features that have association with risk cardiovascular diseases. To

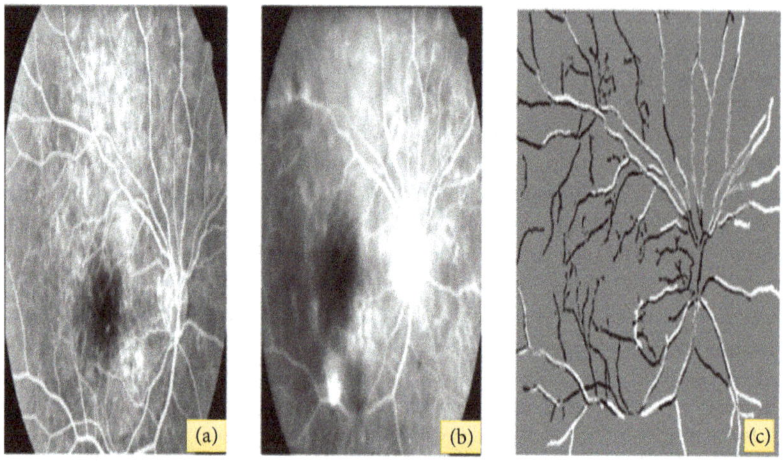

(a) (b) (c)

FIGURE 12: (a) Original fluorescein image. (b) Fluorescein image two years later of temporal registration. (c) Final result of the registration [39].

ensure the effectiveness of the developed systems for clinical diagnosis of diabetic and hypertensive retinopathy, the vessel identification must be accurate. The authors of [123] designed an automated method for the artery and vein identification in dual-wavelength, 570nm and 600nm, retinal images. In this system they utilized the structural feature, relative strength of the vessel central reflex, and the ratio of the vessel optical densities from images at oxygen-sensitive and oxygen-insensitive wavelengths as the functional feature for each vessel segment to differentiate the artery from vein. The study of [123] employed the dual-Gaussian model in which parameter was estimated using a robust M-estimator to compute the relative strength of the central reflex. For the identification of vessel type whether it is artery or vein, the structural and functional features were combined in four classifiers and it was observed that the Support Vector Machine gave the best result with 97% and 90% positive rates for bot arteries and veins, respectively [123].

The study of [124] developed an automatic technique to identify the artery and vein in illumination-corrected retinal images combining Gaussian Mixture Model, Expectation-Maximization unsupervised classifier, and a quadrant-pairwise approach. The obtained specificity and precision of this technique were 0.8978 and 0.9045 for artery, respectively, and 0.9591 and 0.9408 for vein, respectively. The authors of [125] designed a vessel classification system employing the linear discriminant classifier in their proposed method for the measurement of AVR which gave the 92.8% accurate classification results. An automated system for the artery and vein classification was proposed in [126] that provides the classification result for the vasculature analyzing the intersection points (graph nodes) and assigning one of two labels to each vessel segment using graph links. The graph-based labeling results were combined with an intensity features set to perform the final classification and that showed the 89.8%, 8834%, and 87.4% accuracy for the images of VICAVR, INSPIREVR, and DRIVE database, respectively [126].

5. Image Processing-Based Blood Vessel Diameter Measurement for Stroke Risk Detection

Vessel diameter was measured manually by selecting a region of 512 × 512 pixels of a digital image and then calculating the linear distance between two points of opposite edges of a vessel of a digital image. The Gaussian measurement of vessel diameter was performed by analyzing a set of vessel characteristic parameters. These parameters were determined by fitting a double Gaussian model to the intensity cross-section of the vessel. The function below illustrates the intensity cross-section of the vessel and a modified Levenberg-Marquardt [127] least squares method was used to determine the parameters a1 to a5 and a7. The calculated vessel width would be 2.33a3 [128].

$$I(x) = a1e^{-((x-a2)/a3)^2} + a4 - a5e^{-((x-a2)/a7)^2} \text{ [127]}. \quad (2)$$

Sobel edge detection masks were applied to perform Sobel measurements identification of vessel edge position that determines the vessel edge position from the average position of maximum edge strength which was averaged over a three-pixel window. Linear regression within sliding window filter was used to create Sliding Linear Regression Filter (SLRF) measurement identification of edge positions. The SLRF were used to measure the vessel diameter and it was seen that the performance of SLRF was greater than the manual measurement system.

Several studies revealed that there was no association with an arteriolar width decreasing and incident stroke and prevalent stroke rather venular widening was associated with incident stroke [18, 129]. AVR was also analyzed to determine the relation between AVR and incident stroke and prevalent stroke as the arteriolar and venular widening is responsible for changing the value of AVR. No association was found between incident stroke and AVR but the association between

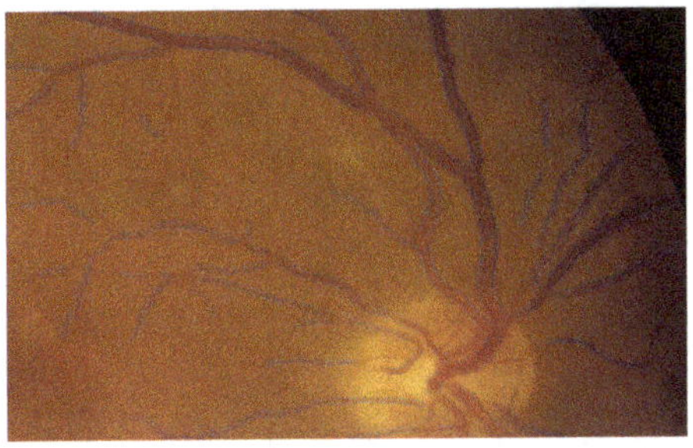

Figure 13: Widened blood vessel of retina [23].

AVR and prevalent stroke was examined [18]. Figure 13 shows the widened vessel of retinal vasculature.

The researches of [130] developed an algorithm for measuring the diameter of the retinal blood vessel to subpixel accuracy by applying the two-dimensional difference of Gaussian model. The researchers came out with 30% more precision in comparison with Zhou's Gaussian model, Brinchmann-Hansen's half height and Gregson's rectangular profile and accuracy of a third of a pixel [130].

A semiautomatic vessel width measuring method named Computer-Aided Image Analysis of Retina (CAIAR) was formulated in [131], in which computer-generated lines, similar to the blood vessel with prefixed frequency, amplitude, and width were used. The resultant width values of retinal blood vessel obtained from this formula were found as less correlated with the ophthalmologist grading [131]. The study of [23] proposed a graph-theoretic algorithm for measuring the vessel width. The work in [20] measures the blood vessel diameter based on thresholding segmentation and training step determining the characteristic point using Douglas-Peucker algorithm. The work in [20] detected vessel contour using active contour and measured vessel diameter using Heron's formula.

An algorithm based on the Graph Tracer method was developed in [132], which can identify the true blood vessels, appropriate bifurcations, and crossover. Multiscale Line Tracing was applied for segmentation and blood vessel annotation tool was used to measure the width of the identified blood vessel and this technique achieved 94.6% accuracy in measuring the diameter of the blood vessel. The study of [133] proposed an algorithm for measuring vessel diameter based on intensity profiles and Dijkstra's shortest path algorithm. The work of [130] also presented a vessel diameter calculating method based on 2D modeling that is more precise than HHFM and Gregson algorithm.

The study of [24] introduced a graph-based algorithm to measure the width of the retinal vessel that segmented both vessel edges following a two-slice, 3D surface segmentation problem model which was converted into a minimum closed set in a node-weighted graph problem in the next step. In their experiment it was shown that more accurate measurement of the vessel width of fundus images can be obtained with larger standard deviation (σ) and the success rates of this algorithm on four datasets, KPIS, CLRIS, VDIS, and HRIS, of REVIEW database are 99.4%, 94.1%, 96%, and 100%, respectively [24]. The authors of [25] proposed a vessel diameter measurement algorithm based on Gaussian modeling combining a series of second-order and higher order Gaussians to design the vessel profile and they used the sigma parameter of generalized Gaussians to the vessel boundaries. The accuracy and precision of this method for CLRIS are −1.574 and 1.691 and for VDIS are −0.443 and 1.182. They compared their result with twin-Gaussian, SLRF, and manual measurement and it was claimed that the accuracy and precision of twin-Gaussian are the least [25].

The authors of [26] designed a hypothesis based method for retinal vessel width measurement applying the theory that Higuchi's dimension of the cross-section is proportional to the vessel diameter. They used REVIEW database to validate their proposed method and came out with the precision and success rate of 0.65 and 99.45%, 1.56 and 98%, 0.45 and 100%, and 1.14 and 97.8% for HRIS, CLRIS, KPIS, and VDIS datasets, respectively. The advantage of this technique is that the results cannot be degraded by edge detection performance and segmentation process as it does not depends on segmentation process [26]. The study of [21] shaped an active contour model named Extraction of Segments Profile (ESP) to measure the retinal blood vessel diameter. Though their algorithm may fail sometimes, they obtained the success rates 99.7%, 99.6%, 93%, and 100% for HRIS, VDIS, CLRIS, and KPIS datasets, respectively, of REVIEW database [21].

An automated method for vessel diameter measurement based on LDA was developed in [53] unsupervised method, which can measure the vessel diameter to subpixel accuracy for all datasets. The study of [52] proposed an algorithm for measuring the retinal vessels widths based on deformable models and which was integrated into an AVR computing framework. This method was robust against different grey color spaces.

The authors of [54] used a novel parametric surface model of the cross-sectional intensities of vessel and combination of bagged decision trees in their algorithm to estimate the retinal vessel width in fundus images. This algorithm was compared with several algorithms such as 1D Gaussian, 2D Gaussian, Gregson, HHFW, Extraction of Segment Profiles (ESP) of Al-Diri, Unsupervised Linear Discriminant Analysis based algorithm of Kumar et al., and graph-based algorithm of Xu et al. and came out with good stability, 100% success rate on all four datasets of REVIEW database [54]. Table 2 displays the performance in terms of accuracy and the applied method to shape the most recent algorithm for measuring the diameter of retinal blood vessels

In Table 2, the most widely used performance measurement of their proposed algorithm is success rate. The success rate can be defined as the ratio of the number of successful runs and the total number of runs. It is a measure of the stability of measurement that was returned [21], along with the mean and standard deviation of measurements and differences. From Table 2 it is seen that the REVIEW database was used in most of the researches, whereas the success rate of the experiment using KPIS datasets achieved 100% with the better standard deviation in maximum cases. This is because the images of this dataset were taken from the clean,

Table 2: Accuracy and applied method of a recently proposed algorithm for measuring the vessel diameter [20, 21, 23–26, 52–54].

No	Paper Info (Authors)	Contribution	Feature	Method	Database REVIEW	Success rate%	Results Accuracy Measurement (in pixels) μ	Results Accuracy Measurement (in pixels) σ	Difference (in pixels) μ	Difference (in pixels) σ
1	[23]	Algorithm for retinal vessel boundary detection and width measurement	Retinal blood vessel width	Graph-Theoretic method	HRIS 90 segments 2368 vessel profile	100	4.54	1.23	0.18	0.47
					VDIS, 79 segments, 2249 vessel profile	96.0	8.59	2.44	−0.29	1.13
2	[52]	A method for measuring the retinal vessels widths and computing AVR in REVIEW database.	Retinal blood vessel width	This algorithm is based on deformable models and integrated into an AVR computing framework	REVIEW 5066 vessel profiles					
					KPIS, SIRIUS					
					G	100	6.20	0.63	−1.28	0.76
					L	100	6.15	0.61	−1.33	0.74
					J	100	6.44	0.63	−1.05	0.73
					I	100	6.23	0.63	−1.26	0.75
					CLRIS, SIRIUS					
					G	91.58	14.69	3.69	0.83	2.17
					L	74.39	16.10	4.74	1.20	4.26
					J	80.35	16.04	3.61	1.52	2.30
					I	75.79	15.85	3.52	1.27	2.51
					VDIS, SIRIUS					
					G	78.70	8.13	2.45	−0.95	1.11
					L	69.68	8.17	2.19	−1.29	1.08
					J	57.80	8.44	2.39	−0.84	1.19
					I	74.34	8.18	2.36	−1.17	1.12
					HRIS, SIRIUS					
					G	78.89	4.26	1.10	−0.13	0.85
					L	81.71	4.16	1.07	−0.22	0.88
					J	73.86	4.35	1.12	−0.08	0.80
					I	83.36	4.25	1.17	−0.14	0.96
3	[53]	An automated vessel diameter measurement technique	Retinal blood vessel width	Unsupervised Linear Discriminant Analysis Diameter Measurement	REVIEW, 5066 Profiles					
					KPIS	100	7.02	0.67	−0.50	0.60
					CLRIS	98.20	13.23	3.55	−0.55	1.79
					VDIS	96.3	8.68	2.82	−0.64	1.18
					HRIS	99.6	4.19	1.35	0.21	0.79

Table 2: Continued.

No	Paper Info (Authors)	Contribution	Feature	Method	Database REVIEW	Success rate%	Results Accuracy Measurement (in pixels) μ	σ	Difference (in pixels) μ	σ
4	[54]	An algorithm for estimating the width of a retinal blood vessel in fundus camera images.	Retinal blood vessel width	Supervised learning is performed by bagged decision trees and an extended multiresolution Hermite model	REVIEW, 5066 Profiles					
					KPIS	100	7.54	0.24	O.015	0.318
					CLRIS	100	13.80	3.89	0.006	1.154
					VDIS	100	8.87	2.22	0.015	1.073
					HRIS	100	4.36	1.13	0.004	0.438
					Tayside data set, 38 fundus images	100	20.43	3.55	0.03	3.168
5	[24]	An algorithm to measure the width of the retinal vessels and find the vessels boundary in fundus photographs	Retinal blood vessel width	Graph-based segmentation method	REVIEW, 5066 profiles					
					KPIS	99.4	6.38	0.59	−1.14	0.67
					CLRIS	94.10	14.05	4.47	0.08	1.78
					VDIS	96.0	8.35	3.00	−0.53	1.43
					HRIS	100	4.56	1.30	0.21	0.567
6	[20]	A novel method for measuring the blood vessel diameter in the retinal image.	Retinal blood vessel width	Thresholding segmentation and thinning step, followed by Douglas-Peucker algorithm. active contours and Heron's Formula	STARE Database		7.73242	0.016	−0.01892	
							4.9278	0.1094	0.06298	
							5.39212	0.1401	0.00544	
					HRF Database		7.7316	0.0234	0.0286	
							15.17584	0.0092	0.00352	
							13.84122	0.0063	−0.00738	
7	[21]	An algorithm for the segmentation and measurement of retinal vessels width, the ESP algorithm.	Retinal blood vessel width	Active contour model	REVIEW, 5066 profiles					
					KPIS	100	6.56			0.328
					CLRIS	93.00	15.70			1.469
					VDIS	99.0	8.80			0.766
					HRIS	99.7	4.63			0.420
8	[26]	An adaptive model to measure the width of retinal vessels in fundus photographs.	Retinal blood vessel width	Adaptive Higuchi's Dimension	REVIEW, 5066 profiles		Precision		Accuracy	
					KPIS	100	0.45		0.72	
					CLRIS	98.00	1.56		0.33	
					VDIS	97.8	1.14		0.45	
					HRIS	99.4	0.65		0.24	
9	[25]	A technique of retinal vessel diameter measurement.	Retinal blood vessel width	Multi-Step Regression Method (Higher order Gaussian modeling)	REVIEW		Precision		Accuracy	
					CLRIS		1.691		−1.574	
					VDIS		1.182		−0.443	

large vessel segments and downsampled such as HRIS [21]. Among the applied methods described here, the [54]'s algorithm based on supervised learning performed by bagged decision trees and an extended multiresolution Hermite model showed the 100% success rate, maximum stability with a good standard deviation of the point to point difference for all the four datasets of REVIEW database. The graph-theoretic method of [23, 24], ESP algorithm of [21], Adaptive Higuchi's Dimension based algorithm of [26], ULDM of [53], and deformable model-based algorithm of [52] exhibited poor accuracy as the mean diameter of retinal blood vessels deviated largely for the CLRIS and VDIS datasets.

The success rate of [53] on HRIS dataset was degraded due to the presence of diabetic abnormalities nearby the vessel boundaries. As the images of VDIS dataset are noisy with lower resolution which is mostly used in pathological purpose, the testing of several algorithms on this dataset led to large deviation and poor accuracy result [53]. The edges of the vessels of CLRIS dataset are highly blurred which can also affect the performance of algorithm [21].

6. Challenges and Future Work

Fundus retinal image processing has become one of the most interesting technologies in diagnosing many cardiovascular diseases such as stroke. According to the medical study, some of the key features of retinal microvasculature convey the symptom of stroke such as the existence of hard exudates, microaneurysm, CWS, and changes in the vessel diameter. To be cooperative with the physicians in terms of detecting the early signs of this lethal condition, researchers from biomedical engineering discipline are being involved more enthusiastically. Numerous quantitative methods for quantifying the abnormal changes in vessel diameter have been developed to provide more precision in medical diagnosing. But, still, there is a scarcity of automatic vascular caliber quantifying methods with more accuracy especially when dealing with images with abnormality. The existing image acquisition technique still has limitations on the basis of automation though there are some ultramodern image acquisition techniques that have been developed which are applicable in the smartphone and the misery in autoalignment of the captured image based on a standard preregistered image. One of the potential ways to avoid this penury is to apply the image registration technique prior to image acquisition or segmentation. Images can be degraded with blur at the time of acquisition and it is either space-variant or space-invariant. Maximum image acquisition algorithm is not effective in a space-invariant blur. It is known that spectral image is more informative for retinal images. Lack of spectral retinal image is also considered as an obstacle of image processing [95].

Edge detection is also important in the workflow of measuring blood vessel diameter as several vessel diameter measuring algorithms go through this technique. Though a lot of edge detection algorithms are available, sometimes the performance of the existing algorithms degraded due to the poor local contrast and wrong illustration of the central light reflection in fundus images [97]. This scenario greatly affects the results of computational technique to measure the vessel diameter especially the smaller vessel diameter measurement. Another worst condition arises in detecting edges and measuring the vessel width if false positive or vessel discontinuities occurred [23]. To avoid this situation edge detecting technology should be developed to a great extent.

Some challenges for image registration technique as mentioned by ETDRS are the feature-based method problems in superfluous landmark points due to small overlaps between adjacent fields. The area-based technique cannot perform well because of irregular intensity distribution from defective data acquisition. Also, both feature and area-based technique can result in less accuracy due to the high-resolution images containing huge homogeneous textureless regions [103].

It is observed that graph-theoretic method has a large contribution in the development of the vessel boundary segmentation and vessel width measurement technique. Some drawback of the vessel width measurement algorithm based on the graph is unable to define the normal direction of the blood vessel that leads them to result in poor accuracy [23]. As the vessel width measurement algorithm depends on the segmentation, the inefficient segmentation method also distorts the outcome of the system.

Another most important issue is that the image quality of the databases which are being used widely to validate the designed algorithm for retinal vessel width measurement. In some cases, it was found that the performance of vessel width measurement method was affected due to the different features and qualities of the images from different datasets. Supposing the diabetic abnormalities are present in the images of HRIS dataset, highly blurred vessel edges of the image of CLRIS dataset and the noisy, pathological images with lower resolution images of the VDIS dataset are found as responsible for the degradation of the performance of the proposed algorithm. For the better accuracy and to create supreme enthusiasm among the researchers the quality of the contents of the testing image sets needs to be updated. Employment of robust retinal blood vessel segmentation algorithm also overcomes the issue of image quality. The proposed method of [134] had shown its effectiveness on the unhealthy dataset. Thus it is believed that the method is meaningful when applying on the abnormal retinal dataset.

With the rapid development of technology, retinal image processing is also marching forward with lots of novel achievements to provide the human beings with a more secured life creating a notable application that can detect the preindication of stroke with more accuracy. The latest technology of retinal image analysis is only for the image acquisition using a smartphone and that is also dependent on a large screen and power consuming devices for further processing. A fully automated image processing technique to extract the features and measure the vessel diameter of the human retinal vasculature with maximum accuracy needs to be developed. The system can be operated by using a smartphone to facilitate the regarding ophthalmologist with the most feasible and expected teleophthalmic system for the diagnosis, supervision, and monitoring diseases state of life-threatening microvascular disease like stroke.

7. Conclusion

The field of digital image processing has a wide variety of concerns of numerous applications. These workable applications of image processing are being used in the largest platform, medical diagnostic system. Fundus retinal image processing is one of the most fertile disciplines of digital image processing that is continuously creating a lot of robust applications with numerous novel features to facilitate the biomedical engineering sector in terms of diagnosing the diseases, planning and supervising the treatment of the diseases, and monitoring the condition of diseases together. The cause of increasing appeal of this widely ranged area of research is that the digital fundus image of human retina can be analyzed noninvasively in vivo and the objects of interest can be visualized with more accuracy.

A lot of life-threatening diseases such as stroke can be diagnosed in their early state analyzing the microvascular structure of human fundus eye image. Some of the features of retinal microvasculature such as hard exudates, haemorrhages, AV nicking, CWS, focal arteriolar narrowing and vessel width change remarkably due to that dangerous disease. Researchers are being involved in this potential field, retinal image processing, to facilitate biomedical engineering sector by developing novel techniques. These techniques can be used to extract and analyze the features of interest with more precision.

Retinal image acquisition is the initial step of retinal image processing to analyze and extract the relevant features. After the acquisition of the retinal image, it goes through several steps such as image enhancement, image restoration, image construction, and image segmentation. These processes are used to figure out the particular objects of interest and some further processing to find out the more specific features and their characteristics. There are a good number of automatic qualitative and quantitative applications that have been embodied based on these image processing techniques. These techniques can detect abnormalities in retinal vasculature such as AV nicking, hard exudates, CWS, and haemorrhages. These techniques also can measure the unexpected changes of vascular caliber such as widening the vessel diameter, which occurred in retinal microvasculature which are directly related to stroke even though the blood pressure and other relevant vascular risk factors are at the tolerable stage. Linear Discriminant Analysis, image gradient segmentation technique (ARG) for vessel segmentation, mask creation, Sobel edge detection, Gaussian Measurement, Sliding Linear Regression Filter (SLRF), and Computer-Aided Image Analysis of Retina (CAIAR) are some of the systems to measure the retinal blood vessel diameter that has been developed using image processing principle. Exploring the advancement of retinal image processing, it can be expected to have a great revolution in modern ophthalmology as this discipline depends on the screened information at a large extent.

Conflicts of Interest

It is hereby declared that there are no conflicts of interest regarding the publication of this paper.

Acknowledgments

The authors would like also to acknowledge Universiti Malaysia Sarawak (UNIMAS). This work has been supported in part by Special Short Term Grant (SpSTG) MyRA F02/SpSTG/1374/16/16.

References

[1] N. Witt, T. Y. Wong, A. D. Hughes et al., "Abnormalities of retinal microvascular structure and risk of mortality from ischemic heart disease and stroke," *Hypertension*, vol. 47, no. 5, pp. 975–981, 2006.

[2] M. James, "Cost effectiveness analysis of screening for sight threatening diabetic eye disease," *British Medical Journal*, vol. 320, no. 7250, pp. 1627–1631.

[3] M. Favali, S. Abbasi-Sureshjani, B. ter Haar Romeny, and A. Sarti, "Analysis of vessel connectivities in retinal images by cortically inspired spectral clustering," *Journal of Mathematical Imaging and Vision*, vol. 56, no. 1, pp. 158–172, 2016.

[4] F. Huang, B. Dashtbozorg, and B. M. Romeny, "Artery/vein classification using reflection features in retina fundus images," *Machine Vision and Applications*, vol. 29, no. 1, pp. 23–34, 2018.

[5] J. Wigdahl, P. Guimaraes, and A. Ruggeri, "A shortest path approach to optic disc detection in retinal fundus images," *Journal for Modeling in Ophthalmology*, vol. 1, no. 4, pp. 29–42, 2017.

[6] L. Bonaldi, E. Menti, L. Ballerini, A. Ruggeri, and E. Trucco, "Automatic generation of synthetic retinal fundus images: vascular network," *Procedia Computer Science*, vol. 90, pp. 54–60, 2016.

[7] S. B. Seidelmann, B. Claggett, P. E. Bravo et al., "Retinal vessel calibers in predicting long-term cardiovascular outcomes: the atherosclerosis risk in communities study," *Circulation*, vol. 134, no. 18, pp. 1328–1338, 2016.

[8] C. S. M. Cheng, Y. F. Lee, C. Ong et al., "Inter-eye comparison of retinal oximetry and vessel caliber between eyes with asymmetrical glaucoma severity in different glaucoma subtypes," *Clinical Ophthalmology*, vol. 10, pp. 1315–1321, 2016.

[9] M. L. Rasmussen, R. Broe, U. Frydkjaer-Olsen et al., "Retinal vascular geometry and its association to microvascular complications in patients with type 1 diabetes: the Danish Cohort of Pediatric Diabetes 1987 (DCPD1987)," *Graefe's Archive for Clinical and Experimental Ophthalmology*, vol. 255, no. 2, pp. 293–299, 2017.

[10] D. A. De Silva, J. J. F. Manzano, E. Y. Liu et al., "Retinal microvascular changes and subsequent vascular events after ischemic stroke," *Neurology*, vol. 77, no. 9, pp. 896–903, 2011.

[11] A. D. Henderson, B. B. Bruce, N. J. Newman, and V. Biousse, "Hypertension-related eye abnormalities and the risk of stroke," *Reviews in Neurological Diseases*, vol. 8, no. 404, pp. 1–9, 2011.

[12] R. Kawasaki, J. Xie, N. Cheung et al., "Retinal microvascular signs and risk of stroke: the Multi-Ethnic Study of Atherosclerosis (MESA)," *Stroke*, vol. 43, no. 12, pp. 3245–3251, 2012.

[13] T. Y. Wong, A. Kamineni, R. Klein et al., "Quantitative retinal venular caliber and risk of cardiovascular disease in older persons: the cardiovascular health study," *Archives of Internal Medicine*, vol. 166, no. 21, pp. 2388–2394, 2006.

[14] H. E. Moss, "Retinal vascular changes are a marker for cerebral vascular diseases," *Current Neurology and Neuroscience Reports*, vol. 15, no. 7, article 40, 2015.

[15] Y. T. Ong, T. Y. Wong, R. Klein et al., "Hypertensive retinopathy and risk of stroke," *Hypertension*, vol. 62, no. 4, pp. 706–711, 2013.

[16] H.-Q. Wu, H. Wu, L.-L. Shi et al., "The association between retinal vasculature changes and stroke: a literature review and Meta-analysis," *International Journal of Ophthalmology*, vol. 10, no. 1, pp. 109–114, 2017.

[17] C. Y.-L. Cheung, W. T. Tay, M. K. Ikram et al., "Retinal microvascular changes and risk of stroke: the Singapore Malay eye study," *Stroke*, vol. 44, no. 9, pp. 2402–2408, 2013.

[18] F. N. Doubal, P. E. Hokke, and J. M. Wardlaw, "Retinal microvascular abnormalities and stroke: a systematic review," *Journal of Neurology, Neurosurgery & Psychiatry*, vol. 80, no. 2, pp. 158–165, 2009.

[19] M. L. Baker, P. J. Hand, J. J. Wang, and T. Y. Wong, "Retinal signs and stroke: revisiting the link between the eye and brain," *Stroke*, vol. 39, no. 4, pp. 1371–1379, 2008.

[20] H. Guedri, M. B. Abdallah, F. Echouchene, and H. Belmabrouk, "Novel computerized method for measurement of retinal vessel diameters," *Biomedicines*, vol. 5, no. 2, p. 12, 2017.

[21] B. Al-Diri, A. Hunter, and D. Steel, "An active contour model for segmenting and measuring retinal vessels," *IEEE Transactions on Medical Imaging*, vol. 28, no. 9, pp. 1488–1497, 2009.

[22] R. Das and P. P. Acharjya, "Edge detection using the magnitude of the gradient," *International Journal of Innovative Research in Computer and Communication Engineering*, vol. 1, no. 2, pp. 180–184, 2013.

[23] X. Xu, M. Niemeijer, Q. Song, M. K. Garvin, J. M. Reinhardt, and M. D. Abràmoff, "Retinal vessel width measurements based on a graph-theoretic method," in *Proceedings of the IEEE International Symposium on Biomedical Imaging: From Nano to Macro*, pp. 641–644, Chicago, Ill, USA, March-April 2011.

[24] X. Xu, M. Niemeijer, Q. Song et al., "Vessel boundary delineation on fundus images using graph-based approach," *IEEE Transactions on Medical Imaging*, vol. 30, no. 6, pp. 1184–1191, 2011.

[25] B. Aliahmad, D. K. Kumar, S. Janghorban, M. Z. C. Azemin, H. Hao, and R. Kawasaki, "Retinal vessel diameter measurement using multi-step regression method," in *Proceedings of the 2012 ISSNIP Biosignals and Biorobotics Conference: Biosignals and Robotics for Better and Safer Living*, pp. 1–4, January 2012.

[26] B. Aliahmad and D. K. Kumar, "Adaptive Higuchi's dimension-based retinal vessel diameter measurement," in *Proceedings of the 38th Annual International Conference of the IEEE Engineering in Medicine and Biology Society (EMBC '16)*, pp. 1308–1311, August 2016.

[27] P. Prentasic, S. Loncaric, Z. Vatavuk et al., "Diabetic retinopathy image database (DRiDB): a new database for diabetic retinopathy screening programs research," in *Proceedings of the 2013 8th International Symposium on Image and Signal Processing and Analysis (ISPA)*, pp. 711–716, Trieste, Italy, September 2013.

[28] B. Al-Diri, A. Hunter, D. Steel, M. Habib, T. Hudaib, and S. Berry, "REVIEW—a reference data set for retinal vessel profiles," in *Proceedings of the 2008 30th Annual International Conference of the IEEE Engineering in Medicine and Biology Society*, vol. 2008, pp. 2262–2265, August 2008.

[29] J. Kur, E. A. Newman, and T. Chan-Ling, "Cellular and physiological mechanisms underlying blood flow regulation in the retina and choroid in health and disease," *Progress in Retinal and Eye Research*, vol. 31, no. 5, pp. 377–406, 2012.

[30] U. T. V. Nguyen, A. Bhuiyan, L. A. F. Park et al., "An automated method for retinal arteriovenous nicking quantification from color fundus images," *IEEE Transactions on Biomedical Engineering*, vol. 60, no. 11, pp. 3194–3203, 2013.

[31] Y. Hatanaka, T. Inoue, S. Okumura, C. Muramatsu, and H. Fujita, "Automated microaneurysm detection method based on double-ring filter and feature analysis in retinal fundus images," in *Proceedings of the 25th IEEE International Symposium on Computer-Based Medical Systems (CBMS '12)*, vol. 7260, pp. 1–8, June 2012.

[32] S. Irshad and M. U. Akram, "Classification of retinal vessels into arteries and veins for detection of hypertensive retinopathy," in *Proceedings of the 7th Cairo International Biomedical Engineering Conference (CIBEC '14)*, pp. 133–136, Giza, Egypt, December 2014.

[33] W. Hsu, P. M. D. S. Pallawala, M. L. Lee, and K.-G. A. Eong, "The role of domain knowledge in the detection of retinal hard exudates," in *Proceedings of the IEEE Computer Society Conference on Computer Vision and Pattern Recognition (CVPR '01)*, pp. II-246–II-251, 2001.

[34] L. E. Downie, L. A. B. Hodgson, C. Dsylva et al., "Hypertensive retinopathy: comparing the Keith-Wagener-Barker to a simplified classification," *Journal of Hypertension*, vol. 31, no. 5, pp. 960–965, 2013.

[35] L. Giancardo, "Docteur de l'université Automated Fundus Images Analysis Techniques to Screen Retinal Diseases in Diabetic Patients," 2011.

[36] G. C. Manikis, V. Sakkalis, X. Zabulis, P. Karamaounas, and A. Triantafyllou, "An image analysis framework for the early assessment of hypertensive retinopathy signs," in *Proceedings of the 3rd International Conference on E-Health and Bioengineering (EHB '11)*, pp. 413–418, 2011.

[37] A. G. Marrugo, M. S. Millán, M. Šorel, and F. Šroubek, "Restoration of retinal images with space-variant blur," *Journal of Biomedical Optics*, vol. 19, no. 1, Article ID 016023, 2014.

[38] Z. Fazekas, A. Hajdu, I. Lßzßr et al., "Influence of Using Different Segmentation Methods on the Fractal Properties of the Identified Retinal Vascular Networks in Healthy Retinas and in Retinas with Vein".

[39] F. Zana and J. C. Klein, "A multimodal registration algorithm of eye fundus images using vessels detection and hough transform," *IEEE Transactions on Medical Imaging*, vol. 18, no. 5, pp. 419–428, 1999.

[40] Henkind P., R. I. Hansen, and J. Szalay, "Ocular circulation," in *Physiology of the Human Eye and Visual System*, Harper and Row, New York, NY, USA, 1979.

[41] B. Pemp, A.-P. Cherecheanu, G. Garhofer, and L. Schmetterer, "Calculation of central retinal artery diameters from non-invasive ocular haemodynamic measurements in type 1 diabetes patients," *Acta Ophthalmologica*, vol. 91, no. 5, pp. e348–e352, 2013.

[42] Q. Wang, O. P. Kocaoglu, B. Cense et al., "Imaging retinal capillaries using ultrahigh-resolution optical coherence tomography and adaptive optics," *Investigative Ophthalmology & Visual Science*, vol. 52, no. 9, pp. 6292–6299, 2011.

[43] L. M. Parver, "Temperature modulating action of choroidal blood flow," *Eye*, vol. 5, no. 2, pp. 181–185, 1991.

[44] A. Grosso, F. Veglio, M. Porta, F. M. Grignolo, and T. Y. Wong, "Hypertensive retinopathy revisited: some answers, more questions," *British Journal of Ophthalmology*, vol. 89, no. 12, pp. 1646–1654, 2005.

[45] S. Chatterjee, S. Chattopadhya, M. Hope-Ross, and P. L. Lip, "Hypertension and the eye: Changing perspectives," *Journal of Human Hypertension*, vol. 16, no. 10, pp. 667–675, 2002.

[46] R. Marcucci, F. Sofi, E. Grifoni, A. Sodi, and D. Prisco, "Retinal vein occlusions: a review for the internist," *Internal and Emergency Medicine*, vol. 6, no. 4, pp. 307–314, 2011.

[47] A. G. Marrugo, M. Šorel, F. Šroubek, and M. S. Millán, "Retinal image restoration by means of blind deconvolution," *Journal of Biomedical Optics*, vol. 16, no. 11, Article ID 116016, 2011.

[48] J. J. Wang, M. L. Baker, P. J. Hand et al., "Transient ischemic attack and acute ischemic stroke: associations with retinal microvascular signs," *Stroke*, vol. 42, no. 2, pp. 404–408, 2011.

[49] T. Y. Wong, R. Klein, D. J. Couper et al., "Retinal microvascular abnormalities and incident stroke: the Atherosclerosis Risk in Communities Study," *The Lancet*, vol. 358, no. 9288, pp. 1134–1140, 2001.

[50] D.-H. Lee, S.-J. Lee, and I.-N. Yoon, "Clinical progress in impending central retinal vein occlusion," *Korean Journal of Ophthalmology*, vol. 24, no. 2, pp. 83–88, 2010.

[51] G. Coll-De-Tuero, S. González-Vázquez, A. Rodríguez-Poncelas et al., "Retinal arteriole-to-venule ratio changes and target organ disease evolution in newly diagnosed hypertensive patients at 1-year follow-up," *Journal of the American Society of Hypertension*, vol. 8, no. 2, pp. 83–93, 2014.

[52] S. G. Vázquez, N. Barreira, M. G. Penedo, M. Pena-Seijo, and F. Gómez-Ulla, "Evaluation of SIRIUS retinal vessel width measurement in REVIEW dataset," in *Proceedings of the IEEE 26th International Symposium on Computer-Based Medical Systems (CBMS '13)*, pp. 71–76, Porto, Portugal, June 2013.

[53] D. K. Kumar, B. Aliahmad, and H. Hao, "Retinal vessel diameter measurement using unsupervised linear discriminant analysis," *ISRN Ophthalmology*, vol. 2012, Article ID 151369, 7 pages, 2012.

[54] C. A. Lupaşcu, D. Tegolo, and E. Trucco, "Accurate estimation of retinal vessel width using bagged decision trees and an extended multiresolution Hermite model," *Medical Image Analysis*, vol. 17, no. 8, pp. 1164–1180, 2013.

[55] D. D. Varma, S. Cugati, A. W. Lee, and C. S. Chen, "A review of central retinal artery occlusion: clinical presentation and management," *Eye*, vol. 27, no. 6, pp. 688–697, 2013.

[56] R. Lee, T. Y. Wong, and C. Sabanayagam, "Epidemiology of diabetic retinopathy, diabetic macular edema and related vision loss," *Eye and Vision*, vol. 2, 17 pages, 2015.

[57] L. Claesson-Welsh, "Vascular permeability—the essentials," *Upsala Journal of Medical Sciences*, vol. 120, no. 3, pp. 135–143, 2015.

[58] E. J. Duh, J. K. Sun, and A. W. Stitt, "Diabetic retinopathy: current understanding, mechanisms, and treatment strategies," *JCI Insight*, vol. 2, no. 14, pp. 1–13, 2017.

[59] D. B. Archer, "Bowman lecture 1998. Diabetic retinopathy: some cellular, molecular and therapeutic considerations," *Eye*, vol. 13, no. 4, pp. 497–523, 1999.

[60] A. Hofman, M. M. Breteler, C. M. van Duijn et al., "The Rotterdam Study: 2010 objectives and design update," *European Journal of Epidemiology*, vol. 24, no. 9, pp. 553–572, 2009.

[61] M. K. Ikram, F. J. De Jong, E. J. Van Dijk et al., "Retinal vessel diameters and cerebral small vessel disease: the Rotterdam Scan Study," *Brain*, vol. 129, no. 1, pp. 182–188, 2006.

[62] S. Nunes, I. Pires, A. Rosa, L. Duarte, R. Bernardes, and J. Cunha-Vaz, "Microaneurysm turnover is a biomarker for diabetic retinopathy progression to clinically significant macular edema: findings for type 2 diabetics with nonproliferative retinopathy," *Ophthalmologica*, vol. 223, no. 5, pp. 292–297, 2009.

[63] M. J. Cree, J. A. Olson, K. C. McHardy, P. F. Sharp, and J. V. Forrester, "A fully automated comparative microaneurysm digital detection system," *Eye*, vol. 11, pp. 622–628, 1997.

[64] M. L. Gomez, F. Mojana, D.-U. Bartsch, and W. R. Freeman, "Imaging of long-term retinal damage after resolved cotton wool spots," *Ophthalmology*, vol. 116, no. 12, pp. 2407–2414, 2009.

[65] M. Niemeijer, B. Van Ginneken, S. R. Russell, M. S. A. Suttorp-Schulten, and M. D. Abràmoff, "Automated detection and differentiation of drusen, exudates, and cotton-wool spots in digital color fundus photographs for diabetic retinopathy diagnosis," *Investigative Ophthalmology & Visual Science*, vol. 48, no. 5, pp. 2260–2267, 2007.

[66] T. Walter, J.-C. Klein, P. Massin, and A. Erginay, "A contribution of image processing to the diagnosis of diabetic retinopathy-detection of exudates in color fundus images of the human retina," *IEEE Transactions on Medical Imaging*, vol. 21, no. 10, pp. 1236–1243, 2002.

[67] C. I. Sánchez, R. Hornero, M. I. López, M. Aboy, J. Poza, and D. Abásolo, "A novel automatic image processing algorithm for detection of hard exudates based on retinal image analysis," *Medical Engineering & Physics*, vol. 30, no. 3, pp. 350–357, 2008.

[68] A. W. Reza, C. Eswaran, and S. Hati, "Automatic tracing of optic disc and exudates from color fundus images using fixed and variable thresholds," *Journal of Medical Systems*, vol. 33, no. 1, pp. 73–80, 2009.

[69] C. I. Sánchez, M. García, A. Mayo, M. I. López, and R. Hornero, "Retinal image analysis based on mixture models to detect hard exudates," *Medical Image Analysis*, vol. 13, no. 4, pp. 650–658, 2009.

[70] M. L. Baker, P. J. Hand, G. Liew et al., "Retinal microvascular signs may provide clues to the underlying vasculopathy in patients with deep intracerebral hemorrhage," *Stroke*, vol. 41, no. 4, pp. 618–623, 2010.

[71] T. Y. Wong, R. Klein, A. R. Sharrett et al., "Retinal arteriolar diameter and risk for hypertension," *Annals of Internal Medicine*, vol. 140, no. 4, pp. 248–255, 2004.

[72] H. E. Moss, "Retinal vascular changes are a marker for cerebral vascular diseases," *Current Neurology and Neuroscience Reports*, vol. 15, no. 7, article 40, 15 pages, 2015.

[73] E. Grisan, *Automatic analysis of retinal images: retinopathy detection and grading [Ph.D. thesis]*, 2005.

[74] U. R. Acharya, C. M. Lim, E. Y. K. Ng, C. Chee, and T. Tamura, "Computer-based detection of diabetes retinopathy stages using digital fundus images," *Proceedings of the Institution of Mechanical Engineers, Part H: Journal of Engineering in Medicine*, vol. 223, no. 5, pp. 545–553, 2009.

[75] R. C. Gonzalez and Woods, *Digital Image Processing*, Pearson, Upper Saddle River, NJ, USA, 2008.

[76] E. Adelson and C. Anderson, "Pyramid methods in image processing," *RCA Engineer*, vol. 29, no. 6, pp. 33–41, 1984.

[77] H. Knutsson, "Filtering and reconstruction in image processing," *Signal Processing*, vol. 7, no. 1, p. 83, 1984.

[78] D. L. G. Hill, P. G. Batchelor, M. Holden, and D. J. Hawkes, "Medical image registration," *Physics in Medicine and Biology*, vol. 46, no. 3, pp. R1–R45, 2001.

[79] J. Suckling, J. Parker, and D. Dance, "The mammographic image analysis society digital mammogram database," in *Exerpta Medica*, vol. 1069 of *International Congress Series*, pp. 375–378, 1994.

[80] M. A. Palomera-Pérez, M. E. Martinez-Perez, H. Benítez-Pérez, and J. L. Ortega-Arjona, "Parallel multiscale feature

extraction and region growing: application in retinal blood vessel detection," *IEEE Transactions on Information Technology in Biomedicine*, vol. 14, no. 2, pp. 500–506, 2010.

[81] N. Patton, T. M. Aslam, T. MacGillivray et al., "Retinal image analysis: concepts, applications and potential," *Progress in Retinal and Eye Research*, vol. 25, no. 1, pp. 99–127, 2006.

[82] K. Kade Mahesh and S. Kashid Nilesh, "Review on fundus image acquisition techniques with data base reference to retinal abnormalities in diabetic retinopathy," *International Journal of Computer Applications*, vol. 68, no. 8, pp. 17–27, 2013.

[83] J. Asmuth, "Mosaicking and enhancement of slit lamp biomicroscopic fundus images," *British Journal of Ophthalmology*, vol. 85, no. 5, pp. 563–565, 2001.

[84] S. Jiao, R. Knighton, X. Huang, G. Gregori, and C. A. Puliafito, "Simultaneous acquisition of sectional and fundus ophthalmic images with spectral-domain optical coherence tomography," *Optics Express*, vol. 13, no. 2, pp. 444–452, 2005.

[85] M. Wojtkowski, V. Srinivasan, J. G. Fujimoto et al., "Three-dimensional retinal imaging with high-speed ultrahigh-resolution optical coherence tomography," *Ophthalmology*, vol. 112, no. 10, pp. 1734–1746, 2005.

[86] B. Lochocki, A. Gambin, S. Manzanera et al., "A single pixel camera video ophthalmoscope," p. 100570M, February 2017.

[87] D. Myung, A. Jais, L. He, M. S. Blumenkranz, and R. T. Chang, "3D Printed smartphone indirect lens adapter for rapid, high quality retinal imaging," *Journal of Mobile Technology in Medicine*, vol. 3, no. 1, pp. 9–15, 2014.

[88] R. C. Gonzalez, R. E. Woods, and S. L. Eddins, *Digital Image Processing Using MATLAB*, Gatesmark Publishing, Knoxville, Tenn, USA, 2nd edition, 2009.

[89] A. K. Jain, "Fundamentals of digital image processing," vol. 14, no. 8, 1989.

[90] C. Saravanan, "Color image to grayscale image conversion," in *Proceedings of the 2010 2nd International Conference on Computer Engineering and Applications (ICCEA '10)*, vol. 2, pp. 196–199, Bali Island, Indonesia, March 2010.

[91] A. Shnayderman, A. Gusev, and A. M. Eskicioglu, "An SVD-based grayscale image quality measure for local and global assessment," *IEEE Transactions on Image Processing*, vol. 15, no. 2, pp. 422–429, 2006.

[92] M. Foracchia, E. Grisan, and A. Ruggeri, "Luminosity and contrast normalization in retinal images," *Medical Image Analysis*, vol. 9, no. 3, pp. 179–190, 2005.

[93] M. J. Cree, E. Gamble, and D. J. Cornforth, "Colour normalisation to reduce inter-patient and intra-patient variability in microaneurysm detection in colour retinal images," in *Proceedings of the APRS Workshop on Digital Image Computing (WDIC '05)*, pp. 163–168, 2005.

[94] E. Romero, N. Lepore, A. G. Marrugo et al., "Improving the blind restoration of retinal images by means of point-spread-function estimation assessment," in *10th International Symposium on Medical Information Processing and Analysis*, vol. 9287 of *Proceedings of SPIE*, p. 92871D, Cartagena de Indias, Colombia, September 2015.

[95] U. Nguyen, L. Laaksonen, H. Uusitalo, and L. Lensu, "Reconstruction of retinal spectra from RGB data using a RBF network," in *Proceedings of the 6th International Conference on Image Processing Theory, Tools and Applications (IPTA '16)*, 5 pages, Oulu, Finland, December 2016.

[96] M. I. Khan, H. Shaikh, and A. M. Mansuri, "A review of retinal vessel segmentation techniques and algorithms," *International Journal of Computer Applications in Technology*, vol. 2, no. 5, pp. 1140–1144, 2011.

[97] S. Chaudhuri, S. Chatterjee, N. Katz, M. Nelson, and M. Goldbaum, "Detection of blood vessels in retinal images using two-dimensional matched filters," *IEEE Transactions on Medical Imaging*, vol. 8, no. 3, pp. 263–269, 1989.

[98] H. Li and O. Chutatape, "Automated feature extraction in color retinal images by a model based approach," *IEEE Transactions on Biomedical Engineering*, vol. 51, no. 2, pp. 246–254, 2004.

[99] A. Aquino, M. E. Gegúndez-Arias, and D. Marín, "Detecting the optic disc boundary in digital fundus images using morphological, edge detection, and feature extraction techniques," *IEEE Transactions on Medical Imaging*, vol. 29, no. 11, pp. 1860–1869, 2010.

[100] T. Chanwimaluang and G. Fan, "An efficient algorithm for extraction of anatomical structures in retinal images," in *Proceedings of the International Conference on Image Processing*, vol. 2, pp. 8–11, Barcelona, Spain, September 2003.

[101] X. Jiang and D. Mojon, "Adaptive local thresholding by verification-based multithreshold probing with application to vessel detection in retinal images," *IEEE Transactions on Pattern Analysis and Machine Intelligence*, vol. 25, no. 1, pp. 131–137, 2003.

[102] G. Molodij, E. N. Ribak, M. Glanc, and G. Chenegros, "Enhancing retinal images by nonlinear registration," *Optics Communications*, vol. 342, pp. 157–166, 2015.

[103] T. Chanwimaluang, G. Fan, and S. R. Fransen, "Erratum to: 'Hybrid retinal image registration', *IEEE Transactions on Information Technology in Biomedicine*, vol. 10, no. 1, pp. 129–142, 2006," *IEEE Transactions on Information Technology in Biomedicine*, vol. 11, no. 1, p. 110, 2007.

[104] G. Ramaswamy and N. Devaney, "Pre-processing, registration and selection of adaptive optics corrected retinal images," *Ophthalmic and Physiological Optics*, vol. 33, no. 4, pp. 527–539, 2013.

[105] M. E. Martínez-Pérez, A. D. Hughes, A. V. Stanton, S. A. Thom, A. A. Bharath, and K. H. Parker, "Retinal blood vessel segmentation by means of scale-space analysis and region growing," in *Medical Image Computing and Computer-Assisted Intervention—MICCAI'99*, vol. 1679, pp. 90–97, Springer, 1999.

[106] J. de Moura, J. Novo, P. Charlón, N. Barreira, and M. Ortega, "Enhanced visualization of the retinal vasculature using depth information in OCT," *Medical & Biological Engineering & Computing*, vol. 55, no. 12, pp. 2209–2225, 2017.

[107] Q. P. Lau, M. L. Lee, W. Hsu, and T. Y. Wong, "Simultaneously identifying all true vessels from segmented retinal images," *IEEE Transactions on Biomedical Engineering*, vol. 60, no. 7, pp. 1851–1858, 2013.

[108] G. Delucta Mary, "Identifying all true vessels from segmented retinal images," *Compusoft*, vol. 3, no. 2, pp. 540–544, 2014.

[109] K. A. Nithya and A. Rajini, "Identifying true vessels using krisch edge segmentation," *Compusoft*, vol. 3, no. 8, pp. 1053–1058, 2014.

[110] J. V. B. Soares, J. J. G. Leandro, R. M. Cesar Jr., H. F. Jelinek, and M. J. Cree, "Retinal vessel segmentation using the 2-D Gabor wavelet and supervised classification," *IEEE Transactions on Medical Imaging*, vol. 25, no. 9, pp. 1214–1222, 2006.

[111] C.-H. Wu, G. Agam, and P. Stanchev, "A general framework for vessel segmentation in retinal images," in *Proceedings of the 2007 International Symposium on Computational Intelligence in Robotics and Automation*, pp. 37–42, Jacksonville, Fla, USA, June 2007.

[112] M. U. Akram and S. A. Khan, "Multilayered thresholding-based blood vessel segmentation for screening of diabetic retinopathy," *Engineering with Computers*, vol. 29, no. 2, pp. 165–173, 2013.

[113] M. M. Fraz, S. A. Barman, P. Remagnino et al., "An approach to localize the retinal blood vessels using bit planes and centerline detection," *Computer Methods and Programs in Biomedicine*, vol. 108, no. 2, pp. 600–616, 2012.

[114] A. M. Mendonça and A. Campilho, "Segmentation of retinal blood vessels by combining the detection of centerlines and morphological reconstruction," *IEEE Transactions on Medical Imaging*, vol. 25, no. 9, pp. 1200–1213, 2006.

[115] M. Vlachos and E. Dermatas, "Multi-scale retinal vessel segmentation using line tracking," *Computerized Medical Imaging and Graphics*, vol. 34, no. 3, pp. 213–227, 2010.

[116] M. Ben Abdallah, J. Malek, A. T. Azar et al., "Automatic extraction of blood vessels in the retinal vascular tree using multiscale medialness," *International Journal of Biomedical Imaging*, vol. 2015, Article ID 519024, 16 pages, 2015.

[117] M. Esmaeili, H. Rabbani, A. Mehri, and A. Dehghani, "Extraction of retinal blood vessels by curvelet transform," in *Proceedings of the 2009 IEEE International Conference on Image Processing (ICIP '09)*, vol. 2, pp. 3353–3356, November 2009.

[118] Y. Yin, M. Adel, and S. Bourennane, "Retinal vessel segmentation using a probabilistic tracking method," *Pattern Recognition*, vol. 45, no. 4, pp. 1235–1244, 2012.

[119] M. Modeling, R. M. Modeling, B. S. Y. Lam, Y. Gao, S. Member, and A. W. Liew, "Using Regularization-Based Author," 2010.

[120] A. A. Mudassar and S. Butt, "Extraction of blood vessels in retinal images using four different techniques," *Journal of Medical Engineering*, vol. 2013, Article ID 408120, 21 pages, 2013.

[121] D. Marín, A. Aquino, M. E. Gegúndez-Arias, and J. M. Bravo, "A new supervised method for blood vessel segmentation in retinal images by using gray-level and moment invariants-based features," *IEEE Transactions on Medical Imaging*, vol. 30, no. 1, pp. 146–158, 2011.

[122] S. Paripurana, W. Chiracharit, K. Chamnongthai, and H. Saito, "Extraction of blood vessels in retinal images using resampling high-order background estimation," *IEICE Transaction on Information and Systems*, vol. E98D, no. 3, pp. 692–703, 2015.

[123] H. Narasimha-Iyer, J. M. Beach, B. Khoobehi, and B. Roysam, "Automatic identification of retinal arteries and veins from dual-wavelength images using structural and functional features," *IEEE Transactions on Biomedical Engineering*, vol. 54, no. 8, pp. 1427–1435, 2007.

[124] D. Relan, T. MacGillivray, L. Ballerini, and E. Trucco, "Retinal vessel classification: sorting arteries and veins," in *Proceedings of the 35th Annual International Conference of the IEEE Engineering in Medicine and Biology Society (EMBC '13)*, pp. 7396–7399, July 2013.

[125] C. Muramatsu, Y. Hatanaka, T. Iwase, T. Hara, and H. Fujita, "Automated selection of major arteries and veins for measurement of arteriolar-to-venular diameter ratio on retinal fundus images," *Computerized Medical Imaging and Graphics*, vol. 35, no. 6, pp. 472–480, 2011.

[126] B. Dashtbozorg, A. M. Mendonca, and A. Campilho, "An automatic graph-based approach for artery/Vein classification in retinal images," *IEEE Transactions on Image Processing*, vol. 23, no. 3, pp. 1073–1083, 2014.

[127] K. Kipli, M. S. Muhammad, S. M. W. Masra, Z. Nurdini, K. Lias, and D. A. A. Mat, "Performance of levenberg-marquardt backpropagation for full reference hybrid image quality metrics," in *Proceedings of the International MultiConference of Engineers and Computer Scientists*, vol. I, pp. 14–17, March 2012.

[128] N. Chapman, N. Witt, X. Gao et al., "Computer algorithms for the automated measurement of retinal arteriolar diameters," *British Journal of Ophthalmology*, vol. 85, no. 1, pp. 74–79, 2001.

[129] R. A. Jensen, S. Xueling, A. V. Smith et al., "Novel genetic loci associated with retinal microvascular diameter," *Circulation. Cardiovascular Genetics*, vol. 9, no. 1, 2016.

[130] J. Lowell, A. Hunter, and D. Steel, "Quantitative measurement of retinal vessel widths from fundus images based on 2D modelling," January 2014.

[131] C. M. Wilson, K. D. Cocker, M. J. Moseley et al., "Computerized analysis of retinal vessel width and tortuosity in premature infants," *Investigative Ophthalmology & Visual Science*, vol. 49, no. 8, pp. 3577–3585, 2008.

[132] M. S. K. Sundaram and K. Gokulakrishnan, "Width measurement from true retinal blood vessels," *International Journal of Emerging Technology and Advanced Engineering*, vol. 4, no. 2, pp. 57–64, 2014.

[133] L. Pedersen, M. Grunkin, B. Ersbøll et al., "Quantitative measurement of changes in retinal vessel diameter in ocular fundus images," *Pattern Recognition Letters*, vol. 21, no. 13-14, pp. 1215–1223, 2000.

[134] M. S. Kuryati Kipli, C. Jiris, S. K. Sahari, R. Sapawi, N. Junaidi, and T. M. A. Z. Kismet Hong Ping, "Morphological and otsu thresholding based retinal blood vessel segmentation for detection of retinopathy".

14

Image Decomposition Algorithm for Dual-Energy Computed Tomography via Fully Convolutional Network

Yifu Xu,[1] Bin Yan,[1] Jingfang Zhang,[2] Jian Chen,[1] Lei Zeng,[1] and Linyuang Wang[1]

[1]*National Digital Switching System Engineering & Technological R&D Centre, Zhengzhou 450002, China*
[2]*153 Central Hospital of Henan Province, Zhengzhou 450002, China*

Correspondence should be addressed to Bin Yan; ybspace@hotmail.com

Academic Editor: Maria E. Fantacci

Background. Dual-energy computed tomography (DECT) has been widely used due to improved substances identification from additional spectral information. The quality of material-specific image produced by DECT attaches great importance to the elaborated design of the basis material decomposition method. *Objective.* The aim of this work is to develop and validate a data-driven algorithm for the image-based decomposition problem. *Methods.* A deep neural net, consisting of a fully convolutional net (FCN) and a fully connected net, is proposed to solve the material decomposition problem. The former net extracts the feature representation of input reconstructed images, and the latter net calculates the decomposed basic material coefficients from the joint feature vector. The whole model was trained and tested using a modified clinical dataset. *Results.* The proposed FCN delivers image with about 60% smaller bias and 70% lower standard deviation than the competing algorithms, suggesting its better material separation capability. Moreover, FCN still yields excellent performance in case of photon noise. *Conclusions.* Our deep cascaded network features high decomposition accuracies and noise robust property. The experimental results have shown the strong function fitting ability of the deep neural network. Deep learning paradigm could be a promising way to solve the nonlinear problem in DECT.

1. Introduction

Conventional single-energy X-ray technique provides information about the examined object which is not sufficient to characterize it precisely. Dual-energy computed tomography (DECT) provides additional information by using two different energy spectra to scan the object, which has been presented as a valid alternative to conventional single-energy X-ray imaging. In recent years, the adoption of DECT has gained increased attention in public security [1] and medical field [2, 3]. The advantage of DECT is the ability for material characterization and differentiation [4]. This decomposition of mixture into two basic materials depends on the principle that the attenuation coefficient is material and energy dependent. Thus, measurements at two distinct energies should permit the separation of the attenuation into its basic components.

The quality of material-specific image produced by DECT attaches great importance to the elaborated design of the basis material decomposition method. The existing decomposition methods can be divided into two main categories: projection-based [5–7] and image-based [8–10]. Projection-based methods pass the projection data through a decomposition function, followed by image reconstruction such as filtered backprojection (FBP). It commonly provides better accuracy and reconstructed image with reduced beam-hardening artifacts in comparison with image-based methods. However, projection-based methods need matched projection datasets. This means that physically the same lines need to be measured for each spectrum, which is usually not the case in today's CT scanners. Image-based methods use linear combinations of reconstructed images to get an image that contains material-selective DECT information. It is an approximative technique, and the resulting images are less quantitative than with projection-based methods. But image-based methods can handle mismatched projection datasets and are applicable to the decomposition of three or more constituent materials, which is more expedient in practice. Thus, they have been employed more frequently in modern DECT implementations.

The material decomposition problem in image domain can be described by the following equation:

$$\begin{pmatrix} \mu_H \\ \mu_L \end{pmatrix} = \begin{pmatrix} \overline{\mu}_{1H} & \overline{\mu}_{2H} \\ \overline{\mu}_{1L} & \overline{\mu}_{2L} \end{pmatrix} \begin{pmatrix} x_1 \\ x_2 \end{pmatrix}, \tag{1}$$

where μ_H and μ_L are the pixels in reconstructed images from low- and high-energy projections, respectively, and x_1 and x_2 are the corresponding points in decomposed basic materials images. The subscripts 1 and 2 indicate two specific materials. $\overline{\mu}_{1L/H}$ and $\overline{\mu}_{2L/H}$ are the average attenuation coefficients of the two basic materials under low/high-energy spectra. These attenuation coefficients are usually obtained by manually selecting two uniform regions of interest (ROIs) on the CT images that contain the basic materials [9, 11, 12]. Direct material decomposition via matrix inversion is a way of calculating the points x_1 and x_2 in the decomposed image, which is written as follows:

$$\begin{pmatrix} x_1 \\ x_2 \end{pmatrix} = \frac{1}{\overline{\mu}_{1H}\overline{\mu}_{2L} - \overline{\mu}_{2H}\overline{\mu}_{1L}} \begin{pmatrix} \overline{\mu}_{2L} & -\overline{\mu}_{2H} \\ -\overline{\mu}_{1L} & \overline{\mu}_{1H} \end{pmatrix} \begin{pmatrix} \mu_H \\ \mu_L \end{pmatrix}. \tag{2}$$

Equation (2) can be easily solved as long as the value of $\Delta = \overline{\mu}_{1H}\overline{\mu}_{2L} - \overline{\mu}_{2H}\overline{\mu}_{1L}$ is not null. However, values of the two terms in Δ do not differ significantly from each other. Therefore, the decomposition result is very sensitive to the noise in the input reconstructed images. Various methods have been proposed to solve this noise suppression problem. Precorrection [13, 14] methods reconstruct two water-precorrected images, followed by a linear combination, to yield images that are free from cupping artifacts usually in water-equivalent materials. The noise reduction techniques after image decomposition include Kalender's correlated noise reduction (KCNR) [15, 16], noise forcing (NOF) [17], and noise clipping (NOC) [18], whose most fundamental strategy is the application of a smoothing filter. Recent advanced iterative methods [9, 10] consider the statistical properties of the decomposition process, producing high-quality edge-preserving images. These methods have shown great success on the decomposition problem. Their well performances rely on the well-handcrafted design of the algorithm.

In recent years, deep learning techniques, which use neural networks having a deep structure with three or more layers, have attracted widespread attention, mainly by outperforming alternative machine learning methods in numerous important applications. The current most popular deep model is the convolutional neural network (CNN) which has emerged as a powerful class of models for image classification [19, 20] and object detection [21]. In the field of computed tomography, some of the recent studies have already attempted to use deep neural networks to solve the problems such as low-dose image denoising [22] and artifact reduction [23]. Wang [24] provides an analytical and global perspective to the combination of tomographic imaging and deep learning. For the material decomposition problem in DECT, several neural network-based methods have also been proposed, but they all decompose the material in the projection domain [7, 25, 26].

Inspired by the recent learning-based methods [27, 28], in this paper, we propose an end-to-end image decomposition algorithm via deep learning techniques. A modified fully convolutional network is applied to extract the feature of reconstructed images and suppress the image noise at the same time. The last layer of the model is a fully connected layer to calculate the decomposed images from the extracted features. We demonstrate the effectiveness of our algorithm by the experiment on a clinical dataset. Two conventional algorithms are implemented and compared to the proposed FCN.

2. Methods

2.1. Fully Convolutional Network. Fully convolutional network (FCN) is one kind of CNN, which is firstly proposed and used for semantic segmentation [29]. The standard CNN generally is composed of a pooling layer and a convolutional layer which are alternately connected. The convolutional layers learn the features of the input. The pooling layers guarantee that the deeper layers can extract higher scale-level features through downsampling. In order to map the feature to the class labels, a fully connected layer will be added to the last output layer, which has fixed dimensions and throws away spatial coordinates. Due to this structural design, the naive CNN requires fixed-sized inputs and produces no-spatial outputs.

The main idea of FCN is transforming the last fully connected layer into a convolution layer with kernels that cover its entire input region. This replacement policy brings about several advantages for FCN. First, the input of the net can be the images of arbitrary sizes, which means that the net can be trained on image patches and then tested on the full-sized images. Second, it can efficiently learn to make dense predictions for per-pixel tasks such as semantic segmentation. Lastly, per-pixel tasks for naive CNN generate a huge amount of redundant convolution computations at adjacent patches. FCN avoids such problems by computing all convolutions in the first layer on the entire input image, leading to significant speedup in the forward-propagation process.

Because of these advantages, FCN is especially suitable for solving the image-based material decomposition problem which can also be regarded as a per-pixel prediction task. In addition, convolution operation to image is interpretable, since it can be seen as a kind of image filtering.

2.2. Image Decomposition Model. For image decomposition, we designed an end-to-end decomposition model based on FCN. The proposed model takes reconstructed images as inputs and predicts the basic material coefficients pixel by pixel in the decomposed image, completing image decomposition and noise suppression at one time.

An overview of our model is illustrated in Figure 1. It is composed of two types of layers: convolutional and fully connected layers. Since the pooling layer may discard important structural details in feature maps, we omit it from the model to avoid losing the quality of result images. But no downsampling process by the pooling layer will lead to the same size of the feature maps at different layers. We hope the model can still catch the multiscale features of the image at different layers, so the strides of the convolutional layers are set to 2 to finish the downsampling operation. The input of the model is the image patch of 65 × 65 size in reconstructed images. There are two independent fully convolutional nets corresponding to the reconstructed images from low- and high-energy projections. The two nets have the same layer structure and are called the L-FCN and H-FCN in short in this study. They are composed of four convolutional layers. The output of layer n can be formulated as follows:

$$\mathbf{C}^{n}(\mathbf{x}_n) = \mathrm{ReLU}(\mathbf{W}_{\mathrm{f}}^{n} * \mathbf{x}_n + \mathbf{b}_{\mathrm{f}}^{n}), \quad n = 1, 2, 3, 4, \tag{3}$$

where $\mathbf{x}_n$ is the input feature map or images and $\mathbf{W}_{\mathrm{f}}^{n}$ and $\mathbf{b}_{\mathrm{f}}^{n}$ represent the convolutional kernel weights and bias parameter, respectively. $*$ is the convolutional operation. $\mathrm{ReLU}(x) = \max(0, x)$ is the nonlinear active function of the neuron. The outputs of L-FCN or H-FCN $(\mathbf{C}^{4}(\mathbf{x}_4))$ are a 512 × 1 vector which represents the feature of the current input patch. The two feature vectors from L-FCN and H-FCN are merged into a joint vector. Then, a fully connected layer is used to calculate the decomposed basic material coefficients from the joint vector, which follows the following equation:

$$\mathbf{X} = \mathbf{W}_{\mathrm{c}}\mathbf{M} + \mathbf{b}_{\mathrm{c}}, \tag{4}$$

where $\mathbf{X} = (x_1, x_2)$ is the predicted material coefficients vector, $\mathbf{W}_{\mathrm{c}}$ and $\mathbf{b}_{\mathrm{c}}$ are the unsolved parameter matrixes, and $\mathbf{M}$ represents the merged vector from L-FCN and H-FCN.

The whole decomposed images can be obtained by traversing all the patches in the input images. The specific information about each layer of the proposed FCN is listed in Table 1.

2.3. The Training Detail. The proposed FCN is implemented via the TensorFlow [30] framework on a computer platform containing two Titan X GPUs (a total of 24 GB video memory). The base learning rate of the model is 5×10^{-3}, which decays by an exponential power of 0.9. There are 1200 training samples in one batch. The mean squared error (MSE) is utilized as the loss function:

$$L(\mathbf{W}_{\mathrm{c}}, \mathbf{b}_{\mathrm{c}}, \mathbf{W}_{\mathrm{f}}, \mathbf{b}_{\mathrm{f}}) = \frac{1}{2}\|\mathbf{X} - \widehat{\mathbf{X}}\|_2, \tag{5}$$

where $\widehat{\mathbf{X}} = (\widehat{x}_1, \widehat{x}_2)$ is the true value of the decomposed image. We used Adam [31] to optimize the loss function in this study. The entire model contains about 64k unsolved parameters and is trained for 40 epochs in 37 hours. The loss curve for training is plotted in Figure S1 in the Supplementary Materials.

3. Experimental Design

3.1. Experimental Dataset. The experimental data are acquired from a real clinical dataset which contains 5987 pleural and cranial cavity 512 × 512 images from 12 patients. These raw data are obtained by one single-energy scan. The tissue and bone regions in the images are all manually sketched out. The images from 10 patients were selected to generate training samples, and the images from the rest of the patients were used for testing. All the images are split up into two partitions. Each partition includes regions of bone or tissue only and is used as the ground truth of the decomposed images. In order to generate dual-energy images, we processed the original raw data and simulated the imaging system. The original image is inconvenient to process for its small value. So, firstly, we amplified the value of the raw data to a proper range via a linear transform that follows the following equation:

$$\begin{cases} x_{\mathrm{t}} = \lambda_{\mathrm{t}}\widetilde{x}_{\mathrm{t}}, \\ x_{\mathrm{b}} = \lambda_{\mathrm{b}}\widetilde{x}_{\mathrm{b}}, \end{cases} \tag{6}$$

where $\widetilde{x}_{\mathrm{t}}$ and $\widetilde{x}_{\mathrm{b}}$ are the pixel values of tissue and bone regions in original images and x_{t} and x_{b} are the corresponding pixel values in transformed images. The values λ_{t} and λ_{b} in the experiment are set to 50 and 15, respectively. Here, the different setting of λ_{t} and λ_{b} is for the purpose of better visual contrast in the transformed images. Secondly, we applied a BM3D [32] algorithm for attenuation of additive white Gaussian noise from the image. Thirdly, we used SpekCalc [33] software to generate 80 kVp and 140 kVp energy spectra, calculated the projection under the simulated scan of dual energy, and obtained the reconstructed images via filtered backprojection (FBP). Lastly, for each patient in the training set, we selected one slice every 10 images. Then, for each image, we extracted 65 × 65 patches with the sliding interval of 5 pixels. The patch size was set to 65 × 65, the same as the input layer of the proposed FCN, getting totally 2,454,300 training patches.

3.2. Evaluation Metrics. The proposed FCN is compared with two other algorithms, direct decomposition (matrix inversion) and iterative decomposition [9]. We choose the bias and standard deviation to evaluate the performance of these methods. Bias shows the difference between the measured value and expected value, which can be a measure of the precision of the result. Standard deviation (SD) reflects the degree of dispersion of the result. They are calculated as follows:

$$\mathrm{Bias} = \frac{1}{N}\sum_{i=1}^{N}|x_i - \widehat{x}_i|,$$
$$\mathrm{SD} = \sqrt{\frac{1}{N}\sum_{i=1}^{N}(x_i - \mu)^2}, \tag{7}$$

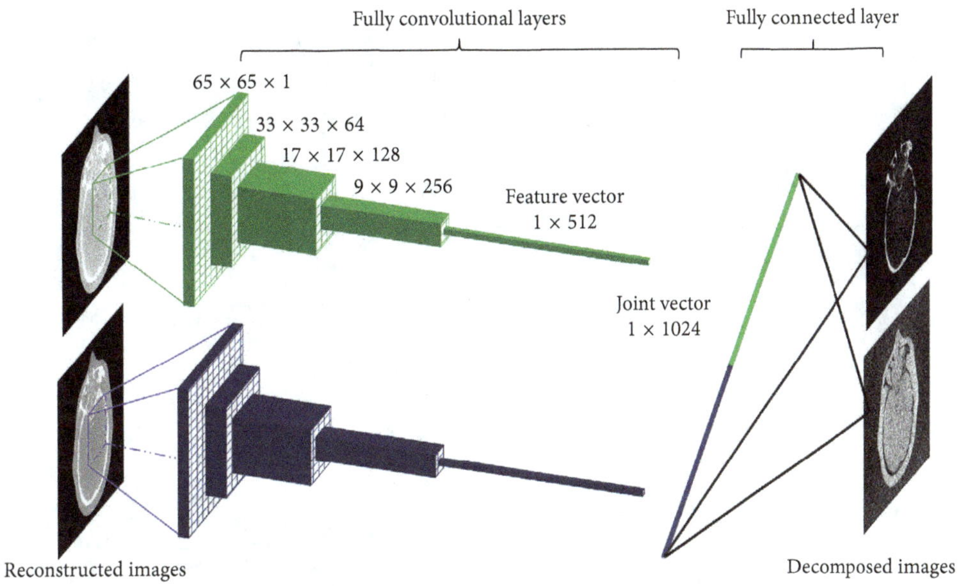

FIGURE 1: Overall architecture of the proposed network.

where x_i and $\hat{x}_i$ are the predicted value and true value at point i of the image, respectively, μ is the mean value of the material, and N is the number of points in ROI.

In order to further investigate the robustness of the proposed FCN, before reconstruction via FBP, photon noise is introduced into the dual-energy projections. There are two major types of noise in X-ray projection images [34]. One type is due to the electrical and roundoff error, which is image independent and can be considered as the Gaussian noise; the other type is due to the statistical fluctuation of the X-ray photons, which is image dependent and can be considered as the Poisson noise. The noise of the first type is small and is omitted in this study. The noise of the second type can be calculated as follows:

$$\begin{cases} \tilde{p}_{\mathrm{L}} = \ln I_{\mathrm{L}} - \ln\left(g\left(I_{\mathrm{L}} e^{-p_{\mathrm{L}}}\right)\right), \\ \tilde{p}_{\mathrm{H}} = \ln I_{\mathrm{H}} - \ln\left(g\left(I_{\mathrm{H}} e^{-p_{\mathrm{H}}}\right)\right), \end{cases} \quad (8)$$

where $\tilde{p}_{\mathrm{L}}$ and $\tilde{p}_{\mathrm{H}}$ are the noise-corrupted low- and high-energy projections, $g(x)$ is a random process according to Poisson's distribution with mean x, and I_{L} and I_{H} are the number of photons of low- and high-energy incident X-rays. We set $I_{\mathrm{L}} = 5 \times 10^5$ and $I_{\mathrm{H}} = 1 \times 10^6$ in the experiments, respectively.

4. Results

We test our model on a cranial image and a pleural image which are excluded from the training dataset. Figure 2 shows the decomposition results by using three algorithms. The first column is the ground truth. Bone and tissue are chosen as the basis materials. Matrix inversion achieves similar results in vision as iterative decomposition. Loss of details and noticeable blocky artifacts are observed for the tissue and bone images from both algorithms. Figure 3 shows the zoom-in images whose areas are indicated in Figure 2 with a dashed rectangle. The iterative decomposition delivers smooth image due to its smoothness regularization term in loss function. It is noticeable that the proposed FCN suppresses most artifacts while preserving the structural features better than the competing algorithms. But there are not distinct improvements in edge-preserving. We guess this is mainly caused by the convolution kernel in the model. The convolutional operation of image can be seen as a kind of filtering.

TABLE 1: Detailed configuration of L-FCN/H-FCN.

Layer name	Kernel size	Stride	Pad	Output size
Input	—	—	—	$65 \times 65 \times 1$
Conv1	5×5	2	1	$33 \times 33 \times 64$
Conv2	5×5	2	1	$17 \times 17 \times 128$
Conv3	5×5	2	1	$9 \times 9 \times 256$
Conv4	9×9	1	0	$1 \times 1 \times 512$

For quantitative evaluation, the bias and SD are calculated on the images generated by using different algorithms inside material's ROI and summarized in Table 2. Generally, the estimate of bone is more accurate than that of tissue. The proposed FCN achieves results closest in values to the ground truth, with about 60% smaller bias and 70% lower standard deviation than the competing algorithms, suggesting its better material separation capability.

To evaluate the potential improvement by FCN, we investigate the effects of photon noise on the material decomposition algorithms. The reconstructed image is generated from noise-corrupted projections as described in Section 3.2. Figure 4 presents the decomposition results on same testing images. It can be seen that direct matrix inversion magnifies the noises both in ROI and background. Iterative decomposition also suffers from serious artifacts. This indicates that both algorithms are more sensitive to the noise. The proposed FCN yields the decomposed images that have not much noticeable change in comparison with the results in Figure 2.

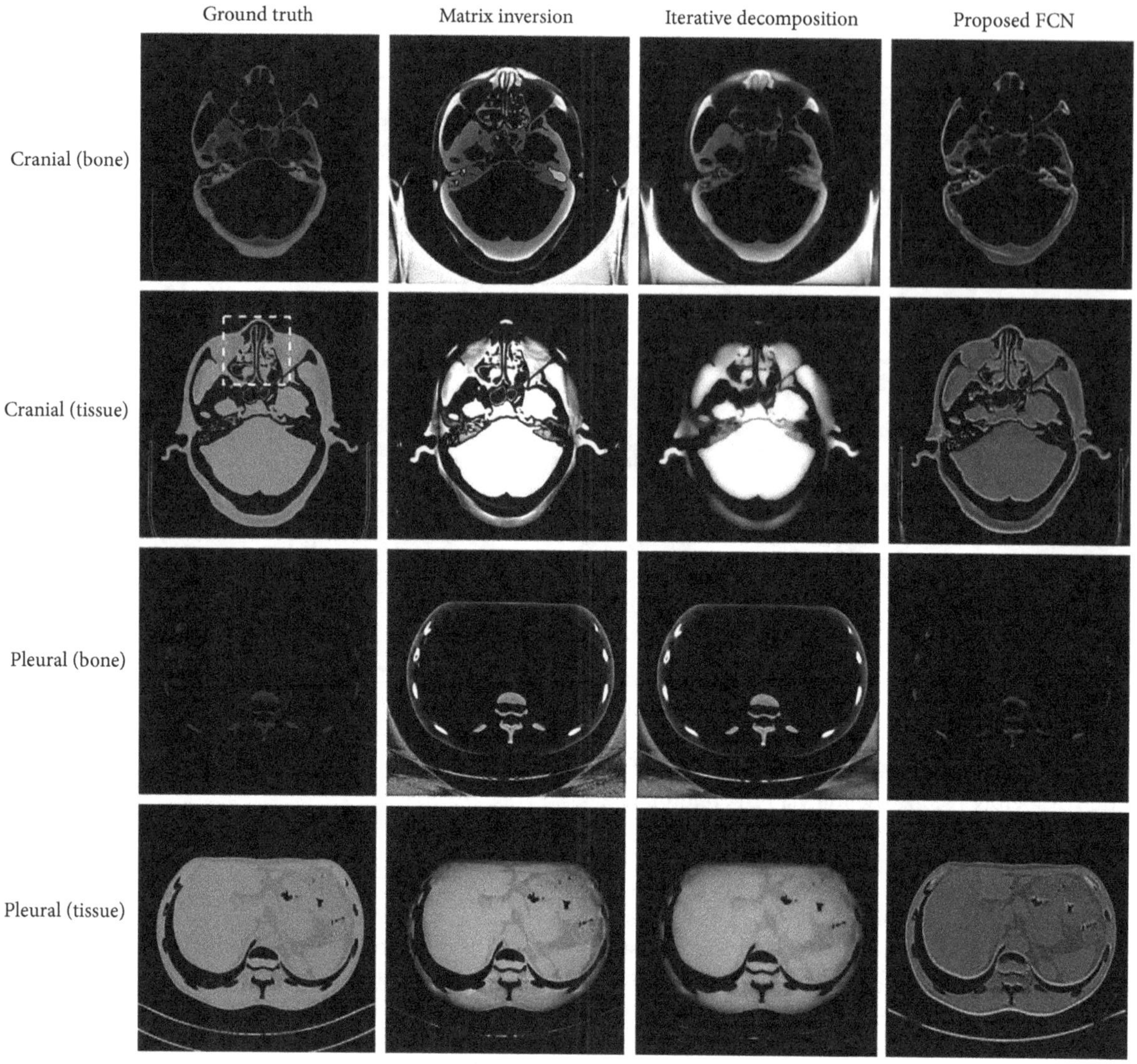

FIGURE 2: The decomposed images by using three methods.

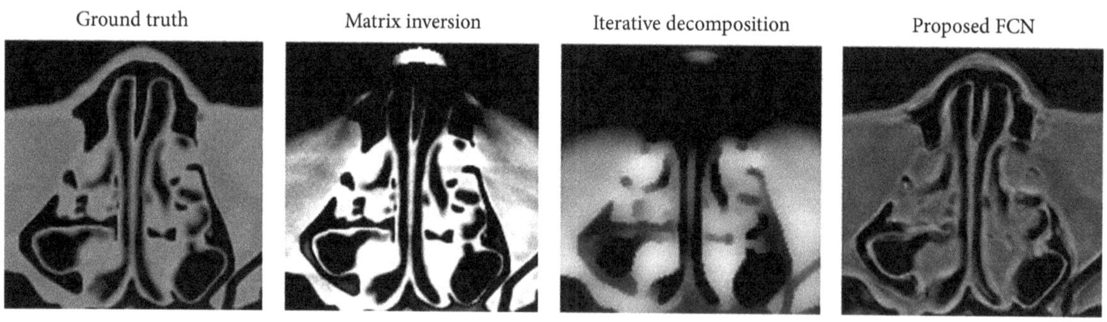

FIGURE 3: Result comparisons in the zoom-in area which is indicated in Figure 2 with a dashed rectangle.

TABLE 2: A list of bias and SD on the images generated by using different algorithms.

Material	Bone (cranial)	Tissue (cranial)	Bone (pleural)	Tissue (pleural)
Matrix inversion	0.410 ± 0.799	0.790 ± 0.930	0.823 ± 1.126	$\mathbf{0.191 \pm 0.348}$
Iterative decomposition	0.330 ± 0.621	0.833 ± 1.221	0.763 ± 0.994	0.220 ± 0.417
Proposed FCN	$\mathbf{0.111 \pm 0.280}$	$\mathbf{0.283 \pm 0.261}$	$\mathbf{0.097 \pm 0.208}$	0.322 ± 0.171

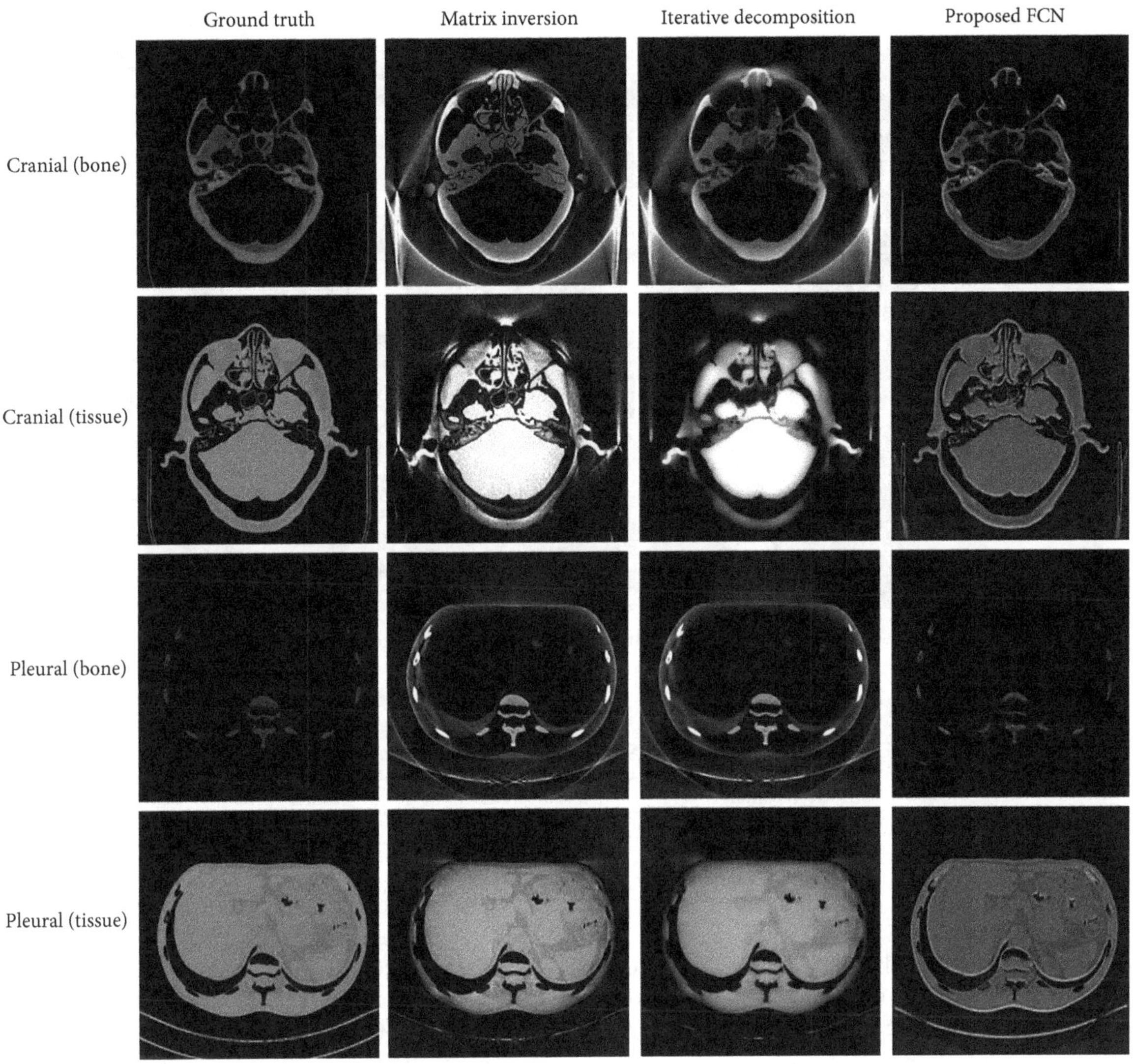

FIGURE 4: The decomposition results on data with photon noise.

Figure 5 illustrates the absolute value of the difference between images in Figures 2 and 4, providing a visual comparison of the performance of noise suppression. For matrix inversion, the noise is statistically independent and evenly distributed in the images because the value of each pixel in decomposed images is calculated by using the corresponding pixel in projections. For iterative decomposition, the noise demonstrates a regional distribution characteristic. The region of tissue and background contain larger amount of noises than bone. In contrast, there are not much obvious differences in the result produced by the proposed FCN. Clearly, it outperforms the other two algorithms, more effectively suppressing image noise while keeping subtle structures.

The quantitative results are listed in Table 3. In the case of photon noise, the bias and SD of the competing algorithms have increased in varying degrees. FCN still demonstrates good agreement to the true value, indicating its advantages on the antinoise capability.

5. Discussion

We have designed a cascaded neural network for the material decomposition problem. The reconstructed images are pixel wisely mapped to decomposed images via several convolutional layers and a fully connected layer. The size of the input layer is 65×65, based on the hypothesis that the value of the material coefficient depends largely on the local region in reconstructed images. The proposed FCN processes data in an end-to-end way, without any needs of precorrected images or other prior knowledge. The experimental results show its strong performance in capturing the localized structural information and suppressing image noise. The decomposed images generated by matrix inversion and iterative decomposition contain relatively a large amount of artifacts. In the robustness-testing experiment, the noise-corrupted inputs will have a negative impact on the performance of the other competing algorithms, but not much on the FCN. The proposed FCN still

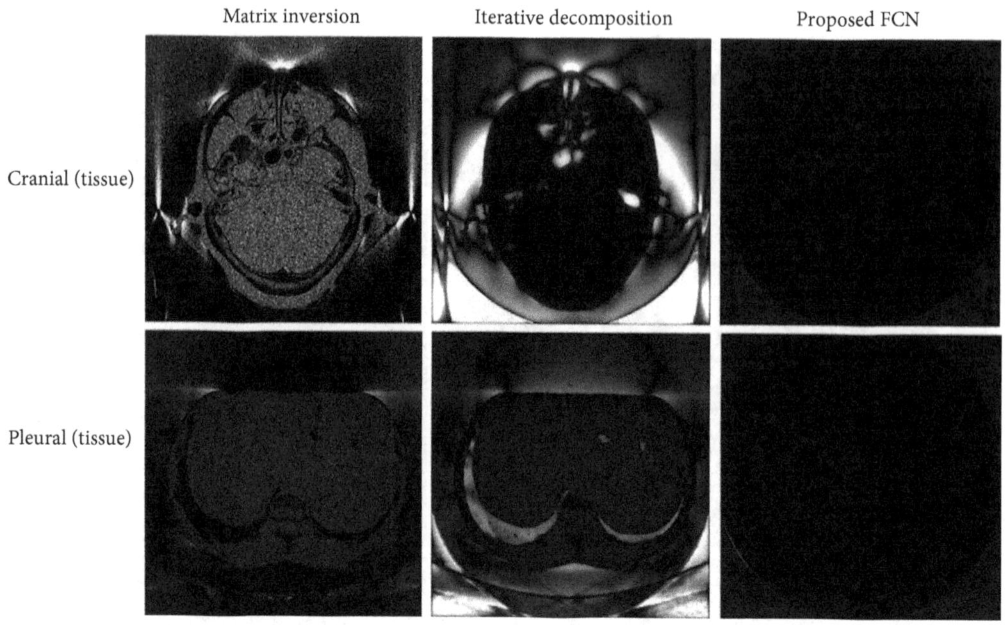

FIGURE 5: The absolute value of the difference between images in Figures 2 and 4.

TABLE 3: A list of bias and SD on the images in case of photon noise.

Material	Bone (cranial)	Tissue (cranial)	Bone (pleural)	Tissue (pleural)
Matrix inversion	0.425 ± 0.807	0.804 ± 0.983	0.840 ± 1.162	$\mathbf{0.241 \pm 0.390}$
Iterative decomposition	0.322 ± 0.608	0.823 ± 1.180	0.773 ± 1.012	0.242 ± 0.423
Proposed FCN	$\mathbf{0.108 \pm 0.284}$	$\mathbf{0.283 \pm 0.260}$	$\mathbf{0.097 \pm 0.208}$	0.290 ± 0.169

achieves excellent results which have low bias and standard deviation. Data augmentation was used in the training process. It brought no boost in performance but costs more training time. We guess the main reason for this issue is that the material decomposition is a regression problem. The value of the label is in a continuous space. Data augmentation assumes that the examples in vicinity share the same class. This hypothesis is usually plausible to the classification problem in which the label is a discrete variable, but unnecessary for the regression problem. The main drawback of our algorithm is the requirement of the specific type of material. Tissue and bone are selected as the basic material in the experiment. The whole model needs to be retrained if one of the materials was changed. So, we hope the proposed algorithm can be used in some applications such as medical diagnosis where the selection of the material is relatively fixed. The amount of training samples is another main factor contributing to the effectiveness of our model. Normally, more data bring better performance of the model. But it may be difficult to collect enough data in some conditions.

6. Conclusions and Further Work

In this study, we present a deep learning approach for the image decomposition problem in DECT. According to the preliminary decomposition results, we successfully prove the feasibility of the proposed algorithm which delivers image with 70% smaller bias and 60% lower standard deviation than the competing algorithms. A deep learning paradigm promises to improve the ability of solving the nonlinear problem in DECT.

We think there are two directions of work that are worth further researching. One is to extend our model to make it applicable to the three-materials decomposition problem. The other is the attempt of using the deconvolutional network which will output the whole decomposed images in a forward-propagation calculation rather than pixel wisely prediction.

Conflicts of Interest

The authors have no conflicts of interest to declare.

Acknowledgments

This work was supported by the National Key R&D Program of China under grant no. 2017YFB1002502, the National Natural Science Foundation of China (nos. 61601518 and 61372172) and the National Natural Science Foundation of Henan Province of China (no. 162300410333).

Supplementary Materials

Figure S1: the proposed model contains about 64k unsolved parameters and is trained for 40 epochs in 37 hours. The training batch size is 1200 reconstructed images from the noise-corrupted low- and high-energy projections. Figures S2 and S3: more testing results to show the superiority of the proposed method. All the testing images are reconstructed from the noise-corrupted low- and high-energy projections. (*Supplementary Materials*)

References

[1] Z. Ying, R. Naidu, and C. R. Crawford, "Dual energy computed tomography for explosive detection," *Journal of X-Ray Science and Technology*, vol. 14, no. 4, pp. 235–256, 2006.

[2] H. W. Goo and J. M. Goo, "Dual-energy CT: new horizon in medical imaging," *Korean Journal of Radiology*, vol. 18, no. 4, pp. 555–569, 2017.

[3] C. Hong, T. Y. Chin, and W. C. G. Peh, "Dual-energy CT in gout-a review of current concepts and applications," *Journal of Medical Radiation Sciences*, vol. 64, no. 1, pp. 41–51, 2017.

[4] A. A. Postma, M. Das, A. A. R. Stadler, and J. E. Wildberger, "Dual-energy CT: what the neuroradiologist should know," *Current Radiology Reports*, vol. 3, no. 5, 2015.

[5] B. Li and Y. Zhang, "Projection decomposition algorithm of x-ray dual-energy computed tomography based on projection matching," *Acta Optica Sinica*, vol. 31, no. 3, article 311002, 2011.

[6] B. Brendel and J. P. Schlomka, "Empirical projection-based basis-component decomposition method," in *Proceedings of Medical Imaging 2009: Physics of Medical Imaging*, vol. 7258, p. 72583Y, International Society for Optics and Photonics, Bellingham, WA, USA, March 2009.

[7] T. G. Schmidt and K. C. Zimmerman, "Material decomposition of multi-spectral x-ray projections using neural networks," US Patent US20150371378, 2015.

[8] J. W. Lambert, Y. Sun, R. G. Gould, M. A. Ohliger, Z. Li, and B. M. Yeh, "An image-domain contrast material extraction method for dual-energy computed tomography," *Investigative Radiology*, vol. 52, no. 4, pp. 245–254, 2017.

[9] T. Niu, X. Dong, M. Petrongolo, and L. Zhu, "Iterative image-domain decomposition for dual-energy CT," *Medical Physics*, vol. 41, no. 4, article 041901, 2014.

[10] X. Dong, T. Niu, and L. Zhu, "Combined iterative reconstruction and image-domain decomposition for dual energy CT using total-variation regularization," *Medical Physics*, vol. 41, no. 5, article 051909, 2014.

[11] T. P. Szczykutowicz and G. H. Chen, "Dual energy CT using slow kVp switching acquisition and prior image constrained compressed sensing," *Physics in Medicine and Biology*, vol. 55, no. 21, pp. 6411–6429, 2010.

[12] P. V. Granton, S. I. Pollmann, N. L. Ford, M. Drangova, and D. W. Holdsworth, "Implementation of dual- and triple-energy cone-beam micro-CT for postreconstruction material decomposition," *Medical Physics*, vol. 35, no. 11, pp. 5030–5042, 2008.

[13] M. Baer, C. Maaß, W. A. Kalender, and M. Kachelrieß, "Image-based dual energy CT using optimized precorrection functions: a practical new approach to material decomposition in the image domain," in *World Congress on Medical Physics and Biomedical Engineering*, pp. 205–208, Springer Berlin Heidelberg, Munich, Germany, September 2009.

[14] R. A. Brooks, "A quantitative theory of the Hounsfield unit and its application to dual energy scanning," *Journal of Computer Assisted Tomography*, vol. 1, no. 4, pp. 487–493, 1977.

[15] W. A. Kalender, E. Klotz, and L. Kostaridou, "An algorithm for noise suppression in dual energy CT material density images," *IEEE Transactions on Medical Imaging*, vol. 7, no. 3, pp. 218–224, 1988.

[16] D. L. Ergun, C. A. Mistretta, D .E. Brown et al., "Single-exposure dual-energy computed radiography: improved detection and processing," *Radiology*, vol. 174, no. 1, pp. 243–249, 1990.

[17] J. T. Dobbins, "Recent progress in noise reduction and scatter correction in dual-energy imaging," in *Proceedings of SPIE-the International Society for Optical Engineering*, pp. 134–142, San Diego, CA, USA, May 1995.

[18] W. W. Peppler, J. T. Dobbins, E. B. Bellers et al., "Dual-energy computed radiography: improvements in processing," in *Proceedings of SPIE-the International Society for Optical Engineering*, vol. 2167, pp. 663–671, Newport Beach, CA, USA, May 1994.

[19] L. Kaiser, A. N. Gomez, N. Shazeer et al., "One model to learn them all," http://arxiv.org/abs/1706.05137, 2017.

[20] K. He, X. Zhang, S. Ren, and J. Sun, "Deep residual learning for image recognition," http://arxiv.org/abs/1512.03385, 2015.

[21] T. Kong, A. Yao, Y. Chen, and F. Sun, "HyperNet: towards accurate region proposal generation and joint object detection," in *Proceedings of Computer Vision and Patter Recognition IEEE*, pp. 845–853, Piscataway, NJ, USA, April 2016.

[22] D. Wu, K. Kim, G. E. Fakhri et al., "A cascaded convolutional neural network for x-ray low-dose CT image denoising," http://arxiv.org/abs/1705.04267, 2017.

[23] H. Zhang, L. Li, K. Qiao et al., "Image prediction for limited-angle tomography via deep learning with convolutional neural network," http://arxiv.org/abs/1607.08707, 2016.

[24] G. Wang, "A perspective on deep imaging," *IEEE Access*, vol. 4, no. 99, pp. 8914–8924, 2016.

[25] K. C. Zimmerman and T. G. Schmidt, "Experimental comparison of empirical material decomposition methods for spectral CT," *Physics in Medicine & Biology*, vol. 60, no. 8, pp. 3175–3191, 2015.

[26] W. J. Lee, D. S. Kim, S. W. Kang, and W. J. Yi, "Material depth reconstruction method of multi-energy X-ray images using neural network," in *Proceedings of International Conference of the IEEE Engineering in Medicine and Biology Society*, pp. 1514–1517, Piscataway, NJ, USA, August–September 2012.

[27] Y. Lu, M. Berger, M. Manhart et al., "Bridge to real data: empirical multiple material calibration for learning-based material decomposition," in *Proceedings of IEEE International Symposium on Biomedical Imaging*, pp. 457–460, Prague, Czech Republic, April 2016.

[28] Y. Lu, J. Geret, M. Unberath et al., "Projection-based material decomposition by machine learning using image-based features for computed tomography," in *Proceedings of the 13th International Meeting on Fully Three-Dimensional Image Reconstruction in Radiology and Nuclear Medicine*, pp. 448–452, Newport, RI, USA, July 2015.

[29] E. Shelhamer, J. Long, and T. Darrell, "Fully convolutional networks for semantic segmentation," in *Proceedings of IEEE Transactions on Patter Analysis & Machine Intelligence*, vol. 39, no. 4, pp. 640–651, Washington, DC, USA, 2014.

[30] M. Abadi, A. Agarwal, P. Barham et al., "TensorFlow: large-scale machine learning on heterogeneous distributed systems," http://arxiv.org/abs/1603.04467, 2016.
[31] D. P. Kingma and J. Ba, "Adam: a method for stochastic optimization," http://arxiv.org/abs/1412.6980, 2014.
[32] K. Dabov, A. Foi, and V. Katkovnik, "Image denoising by sparse 3-D transform-domain collaborative filtering," *IEEE Transactions on Image Processing*, vol. 16, no. 8, pp. 2080–2095, 2007.
[33] G. Poludniowski, G. Landry, F. Deblois, P. M. Evans, and F. Verhaegen, "SpekCalc: a program to calculate photon spectra from tungsten anode x-ray tubes," *Physics in Medicine & Biology*, vol. 54, no. 19, pp. N433–N438, 2009.
[34] L. Zhu, J. Wang, and L. Xing, "Noise suppression in scatter correction for cone-beam CT," *Medical Physics*, vol. 36, no. 3, pp. 741–752, 2009.

15

Shape Analysis of the Patellar Bone Surface and Cutting Plane for Knee Replacement Surgery

E. L. Rex,[1,2] J. Werle,[2,3] B. C. Burkart,[2,3] J. R. MacKenzie,[2,3] K. D. Johnston,[2,3] and C. Anglin[1,2,4]

[1]*Biomedical Engineering, University of Calgary, Calgary, Canada*
[2]*McCaig Institute for Bone and Joint Health, University of Calgary, Calgary, Canada*
[3]*Section of Orthopaedic Surgery, University of Calgary, Calgary, Canada*
[4]*Dept. of Civil Engineering, University of Calgary, Calgary, Canada*

Correspondence should be addressed to C. Anglin; carolyn.anglin@gmail.com

Academic Editor: Hiro Yoshida

Geometry of the patella (kneecap) remains poorly understood yet is highly relevant to performing the correct patellar cut to reduce pain and to improve function and satisfaction after knee replacement surgery. Although studies routinely refer to "parallel to the anterior surface" and "the patellar horizon," a quantitative definition of these is lacking and significant variability exists between observers for this irregularly-shaped bone. A 2D-3D shape analysis technique was developed to determine the optimal device configuration for contacting the patellar surface. Axial and sagittal pseudo-X-rays were created from 18 computed tomography (CT) scans of cadaveric knees. Four expert surgeons reviewed three repetitions of the X-rays in randomized order, marking their desired cut plane and their estimate of the anterior surface. These 2D results were related back to the 3D model to create the desired plane. There was considerable variability in perceptions, with intra- and intersurgeon repeatability (standard deviations) ranging from 1.3° to 2.4°. The best configuration of contact points to achieve the desired cutting plane was three pegs centred on the patellar surface, two superior and one inferior, forming a 16 mm equilateral triangle. This configuration achieved predicted cut planes within 1° of the surgeon ranges on all 18 patellae. Implementing this, as was done in a subsequent prototype surgical device, should help improve the success and satisfaction of knee replacement surgery.

1. Introduction

The patella has a critical impact on the outcome of knee replacement surgery (also called total knee arthroplasty or TKA) (Figure 1). An incorrect cut of the patella, meaning too thick or too thin in one or more of the quadrants, can lead to pain [1–3], improper tracking of the patella within the femoral groove [4], and decreased range of motion [5–7], directly affecting the person's activities of daily living and quality of life. Patellar asymmetry occurs in at least 10% of cases, even amongst expert surgeons [8], contributing to the approximately 20% of TKA patients who are not satisfied with the result of their surgery [9, 10].

A variety of devices have been developed to aid surgeons with cutting (resecting) the patella during TKA. Unfortunately, using these devices can still lead to asymmetry and incorrect thickness, resulting in the clinical complications listed above. Computer-assisted surgery (CAS) can be used to improve accuracy [11] but remains invasive due to attaching a marker array onto the bone and is still only as accurate as the placement of the attachment plate. Many surgeons prefer to resect the patella freehand, since they consider the current devices to be inaccurate. A new approach is needed to create a device that is accurate, fast, and noninvasive.

Fundamental to designing a new, noninvasive instrument for patellar resection is determining suitable contact points on the front (anterior) surface. Some devices, such as the standard pliers-like sawguide, which wraps around the patella, have no contact with the anterior surface,

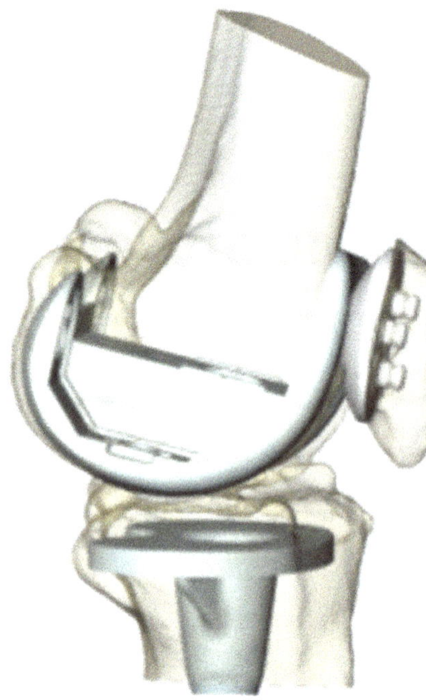

FIGURE 1: In TKA, the arthritic surfaces of the femur, tibia, and patella are replaced with artificial components to reduce pain and improve function. The patella is normally cut "parallel" to the anterior (front) surface. Holes are then drilled into the flat cut surface to insert the patellar component. An incorrect cut of the patella can result in postoperative pain and complications.

relying on the surgeon's judgment to align it with the anterior surface [12]. Others, such as reamers, have a circle of spikes that contact the anterior surface. However, depending on the ridges and hollows on the surface, the plate is sometimes not parallel to the anterior surface, resulting in an asymmetric cut [12, 13]. Symmetry and thickness are also affected when the spikes enter the soft tissue and bone to varying depths [12, 13].

Although various landmark guidelines, such as the medial-lateral (side-to-side) extents, have been suggested for performing the patellar cut, the ultimate goal is to be parallel to the anterior surface, producing a visually rectangular cut, with equal thicknesses in all quadrants [13]. Since asymmetry with respect to the anterior surface, considered to be within ±7° [11], has been correlated with anterior knee pain [2, 3], it is essential to reference a device off of the anterior surface.

Two surprising challenges are to determine what the anterior surface plane is (although the visual goal is a "rectangular" cut, the anterior surface is not flat, and is rarely considered in three dimensions) and to determine whether parallel to this is indeed what looks right to surgeons on the postoperative X-rays. Due to the ellipsoidal and variable shape of the patella, what looks obvious to one person as the anterior surface and the desired resection line is often different from what another person sees [8]. A mathematical definition of the anterior surface plane could be generated, but it was nonetheless essential to confirm that the estimated surface plane corresponds to the resection plane desired by surgeons, thus requiring surgeon input. Statistical shape models have been created of the patellar bone [14], but these provide only a generalized shape of the anterior surface, without a detailed investigation of the surface's hills and valleys. We are unaware of any other study that has analysed the three-dimensional geometry of the anterior surface of the patella and its relationship to the cutting plane.

The purpose of this study was therefore (a) to determine the desired resection plane by having surgeons draw virtual resection lines on axial and sagittal preoperative X-rays, (b) to determine the corresponding estimated anterior surface plane, and (c) to determine a contact configuration on the anterior patellar surface that can achieve the desired patellar resection plane.

2. Materials and Methods

There were four stages to this study: preparing the images, obtaining surgeon input, analysing the surgeon data, and evaluating different contact geometries to achieve the desired resection plane.

2.1. Preparing the Images. Following institutional review board approval, we imaged 18 cadaveric knee specimens (9 left, 9 right; 8 male, 10 female; average age 76; range, 34–91 years) using computed tomography (CT), with a slice thickness of 0.6 mm. All of the specimens had largely normal anatomy, without obvious osteophytes, although some showed signs of early osteoarthritis.

Pseudoradiographs were created from the CT scans in the axial and sagittal directions, using Amira image analysis software (version 5.3.1, Visage Imaging, Andover, MA). The CT was viewed using the volume rendering (Volren) function, which was adjusted so that the bone appeared white and the background black. The threshold was set to remove most of the soft tissue in the view to look as similar as possible to a clinical X-ray. The image was rotated to align the knee similar to how a radiation technologist would align a patient: for the axial view, the image was aligned to provide a skyline view, where the patella is viewed from the superior looking downward (Figure 2(a)) (in patients, this requires flexing the knee to be able to view the patella from the top; in creating the pseudo-X-ray, the viewpoint was rotated about the patella until the patellar ridge was aligned with the *z*-axis, i.e., coming directly in/out of the image) (Figure 2(a)); for the sagittal view, the image was rotated until the femoral condyles were aligned, i.e., overlapping, as they would in a typical sagittal X-ray (Figure 2(b)). Screenshots of each view were taken and the orientation saved for later analysis. Three repetitions of each of the projections were saved in randomized order, for a total of 54 axial projections and 54 sagittal projections.

2.2. Obtaining Surgeon Input. Four experienced, fellowship-trained orthopaedic surgeons assisted in determining the desired resection plane and the corresponding anterior surface plane. Using a custom macro in ImageJ (version 1.42q, National Institutes of Health (NIH), USA), the

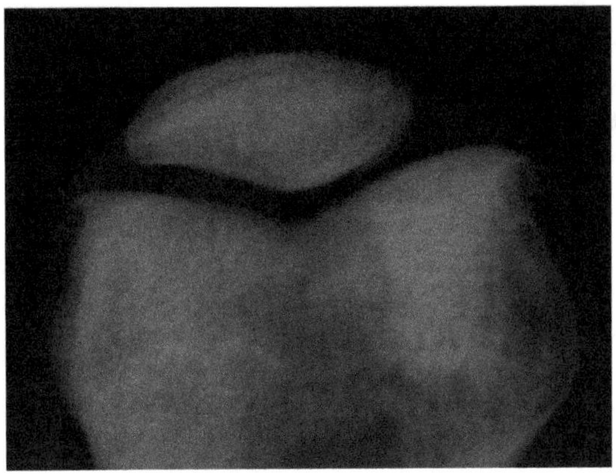

(a)

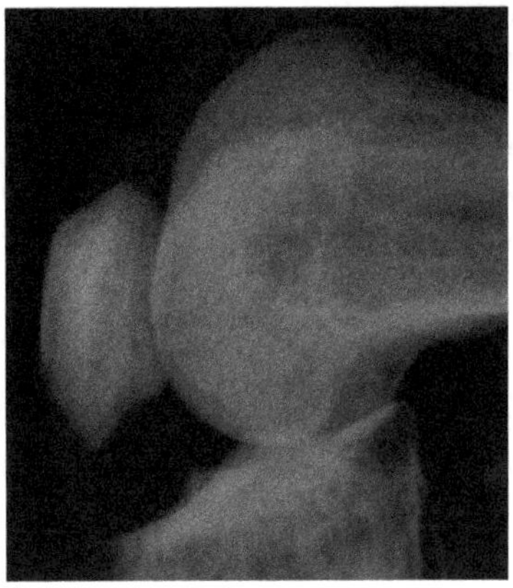

(b)

FIGURE 2: Pseudo-X-rays of a left knee created from a CT scan: (a) axial (skyline) and (b) sagittal (lateral).

surgeons marked the lines electronically on each X-ray; previous lines did not appear. The (x, y) coordinates of the endpoints of the line were recorded in a text file along with the surgeon's initials and date.

X-ray input was performed in four passes. In the first pass, the surgeon went through the axial projections, drawing the line at which they would ideally resect the patella. In the second pass, they drew a line to represent their best estimate of the anterior surface. Third, they went through the sagittal projections, drawing their ideal resection line, and fourth, they went through the sagittal projections to estimate the anterior surface. All of this was done without concern for thickness: while patellar thickness is an important aspect of the resection, it is set independently by the surgeon during surgery, after aligning to the anterior surface, which is the focus of this study. Mediolateral positioning of the patellar component on the flat cut surface, while also an important aspect of patellar resurfacing (for which slight medialization is recommended [15]), is a separate surgical consideration from the resection plane itself and was therefore not included in this study. Altogether 864 lines were drawn: 4 surgeons × 18 patellae × 2 views × 2 lines per view × 3 repetitions.

2.3. Analysing the Surgeon Data. The first step in analysing the data was to understand them visually, which was done by displaying the drawn lines on circumferences of the patellae.

For the quantitative analysis, line angles were measured from the horizontal (mediolateral, ML) axis in the axial views and from the vertical (superoinferior, SI) axis in the sagittal views. For plotting purposes, an average of all the surgeon angles was computed for each patella and this result subtracted from each of the measured angles so that all of the angles were centred about zero. PASW Statistics software (version 17.0, Statistical Package for Social Sciences (SPSS) Inc., Chicago, IL) was used to calculate the interobserver and intraobserver repeatability.

2.4. Evaluating Contact Geometries. To establish a 3D plane from the two sets of 2D data (axial and sagittal), the patellae were segmented from the CT scans using Amira. The surface model was then exported as a dxf file to AutoCAD (version 2010, AutoDesk, San Rafael, CA) to create planes along the surgeon-identified resection lines. The alignment of the model was crucial to transferring the surgeon lines accurately to the virtual 3D model; by saving the alignment of the CT earlier, when the 2D projections were created, the 3D model created from the CT was already aligned.

For each patella and each surgeon, a plane was created based on the average of the three repetitions. This was done by drawing a line along the axial view, a line along the sagittal view, and creating a plane along the two lines.

The resection planes, created from the surgeons' input, were visually compared to each other as well as the patellar surface. Without changing alignment, the planes were moved anteroposteriorly (representing different resection thicknesses) to visually inspect how they looked on the models and if there were any landmarks to which they specifically aligned. This visual inspection, along with the surgeons' comments about desiring a symmetric resection, was used to identify a contact configuration to align to the surgeons' desired resection plane.

A variety of contact configurations were tested and compared to the resection planes by modelling the geometry of the contact plane in comparison to the anterior surface and resection planes. Possible contact configurations included different distances and layouts of pegs, a flat plate corresponding to the previously-developed CAS system [11, 12] and a ring corresponding to the existing reamer. Applying flat-jawed callipers, i.e., the standard clinical measurement instrument, to the patella has been shown previously to cause it to rotate [8], showing that this is not a suitable contact mechanism. For each configuration tested, the simulated geometry of the contact model was applied to the anterior surface of the 18 patella models, creating a plane that was compared visually to the surgeon input in the ML and SI directions. Most configurations could be discarded based on a visual comparison between the proposed plane and the desired plane. To quantify the discrepancies of the final selected configuration, the axial resection (AR), sagittal resection (SR), axial surface (AS), and sagittal surface (SS) results were each sorted into three categories: (1) patellae for which the contact-based plane was within the range of surgeon averages for that specimen, (2) patellae for which this plane was within the overall range of surgeon angles for that specimen, and (3) patellae for which the plane fell outside the range of surgeon input for that specimen.

3. Results

3.1. Surgeon-Defined Resection Planes and Anterior Surface Planes. There was high variability in the estimated resection lines and anterior surface lines within and between the surgeons. As a visual representation, the most and least consistently drawn lines are shown on the patellar circumferences in Figure 3 (axial) and Figure 4 (sagittal). In general, the flatter the patella, the more consistently the lines were drawn, while the more rounded or irregular the patella, the more inconsistently they were drawn, although this did not hold true in all cases. In general, the more well-defined the medial and lateral extents were on the axial views, the more consistently the lines were drawn, although this also did not hold true in all cases. The surgeons appeared to be taking a mental least-squares fit to the anterior surface but varied in which part of the anterior surface they included.

Intrasurgeon repeatability, as judged by the standard deviation, was best for Surgeon 1 (SD = 1.3°), followed by Surgeon 4 (SD = 1.7°), Surgeon 2 (SD = 2.2°), and Surgeon 3 (SD = 2.4°) (Figures 5 and 6). Intersurgeon repeatability was best for the *axial resection line* (SD = 1.5°), followed by the *sagittal resection line* (SD = 1.7°), *axial anterior surface line* (SD = 2.2°), and *sagittal anterior surface line* (SD = 2.2°) (Figures 4 and 5). Perceptions of the resection lines varied by as much as ±9°, and perceptions of the anterior surface varied by as much as ±11°.

The mean angular difference between the *axial resection line* and the *anterior surface* line was 2.4° (SD = 2.1°). The mean angular difference between the *sagittal resection line* and the *anterior surface* line was 3.5° (SD = 3.3°). From the visual display of the lines on the patellar circumferences (see Figures 2 and 3 for examples), it appeared that, when asked to identify the anterior surface, the surgeons focused on a more localized portion of the anterior surface whereas the estimated resection plane approximated a symmetric resection; if the resection plane were to align with the drawn anterior surface plane, it would incorrectly result in thicker medial and superior sides.

3.2. Anterior Surface Contact Configuration Selection. The first decision in the process of defining the desired configuration was to use a system of three contact points since three points always find a stable orientation, as with a three-legged stool. With four or more points, a ring, or a flat plate, there is a risk of instability, as with a chair with one shorter leg. In general, the patellar geometry was higher in the middle and sloped off inferiorly, which explains why a ring of contacts (as with the reamer) or a flat plate can create a tilt of several degrees on some patellae [8, 12, 13]. Although early investigations revealed that the flattest portion of the anterior surface was consistently in the superolateral quadrant, the pegs were too close together in this region to provide good stability; furthermore, after analysis, it was discovered that focusing the pegs in this region did not provide a symmetric resection.

The final configuration that fit best with the surgeon-defined resection planes was a 16 mm equilateral triangle, with two points superior (medial and lateral) and one point inferior, centred on the patellar centre. For the cut to be symmetric, the device needed to be central on the patella, and due to the geometry where the inferior edge begins to slope off and narrow, 2 pegs fit best superiorly, with the 3rd peg inferior. The average patella is 46 mm in the mediolateral direction [16] and symmetry is measured 15 mm from the patellar extents [2], which leaves 16 mm in the centre of the patella; thus a 16 mm spacing about the centre was used. This resulted in a good estimate of the desired resection plane.

The patellar centre was defined halfway between the superior and inferior extents and medial and lateral extents (Figure 7). Having symmetry about the centrepoint further simplifies the design of a surgical instrument. We considered a 15 mm triangle for smaller patellae, but this altered the resulting angles by at most 0.3°, which was too small to be worth the practical consequences. With 16 mm sides, the distance between the inferior and superior points is 13.9 mm.

Using this peg configuration, most of the resection plane and anterior surface plane angles (53 out of 72; 15 AR, 11 SR, 14 AS, and 13 SS) fell into the first category, within the surgeon averages, as expected and desired (Figures 4 and 5); 15 lines (3 AR, 4 SR, 4 AS, and 4 SS) fell into the second category, within the surgeon ranges; and 4 (3 SR, 1 SS) fell into the third category, outside of this range, but only by 1.0°, 0.5°, 0.4°, and 0.6°, respectively.

4. Discussion

This study analysed the relationship between the anterior surface of the patella and the surgeons' desired resection planes. On the basis of this shape analysis study, a novel peg configuration was generated for a patellar resection device that was within 1° of the range of angles that the surgeons defined as their desired resection line for all 18 patellae studied, with most falling within the range of surgeon averages. Subsequent use of this configuration in a device designed to detect patellar asymmetry intraoperatively resulted in good accuracy in comparison to CT scans [17]. Note that if later clinical evidence supports patellar resection at an angle other than parallel to the anterior surface, extending one or two of the pegs can achieve this angle easily.

The high variability in surgeon input confirms the difficulty of defining the anterior surface plane and of defining an ideal resection plane. It also highlights the difficulty of reporting patellar asymmetry or patellar tilt clinically since the definition of the patellar horizon is so variable, despite the fact that studies describe it as a fixed reference. Conversely, the high variability provides some leeway in achieving an exact resection.

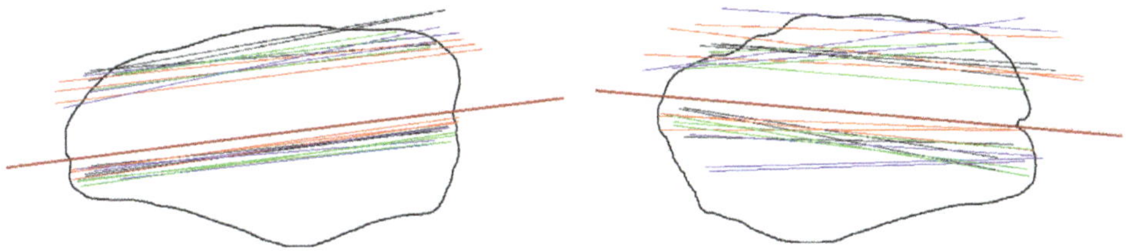

(a) (b)

FIGURE 3: Patellae with the most and least consistently drawn axial resection lines. The four colours represent the four different surgeons, with three repetitions each. The solid black line represents the corresponding peg plane that was later defined (Section 3.2). Thickness is unimportant in the representation since this is decided intraoperatively based on the patellar thickness and component thickness.

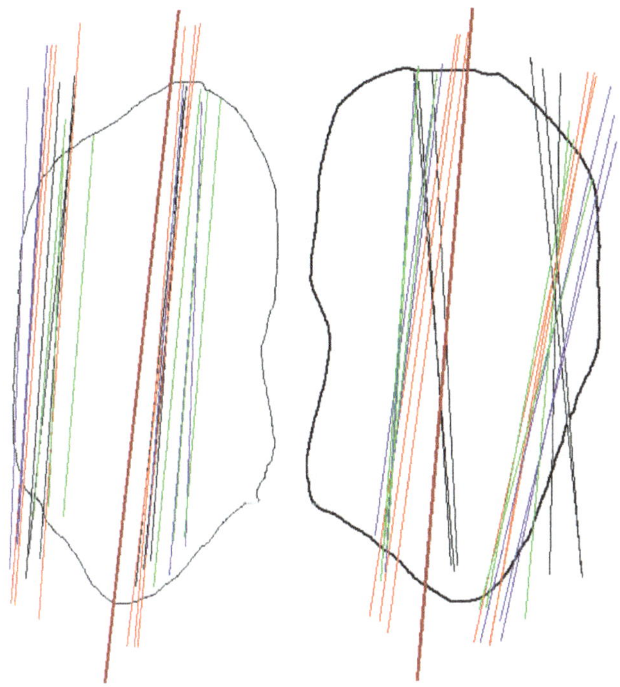

(a) (b)

FIGURE 4: Patellae with the most and least consistently drawn sagittal resection lines. The four colours represent the four different surgeons, with three repetitions each. The solid black line represents the corresponding peg plane defined later (Section 3.2). The thickness is unimportant in the representation since this is decided intraoperatively based on the patellar thickness and component thickness.

The intrasurgeon and intersurgeon repeatability were similar to a previous study with three different surgeons, in which the AR intrasurgeon repeatability using the medial-lateral extents method was 1.6°, and the AR intersurgeon repeatability was 2.0°, although the maximum differences were larger in the present study [8]. The present study is novel in evaluating sagittal radiographs and the anterior surface, as well as in investigating the relationship between the 3D anterior surface geometry and the desired resection plane. Sagittal symmetry is important because SI asymmetry has an even stronger correlation with anterior knee pain than ML asymmetry [2]. The same 2D-3D analysis techniques could be used for any joint in which surgeons routinely use plain X-rays.

The only quantitative definition of asymmetry that we are aware of is a difference of greater than 2 mm between the medial and lateral or superior and inferior patellar thicknesses, measured 15 mm in from the patellar extents [2, 3]. On an average sized patella [16], this results in a similar ML spacing to our peg configuration, although as previously mentioned, a 3-peg configuration was chosen since it is more stable and predictable than a 4-peg configuration.

The main limitation of this study is that the surgeons' lines drawn on X-rays may not be the same plane as they would choose to resect intraoperatively. Nevertheless, it reflects what they would like to see on the X-ray post-operatively and is the basis for measuring asymmetry. Only one of the surgeons (Surgeon 1) routinely does preoperative planning for the patella (axial only) and was therefore most familiar with drawing lines on the X-rays; he had the smallest standard deviation and his resection lines tended to be closer to the overall surgeon average. The particular projection of the patella could also impact the resulting plane; however, all

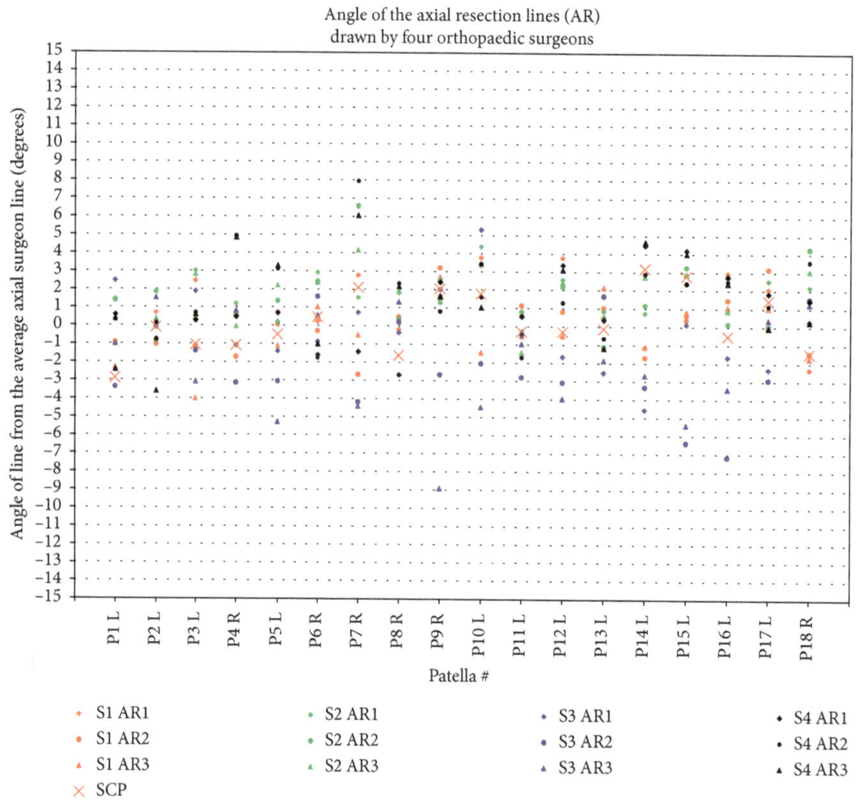

(a)

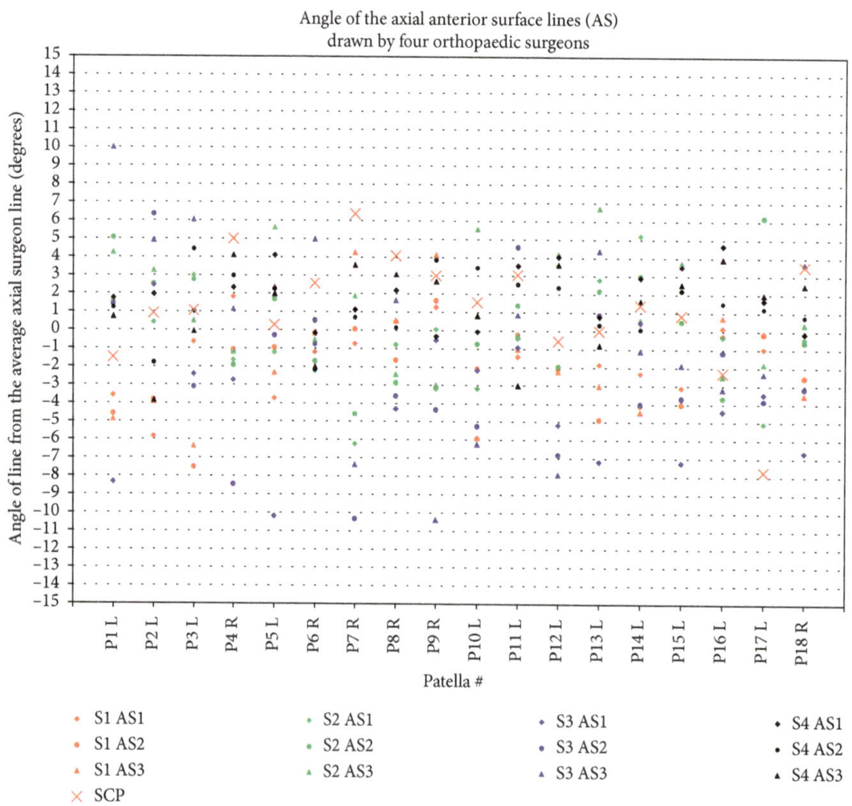

(b)

Figure 5: Angle of drawn lines on axial projections: (a) axial resection (AR) lines and (b) axial anterior surface (AS) lines. Positive angles mean the patella is thicker medially. Results are centred around zero for each patella. SCP = symmetric contact points, i.e., the calculated plane based on the final selected configuration (Section 3.2), which is symmetric about the patellar centre. While the SCP planes did not always match the anterior surfaces drawn (b), they were within a few degrees of the desired resection plane (a), which is the primary clinical goal. S1–S4 = surgeons 1–4.

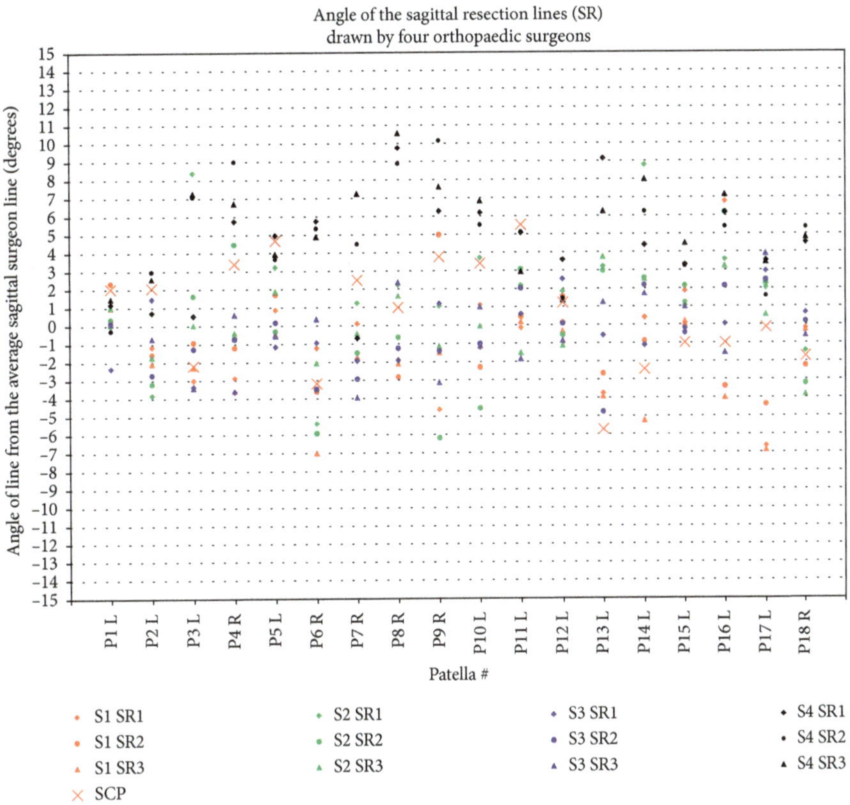

(a)

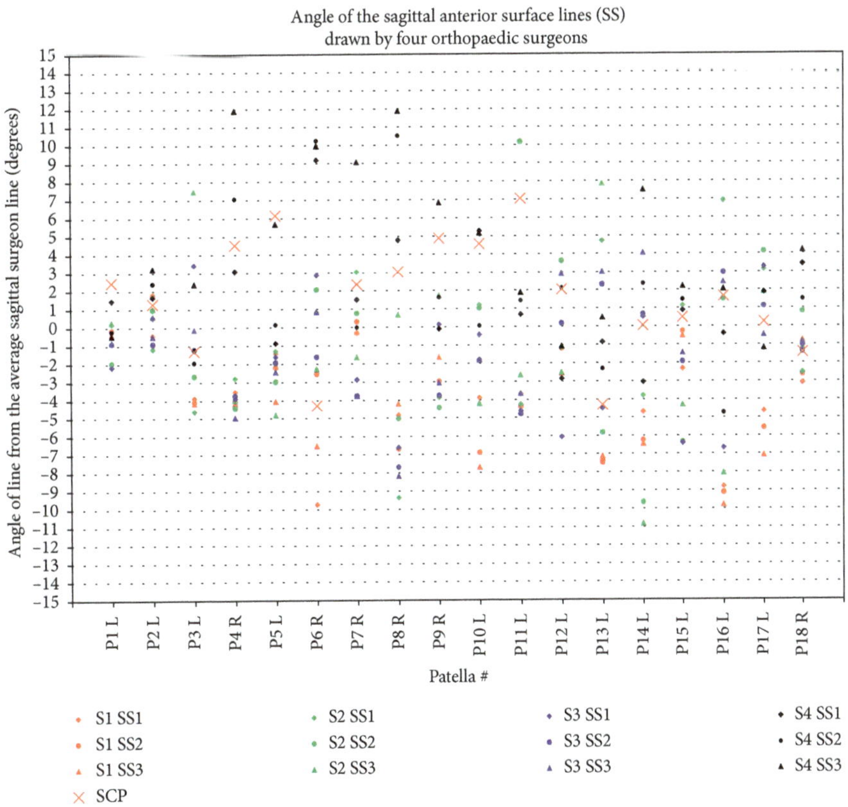

(b)

FIGURE 6: Angle of lines drawn on sagittal projections: (a) sagittal resection (SR) lines and (b) sagittal anterior surface (SS) lines. Positive angles mean the patella is thicker superiorly. Results are centred around zero, i.e., the average, for each patella. SCP = symmetric contact points, i.e., the calculated plane based on the final selected configuration (Section 3.2), which is symmetric about the patellar centre. S1–S4 = surgeons 1–4.

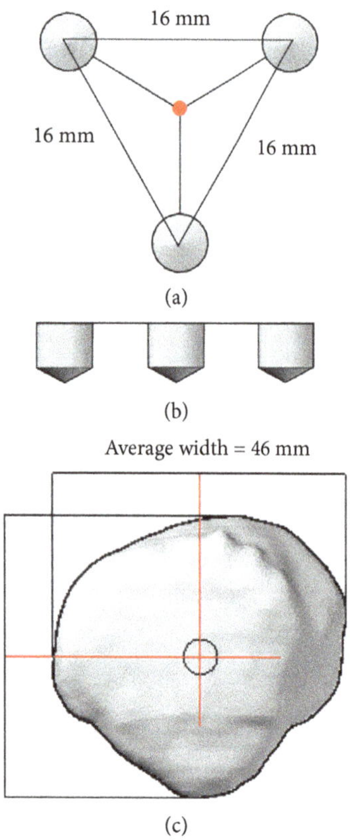

Figure 7: Final peg configuration proposed to align the resection device to the desired resection plane: (a) shows the top view of peg configuration, which is aligned (red dot) to the centre of the patella; (b) shows the side view of the equal-length pegs; and (c) shows the patella view, whereby the patellar centre is defined midway between the mediolateral and superoinferior extents. Patellar width averages 46 mm [16].

surgeons saw the same projections, so the variability within and between surgeons is unrelated to the projection. Another potential limitation is that the patellae investigated were largely healthy patellae, but since arthritis affects the posterior surface rather than the anterior surface, this should not represent a problem: a benefit of the device is that the peg configuration should produce the desired resection on a diseased patella equally well to a healthy one. The shape of the articulating patellar surface and whether it is a male or female patella, which have previously been shown to affect resection accuracy clinically [8], should likewise have no impact on the result since the device references solely off of the anterior surface.

5. Conclusions

Based on a rigorous and extensive shape analysis of the patella, a peg configuration was derived to achieve the surgeons' desired resection plane relative to specific contact points on the anterior surface. By analysing numerous alternative configurations on the 18 patellae, the best peg configuration consisted of a 16 mm equilateral triangle centred on the patella, which has been adopted into and successfully tested on two prototype surgical devices [17, 18]. The resulting configuration could also benefit custom rapid-prototyped surgical guides (which have often been used for the femur and tibia), by working from similar contact points. Furthermore, existing surgical devices could be modified to incorporate the proposed contact configuration.

Improving patellar symmetry will help reduce the number of patients who experience postoperative pain and reduced function. Using the proposed peg configuration in a patellar resection device should improve accuracy while remaining simple, fast, and noninvasive to use.

Conflicts of Interest

The authors declare that there are no conflicts of interest regarding the publication of this paper.

Acknowledgments

We would like to thank Dr. Tak Fung for his statistical advice. We gratefully acknowledge funding support from the Natural Sciences and Engineering Research Council of Canada (NSERC) and Alberta Innovates – Technology Futures (AITF).

References

[1] A. Baldini, J. A. Anderson, P. Zampetti, H. Pavlov, and T. P. Sculco, "A new patellofemoral scoring system for total knee arthroplasty," *Clinical Orthopaedics and Related Research*, vol. 452, pp. 150–154, 2006.

[2] A. Baldini, J. A. Anderson, P. Cerulli-Mariani, J. Kalyvas, H. Pavlov, and T. P. Sculco, "Patellofemoral evaluation after total knee arthroplasty: validation of a new weight-bearing axial radiographic view," *Journal of Bone and Joint Surgery*, vol. 89, no. 8, pp. 1810–1817, 2007.

[3] M. W. Pagnano and R. T. Trousdale, "Asymmetric patella resurfacing in total knee arthroplasty," *American Journal of Knee Surgery*, vol. 13, pp. 228–233, 2000.

[4] C. Anglin, J. M. Brimacombe, A. J. Hodgson et al., "Determinants of patellar tracking in total knee arthroplasty," *Clinical Biomechanics*, vol. 23, no. 7, pp. 900–910, 2008.

[5] B. C. Bengs and R. D. Scott, "The effect of patellar thickness on intraoperative knee flexion and patellar tracking in total knee arthroplasty," *Journal of Arthroplasty*, vol. 21, no. 5, pp. 650–655, 2006.

[6] W. Mihalko, Z. Fishkin, and K. Krackow, "Patellofemoral overstuff and its relationship to flexion after total knee arthroplasty," *Clinical Orthopaedics and Related Research*, vol. 449, pp. 283–287, 2006.

[7] Y. S. Youm, W. S. Cho, J. H. Woo, and B. K. Kim, "The effect of patellar thickness changes on patellar tilt in total knee arthroplasty," *Knee Surgery, Sports Traumatology, Arthroscopy*, vol. 18, no. 7, pp. 923–927, 2010.

[8] C. Anglin, C. Fu, A. J. Hodgson, N. Helmy, N. V. Greidanus, and B. A. Masri, "Finding and defining the ideal patellar resection plane in total knee arthroplasty," *Journal of Biomechanics*, vol. 42, no. 14, pp. 2307–2312, 2009.

[9] R. Gunaratne, D. N. Pratt, J. Banda, D. P. Fick, R. J. K. Khan, and B. W. Robertson, "Patient dissatisfaction following total knee arthroplasty: a systematic review of the literature," *Journal of Arthroplasty*, vol. 32, no. 12, pp. 3854–3860, 2017.

[10] D. Shervin, K. Pratt, T. Healey et al., "Anterior knee pain following primary total knee arthroplasty," *World Journal of Orthopedics*, vol. 6, no. 10, pp. 795–803, 2015.

[11] C. Fu, J. Wai, E. Lee et al., "Computer-assisted patellar resection for total knee arthroplasty," *Computer Aided Surgery*, vol. 17, no. 1, pp. 21–28, 2012.

[12] C. Fu, J. Wai, E. Lee et al., "Computer-assisted patellar resection system: development and insights," *Journal of Orthopaedic Research*, vol. 30, no. 4, pp. 535–540, 2012.

[13] M. Ledger, D. Shakespeare, and M. Scaddan, "Accuracy of patellar resection in total knee replacement: a study using the medial pivot knee," *The Knee*, vol. 12, no. 1, pp. 13–19, 2005.

[14] L. M. Smoger, K. B. Shelburne, A. J. Cyr, P. J. Rullkoetter, and P. J. Laz, "Statistical shape modeling predicts patellar bone geometry to enable stereo-radiographic kinematic tracking," *Journal of Biomechanics*, vol. 58, pp. 187–194, 2017.

[15] C. Anglin, J. M. Brimacombe, D. R. Wilson et al., "Biomechanical consequences of patellar component medialization in total knee arthroplasty," *Journal of Arthroplasty*, vol. 25, no. 5, pp. 793–802, 2009.

[16] J. L. Baldwin and C. K. House, "Anatomic dimensions of the patella measured during total knee arthroplasty," *Journal of Arthroplasty*, vol. 20, no. 2, pp. 250–257, 2005.

[17] E. L. Rex, E. M. Illical, C. Gaudelli et al., "Device for verifying the patellar cut during knee replacement surgery," *Journal of Medical Devices*, vol. 10, no. 2, article 024502, 2016.

[18] E. L. Rex, C. Gaudelli, E. M. Illical et al., "Guiding device for the patellar cut in total knee arthroplasty: design and validation," *Bioengineering*, vol. 5, no. 2, p. E38, 2018.

16

Airflow and Particle Deposition in Acinar Models with Interalveolar Septal Walls and Different Alveolar Numbers

Jinxiang Xi,[1,2,3] Mohamed Talaat,[2] Hesham Tanbour,[3] and Khaled Talaat[4]

[1]*Department of Biomedical Engineering, California Baptist University, Riverside, CA 92504, USA*
[2]*Department of Aerospace, Industrial, and Mechanical Engineering, California Baptist University, Riverside, CA 92504, USA*
[3]*School of Engineering and Technology, Central Michigan University, Mt. Pleasant, MI 48858, USA*
[4]*Department of Nuclear Engineering, The University of New Mexico, Albuquerque, NM 87131, USA*

Correspondence should be addressed to Jinxiang Xi; xi1j@cmich.edu

Academic Editor: Yu Feng

Unique features exist in acinar units such as multiple alveoli, interalveolar septal walls, and pores of Kohn. However, the effects of such features on airflow and particle deposition remain not well quantified due to their structural complexity. This study aims to numerically investigate particle dynamics in acinar models with interalveolar septal walls and pores of Kohn. A simplified 4-alveoli model with well-defined geometries and a physiologically realistic 45-alveoli model was developed. A well-validated Lagrangian tracking model was used to simulate particle trajectories in the acinar models with rhythmically expanding and contracting wall motions. Both spatial and temporal dosimetries in the acinar models were analyzed. Results show that collateral ventilation exists among alveoli due to pressure imbalance. The size of interalveolar septal aperture significantly alters the spatial deposition pattern, while it has an insignificant effect on the total deposition rate. Surprisingly, the deposition rate in the 45-alveoli model is lower than that in the 4-alveoli model, indicating a stronger particle dispersion in more complex models. The gravity orientation angle has a decreasing effect on acinar deposition rates with an increasing number of alveoli retained in the model; such an effect is nearly negligible in the 45-alveoli model. Breath-holding increased particle deposition in the acinar region, which was most significant in the alveoli proximal to the duct. Increasing inhalation depth only slightly increases the fraction of deposited particles over particles entering the alveolar model but has a large influence on dispensing particles to the peripheral alveoli. Results of this study indicate that an empirical correlation for acinar deposition can be developed based on alveolar models with reduced complexity; however, what level of geometry complexity would be sufficient is yet to be determined.

1. Introduction

Alveolar sacs are the ends of the respiratory tree and the smallest respiration units. The alveoli have a complex framework and are supported by interalveolar septa [1]. Pores of Kohn are apertures in the alveolar septum, which are circular or oval in shape and allow communications among adjacent alveoli [2]. This collateral ventilation helps equalize pressures across alveoli and plays an important role in preventing lung collapse (i.e., atelectasis) [3] and promoting alveolar recruitment [4]. In the case of emphysema, the number and size of pores increase in early stages [5–7]. In later stages, destruction of septa and even framework breakdown have been observed, leading to a decrease of the elasticity and an increase of the sac airspace [8]. Aging can also cause the breakdown of interalveolar septa and increase the number and sizes of alveolar pores of Kohn, thereby decreasing the collateral flow resistance among alveoli [9, 10]. Understanding the effects of alveolar septum and pore on alveolar airflow and particle dynamics is fundamental for understanding the pathology of pulmonary diseases, improving ventilator management, and devising more effective treatment strategies.

Due to the extremely large number of alveoli in the lungs (~480 million), it is impossible to construct a complete model of the acinar airspace. Even to construct a single acinar unit will be a formidable task, considering that each acinar unit consists of more than 10,000 alveoli [1]. As a result, alveolar models with varying realism and complexities had been proposed to understand particle dynamics inside the alveoli. These included alveolus models comprising a single hemisphere attached to a duct (i.e., respiratory bronchiole) [11–16], an alveolar duct with multiple alveoli (i.e., alveolated duct) [17, 18], and space-filling-based models with honeycomb or polygonal structures (i.e., terminal alveolar sacs) [19–23]. Based on *in vivo* microscopy observations, Kitaoka et al. [19] proposed that the alveolar mouth closes at minimum volume and gradually opens during inhalation. Based on a series of acinar models, Khajeh-Hosseini-Dalasm and Longest [22] investigated the effect of geometry complexity on acinar deposition. It was observed that when the acinar models consisted of more than three alveolar duct generations, the total acinar deposition rates were similar among models and were not affected by gravity orientation either. However, the effects of pores of Kohn on acinar deposition remain unclear and underinvestigated. More recently, Hofemeier et al. [23] proposed an algorithm to construct generic heterogeneous acinar models ranging from 372 to 2361 alveoli. They observed that variance in acinar heterogeneity played a minor role in determining alveolar deposition while the deposition rate increased for deeper inhalations.

General findings from previous studies are summarized as follows: classical alveolar deposition correlations are less accurate and need to be improved; wall motion is essential in determining the alveolar flows and aerosol dynamics; geometry complexity and realism strongly affect the predictions of alveolar deposition. Classical alveolar deposition correlations were typically based on particle sedimentation in steady tubular flows [24, 25] and used analytical approximations of particle deposition mechanisms [26]. Such corrections neglected a number of factors that are crucial in accurately determining particle transport and deposition, such as tidal breathing, wall motion, and geometry details. Kojic and Tsuda [27] showed that using steady-flow solutions to approximate oscillatory flows underestimated local particle deposition densities, and this error increased quickly for increasing oscillation frequencies. When the oscillation period became equivalent to the characteristic time for gravitational sedimentation, particle deposition would no longer be approximated by the classical solution based on steady flows as proposed in [28]. Alveolar wall motion has been demonstrated to be essential to match single-path-transport model predictions with *in vivo* alveolar deposition data [29, 30]. It is noted that wall kinematics, such as symmetric vs. anisotropic oscillations, played a negligible role in the alveolar deposition [15, 31]. On the contrary, the type and complexity of the acinar airway yield unique features of airflow and particle transport patterns. Kumar et al. [20] simulated airflow in acinar models with honeycomb structures and reported recirculation inside the alveoli induced by oscillatory wall motions. Talaat and Xi [15] numerically investigated aerosol deposition in a single terminal alveolus with rhythmical oscillations and found significantly different particle dynamics in comparison to that in alveolated ducts or respiratory bronchioles [12, 21, 32, 33]. Particles move back and forth driven by the oscillating walls of the terminal alveolus and form multifolding trajectories [15]; by contrast, particles in an alveolated duct or respiratory bronchiole geometry remain suspended in the alveolus for several breathing cycles, rotating clockwise during exhalation and counterclockwise during inhalation [12, 21].

Several controversial observations have been reported regarding the influences from gravity orientation angle, airway realism, and breathing depth. Haber et al. [12] and Sznitman et al. [21] reported that particle deposition efficiencies are strongly related to the gravity orientation in both alveolated ducts and space-filling geometries. By contrast, Khajeh-Hosseini-Dalasm and Longest [22] suggested that total acinar deposition rates were insensitive to the gravity orientation when the geometry had more than three alveolar duct generations. While some studies [18, 34, 35] found that the geometry complexity significantly affected acinar aerosol deposition, Hofemeier et al. [23] recently observed that heterogeneity in acinar geometry had little effect on alveolar deposition. Similarly, while Hofemeier et al. [23] reported that the acinar deposition rate increased for deeper inhalations, Talaat and Xi [15] reported that the deposition was relatively insensitive to the breathing depth in single terminal alveolar models.

Unique features exist in the acinar airspace. Structurally, there are septal walls and pores of Kohn between neighboring alveoli, which are expected to strongly affect airflow and particle deposition, but whose influences have rarely been studied. Dynamically, the contribution of particle-wall interception to deposition can be important due to the geometrical complexity. But its importance relative to other deposition mechanisms, such as oscillatory convection, gravitational sedimentation, and particle dispersion, is unclear. Moreover, interalveolar septa and pores of Kohn can change in shape and size due to diseases or aging. As a result of these complexities, most space-filling-based honeycomb or polyhedral models to date have neglected the interalveolar septa and pores.

The objective of this study is to investigate the effects of acinar geometrical details, such as the interalveolar septum and pore size, on acinus airflow and particle dynamics in both a simplified 4-alveoli model and a 45-alveoli model. There are four specific aims in this study: (1) to develop acinar models with different number of alveoli and with septa and pores of different sizes, (2) to characterize airflows and particle motions in these acinar geometries, (3) to quantify the effect of pore size on surface doses both temporally and spatially, and (4) to evaluate the acinar deposition in simplified and complex models under the influence of the following factors: particle size, geometrical complexity, gravity orientation angle, and inhalation depth.

2. Methods

2.1. Study Design. Pulmonary alveoli are composed of a myriad of alveolar sacs arranged in a fractal manner. To study particle deposition in these regions, we started with a simplified four-sac alveolar model, with septal walls and pores of Kohn between adjacent sacs. Due to its well-defined shape and dimension, this model allowed controlled testing of influential parameters such as particle diameter, breathing conditions, and pore sizes. As a result, a comprehensive understanding of deposition mechanisms in alveolar sacs can be obtained. By doing so, airflow and particle dynamics in a control case with rhythmic wall motions were examined; the resulting particle deposition was characterized both temporally (dynamic deposition growth) and spatially (in each alveolus). Six controlled tests were then conducted. To study the effects of rhythmic wall motions, Test 1 compared particle deposition in alveolar models with dynamic and rigid walls. For the model with rigid walls, the flow was stagnant (i.e., zero velocity), and particles moved due to gravity. Test 2 investigated the effect of particle size on alveolar deposition, which was 0.5 μm, 1 μm, 2 μm, and 3 μm. Test 3 investigated alveolar deposition under four gravity orientation angles (0°, 45°, 90°, and 135° from the gravity). The effects of breath-holding and respiration depth were studied in Tests 4 and 5, respectively. The breath holding durations included 0.5, 1, 2, 3, and 4 seconds following the inhalation. The respiration depths included 1, 2, 3, and 4 times of the standard tidal volumes, which is 23.3% of the functional residual capacity (FRC). The last test (Test 6) studied the effects of pore sizes on particle deposition that included three different pore sizes and one model without the septal wall.

A more complex and physiologically realistic model was then developed that contained 45 alveolar sacs. Airflow and particle deposition in this model were compared to the simplified 4-alveoli (or 4-sac) model to determine the feasibility of using simplified alveolar models for inhalation dosimetry predictions and to evaluate the impact from gravitational orientation and inhalation depth. Model development, fluid-particle tracking algorithms, and numerical methods are detailed below.

2.2. Computational Acinar Model and Kinematics. To model an acinar cluster, four alveoli were retained in the simplified model. Individual alveolus was approximated using a 0.3 mm diameter sphere. The alveolar cluster was joined to a duct with a length of 0.2 mm and a diameter of 0.1 mm. The left panel of Figure 1(a) shows the air-filled geometry of the above duct-alveoli model, while the middle panel of Figure 1(a) shows the cut-open view of the hollow duct-alveoli model. There were openings (i.e., pores of Kohn) connecting any two neighboring alveoli, and thus totally five pores existed in this model geometry (Figure 1(a), middle panel). To facilitate later reference to the four alveolar sacs, the upper alveolus was termed as Sac 1, the lower alveolus as Sac 4, and the left and right alveoli as Sac 2 and Sac 3, respectively (Figure 1(a), left panel).

The wall kinematics of the acinar model followed the anisotropic motion of the chest [36, 37], which reported a smaller expansion in the arm-arm (z) direction than the head-foot (x) and back-front (y) directions (i.e., $z:y:x = 0.375:1:1$). Under normal breathing conditions, the volume expansion was assumed to be the standard tidal volume, that is, $V_T/\text{FRC} = 0.233$, where V_T represents the standard tidal volume [38]. A user-defined function (UDF) was written that specified the oscillatory wall motions (right panel of Figure 1(a)). More details of the UDF can be found in Talaat and Xi [15].

To investigate the effects of pore sizes on alveolar deposition, three geometrically similar models with different pore sizes were developed, that is, 40 μm, 100 μm, and 160 μm, as shown in Figures 1(a), 1(b), and 1(c), respectively. For comparison purposes, an alveolar model with no septal wall was also developed (Figure 1(d)).

To investigate the model complexity effects, the second model consisted of 45 alveoli. Similar to the 4-alveoli model, septal walls existed between contiguous alveoli and pores of Kohn existed in the septal walls (Figure 1(e)). Even though spheres were initially used to approximate the alveoli, they naturally evolved into polygonal shapes when multiple spheres intersected each other (Figure 1(e)). The airway volume is 4.32×10^{-11} m^3 for the 4-alveoli model and 4.20×10^{-10} m^3 for the 45-alveoli model in comparison to 1.48×10^{-11} m^3 for the single alveolus model in Talaat and Xi [15].

2.3. Airflow and Particle Transport Models. Particles ranging from 0.5 to 3 μm in diameter were investigated because smaller submicron particles deposited in the pulmonary region with different mechanisms (i.e., diffusion), and larger micrometer particles were captured by the upper respiratory tract and could not reach the pulmonary acinus. For each numerical test, multiple (3–24) respiration cycles were modeled, with the first cycle to create the unsteady flow field. An amount of 10,000 particles was inhaled at 0.20 s of the second cycle to simulate the inhalation of a bolus of pharmaceutical particles and was tracked until all particles deposited or exited the geometry with the expiratory flow.

The airflow was isothermal and incompressible in this study. The flow regime is laminar because the Reynolds number is much smaller than one even during peak inhalations [39]. Therefore, the laminar flow model was used to solve the airflow field. A well-tested direct Lagrangian algorithm was used to track particle motions [40, 41]. This algorithm, enhanced by the near-wall treatment algorithm [42], has been shown in our previous studies to agree with *in vitro* deposition results in human upper airways for both nanoparticles [43] and micrometer particles [44, 45].

2.4. Numerical Methods. ANSYS Fluent (Canonsburg, PA) with dynamic mesh and discrete phase models was used to simulate the transient airflow and particle deposition. User-defined Fortran and C modules were used to specify alveolar wall kinematics and calculate temporal and spatial surface deposition rates [46, 47]. ANSYS ICEM CFD (Ansys, Inc) was utilized for computational mesh generation. One-way coupling from the airflow to particles was assumed. A grid sensitivity analysis was conducted by testing a range of mesh densities, and grid independent result was considered to be achieved when the difference in total particle deposition was less than 1%. The final mesh was chosen to be 1.2 million cells for the 4-alveoli model and 6.0 million for the 45-alveoli model.

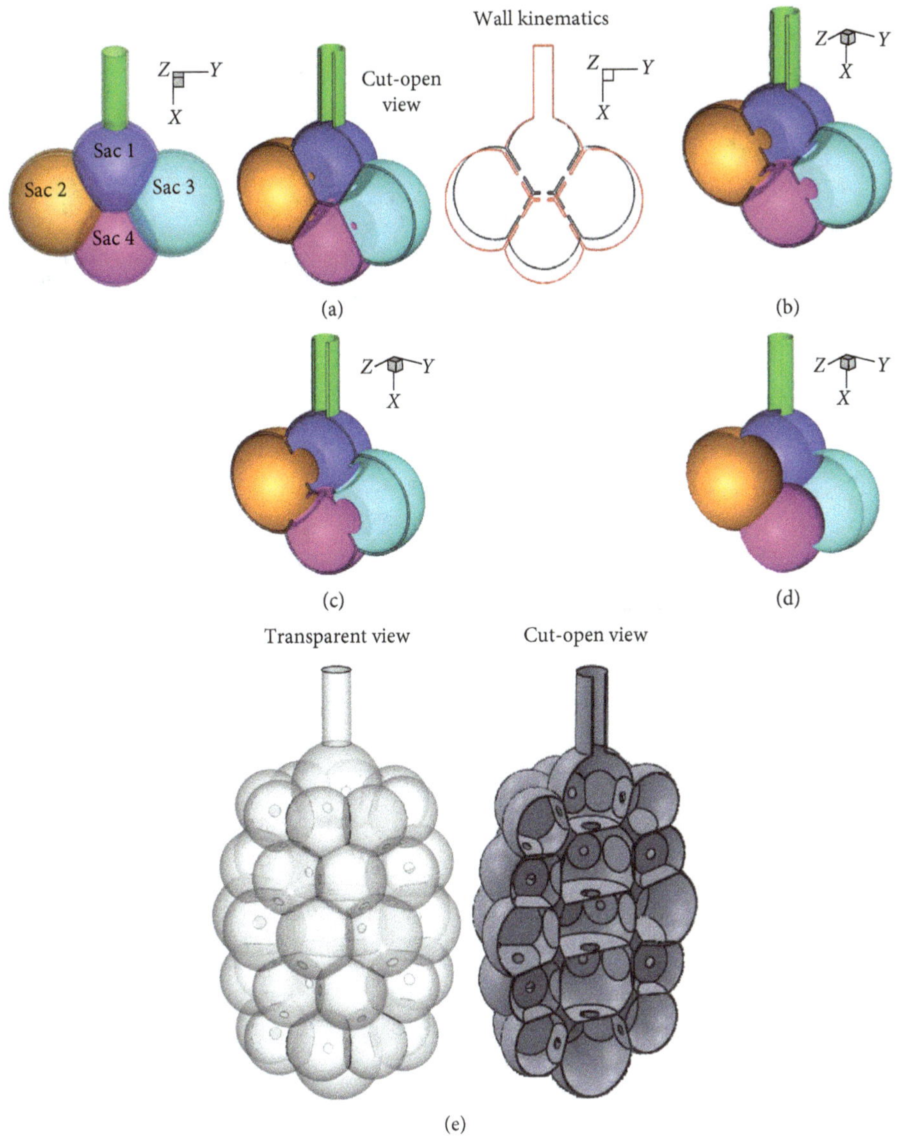

FIGURE 1: Simplified and complex multiple-alveoli models with septa and pores: (a) simplified 4-alveoli (or 4-sac) model with pores (septal apertures) of 40 μm in diameter; (b) 4-alveoli model with 100 μm pores; (c) 4-alveoli model with 160 μm pores; (d) 4-alveoli model with no spectrum; and (e) 45-alveoli model with 40 μm pores.

3. Results

3.1. Airflow Field and Particle Motion. Instantaneous airflow fields at the middle of inhalation and exhalation cycles are shown in Figure 2 in the 4-alveoli model with different pore sizes. Under the normal tidal breathing condition (23.3% FRC, or 1 V_T), the peak velocity at the inlet is around 1 mm/s. In the model with a pore size of 40 μm ((A) in Figure 2), Venturi effect at the pores was observed, which increased the penetration depth of particles into the peripheral alveoli. This effect, however, was absent when the pore sizes were large ((B) in Figure 2) or when the septal wall was missing ((C) in Figure 2). Overall symmetric flow patterns were noted for all of the three models considered (at their peak inhalation and exhalation speeds herein). Interesting discrepancies were also discerned among the three models. In (A) in Figure 2, streamlines flowed from the lower alveolus to the two lateral alveoli during inhalation and reversed their directions during exhalation. By contrast, streamlines in (B) in Figure 2 flowed from the two lateral alveoli to the lower one during inhalation and vice versa. These collateral ventilations were presumably associated with the pressure imbalance between neighboring alveoli.

Particle dynamics in the oscillating alveoli are visualized in Figures 3 and 4. Figure 3 shows the snapshots of particle positions during the first cycle at ten different instants. Particles were inhaled into the geometry approximately at the beginning of inhalation. Depending on local velocities, the swarm of particles exhibits a parabolic pattern in the alveolar duct ($T = 0.25$ s) and a spherical shape after entering the top alveolus (i.e., Sac 1, $T = 0.50$ s). Around the middle of inhalation cycle ($T = 0.75$ and 1.0 s), particles start entering

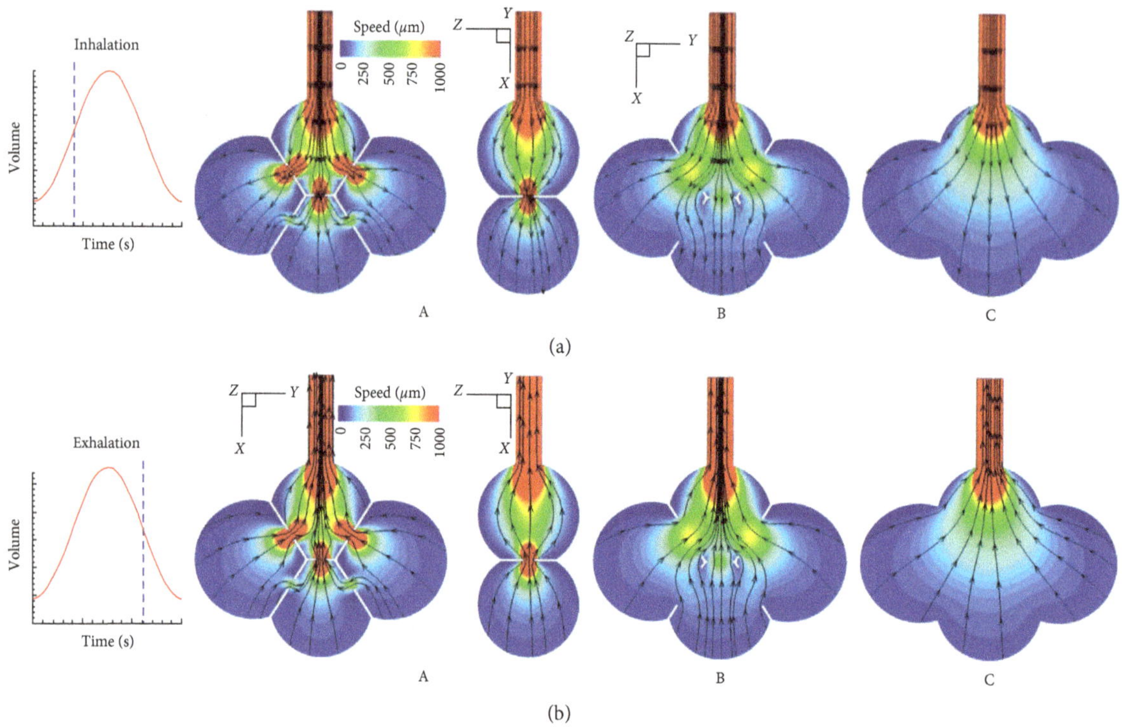

FIGURE 2: Airflow contour and stream traces in the 4-alveoli models with a pore size of (A) 40 μm, (B) 160 μm, and (C) no septal wall during (a) inhalation and (b) exhalation.

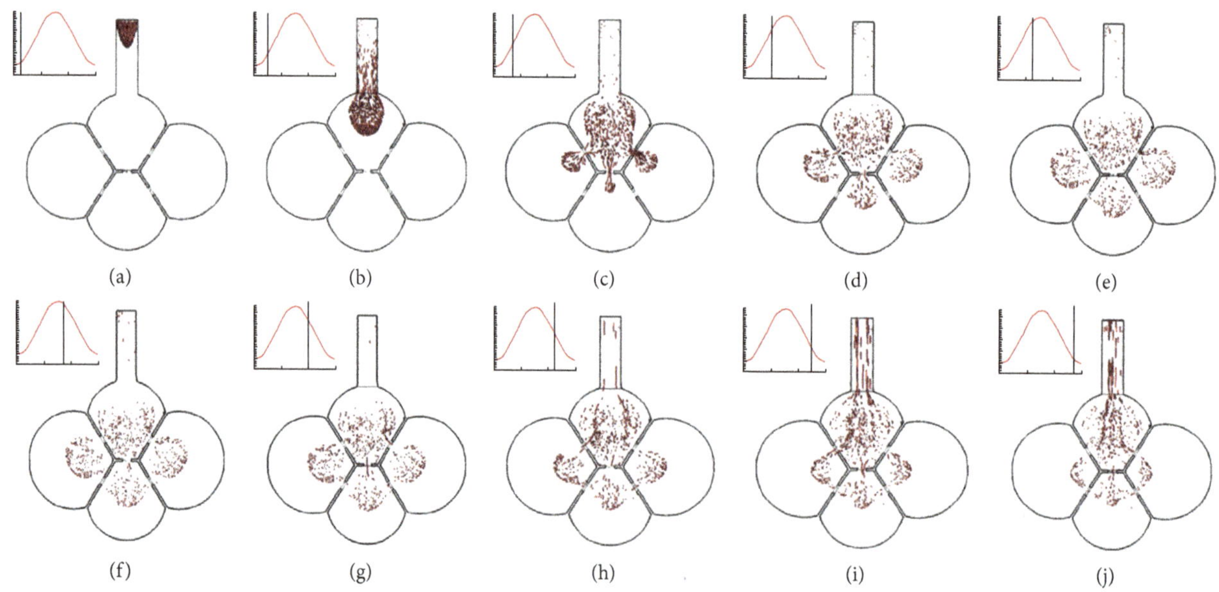

FIGURE 3: Instantaneous snapshots of particle positions in the 4-alveoli model during the first cycle after particle inhalation. Particles were 1 μm in diameter. Due to particle dispersion, some particles exited the geometry with expiratory airflow. (a) $T = 0.25$ s. (b) $T = 0.50$ s. (c) $T = 0.75$ s. (d) $T = 1.00$ s. (e) $T = 1.25$ s. (f) $T = 1.75$ s. (g) $T = 2.00$ s. (h) $T = 2.25$ s. (i) $T = 2.50$ s. (j) $T = 2.75$ s.

the adjacent alveoli through the pores of Kohn and deposit on the septal walls in Sac 1. In contrast, the particle fronts in the three adjacent alveoli only reached half of the airspace at the end of the first inhalation cycle ($T = 1.25$ s).

During expiration, particles gradually reversed their direction and started to move upward ($T = 1.75$–2.75 s). Particles that returned to Sac 1 from the three peripheral alveoli (Sacs 2–4) gained momentum due to the pores' accelerating effect. They mixed with the relativity slow-moving particles in the top alveoli, which enhanced particle dispersion. At the end of the first cycle, some particles were exhaled out of the alveoli, as displayed at $T = 2.50$–2.75 s. In

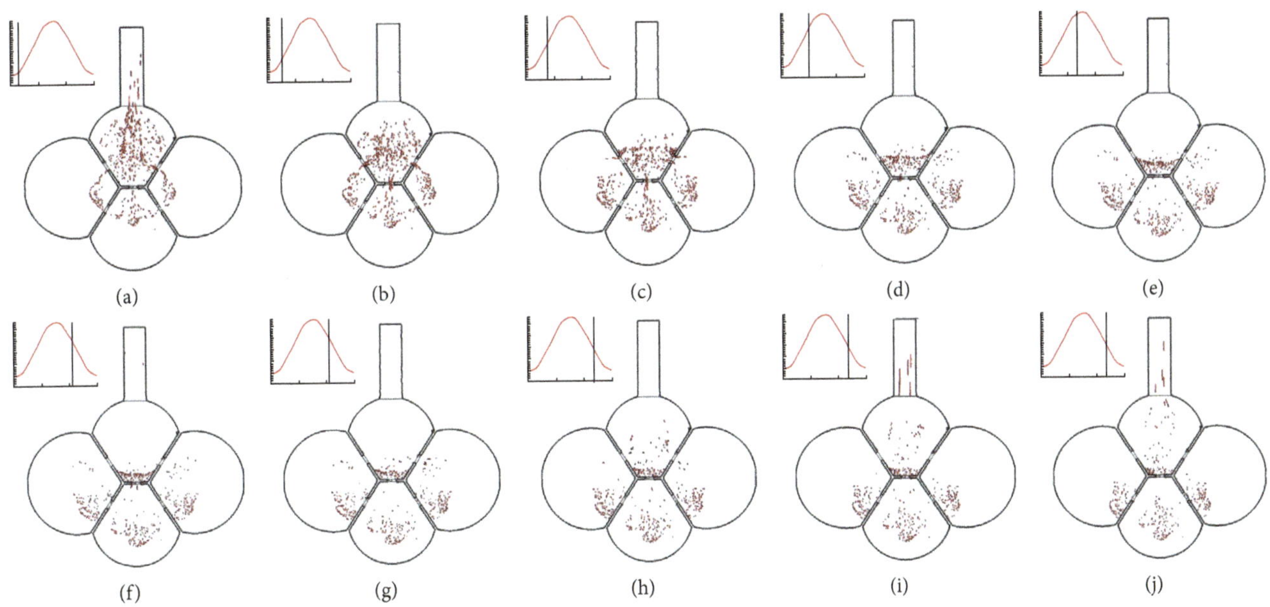

FIGURE 4: Instantaneous snapshots of particle positions in the 4-alveoli model during the second cycle after particle inhalation. All the remaining particles will deposit eventually due to particle interception or gravitational sedimentation. (a) $T = 3.25$ s. (b) $T = 3.50$ s. (c) $T = 3.75$ s. (d) $T = 4.00$ s. (e) $T = 4.25$ s. (f) $T = 4.75$ s. (g) $T = 5.00$ s. (h) $T = 5.25$ s. (i) $T = 5.50$ s. (j) $T = 5.75$ s.

principle, these particles would not be able to re-enter the geometry.

The instantaneous snapshots of particles during the second respiration cycle are shown in Figure 4 at ten instants. Similar as in the first cycle, the particle swarm oscillated with the expanding-contracting wall motion. However, a negligible number of particles was observed to leave the geometry. Instead, all these particles eventually deposited on the septal walls due to oscillatory convection and gravitational sedimentation. In the following sections, we would study the distribution of particles among individual alveoli as well as the influences of breathing conditions and pore size on these distributions.

Figure 5 shows the surface deposition pattern of 1 μm particles. Most particles deposited on the septal walls of the top and two lateral alveoli, indicating that gravitational sedimentation is still the predominant deposition mechanism. It was noted that the deposition pattern in Figure 5 displayed the positions of the particles when they deposited on the moving walls of the alveoli; therefore, particle deposition was not limited to one surface but was possible on any surface in a range spanned by the oscillating wall motion.

From Figure 5, a large portion of particles deposited at the end of exhalation (wall contraction) when alveoli have the smallest volume. This was because that, near the end of the exhalation, gravitational sedimentation overtook the upward-moving momentum from the contracting wall. As a result, particles moved downward and collided with the upward-moving wall. Moreover, at the start of the subsequent inhalation, particles still outran the expanding walls and deposited there. As a result, a seemingly suspending layer of deposited particles was observed in the top and lower alveoli (solid arrow in Figure 5(a)). Considering the two lateral alveoli, no particles were observed on the distal walls, indicating a limited contribution from convective deposition. By contrast, most particles deposited on the interalveolar septal walls that bordered with the lower alveolus (i.e., Sac 4), with even more concentrated deposition in the lower position of the septal walls (dashed ellipse in Figure 5(b)).

Three major differences were observed when comparing temporal deposition profiles between dynamic and static models (Figure 6). First, the cumulative deposition rates were different, with 100% in the static model versus 76.2% in the dynamic model for 1 μm particles. Second, the spatial distributions of particle deposition among alveoli were different. Heterogeneous doses were predicted in the dynamic model, with 26.8% in the upper alveolus (Sac 1), 14.4% in the lower alveolus (Sac 4), and about 12.3 ± 1% in each of the two lateral alveoli (Sac 2 and 3). In addition, there was also an appreciable fraction of particles that deposited in the alveolar duct (8.9%) and the canals of the pores (1.4%), as shown in Figure 6(a). By contrast, no particle deposited in the two lateral alveoli of the static model (Figure 6(b)). The third difference was the time required for deposition. It took about 24 seconds to start deposition in the static model, which was dictated by the alveolar size over particle settling velocity; while in the dynamic model, deposition started almost immediately after particles enter the alveoli. Furthermore, once the deposition started, it took only a short period time for the deposition to complete in the static model; in contrast, the deposition in the dynamic model took about three cycles to complete in the upper alveolus and more than four cycles in the lower and lateral alveoli. Further, a multistage staggering profile of the deposition fraction (DF) versus time was observed in each alveolus of the dynamic model. These three differences clearly demonstrated that

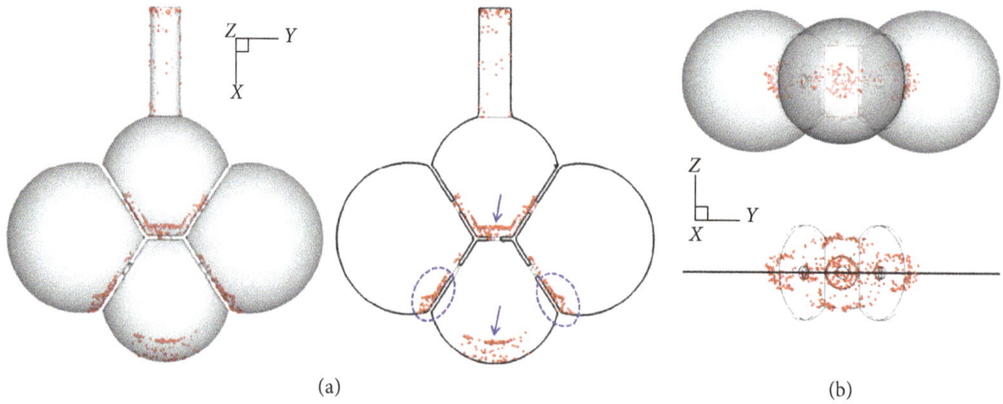

FIGURE 5: Surface deposition of 1 μm particles in the 4-alveoli model with a pore size of 40 μm: (a) side view and (b) bottom view.

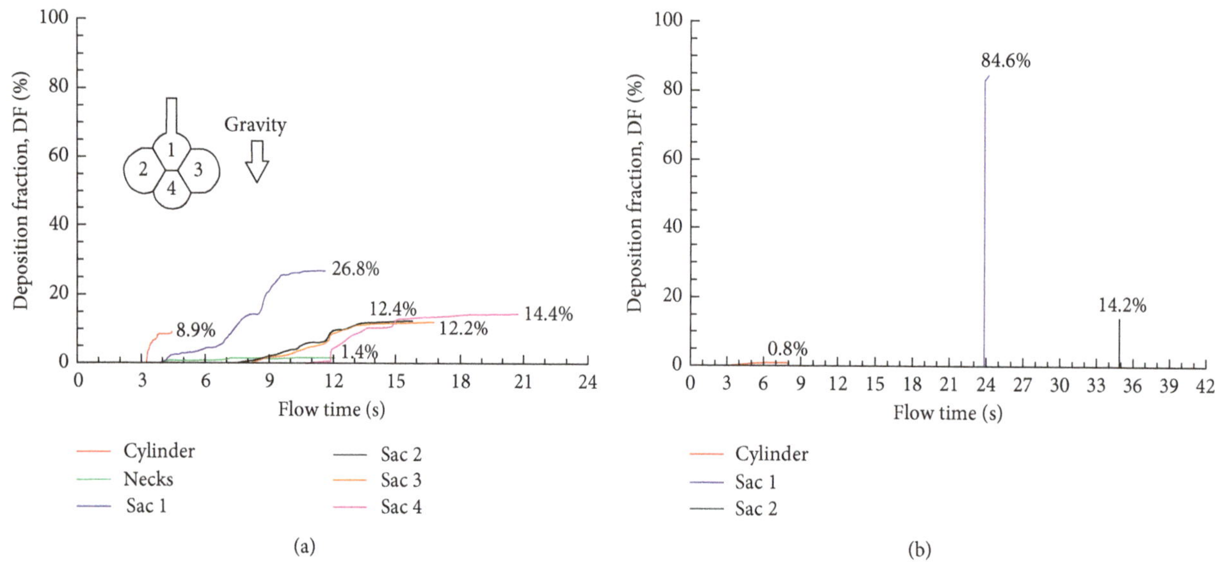

FIGURE 6: Cumulative deposition rate with time for 1 μm particles in the 4-alveoli model: (a) dynamic model with rhythmic wall motions and (b) static model. Each respiration cycle has a period of 3.0 s.

analyses that neglected the dynamic nature of the alveoli would miss the particle behaviors in the acinar region.

Deposition fractions for different sized particles are shown in Figure 7, which increases quickly with particle sizes. It is about 51.5% for 0.5 μm particles and 76.1% for 1 μm particles. For particles larger than 2 μm, nearly all deposits are in the alveolar airspace (i.e., 98.4% for 2 μm particles and 100% for 3 μm particles). It is noted that, in this study, a particle bolus was inhaled into the geometry at the start of the inhalation. Particle boluses inhaled at later instants will have lower deposition rates [16].

To gain a better understanding of particle behaviors at different sizes, temporal variation of DFs was quantified in each section of the alveolar sacs (i.e., alveolar duct, four alveoli, and interalveolar pores, Figures 7(b)–7(d)). Similar to the dynamic case in Figure 6, heterogeneous and continuous particle deposition was observed for all sized particles, with the upper alveolus receiving the highest deposition. The time required for deposition was very different between different particle sizes. It took around 45 s (i.e., 48–3 in Figure 7(b)) for 0.5 μm particles to complete the deposition, while it took 12 s (i.e., 15–3 in Figure 7(c)) for 1 μm particles, 4 s for 2 μm particles, and 2 s for 3 μm particles. These decreases are due to the increasing particle settling velocity, which is proportional to the square of the particle diameter (i.e., $v_{\text{settling}} = 18(\rho_p - \rho_f)gd_p^2/18\mu$). Furthermore, deposition of 0.5 μm particles started 21 s after administration in the two lateral alveoli and 33 s in the lower alveoli. By contrast, much shorter periods of time were needed for larger particles. For instance, it took 1 s for 2 μm particles to start deposition in the lateral alveoli and 1.5 s in the lower alveoli (Figure 7(d)).

The effects of gravity orientation angle on particle deposition in the alveolar sacs are shown in Figure 8. For the four cases considered (0°, 45°, 90°, and 135°), the total DF ranged from 60.3 to 77.2%. The highest subregional DF still occurred in Sac 1 (26.8–31.5%). However, the distribution of DFs in the other three alveoli changed significantly,

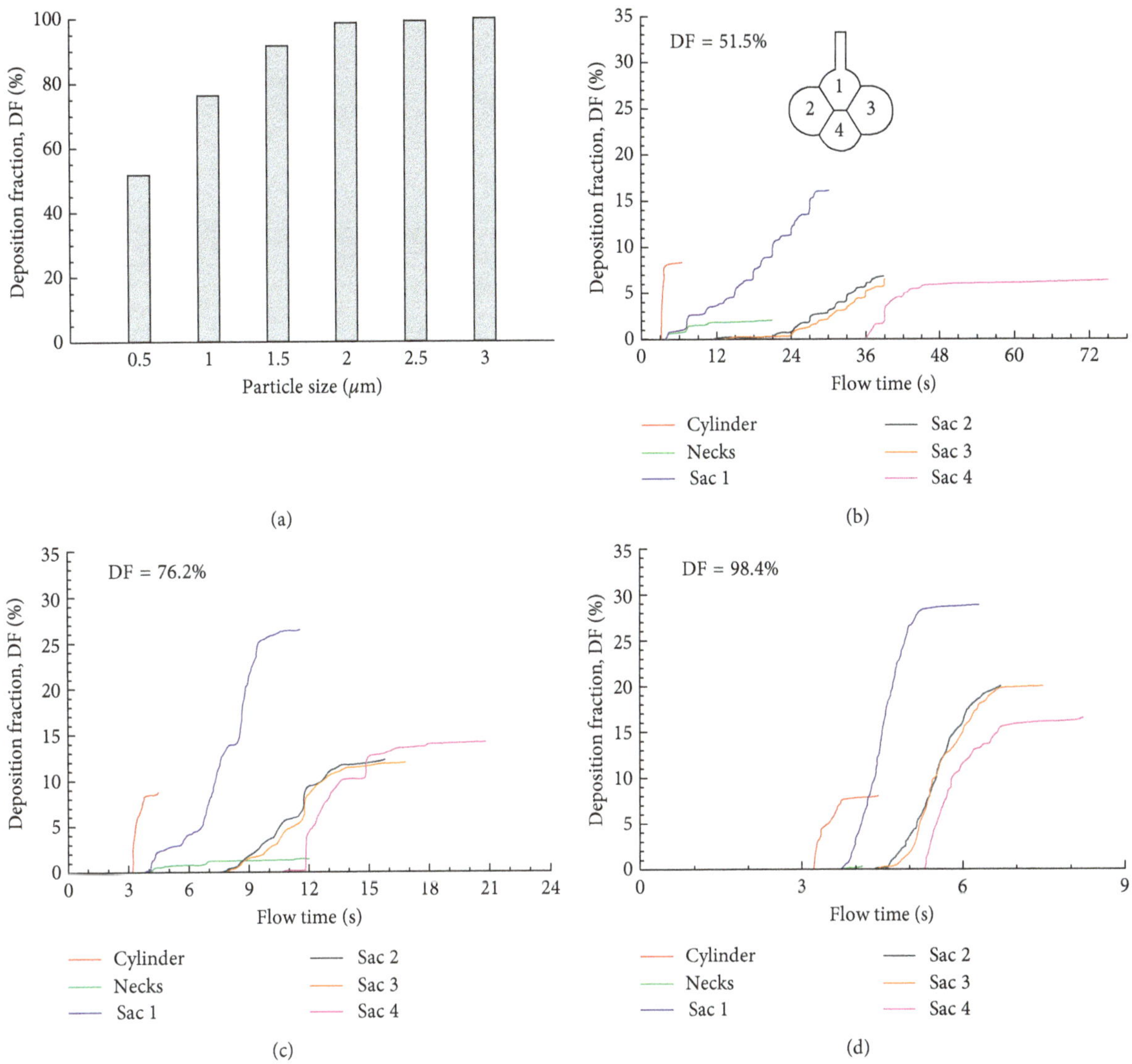

FIGURE 7: Comparison of particle deposition between different particle sizes in the 4-alveoli model: (a) total deposition fractions for 0.5–3 μm particles and the time resolution of sectional cumulative deposition for (b) 0.5 μm, (c) 1 μm, and (d) 2 μm. Deposition was quantified in each section of the model, that is, cylinder, the four alveoli (sacs), and the interalveolar pores (necks).

depending on the alveolar orientation relative to the gravity. For instance, the DF in Sac 2 changed from 12.4% at 0° to ~17% at 45° and 90° and 11.8% at 135°, while the DF in Sac 3 changed from 12.2% at 0° to 2~4% at 45°–135°. Considering that the major deposition mechanisms herein were gravitational sedimentation and oscillatory convection, the orientation-induced DF variation was closely associated to the projected area normal to the gravity, as well as the convective penetration depth of particles into the distal alveoli. Similarly, DF in the alveolar duct was the highest at 90° (16.3%, Figure 8(c)) due to its largest projected area normal to the gravity and was lowest at 0° (8.9%, Figure 8(a)).

Breath-holding after inhalation can significantly enhance alveolar deposition. Figure 9(a) shows the predicted DF of 1 μm particles with different periods of breath-holding. Four seconds or longer hold of breath allowed all inhaled particles to deposit. Figure 9(b) shows the temporal and spatial variation of DFs in the alveolar sacs for breath-holding of 4 seconds. Compared to the case without breath-holding (Figure 8(a)), subregional DF in Sac 1 increased significantly, for example, from 26.8% to 36.5%. Subregional DFs in the three peripheral alveoli also increased with the breathing-holding due to the prolonged action period of gravitational sedimentation relative to the particle dispersion from oscillatory convection.

The effects of inhalation depth on alveolar deposition and its distribution in the 4-alveoli model are shown in Figure 10. Increasing the inhalation depth had a negligible effect on total DF. However, it notably changed particle distributions among alveoli, with deeper inhalations being associated with less heterogeneous subregional depositions. For instance, at higher tidal volumes, the DF in Sac 1 decreased while DF in Sac 4 increased, reducing the difference between them (Figures 10(b) vs. 10(d)). A subtle variation in

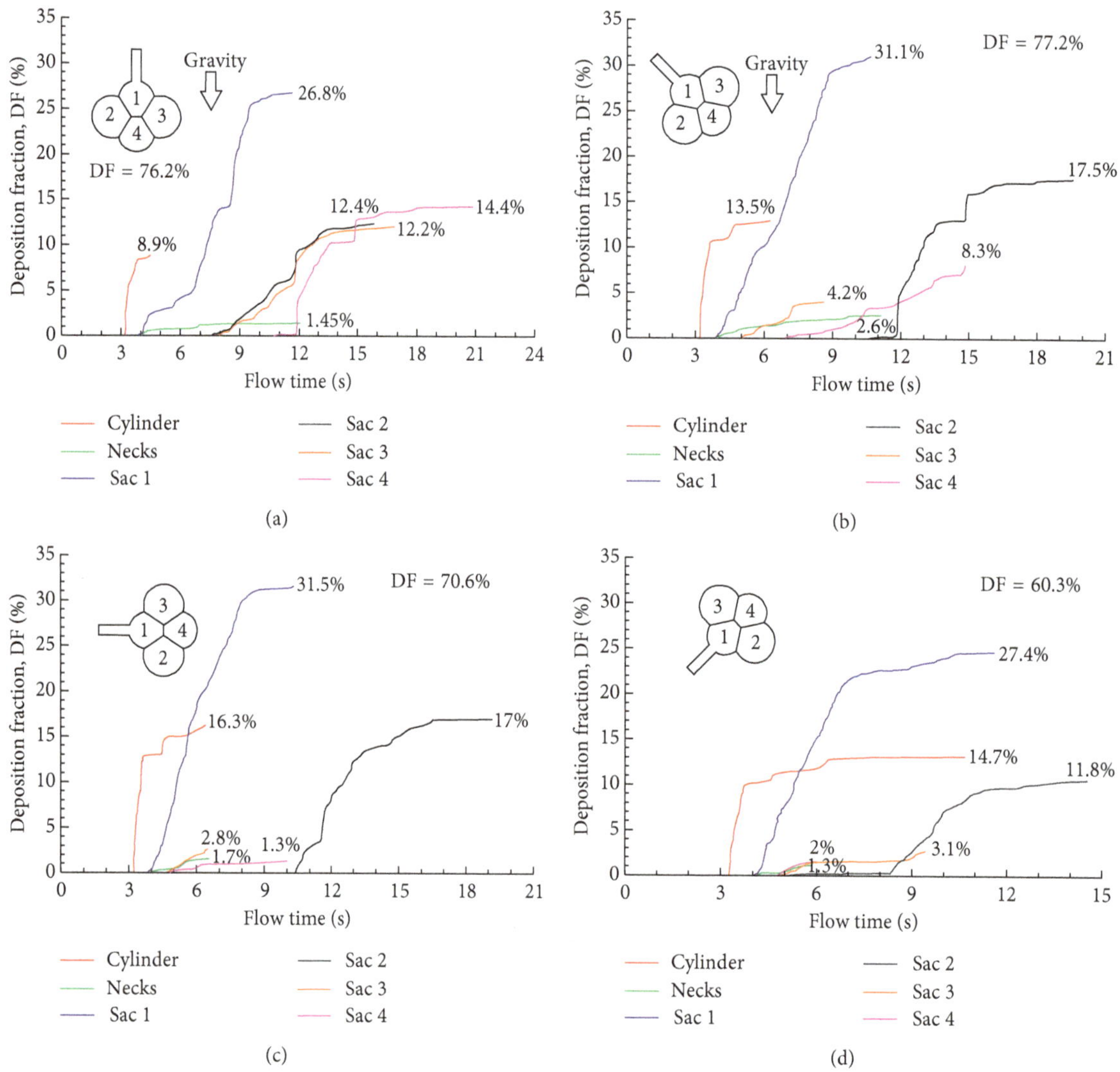

FIGURE 8: Comparison of the cumulative deposition with time for different gravity orientation angles in the 4-alveoli model: (a) 0°, (b) 45°, (c) 90°, and (d) 135° counterclockwise from the gravity.

DF was also noted in the alveolar duct, which increased slightly but persistently with increasing tidal volumes.

One significant issue of pulmonary pathology is the breakdown of interalveolar septal walls and associated collateral ventilation. To investigate their effects on acinar deposition, models with different sizes of septal apertures (pores) were simulated and compared in Figure 11(a). Surprisingly, total DF was found to be insensitive to the pore size. For all pore sizes considered (40 μm, 100 μm, 160 μm, and no septal wall), the total DF for 1 μm particles varied within a narrow range of 71.3~74.4% (Figures 11(b)–11(e)). Dramatic changes in the spatial deposition distribution with the pore size were predicted. Considering the cases of pore size from 40 μm (Figure 8(a)) to 160 μm (Figures 11(c) and 11(d)), the DF in Sac 4 constantly increased with pore size, while DFs in the two lateral alveoli (Sac 2 and 3) decreased with pore size. This trend was reasonable because a larger pore allowed more particles to enter the lower alveolus (due to gravity) but at the same time, decreased the Venturi effect of pore aperture and therefore reduced the particle penetration depth to the lateral alveoli. When there was no septal wall, the alveolar sacs behave more like a single alveolus, as in Talaat and Xi [15], where particle deposition concentrated at the bottom of the alveolus (Figures 11(b) and 11(e)). This phenomenon was even more pronounced for 3 μm particles, where the majority of particles deposited in the lower alveolus (Sac 4, Figures 11(b) and 11(f)).

Airflow and particle deposition in the 45-alveoli model was also simulated. Figure 12(a) shows the inspiratory airflow in the 45-alveoli model. Complex flow fields are observed as the inhaled airflow enters the alveolar sacs sequentially from top to bottom and from central to peripheral. Airflow is stronger in the axial direction (i.e., positive x-direction, Figure 12(a)) and decreases progressively in the lateral compartments. Surface deposition of inhaled particles is shown in Figure 12(b). Overall symmetric surface deposition patterns are observed. Due to the multiple alveolar generations retained in this model, particle deposition exhibits a cascading pattern for all particle

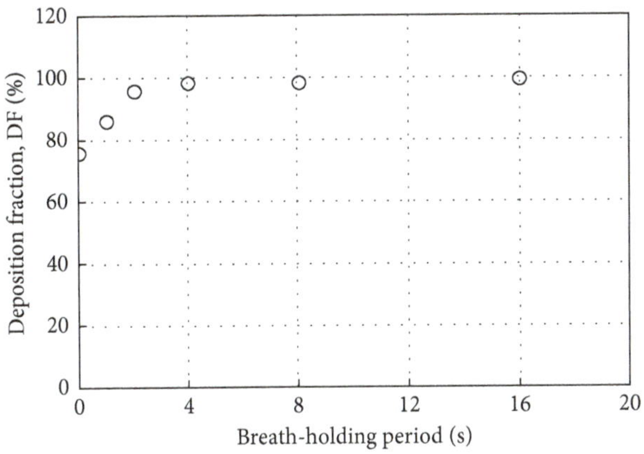

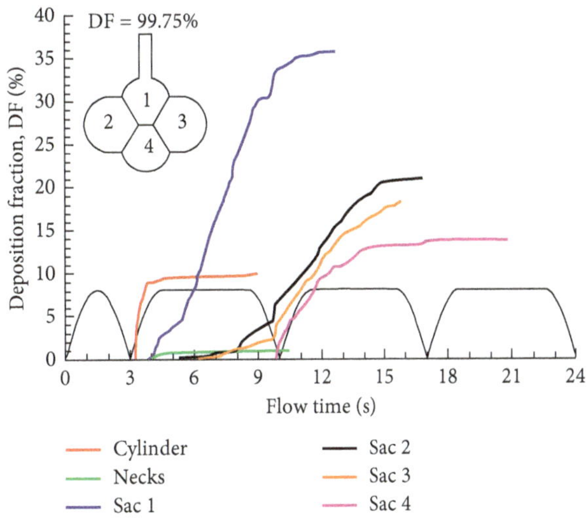

FIGURE 9: Comparison of total deposition rates for different breath-holding periods in the 4-alveoli model: (a) total deposition and (b) temporal deposition for 4 s breath-holding.

sizes (0.5–3.0 μm) considered (Figure 12(b)). However, deposition is more dispersed for small particles and more concentrated for large particles.

Figure 12(c) compares the deposition fractions as a function of particle size between the 45-alveoli and 4-alveoli models, as well as the single-alveolus model in [15]. Surprisingly, for 0.5–1.5 μm particles, lower deposition fractions were predicted in the complex 45-alveoli model than the highly simplified 4-alveoli and single-alveolus models, indicating a negative correlation of acinar deposition with the number of alveoli retained in the model. The cumulative deposition profiles with time were displayed in Figure 12(d) for 0.5–3.0 μm particles. The time required for complete deposition decreased dramatically from 0.5 μm to 3.0 μm (Figure 12(d)).

Effects of gravity orientation angle on particle deposition in the 45-alveoli model are shown in Figure 13. It appears that the total DF is not sensitive to the gravity orientation angle, as demonstrated by the close similarity of temporal deposition profiles among the four angles considered (i.e., 0°, 45°, 90°, and 135° counterclockwise from the gravity, Figure 13(e)). The spatial deposition distributions, however, exhibit high levels of heterogeneity. In contrast to the symmetric distribution in the 0° case (Figure 13(a)), particles in the other three models deposit preferentially in the central alveoli or alveoli along the gravity (Figures 13(b)–13(d)). Particle deposition decreases quickly in the distal alveoli; very few particles are observed in the alveoli opposite to the gravity direction (Figures 13(b)–13(d)).

Effects of inhalation depth on particle deposition in the 45-alveoli model are shown in Figure 14. The inhalation depth ranges from 0.5 to 1.05 standard V_T (i.e., one standard $V_T = 0.233$ FRC), and the particle size is 1 μm. Very different deposition patterns are noted for different inhalation depths. At very shallow breath (i.e., 0.5 standard V_T or volume expansion ratio = 0.117), particles concentrate in the central alveoli along the gravity direction while very few particles deposit in the peripheral alveoli. As the breath depth increases, particle deposition becomes progressively more dispersed (Figures 14(a)–14(d)). This is caused by the deeper ventilation and stronger flow irregularity at a higher flow rate, both of which will lead to enhanced particle mixing. This phenomenon is most pronounced in the highest breath depth considered (1.05 V_T), where particles are spread in all compartments of the 45-alveoli model geometry (Figure 14(d)). Figure 14(e) shows the cumulative deposition with time for 1 μm particles at varying inhalation depths. Similar temporal profiles are observed among the four inhalation depths considered, all with a steep slope during the first two cycles and approaching asymptotic thereafter. The final DF slightly increases with the breath depth, for instance, from 62.6% at 0.5 V_T to 70.5% at 1.05 V_T (Figure 14(e)).

4. Discussion and Summary

A systematic study of acinar deposition of inertial particles was conducted in a well-defined 4-alveoli model and a more realistic 45-alveoli model. Airflow and particle dynamics driven by oscillating wall motions were visualized. A parametric study of various respiratory and structural factors was conducted, which included alveolar wall kinematics, particle size, model orientation, breath-holding, inhalation depth, and size of interalveolar pores.

Similar to the single-alveolus model as considered in [15], oscillatory wall motion was essential for the multi-alveoli models to capture particle dynamics and acinar deposition. A static multialveoli model that neglected oscillatory wall motions failed to predict neither the deposition fraction nor the particle lifetime before deposition. Much longer time was needed for particles to start deposition in a static alveolar model (about seven respiration cycles, Figure 5) than in a dynamic model (less than three cycles) [15]. This observation called into question of the using aerosols to estimate the alveolar size *in vivo* [48].

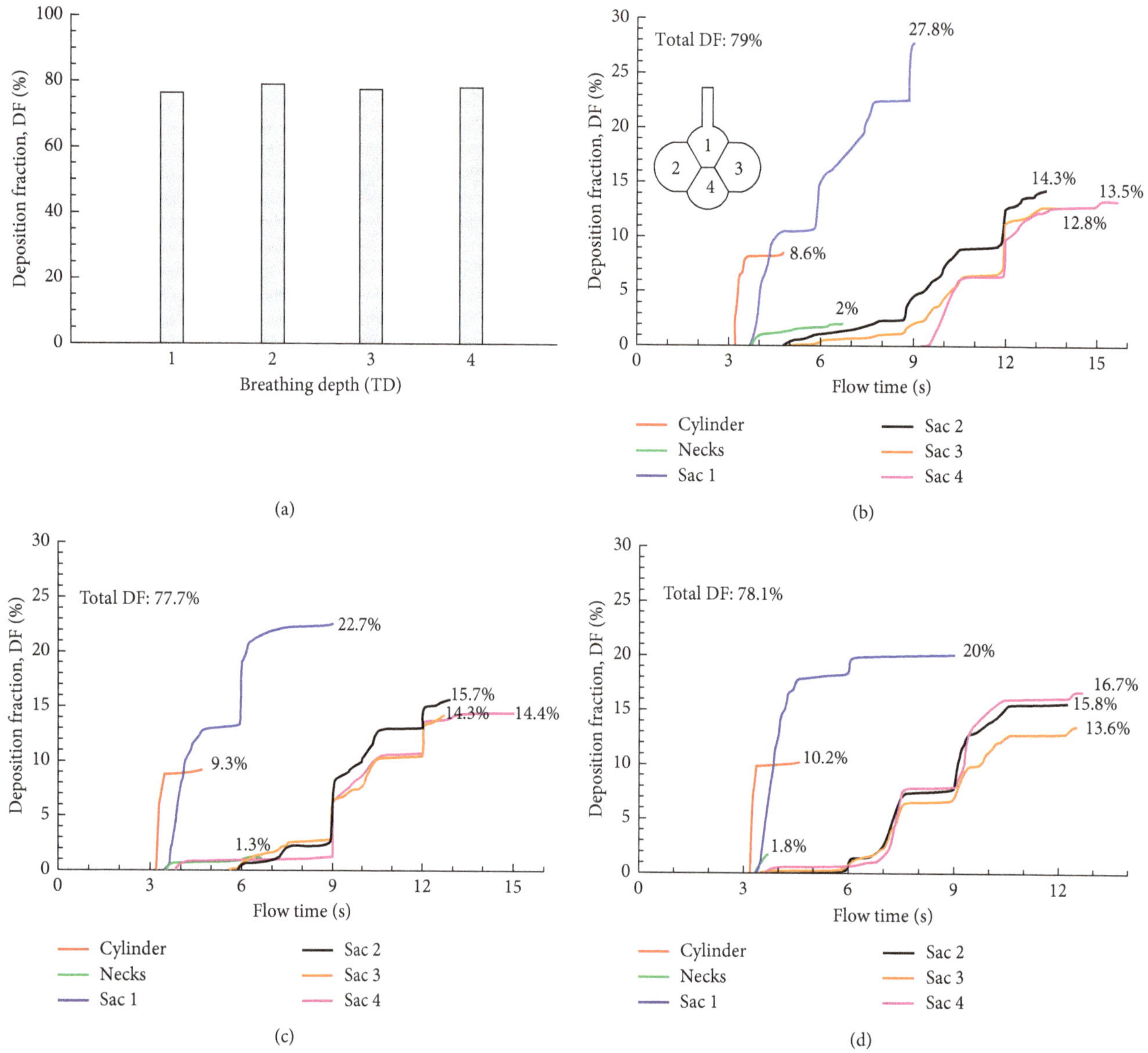

Figure 10: Comparison of particle deposition rates between different inhalation depths in the 4-alveoli model: (a) total deposition at different tidal volumes (V_T) and the time resolution of sectional cumulative deposition for (b) 2 standard V_T (i.e., volume expansion factor: 0.466), (c) 3 standard V_T (i.e., volume expansion factor: 0.699), and (d) 4 V_T (i.e., volume expansion factor: 0.932). Each respiration cycle has a period of 3.0 s.

In comparison to a single-alveolus model [15], interesting differences were observed in multiple-alveoli (i.e., 4-alveoli and 45-alveoli) models. One difference between the single-alveolus model and multialveoli model is the time to start deposition after inhalation. In a terminal single-alveolus model, inhaled particles cannot reach the airway wall during the first inhalation cycle due to the resident air, as observed both experimentally by Berg et al. [49] and in numerical studies by Sera et al. [50] and Talaat and Xi [15]. It took about three breathing cycles for 1 μm particles to start deposition and even longer for submicron particles [15]. In contrast, particle deposition in multialveoli models started during the first inhalation cycle. As shown in Figures 3(d) and 3(e), inhaled particles reached the septal walls of the upper alveolus (Sac 1) around the middle of the cycle. Due to the interalveolar communication in multialveoli models, all residual air in Sac 1 was displaced into the peripheral alveoli, transporting particles to Sac 1's septa; while the residual air in the single-alveolus model remained in the airspace during wall expansion, which restrained particles' forward motion and kept particles from reaching the alveolar wall.

The acinar deposition was found to be considerably sensitive to the number of alveoli retained in the model. In this study, the total DF for 0.5–1.5 μm particles was lower in the 45-alveoli model than that in the 4-alveoli model, which in turn was slightly lower than that in the single-alveolus model. This might be counterintuitive at first sight, as complex structures are generally expected to capture more

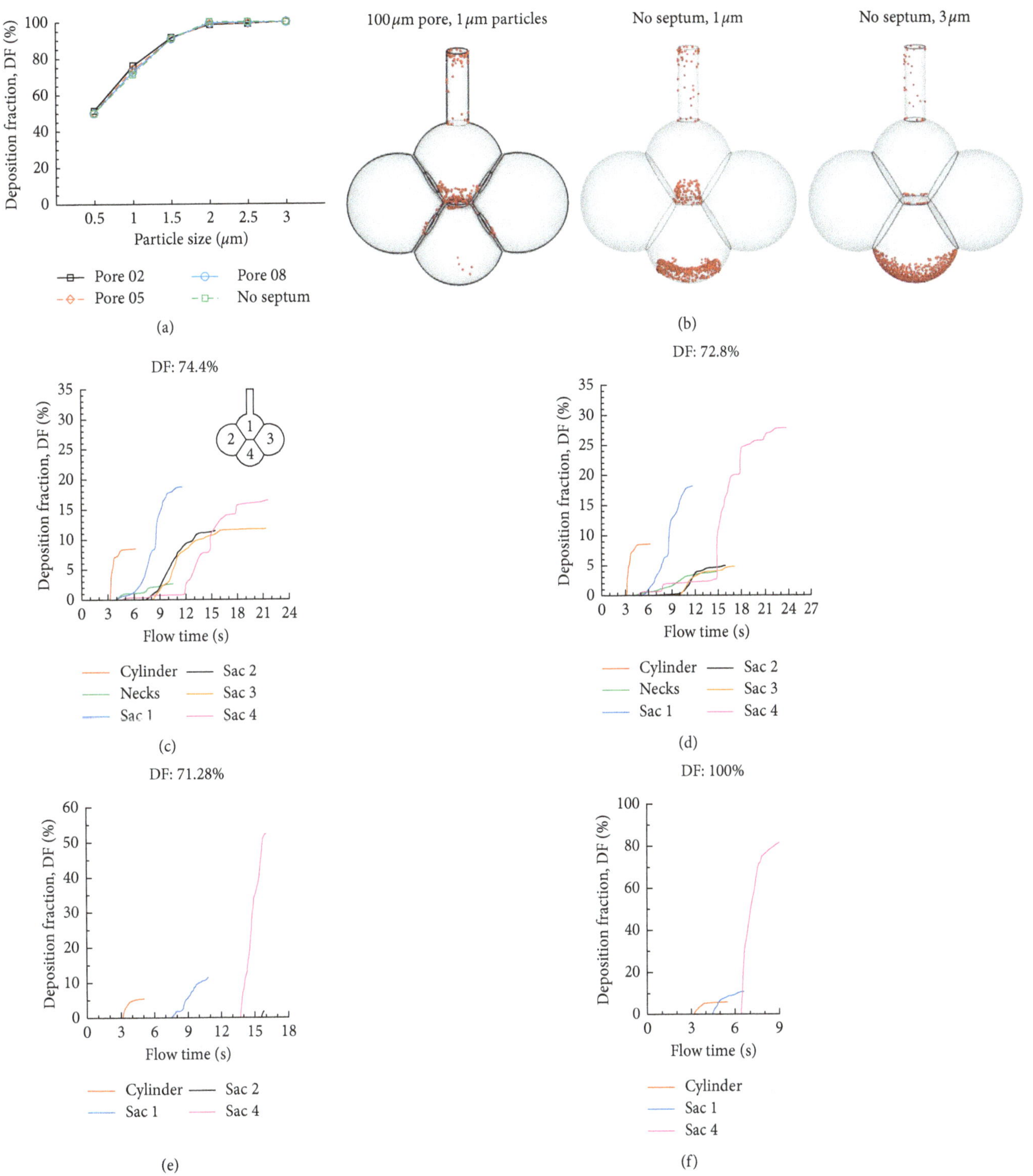

FIGURE 11: Pore size effects on alveolar deposition in the 4-alveoli model: (a) total deposition rates in the alveoli with different pore sizes, (b) surface deposition in different alveoli models, and the time resolution of sectional cumulative deposition for (c) pore size: 100 μm, particles: 1 μm, (d) pore size: 160 μm, particles: 1 μm, (e) no septum, particles: 1 μm, and (f) no septum, particles: 3 μm.

inhaled particles. This is mostly true for throughout flows with inlets and outlets. However, the terminal acinar region is a blind-end airspace, where airflow is driven by the rhythmical wall motion, which enters and exits the geometry through the same inlet. Some particles will be exhaled out of the geometry during expiration, leading to incomplete deposition. The escaping particles are primarily due to dispersion. Particles that remain in the geometry will eventually deposit either by convection impaction or gravitational sedimentation. Concerning the geometrical complexity

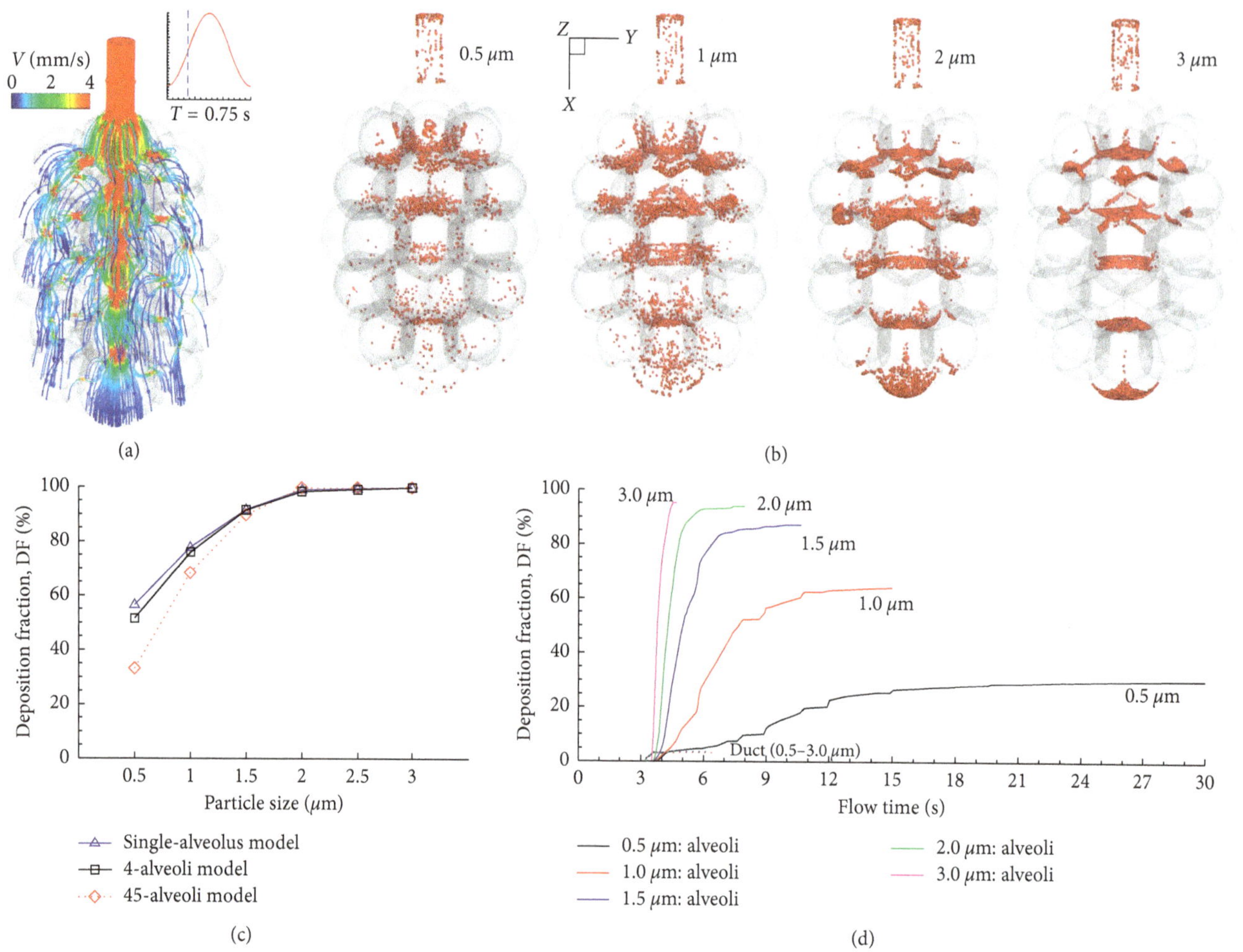

FIGURE 12: Airflow and particle deposition in the 45-alveoli model: (a) airflow, (b) surface deposition, (c) comparison of deposition rate between the 45-alveoli model and 4-alveoli and single-alveolus models, and (d) deposition with time.

effect, the 45-sac model will filter more particles for a given period of time, but meanwhile causes much stronger dispersion, overall giving rise to a lower total DF.

Second, self-similarities exist of particle dynamics in different generations of alveolar sacs. Particles in the central alveoli are more likely to deposit via convective impaction or interception, whereas particles in the peripheral alveoli will deposit via sedimentation, where the flow is slower. For a given tidal volume (i.e., 23.3% FRC), the 45-alveoli model draws in about ten times of air than the 4-alveoli model, inducing stronger inertia impaction in the central alveoli and higher particle deposition. As the inhaled flow bifurcates into the more distal compartments, the inertial impaction effect quickly decreases yielding much lower deposition. Overall, the total DF in the 45-alveoli model can be lower than a simpler 4-alveoli or single-alveolus model.

Particle deposition of these models (i.e., 1-, 4-, and 45-alveoli) reacted differently to the model orientation relative to the gravity. The DF was found to be relatively sensitive to the gravity orientation angle in the single-alveolus model (62.6%–80.0% from 0° to 135°) [15] and the 4-alveoli model (60.3%–77.2% from 0° to 135°), but was insensitive in the 45-alveoli model (67.5%%–71.6% from 0° to 135°). The lowest DF occurred at 135° for all of the three models considered. Considering that gravitational sedimentation was one dominating deposition mechanism, the above variation was most likely attributed to area ratio of the duct inlet over the projected area of the alveoli normal to the gravity, which decreased as the model become more complex. Accordingly, it was anticipated that particle dose in more complex acinus should be orientationally insensitive too. A similar observation was also reported in Khajeh-Hosseini-Dalasm and Longest [22] that acinar deposition was not affected by gravity orientation angle when the number of alveolar duct generations was more than three.

The inhalation depth was found to have an insignificant effect on the acinar DF in both the single-alveolus model [15] and the two multialveoli models herein. However, inhalation depth significantly altered the spatial distribution of the particle deposition, with more deposition rates in peripheral alveoli at deeper inhalations. A higher inhalation depth means a higher airflow speed and stronger wall-flow-particle interaction, which further lead to an enhanced deposition from convection and interception, as well as an intensified

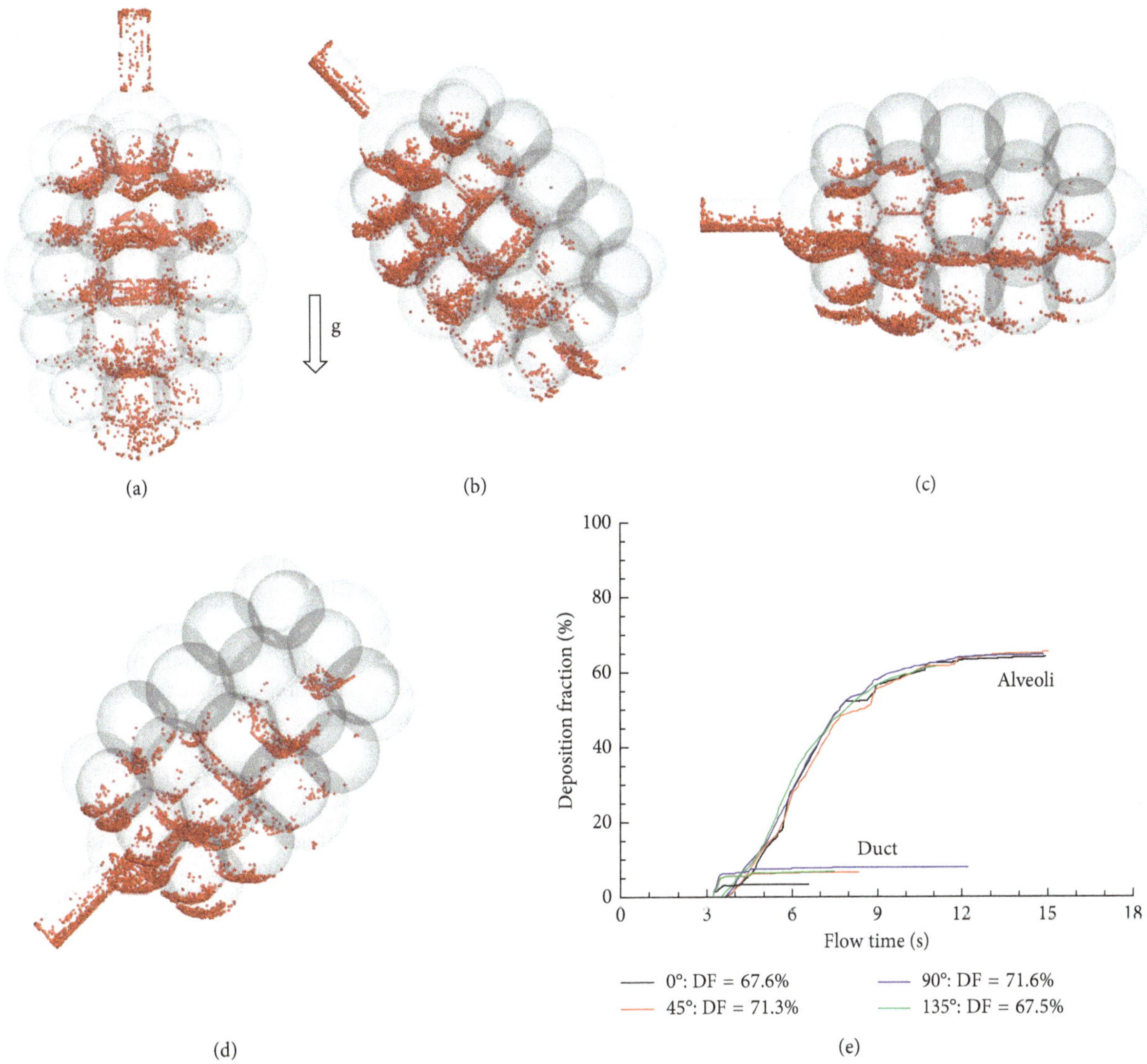

FIGURE 13: Comparison of the cumulative deposition with time for different gravity orientation angles in the 45-alveoli model: (a) 0°, (b) 45°, (c) 90°, and (d) 135° counterclockwise from the gravity, and (e) deposition.

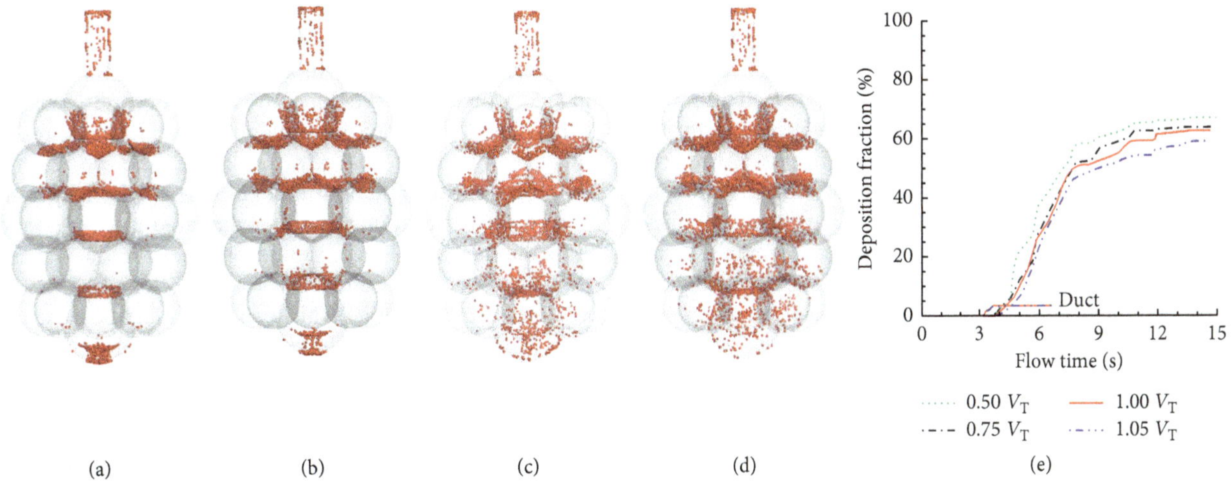

FIGURE 14: Comparison of particle deposition with different breath depths in the 45-alveoli model: (a) total deposition vs. inhalation depth and the time resolution of sectional cumulative deposition for (b) 0.5 standard V_T (i.e., volume expansion factor: 0.117), (c) 0.75 standard V_T (i.e., volume expansion factor: 0.175), (d) 1.05 standard V_T (i.e., volume expansion factor: 0.245), and (e) 1.1 standard V_T (i.e., volume expansion factor: 0.256).

particle dispersion. As discussed earlier, particle dispersion can cause more particles to escape the geometry during exhalation. Overall, the acinar DFs remain similar between different inhalation depths for a given acinar model. It is acknowledged that the acinar deposition fraction presented in this study was calculated as the number of particles deposited in the alveoli over the number of particles entering the alveolar model. The fraction of orally inhaled particles that entered the alveolar airways was not considered. Increasing inhalation depth will convey more particles to the alveolar airways and lead to higher doses in the acinar region.

In a recent study, Hofemeier et al. [23] also reported that that variance in acinar heterogeneity had little effect on total deposition. When comparing the DF between different acinar models (Figure 12(c)), the DF-particle profiles exhibit a similar asymptotic pattern among the three models, despite the difference in DF magnitude for particles ranging from 0.5 to 1.5 μm. This similarity was to a large extent due to the similar oscillating flows and the associated particle dynamics. Taken other similarities altogether, this agreement suggests a possibility of a generic deposition correlation to estimate the dosimetry in the intricate alveoli using relatively simple geometries even though what level of geometry complexity would be sufficient is yet to be determined. It is emphasized that Khajeh-Hosseini-Dalasm and Longest [22] pursued this question in multigeneration space-filling models with no septal walls and reported a relatively constant acinar deposition rate when more than three alveolar duct generations were retained in the acinar model. Furthermore, correlations of alveolar dose were proposed for different ventilation waveforms such as quick-and-deep and slow-and-deep inhalations [22].

Results of the airflow and particle deposition in alveoli with different pore sizes have meaningful implications in emphysema, which is featured by pore size increase and septum destruction. Varying the pore size was found to exert an insignificant impact on the alveolar deposition, indicating that the emphysematous terminal sacs might receive a similar amount of inhaled aerosols. It is noted that the above results were obtained with the remaining parameters being kept constant, while an emphysematous alveolus might also experience other changes, such as size increase, more compliant (i.e., longer exhalation time), and structure remodeling.

Besides interalveolar pores, collateral ventilation can also occur via bronchiole-alveolar communications (Lambert's channels), interbronchiolar communications (Martin's channels), and even interlobular respiratory bronchioles, depending on the alveolar location and emphysema severity [3]. This study focused on the terminal alveolar sac, where the bronchiole-alveolar and interbronchiolar channels are scarce and the interalveolar pores are prevalent [3, 51]. As a result, results of this study should only be applicable in the terminal alveoli or apical acinar regions where pores of Kohn were found in greatest numbers [3].

Other limitations of this study include simplified model geometry, ideal breathing conditions, noncontinuous aerosol inhalation, and one-way (wall-flow-particle) interaction. For simplicity, particle charge [52, 53], size [54], and hygroscopy effect [55] were also excluded. *In vivo* pulmonary alveoli have complicated morphology as revealed by histology and microscopy studies [49, 56–58], which appear as a polyhedral complex with varying-sized alveoli grouped in a fractal form [59]. In this study, both the 4-alveoli and 45-alveoli models were constructed from regular geometries such as a cylindrical duct, spheres, and circular pores. Anatomically, the interalveolar septal wall was demonstrated to have a variable thickness from the alveolar mouth to distal walls [60]; a constant wall thickness was assumed herein for computational simplicity. Likewise, there was no more than one pore in one septum in this study, while there can be one to seven pores in life conditions [61]. Even though scanning electron microscope (SEM) evidence has confirmed that pores of Kohn are normal structures in healthy lungs, it is not clear whether they are open all the time or are mostly covered by surfactant that ruptures during expansion or at high differential pressures. Because SEM samples are *ex vivo*, to the authors' knowledge, there is no literature that tracked the size and shape of *in vivo* pores in human lungs, despite recent attempts that utilized confocal microscopy [4] and optical coherence tomography (OCT) imaging to visualize alveolar structure dynamics [62] in mouse models. Moreover, surface tension of the fluids/surfactant mixture lining the alveolar wall varies during the expansion and contraction of the alveoli (roughly proportional to the surface area) [63]. Along with the nonlinear alveolar tissue elasticity, this surface tension variation further complicates the alveolar kinematics, for instance, by slowing down or constraining the wall stretching as it approaches the end of inhalation [64]. However, essential geometrical features of the alveoli were retained in these two models, such as the septa between alveoli and the pores in the septa, both of which had been neglected in previous numerical studies. The sphere was also naturally reshaped into a polyhedron when several spheres cross-cut each other, as displayed by the polyhedrons inside the 45-alveoli model, as well as the semipolyhedrons in the outer layers of both models (Figure 1). More importantly, well-defined shape and size allowed controlled parametric studies and hence identification of major factors that dictate airflow and particle deposition in alveolar sacs. The second physiological parameter to be improved is the breathing profile, which can have different waveforms and inhalation : exhalation (I : E) ratios. In life conditions, a normal breathing in a healthy subject generally has an I: E ratio of 1 : 2. [65] In emphysematous patients, the loss of alveolar elasticity leads to even longer exhalation time [8]. Compared to an I : E ratio of 1 : 1, this means a longer period of exhalation, a slower expiratory speed, a further downward motion, and therefore a higher deposition rate. This scenario should be close to the case with a postinhalation breath-holding of 1 or 2 seconds. Thirdly, particles were inhaled only at the beginning of the inhalation, and results of this study cannot be applied in scenarios with continuous aerosol exposures. Lastly, kinematics of the alveolar wall was modeled based on experimentally measured chest motions [37], and the amplitude of the alveolar wall motion was based on the tidal volume [66]. Direct measurements of *in vivo* alveolar motion using the imaging method should be used in future studies. It is also noted that pores of Kohn is 2–15 μm in diameter in normal healthy lungs [2, 61], but can constantly increase in size till septal wall breakdown in emphysematous alveoli [59]. Further studies of the influences of smaller and multiple pores on acinar flow and particle deposition are needed.

In Summary, temporal and spatial deposition variations in multialveoli pore-communicated acinar models were numerically investigated under the influences of various physiological factors. Specific findings are listed as follows:

(1) Collateral ventilation existed in multialveoli acinar models

(2) Heterogeneous deposition distributions were found among alveoli, with the highest deposition in the central alveoli and decreasing deposition in more peripheral alveoli

(3) The acinar deposition was highly sensitive to particle size; for particles that were larger than 2 μm and administered at the beginning of the inhalation, nearly 100% alveolar deposition fraction (i.e., particles deposited in the alveolar model over particles entering the alveolar model) was predicted

(4) The number of alveoli retained in the model affected the total deposition, with the 45-alveoli model having lower deposition than the 4-alveoli and single-alveolus models

(5) The size of the pores of Kohn, inhalation depth, and gravity orientation angle had insignificant effects on the acinar deposition fraction but had a dramatic impact on the spatial distribution of particle deposition among alveoli

Conflicts of Interest

The authors report no conflicts of interest in this work.

Acknowledgments

This study was funded by the NSF Grant (CBET 1745602).

References

[1] M. Ochs, J. R. Nyengaard, A. Jung et al., "The number of alveoli in the human lung," *American Journal of Respiratory and Critical Care Medicine*, vol. 169, no. 1, pp. 120–124, 2004.

[2] C. Desplechain, B. Foliguet, E. Barrat, G. Grignon, and F. Touati, "The pores of Kohn in pulmonary alveoli," *Bulletin Européen de Physiopathologie Respiratoire*, vol. 19, no. 1, pp. 59–68, 1983.

[3] P. B. Terry and R. J. Traystman, "The clinical significance of collateral ventilation," *Annals of the American Thoracic Society*, vol. 13, no. 12, pp. 2251–2257, 2016.

[4] E. Namati, J. Thiesse, J. de Ryk, and G. McLennan, "Alveolar dynamics during respiration: are the pores of Kohn a pathway to recruitment?," *American Journal of Respiratory Cell and Molecular Biology*, vol. 38, no. 5, pp. 572–578, 2008.

[5] A. Nagai, H. Inano, K. Matsuba, and W. M. Thurlbeck, "Scanning electronmicroscopic morphometry of emphysema in humans," *American Journal of Respiratory and Critical Care Medicine*, vol. 150, no. 5, pp. 1411–1415, 1994.

[6] C. M. Van Allen, G. E. Lindskog, and H. G. Richter, "Gaseous interchange between adjacent lung lobules," *Yale Journal of Biology and Medicine*, vol. 2, no. 4, pp. 297–300, 1930.

[7] P. B. Terry, R. J. Traystman, H. H. Newball, G. Batra, and H. A. Menkes, "Collateral ventilation in man," *New England Journal of Medicine*, vol. 298, no. 1, pp. 10–15, 1978.

[8] J. C. Hogg, "Pathophysiology of airflow limitation in chronic obstructive pulmonary disease," *The Lancet*, vol. 364, no. 9435, pp. 709–721, 2004.

[9] S. Shimura, E. S. Boatman, and C. J. Martin, "Effects of ageing on the alveolar pores of Kohn and on the cytoplasmic components of alveolar type II cells in monkey lungs," *Journal of Pathology*, vol. 148, no. 1, pp. 1–11, 1986.

[10] N. A. Gillett, R. F. Gerlach, B. A. Muggenburg, J. R. Harkema, W. C. Griffith, and J. L. Mauderly, "Relationship between collateral flow resistance and alveolar pores in the aging beagle dog," *Experimental Lung Research*, vol. 15, no. 5, pp. 709–719, 1989.

[11] I. Balashazy, W. Hofmann, A. Farkas, and B. G. Madas, "Three-dimensional model for aerosol transport and deposition in expanding and contracting alveoli," *Inhalation Toxicology*, vol. 20, no. 6, pp. 611–621, 2008.

[12] S. Haber, D. Yitzhak, and A. Tsuda, "Gravitational deposition in a rhythmically expanding and contracting alveolus," *Journal of Applied Physiology*, vol. 95, no. 2, pp. 657–671, 2003.

[13] D. Y. Lee and J. W. Lee, "Characteristics of particle transport in an expanding or contracting alveolated tube," *Journal of Aerosol Science*, vol. 34, no. 9, pp. 1193–1215, 2003.

[14] J. Sznitman, F. Heimsch, T. Heimsch, D. Rusch, and T. Rösgen, "Three-dimensional convective alveolar flow induced by rhythmic breathing motion of the pulmonary acinus," *Journal of Biomechanical Engineering*, vol. 129, no. 5, pp. 658–665, 2007.

[15] K. Talaat and J. Xi, "Computational modeling of aerosol transport, dispersion, and deposition in rhythmically expanding and contracting terminal alveoli," *Journal of Aerosol Science*, vol. 112, pp. 19–33, 2017.

[16] J. Xi, K. Talaat, and X. Si, "Deposition of bolus and continuously inhaled aerosols in rhythmically moving terminal alveoli," *Journal of Computational Multiphase Flows*, 2018.

[17] C. Darquenne and M. Paiva, "Two- and three dimensional simulations of aerosol transport and deposition in alveolar zone of human lung," *Journal of Applied Physiology*, vol. 80, no. 4, pp. 1401–1414, 1996.

[18] A. Karl, F. S. Henry, and A. Tsuda, "Low reynolds number viscous flow in an alveolated duct," *Journal of Biomechanical Engineering*, vol. 126, no. 4, pp. 420–429, 2004.

[19] H. Kitaoka, G. F. Nieman, Y. Fujino, D. Carney, J. DiRocco, and I. Kawase, "A 4-dimensional model of the alveolar structure," *Journal of Physiological Sciences*, vol. 57, no. 3, pp. 175–185, 2007.

[20] H. Kumar, M. H. Tawhai, E. A. Hoffman, and C. L. Lin, "The effects of geometry on airflow in the acinar region of the human lung," *Journal of Biomechanics*, vol. 42, no. 11, pp. 1635–1642, 2009.

[21] J. Sznitman, T. Heimsch, J. H. Wildhaber, A. Tsuda, and T. Rösgen, "Respiratory flow phenomena and gravitational deposition in a three-dimensional space-filling model of the

pulmonary acinar tree," *Journal of Biomechanical Engineering*, vol. 131, no. 3, article 031010, 2009.

[22] N. Khajeh-Hosseini-Dalasm and P. W. Longest, "Deposition of particles in the alveolar airways: inhalation and breath-hold with pharmaceutical aerosols," *Journal of Aerosol Science*, vol. 79, no. 1, pp. 15–30, 2015.

[23] P. Hofemeier, K. Koshiyama, S. Wada, and J. Sznitman, "One (sub-)acinus for all: fate of inhaled aerosols in heterogeneous pulmonary acinar structures," *European Journal of Pharmaceutical Sciences*, vol. 113, no. 2, pp. 53–63, 2018.

[24] W. H. Finlay, *The Mechanics of Inhaled Pharmaceutical Aerosols*, Academic Press, San Diego, CA, USA, 2001.

[25] G. Rudolf, J. Gebhart, J. Heyder, C. F. Schiller, and W. Stahlhofen, "An empirical formula describing aerosol deposition in man for any particle size," *Journal of Aerosol Science*, vol. 17, no. 3, pp. 350–355, 1986.

[26] K. K. Isaacs, J. A. Rosati, and T. B. Martonen, "Mechanisms of particle deposition," in *Aerosols Handbook*, L. S. Ruzer and N. H. Harley, Eds., pp. 75–99, CRC Press, New York, NY, USA, 2005.

[27] M. Kojic and A. Tsuda, "A simple model for gravitational deposition of non-diffusing particles in oscillatory laminar pipe flow and its application to small airways," *Journal of Aerosol Science*, vol. 35, no. 2, pp. 245–261, 2004.

[28] J. Pich, "Theory of gravitational deposition of particles from laminar flows in channels," *Journal of Aerosol Science*, vol. 3, no. 5, pp. 351–361, 1972.

[29] J. I. Choi and C. S. Kim, "Mathematical analysis of particle deposition in human lungs: an improved single path transport model," *Inhalation Toxicology*, vol. 19, no. 11, pp. 925–939, 2007.

[30] C. S. Kim, "Deposition of aerosol particles in human lungs: in vivo measurement and modelling," *Biomarkers*, vol. 14, no. 1, pp. 54–58, 2009.

[31] P. Hofemeier and J. Sznitman, "The role of anisotropic expansion for pulmonary acinar aerosol deposition," *Journal of Biomechanics*, vol. 49, no. 14, pp. 3543–3548, 2016.

[32] S. Haber, J. P. Butler, H. Brenner, I. Emanuel, and A. Tsuda, "Shear flow over a self-similar expanding pulmonary alveolus during rhythmical breathing," *Journal of Fluid Mechanics*, vol. 405, no. 2, pp. 243–268, 2000.

[33] S. Chhabra and A. K. Prasad, "Flow and particle dispersion in a pulmonary alveolus-part I: velocity measurements and convective particle transport," *Journal of Biomechanical Engineering*, vol. 132, no. 5, article 4001112, 2010.

[34] E. J. Berg and R. J. Robinson, "Stereoscopic particle image velocimetry analysis of healthy and emphysemic alveolar sac models," *Journal of Biomechanical Engineering*, vol. 133, no. 6, article 061004, 2011.

[35] B. Ma and C. Darquenne, "Aerosol bolus dispersion in acinar airways--influence of gravity and airway asymmetry," *Journal of Applied Physiology*, vol. 113, no. 3, pp. 442–450, 2012.

[36] C. Plathow, S. Ley, C. Fink et al., "Evaluation of chest motion and volumetry during the breathing cycle by dynamic MRI in healthy subjects: comparison with pulmonary function tests," *Investigative Radiology*, vol. 39, no. 4, pp. 202–209, 2004.

[37] A. De Groote, M. Wantier, G. Cheron, M. Estenne, and M. Paiva, "Chest wall motion during tidal breathing," *Journal of Applied Physiology*, vol. 83, no. 5, pp. 1531–1537, 1997.

[38] R. L. Jones and M. M. Nzekwu, "The effects of body mass index on lung volumes," *Chest*, vol. 130, no. 3, pp. 827–833, 2006.

[39] E. M. Harding and R. J. Robinson, "Flow in a terminal alveolar sac model with expanding walls using computational fluid dynamics," *Inhalation Toxicology*, vol. 22, no. 8, pp. 669–678, 2010.

[40] J. Xi, X. Si, J. Kim, G. Su, and H. Dong, "Modeling the pharyngeal anatomical effects on breathing resistance and aerodynamically generated sound," *Medical & Biological Engineering & Computing*, vol. 52, no. 7, pp. 567–577, 2014.

[41] X. Si, J. Xi, and J. Kim, "Effect of laryngopharyngeal anatomy on expiratory airflow and submicrometer particle deposition in human extrathoracic airways," *Open Journal of Fluid Dynamics*, vol. 3, no. 4, 2013.

[42] P. W. Longest and J. Xi, "Computational investigation of particle inertia effects on submicron aerosol deposition in the respiratory tract," *Journal of Aerosol Science*, vol. 38, no. 1, pp. 111–130, 2007.

[43] P. W. Longest and J. Xi, "Effectiveness of direct Lagrangian tracking models for simulating nanoparticle deposition in the upper airways," *Aerosol Science and Technology*, vol. 41, no. 4, pp. 380–397, 2007.

[44] J. Kim, J. Xi, X. Si, A. Berlinski, and W. C. Su, "Hood nebulization: effects of head direction and breathing mode on particle inhalability and deposition in a 7-month-old infant model," *Journal of Aerosol Medicine and Pulmonary Drug Delivery*, vol. 27, no. 3, pp. 209–218, 2014.

[45] Z. Li, C. Kleinstreuer, and Z. Zhang, "Simulation of airflow fields and microparticle deposition in realistic human lung airway models. Part II: particle transport and deposition," *European Journal of Mechanics-B/Fluids*, vol. 26, no. 5, pp. 650–668, 2007.

[46] J. Xi and P. W. Longest, "Transport and deposition of microaerosols in realistic and simplified models of the oral airway," *Annals of Biomedical Engineering*, vol. 35, no. 4, pp. 560–581, 2007.

[47] J. Xi, J. E. Yuan, M. Yang, X. Si, Y. Zhou, and Y.-S. Cheng, "Parametric study on mouth–throat geometrical factors on deposition of orally inhaled aerosols," *Journal of Aerosol Science*, vol. 99, pp. 94–106, 2016.

[48] W. D. Bennett and G. C. Smaldone, "Use of aerosols to estimate mean air-space size in chronic obstructive pulmonary disease," *Journal of Applied Physiology*, vol. 64, no. 4, pp. 1554–1560, 1985.

[49] E. J. Berg, J. L. Weisman, M. J. Oldham, and R. J. Robinson, "Flow field analysis in a compliant acinus replica model using particle image velocimetry (PIV)," *Journal of Biomechanics*, vol. 43, no. 6, pp. 1039–1047, 2010.

[50] T. Sera, K. Uesugi, N. Yagi, and H. Yokota, "Numerical simulation of airflow and microparticle deposition in a synchrotron micro-CT-based pulmonary acinus model," *Computer Methods in Biomechanics and Biomedical Engineering*, vol. 18, no. 13, pp. 1427–1435, 2015.

[51] E. J. Cetti, A. J. Moore, and D. M. Geddes, "Collateral ventilation," *Thorax*, vol. 61, no. 5, pp. 371–373, 2006.

[52] J. Xi, X. Si, and P. W. Longest, "Electrostatic charge effects on pharmaceutical aerosol deposition in human nasal-laryngeal airways," *Pharmaceutics*, vol. 6, no. 1, pp. 26–35, 2013.

[53] J. Xi, X. A. Si, and R. Gaide, "Electrophoretic particle guidance significantly enhances olfactory drug delivery: a feasibility study,"PLoS One, vol. 9, no. 1, article e86593, 2014.

[54] X. A. Si, J. Xi, J. Kim, Y. Zhou, and H. Zhong, "Modeling of release position and ventilation effects on olfactory aerosol drug delivery," *Respiratory Physiology & Neurobiology*, vol. 186, no. 1, pp. 22–32, 2013.

[55] J. W. Kim, J. Xi, and X. A. Si, "Dynamic growth and deposition of hygroscopic aerosols in the nasal airway of a 5-year-old

child," *International Journal for Numerical Methods in Biomedical Engineering*, vol. 29, no. 1, pp. 17–39, 2013.
[56] G. C. Smaldone and W. Mitzner, "Viewpoint: unresolved mysteries," *Journal of Applied Physiology*, vol. 113, no. 12, pp. 1945–1947, 2012.
[57] B. Haefeli-Bleuer and E. R. Weibel, "Morphometry of the human pulmonary acinus," *Anatomical Record*, vol. 220, no. 4, pp. 401–414, 1988.
[58] H. Kumar, D. M. Vasilescu, Y. Yin, E. A. Hoffman, M. H. Tawhai, and C.-L. Lin, "Multiscale imaging and registration-driven model for pulmonary acinar mechanics in the mouse," *Journal of Applied Physiology*, vol. 114, no. 8, pp. 971–978, 2013.
[59] A. Yoshikawa, S. Sato, T. Tanaka et al., "Breakdown of lung framework and an increase in pores of Kohn as initial events of emphysema and a cause of reduction in diffusing capacity," *International Journal of Chronic Obstructive Pulmonary Disease*, vol. 11, no. 1, pp. 2287–2294, 2016.
[60] S. F. Ryan, "The structure of the interalveolar septum of the mammalian lung," *Anatomical Record*, vol. 165, no. 4, pp. 467–483, 1969.
[61] J. L. Cordingley, "Pores of Kohn," *Thorax*, vol. 27, no. 4, pp. 433–441, 1972.
[62] L. Kirsten, M. Gaertner, C. Schnabel, S. Meissner, and E. Koch, "Four-dimensional imaging of murine subpleural alveoli using high-speed optical coherence tomography," *Journal of Biophotonics*, vol. 6, no. 2, pp. 148–152, 2013.
[63] C. B. Daniels, S. Orgeig, A. W. Smits, and J. D. Miller, "The influence of temperature, phylogeny, and lung structure on the lipid composition of reptilian pulmonary surfactant," *Experimental Lung Research*, vol. 22, no. 3, pp. 267–281, 1996.
[64] F. J. A. Veldhuizen and H. P. Haagsman, "Role of pulmonary surfactant components in surface film formation and dynamics," *Biochimica et Biophysica Acta (BBA)–Biomembranes*, vol. 1467, no. 2, pp. 255–270, 2000.
[65] I. Van Diest, K. Verstappen, A. E. Aubert, D. Widjaja, D. Vansteenwegen, and E. Vlemincx, "Inhalation/exhalation ratio modulates the effect of slow breathing on heart rate variability and relaxation," *Applied Psychophysiology and Biofeedback*, vol. 39, no. 3-4, pp. 171–180, 2014.
[66] L. Porra, S. Monfraix, G. Berruyer, G. Le Duc, C. Nemoz, and S. Bayat, "Effect of tidal volume on distribution of ventialtion assessed by synchrotron radiation CT in rabbit," *Journal of Applied Physiology*, vol. 96, pp. 1899–1908, 2004.

17

A New Approach for Advertising CTR Prediction Based on Deep Neural Network via Attention Mechanism

Qianqian Wang,[1] Fang'ai Liu,[1] Shuning Xing,[1] and Xiaohui Zhao[2]

[1]*School of Information Science and Engineering, Shandong Normal University, Jinan, China*
[2]*School of Mathematical Science, Shandong Normal University, Jinan, China*

Correspondence should be addressed to Fang'ai Liu; lfa@sdnu.edu.cn

Academic Editor: Martti Juhola

Click-through rate prediction is critical in Internet advertising and affects web publisher's profits and advertiser's payment. The traditional method of obtaining features using feature extraction did not consider the sparseness of advertising data and the highly nonlinear association between features. To reduce the sparseness of data and to mine the hidden features in advertising data, a method that learns the sparse features is proposed. Our method exploits dimension reduction based on decomposition, takes advantage of the attention mechanism in neural network modelling, and improves FM to make feature interactions contribute differently to the prediction. We utilize stack autoencoder to explore high-order feature interactions and use improved FM for low-order feature interactions to portray the nonlinear associated relationship of features. The experiment shows that our method improves the effect of CTR prediction and produces economic benefits in Internet advertising.

1. Introduction

Click-through rate (CTR) prediction is critical to many web applications including web search, recommender systems [1, 2], sponsored search, and display advertising. Search advertising, known as sponsored search, refers to advertisers identifying relevant keywords based on their product or service for advertising. When the user retrieves the keyword purchased by the advertiser, the corresponding advertisement is triggered and displayed. In the cost-per-click model, the advertiser pays the web publisher only when a user clicks their advertisements and visits the advertiser's site. The CTR prediction is defined to estimate the ratio of clicks to impressions of advertisements that will be displayed [3].

With the rapid development of the mobile Internet and its wide range of applications, advertising has become one of the most successful business models in the world. Internet text advertising is regarded as a more effective advertising communication method due to its strong targeted communication and convenience of user clicking and has become an important income resource for many Internet companies. Some electronic commerce companies and search engine companies are seeking targeted advertising to increase their revenue.

In general, the display of online advertising can be seen as a three-party game between media, advertisers, and users. How to advertise to specific user groups is a key issue in the field of online advertising. Inappropriate advertising can lead to a decline in user experience. Advertising cannot achieve the desired effect, and the media can also be affected. Internet text advertising is usually in the form of text, and the advertisers get the opportunity to buy media ads through cost-per-click (CPC) [4]. In the CPC model, the click-through rate (CTR) is an important indicator to measure the effectiveness of advertising display and is a key factor in the three-party game. Therefore, the CTR estimation of advertising is a hot research direction in the field of computing advertising. In this paper, the click-through rate prediction of Internet text advertising shows the probability of predicting a user's click on a text under the current context environment. Due to the three-party information of advertising properties, user properties, and context environment, the CTR prediction is very complicated.

At present, the prediction of click-through rate for online advertising has attracted widespread attention from researchers in industry and academia. Researchers have proposed many models that are usually based on machine learning methods. We can divide them into three categories: linear, nonlinear, and fusion models. Typically, a predictive task is formulated as estimating a function that maps predictor variables to some target. To build predictive models with these predictor variables, a common solution is to convert them to a set of binary features (a.k.a. feature vector) via one-hot encoding [5]. McMahan et al. [6] used the logistic regression [7] model to solve the CTR problems of Google Advertising. They adopted user information, advertising data, search keywords, and other features as the input of the model and proposed an online sparse learning algorithm to train the model. Chapelle [8] proposed a machine-learning framework based on the logistic regression in which advertisers, web publishers, users, and time characteristics were used as input to the model to solve the advertising CTR prediction for Yahoo. Dave and Varma [9] used the gradient boosting decision tree (GBDT) to predict the advertising CTR. They extracted similar features from advertising data and discovered implicit relationships between different features. Finally, they found out the nonlinear relationships between the predicted target and features. He et al. [10] introduced a fusion model which combines decision trees with logistic regression for predicting clicks on Facebook ads. The traditional CTR prediction model mainly depends on the design of features. The features of data are artificially selected and processed. The data have a complex mapping relationship, especially for meaningful data, and it is crucial to account for the interactions between features. Many successful solutions in both industry and academia largely rely on manually crafting combinatorial features [11], i.e., constructing new features by combining multiple predictor variables, also known as cross features. However, the power of such features comes at a high cost since it requires heavy engineering efforts and useful domain knowledge to design effective features. Factorization machines (FMs) [12] are a supervised learning approach that embed features into a latent space and model the interactions between features via inner product of their embedding vectors. Models based on degree-2 polynomial mapping and factorization machines are widely used for CTR prediction. The factorization-based prediction method field-aware factorization machines [13] were developed by Juan et al.

In recent years, deep learning [14, 15] has achieved very good results in the fields of speech recognition [16], image data processing [17], and natural language processing [18]. As a powerful approach to learning feature representation, deep neural networks have the potential to learn sophisticated feature interactions. Liu et al. [19] extended CNN for CTR prediction, but CNN-based models are biased towards the interactions between neighboring features. Zhang et al. [20] studied feature representations and proposed factorization machine-supported neural network (FNN). This model pretrained FM before applying DNN and thus limited by the capability of FM. He and Chua [21] proposed a novel neural factorization machine (NFM) for prediction under sparse setting. NFM combines the linearity of FM in modelling second-order feature interactions and the nonlinearity of neural network in modelling higher-order feature interactions. Despite great promise, we argue that FM can be hindered by its modelling of all factorized interactions with the same weight. In real-world applications, different predictor variables often have different predictive power. Not all features contain useful information for predicting the target. Therefore, the interaction of features with less useful information should be assigned a lower weight indicating that they contribute less to the prediction. However, FM lacks the ability to distinguish the importance of feature interactions, which will lead to suboptimal prediction.

Considering the high-dimensional sparsity of advertising data and the highly nonlinear association between features [22], a hybrid model for advertising CTR estimation based on stacked autoencoder, named Attention Stacked Autoencoder (ASAE), is proposed. Our model takes advantage of the attention mechanism in neural network modelling [23, 24] and improves FM to make feature interactions contribute differently to the prediction. More importantly, the importance of feature interactions is automatically learned from the data with any human domain knowledge. We explore data dimension reduction and identify the relationship between features. Additionally, many experiments are conducted to show that this method improves the accuracy of CTR estimation.

The rest of this paper is organized as follows. Section 2 provides the factorization machines. In Section 3, the sparse feature learning method for advertising data based on the ASAE model is proposed. In Section 4, we design the experiment and verify the prediction effect of the method by comparison experiment. We also analyze the experimental results in this section. Section 5 concludes the paper and lists possible future work.

2. Factorization Machines

The factorization machines are originally proposed for learning feature interactions in the recommendation system. Given a real-valued feature vector $X \in \mathbb{R}^n$ where n denotes the number of features, FM estimates the target by modelling all interactions between each pair of features:

$$\hat{y}\mathrm{FM}(x) = w_0 + \sum_{i=1}^{n} w_i x_i + \sum_{i=1}^{n} \sum_{j=i+1}^{n} \hat{w}_{ij} x_i x_j, \qquad (1)$$

where w_0 is the global bias, w_i denotes the weight of the ith feature, and w_{ij} denotes the weight of the cross feature $x_i x_j$, which is factorized as $\hat{w}_{ij} = v_i^T v_j$, where $v_i \in \mathbb{R}^k$ denotes the embedding vector for feature i and k denotes the size of the embedding vector. Besides linear (order-1) interactions among features, FM models pairwise (order-2) feature interactions as inner product of respective feature latent vectors. It can capture order-2 feature interactions much more effectively than previous approaches especially when the dataset is sparse. It is worth noting that FM models all feature interactions in the same way: first, a latent vector v_i is shared in estimating all feature interactions that the ith feature involves; second, all feature interactions have the same weight of 1. However, it is common that not all features are relevant to the prediction. These interactions of irrelevant features can be considered as noise that does not contribute to the prediction. FM models all features using the same weights for interaction and may have a negative impact on generalization performance.

3. Click-through Rate Estimation Based on Deep Neural Network

One of the necessary steps in the click rate prediction system is to mine features that are highly correlated with the estimated task. To reduce the high sparseness of features and characterize the nonlinear association between features, we propose a sparse feature learning method for advertising data based on deep learning (DLSAE).

3.1. Data Dimensionality Reduction.

Click log data contain many types of objects, such as users, queries, and advertisements. The relationship between these objects is very complex. The same objects have similarity, and there are complex relationships between different types of objects. For instance, given a particular user and the query submitted by the user, it is necessary to predict whether the user will click on the advertisement and the probability. There is a complex implicit relationship between users, queries, and advertising. Based on the characteristics of the click log data, dimension reduction is achieved in the following two aspects: the similarity between the internal objects and the association between different objects.

In this paper, the k-means clustering algorithm [25] based on distance is adopted. We cluster queries, advertisements, and users separately, and the similar objects are aggregated into the same cluster. We use advertising frequency as the weight of the advertisement A_i and query Q_j and create a matrix $W_{M_a \times M_q}$ of the ad-query (where M_a is the number of ads and M_q indicates the number of queries), using the k-means algorithm to cluster the ad-query matrix. We scan the ad-query matrix to obtain the ad sets and query sets, as $A = \{a_1, a_2, \ldots, a_m\}$ and $Q = \{q_1, q_2, \ldots, q_N\}$. Then, we take K samples from the advertising set randomly as the initial point of the cluster center, record as $T = \{t_1, t_2, \ldots, t_k\}$. Next, Equation (2) is used to calculate the distance between ad a_i and each cluster center point t_j. The number of clusters of users, ads, and queries is represented by K_u, K_a, and K_q, respectively. Finally, the number of users, ads, and queries in the dataset is reduced from M_u, M_a, M_q to K_u, K_a, K_q:

$$\mathrm{Dis}(a_i, t_j) = \sqrt{\sum \left(W_{a_i} - W_{q_j}\right)^2}, \quad (2)$$

where W_{a_i} is the weight of a_i, W_{q_j} is the weight of t_j, and $\mathrm{Dis}(a_i, t_j)$ is the distance between a_i and t_j.

There is a ternary relationship between the user-query-ad in the click log data. In this paper, we use the three-dimensional tensor structure model [26, 27] to represent the user, query, and advertisement. Then, the tensor decomposition method is used to reduce the dimensions. The sum of the display number of ads in the cluster is used as the weight of the elements in 3D space. The three-dimensional tensor model is constructed and represented by $X \in R^{K_u \times K_q \times K_a}$. In this paper, tensor X is decomposed using the Tucker factorization. Equation (2) is the decomposition formula.

$$\begin{aligned} X &= [G; A, B, C] = G \times_u A \times_q B \times_a C \\ &= \sum_{p=1}^{P} \sum_{m=1}^{M} \sum_{r=1}^{R} g_{pmr} u_p \circ q_m \circ a_r, \end{aligned} \quad (3)$$

where G represents the core tensor of tensor X. We use A, B, and C to represent the feature matrix of the tensor X on the dimension K_u, K_q, K_a.

Figure 1 is a schematic diagram of the Tucker decomposition. The purpose of the Tucker decomposition is to find an approximate tensor $\widehat{X}$ with the original tensor X and to retain the original tensor information and structural information to the greatest extent. The minimization formula is shown below:

$$\min\left\|X - \widehat{X}\right\|, \quad \widehat{X} = G \times_u A \times_q B \times_a C = [G; A, B, C]. \quad (4)$$

Equation (4) is the objective optimization function. According to Equation (3), the expression of the core tensor can be obtained as follows:

$$G = X \times_u A^T \times_q B^T \times_a C^T, \quad (5)$$

and the objective function can be written in a squared form:

$$\begin{aligned} \|X - [G; A, B, C]\|^2 &= \|X\|^2 - 2\langle X \times_u A^T \times_q B^T \times_a C^T, G\rangle \\ &\quad + \|G\|^2, \\ &= \|X\|^2 - 2\langle G, G\rangle + \|G\|^2, \\ &= \|X\|^2 - \|G\|^2, \\ &= \|X\|^2 - \left\|X \times_u A^T \times_q B^T \times_a C^T\right\|^2. \end{aligned} \quad (6)$$

Therefore, the objective function is transformed to

$$\begin{aligned} &\max\left\|X \times_u A^T \times_q B^T \times_a C^T\right\|^2, \\ &\left\|A^T W\right\|, \quad W = X \times_q B^T \times_a C^T, \\ &\left\|B^T W\right\|, \quad W = X \times_u A^T \times_a C^T, \\ &\left\|C^T W\right\|, \quad W = X \times_u A^T \times_q B^T. \end{aligned} \quad (7)$$

In the process of solving the optimal solution, we need to fix the matrix of the other dimensions W, solve for A^T, B^T, C^T, and then perform a singular value decomposition (SVD) of A^T, B^T, C^T. Next, expand the tensor X into a matrix on the user, query, and advertising dimensions, respectively, as X_1, X_2, X_3 and apply SVD on X_1, X_2, X_3:

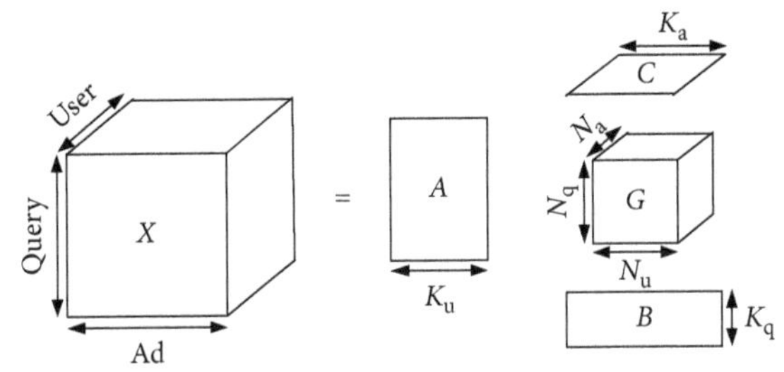

FIGURE 1: Schematic diagram of the Tucker decomposition.

$$\begin{aligned} X_1 &= A \cdot G_1 \cdot V_1^T, \\ X_2 &= B \cdot G_2 \cdot V_2^T, \\ X_3 &= C \cdot G_3 \cdot V_3^T, \end{aligned} \tag{8}$$

where G_1, G_2, G_3 are the diagonal singular value matrices obtained using singular value decomposition of the matrices X_1, X_2, X_3. g_1, g_2, g_3 are the dimensions of the singular value matrix A, B, C. The dimensions g_1, g_2, g_3 are obtained by calculating the diagonal singular values of G_1, G_2, G_3 in proportion. In the process of reducing the dimensions, the proportion of excluded singular values is set to 50% in this paper. Therefore, the calculation of the core tensor after dimension reduction is as follows:

$$\begin{aligned} G' &= X \times_{\mathrm{u}} A_{r1}^T \times_{\mathrm{q}} B_{r2}^T \times_{\mathrm{a}} C_{r3}^T, \\ X' &= G' \times_{\mathrm{u}} A_{r1} \times_{\mathrm{q}} B_{r2} \times_{\mathrm{a}} C_{r3}. \end{aligned} \tag{9}$$

The three dimensions of the initial tensor X are $K_{\mathrm{u}}, K_{\mathrm{q}}, K_{\mathrm{a}}$, and the three dimensions of the approximate tensor X' after decreasing dimension are denoted by $N_{\mathrm{u}}, N_{\mathrm{q}}, N_{\mathrm{a}}$. The time complexity of the Tucker decomposition algorithm is proportional to the tensor dimension, which is expressed as $O(K_{\mathrm{u}} K_{\mathrm{q}} K_{\mathrm{a}})$. We previously used the clustering method to achieve the reduction of the original matrix, which reduced the cost of the Tucker decomposition greatly and improved the efficiency and precision.

3.2. Feature Composition Analysis of the Input Layer. There is a high degree of nonlinear correlation between the features in advertising data. Although the approximate tensor of the original tensor is reduced by the Tucker decomposition, it only reflects the information between the three characteristic dimensions of user, query, and ad. Other useful information in the data is not fully utilized for click-through rate estimates, such as the position of the advertisement on the page, the number of ads, and the age and gender of the user. This paper combines the features of <user, query, ad> after tensor reduction and other valid information in the log data as the object of feature learning. The composition of the input layer features is summarized as follows:

(1) *ID Feature.* ID feature uniquely identifies a class of entities in the actual click log, usually using a set of numeric strings to represent variables. For instance, "10110" can identify only one user group. The ID class used in this article has the UserID, QueryID, AdID, position, and the number of advertisements on the return page. UserID, QueryID, and AdID are collections of "virtual" ID classes that are obtained using k-means clustering and tensor dimension reduction.

(2) *Attribute Characteristics.* The ID class feature is a symbol that cannot be obtained from the new entity data and has weak generalization ability. Attribute features are used to describe a set of users, ad collections, etc., and have better generalization ability and apply to multiple instances. Therefore, it is necessary to attribute the property class as learning the input layer feature further. Commonly used attribute class features are user's URL, user's gender, user's age, and advertising time to trigger and query keywords.

(3) *Statistical Characteristics.* The statistical feature uses historical data statistics information to provide an estimate for the forecasting model. The statistical characteristics of the text consist of the number of advertising histories, the number of clicks on the advertising history, and the click-through rate after the advertising position normalization, denoted by Shows, Clicks, and COEC In the experiment, the input layer feature of the ASAE model is shown in Figure 2.

3.3. Study on CTR Prediction via Attention Mechanism Based on the Stacked Autoencoder

3.3.1. Attentional Factorization Machines. Since the attention mechanism has been introduced into neural network modelling, it has been widely used in many tasks. On the basis of FM, Figure 3 shows the neural network structure of attentional factorization machines (AFM). The input layer and the embedding layer are the same as the FM; the input features are represented with sparse features, and each nonzero feature item is embedded in the dense vector. Formally, let the set of nonzero features in the feature vector x be χ and the output of the embedding layer be $\lambda = \{v_i x_i\}_{i \in \chi}$. In the interaction layer, we can represent the output as a set of vectors:

$$f_{\mathrm{in}}(\lambda) = \left\{ \left(v_i \odot v_j \right) x_i x_j \right\}_{(i,j) \in R_x}, \tag{10}$$

where $\odot$ denotes the element wise product of two vector and $R_x = \{(i, j)\}_{i \in \chi, j \in \chi, j > i}$. By defining the interaction layer, we express FM under the neural network architecture. We compress $f_{\mathrm{in}}(\lambda)$ with a sum pooling. Then use the full connection layer to establish it and get the prediction score:

$$\hat{y} = p^T \sum_{(i,j) \in R_x} \left(v_i \odot v_j \right) x_i x_j + b, \tag{11}$$

where $p \in \mathbb{R}^k$ denotes the weights and $b \in \mathbb{R}$ denotes the bias for the prediction layer.

The attention mechanism has been widely used in many tasks. The idea is to allow different parts to contribute differently when compressing them to a single

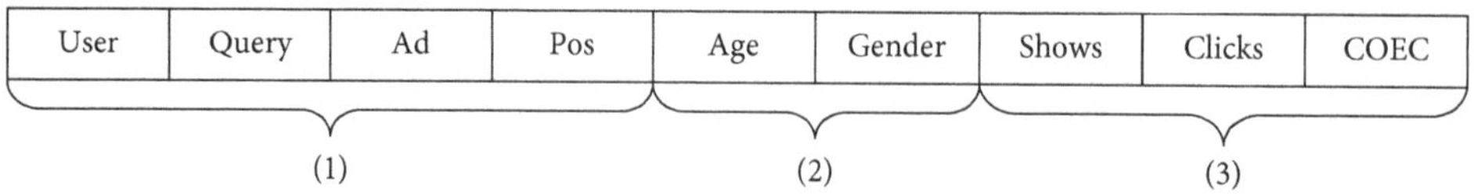

FIGURE 2: The features of the input layer.

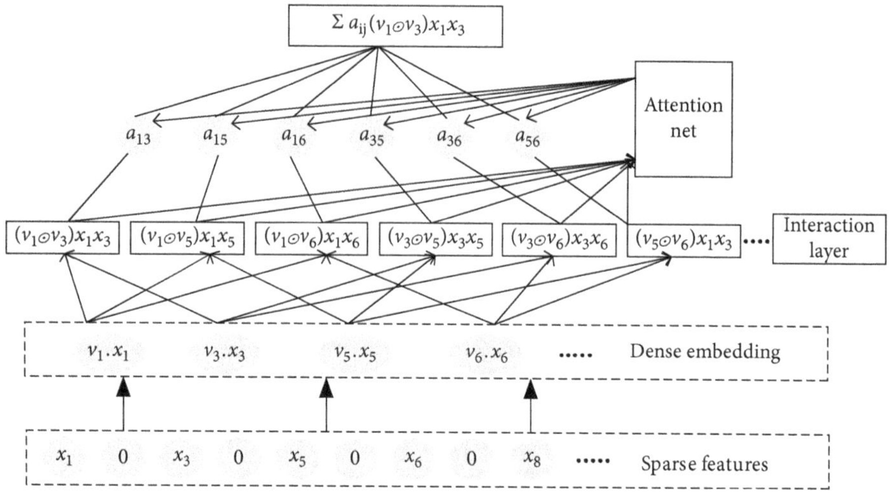

FIGURE 3: The structure of attentional factorization machines.

representation. We use attention mechanisms for feature interaction:

$$f_{\text{att}}(f_{\text{in}}(\lambda)) = \sum_{(i,j)\in R_x} a_{ij}\left(v_i \odot v_j\right)x_i x_j, \tag{12}$$

where a_{ij} is the attention score for feature interaction $\widehat{w}_{ij}$, and it can be interpreted as the importance of $\widehat{w}_{ij}$ in predicting goals. a_{ij} can be learned by minimizing the loss function, but the attention scores of interactions that never occur in training data cannot be estimated. In order to solve the generalization problem, we use the multilayer perceptron (MLP) to further parameterize the attention score, which we call the attention network. The input of the attention network is an interaction vector of two features and can encode their interaction information in the embedding space. In general, the attention network is defined as

$$a'_{ij} = h^T \operatorname{Re}LU\left(W\left(v_i \odot v_j\right)x_i x_j + b\right),$$
$$a_{ij} = \frac{\exp\left(a'_{ij}\right)}{\sum_{(i,j)\in R_x}\exp\left(a'_{ij}\right)}, \tag{13}$$

where $W \in \mathbb{R}^{s\times k}, b \in \mathbb{R}^s$, and $h \in \mathbb{R}^s$ are the model parameters and s is the hidden layer size of the attention network, which we call the attention factor. Rectifiers are used as the activation function for attention scores and show good performance empirically. The output of the attention layer is a k-dimensional vector that compresses all feature interactions in the embedding space by differentiating their importance. We give the overall formulation of attentional factorization machines as

$$y_{\text{AFM}}(x) = w_0 + \sum_{i=1}^{n} w_i x_i + p^T \sum_{i=1}^{n}\sum_{j=i+1}^{n} a_{ij}\left(v_i \odot v_j\right)x_i x_j, \tag{14}$$

where a_{ij} has been defined in Equation (13).

For the part of the attention network, which is a single-layer MLP, we apply L_2 regularization on the weight matrix W to prevent possible overfitting. In other words, the actual objective function we optimize is

$$L = \sum_{x\in \Upsilon} \left(\widehat{y}_{\text{AFM}}(x) - \widehat{y}(x)\right)^2 + \lambda \|W\|^2, \tag{15}$$

where Υ denotes the set of training instances and λ controls the regularization strength.

3.3.2. Stacked Autoencoder. The autoencoder (AE) [28] is a kind of the neural network model that automatically learns features from data without supervision. It consists of three network layers. The bottom is the input layer I, the middle of the hidden layer H, and the output layer O or reconstruction layer. The autoencoder architecture is shown in Figure 4. In Figure 4, w is the connection weight of the two layers and b is the bias. In the input layer and hidden layer, the AE model will convert input data to each node of the hidden layer. In the hidden layer and the reconstruction layer, the value of the nodes in the hidden layer is reconstructed and the output data are obtained.

The stacked autoencoder (SAE) [29] is a kind of network that consists of n AE stacks from the bottom to the top, as shown in Figure 5. The input data of the bottom AE are x. When the training of the bottom AE is finished, the feature

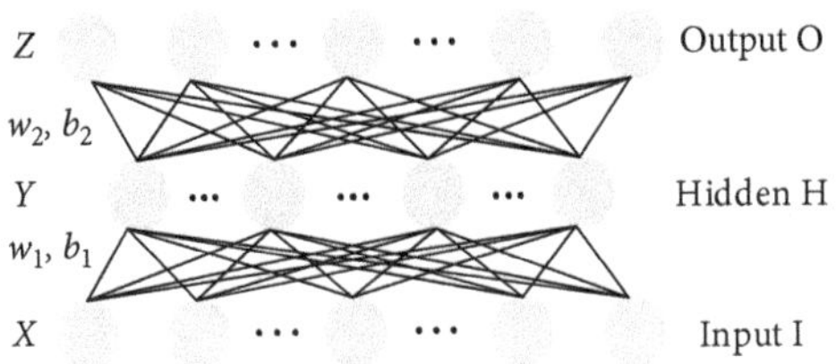

FIGURE 4: The architecture of the autoencoder.

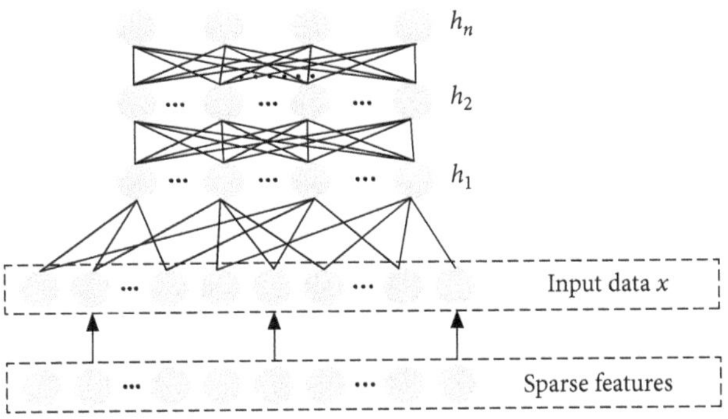

FIGURE 5: The structure of the stacked autoencoder.

of the hidden layer is obtained and can be represented by h_1. Then, h_1 is regarded as the input data of the second AE layer, which is trained and provides the features of the hidden layer and is represented by h_2. This process is repeated until h_n is obtained.

The related definition of the jth node in the hidden layer of AE can be described as follows: s_h is the number of nodes in the hidden layer (H) of the AE. w_{ji}^h is the connection weight between the jth node of hidden layer (H) and the ith node of input layer (I). b_j^h is the bias of the jth node in the hidden layer (H). $\mathrm{net}_j^h = b_j^h + \sum_{i=1}^{s_x} w_{ji}^h o_i^x$ is the weight sum of the input of the jth node in the hidden layer (H). o_j^h is the output value of the jth node in the hidden layer (H). The activation function of every neuron node is $\sigma(x) = 1/(1+e^{-x})$.

The output value of the jth node in the hidden layer (H) can be represented by the following formula:

$$o_j^h = f\left(\mathrm{net}_j^h\right) = \sigma\left(b_j^h + \sum_{i=1}^{s_x} w_{ji}^h o_i^x\right). \tag{16}$$

When the feature of the hidden layer (H) is decoded, the feature of the reconstruction layer O is obtained. The output value of the jth node in the reconstruction layer O can be represented by the following formula:

$$\begin{aligned} o_j^o &= g\left(\mathrm{net}_j^o\right) = g\left(b_j^o + \sum_{i=1}^{s_h} w_{ji}^o o_i^h\right), \\ &= \sigma\left(b_j^o + \sum_{i=1}^{s_h} w_{ji}^o\left(\sigma\left(b_i^h + \sum_{k=1}^{s_o} w_{ik}^h o_k^x\right)\right)\right). \end{aligned} \tag{17}$$

To easily calculate and deduce the formulae, we define the residual error δ_j^l of the jth node in the lth layer. The residual error δ_j^h of the neuron node of the reconstruction layer can be calculated using the following formula according to the chain rule:

$$\begin{aligned} \delta_j^h &= \frac{\partial J}{\partial \mathrm{net}_j^h} = \frac{\partial J}{\partial o_j^h}\frac{\partial o_j^h}{\partial \mathrm{net}_j^h}, \\ &= \sum_{i=1}^{s_o}\left(\frac{\partial J}{\partial \mathrm{net}_i^o}\frac{\partial \mathrm{net}_i^o}{\partial o_j^h}\right)\cdot\frac{\partial o_j^h}{\partial \mathrm{net}_j^h}, \\ &= \left(\sum_{i=1}^{s_o}\delta_i^o w_{ji}^o\right)\frac{\partial \delta\left(\mathrm{net}_j^h\right)}{\partial \mathrm{net}_j^h}, \\ &= \left(\sum_{i=1}^{s_o}\delta_i^o w_{ji}^o\right)\delta\left(\mathrm{net}_j^h\right)\left(1-\delta\left(\mathrm{net}_j^h\right)\right), \\ &= \left(\sum_{i=1}^{s_o}\delta_i^o w_{ji}^o\right)o_j^h\left(1-o_j^h\right). \end{aligned} \tag{18}$$

The parameters w_{ji}^l and b_j^l can be calculated by formulae (19) and (20):

$$\frac{\partial J}{\partial w_{ji}^l} = \frac{\partial J}{\partial \mathrm{net}_j^l}\frac{\partial \mathrm{net}_j^l}{\partial w_{ji}^l} = \delta_j^l \cdot o_i^{l-1}, \tag{19}$$

$$\frac{\partial J}{\partial b_j^l} = \delta_j^l. \tag{20}$$

The parameters w_{ji}^l and b_j^l can be updated as the following formulae, where ε is the learning rate:

$$\begin{aligned} w_{ji}^l &= w_{ji}^l - \beta\frac{\partial J}{\partial w_{ji}^l} = w_{ji}^l - \beta\cdot\delta_j^l\cdot o_i^{l-1}, \\ b_j^l &= b_j^l - \beta\frac{\partial J}{\partial b_j^l} = b_j^l - \beta\cdot\delta_j^l. \end{aligned} \tag{21}$$

The SAE is a generative model that is composed of a stack of autoencoders. This method relies on the training algorithm of the autoencoder to initialize the parameters of a stacked autoencoder. Each new layer is stacked on top of the current autoencoder. The process gradually refines the previously learned information and further discovers more complex features. After this, a dense real-value feature vector is generated, which is finally fed into the sigmoid function for CTR prediction:

$$y_{\mathrm{SAE}} = \sigma\left(W^{h+1}\cdot X^h + b^{h+1}\right), \tag{22}$$

where W is the model weight, b is the bias, and h is the number of hidden layer.

This paper selects the square error as the objective function and adopts the gradient descent [30, 31] to train the parameters, and the objective function can be described by the following formula:

$$J(X,O) = \frac{1}{2}\sum_{i=1}^{n}\left\|X^{(i)} - O^{(i)}\right\|^2. \tag{23}$$

3.3.3. ASAE Model. The ASAE model consists of two components, AFM component and SAE component, which share the same input. The graphical model of the ASAE model is shown in Figure 6. It is able to learn feature interactions of all orders in an end-to-end manner, without any feature engineering besides raw features. x_i is fed in AFM component to model order-2 feature interactions and distinguish their importance. x_i is fed in SAE component to model high-order feature interactions, and it can generalize better to unseen feature combinations through low-dimensional dense embedding learned for the sparse features. All parameters are trained jointly for the combined prediction model:

$$\hat{y} = \text{sigmoid}\left(y_{\text{AFM}} + y_{\text{SAE}}\right), \tag{24}$$

where $\hat{y} \in (0, 1)$ is the predicted CTR, y_{AFM} is the output of the AFM component, and y_{SAE} is the output of the SAE component.

4. Experiments

4.1. Datasets. We perform experiments with two publicly accessible datasets: Frappe [32] and SIGKDD Cup2012 track2. The Frappe dataset has been used for context-aware recommendation. It contains 96,215 app usage logs of users under different contexts. Each log contains 8 context variables, including app ID, user ID, city, and daytime. We convert each log into a feature vector with one-hot encoding, resulting in 5,479 features in total. We split dataset into the training set and testing set using a random partition method by the ratio of 8 to 1. The target value of 1 indicates that the user has used the application in context.

The KDD2012 CUP track2 corresponding research question is based on the actual click data information to predict the click rate of the advertisement. The training dataset provided by the competition has a total of 149,639,105 records, and the size of 9.8 GB. In addition to the number of click and the number of displays, the test dataset is consistent with the training dataset, a total of 20,257,594 records, 1.28 GB in size. After data cleaning and data preprocessing, a total of 3.5 million samples were randomly selected from the candidate dataset for the experiment. Table 1 summarizes the statistics of the final evaluation datasets.

In the KDD2012 CUP track2 dataset, the samples of seven different scale datasets are 150000, 200000, 300000, 500000, 600000, 750000, and 1 million. The training data are grouped randomly, and the final result is the average of all the experimental results to ensure the reliability of the experimental results.

4.2. Evaluation Index. We use two evaluation metrics in our experiments: AUC (area under ROC) and Logloss (cross entropy). The curve in AUC usually means the receiver operating characteristic (ROC) [33], which is usually used to measure performance of two-class classifier. The CTR prediction is a classic binary classification method based on whether the advertising is clicked. The value of AUC is usually between [0.5, 1). The larger the value of AUC becomes, the more accurate the advertising CTR prediction is.

4.3. Baseline Models. We compare the ASAE model with the following methods that are designed for sparse data prediction:

FM [34]: FM is successfully applied to the recommended system and user response prediction task. FM explores feature interaction, which is effective on sparse data

FNN [20]: FNN is a FM-initialized feedforward neural network. It is able to capture high-order latent patterns of multifield categorical data.

CCPM [19]: convolutional click prediction model (CCPM) is based on convolution neural network. It can extract local-global key features from an input instance with varied elements, which can be implemented for single advertising impression and sequential advertising impression.

Deep cross [11]: it applies a multilayer residual network on a feature embedding cascade for learning feature interactions. This model is a deep neural network that automatically combines features to produce superior models.

Wide and deep [35]: this model combines a linear ("wide") model and a deep model. The deep part is a three-layer MLP that first concatenates feature embedding. The wide part (which is a linear regression model) is subject to design to incorporate cross features.

4.4. Analysis of Experimental Results. This section evaluates the ASAE model from two perspectives: (1) discussing the impact of relevant parameters and (2) comparing the ASAE model with five existing prediction models.

4.4.1. Impact of Parameters. Dropout [36] refers to the probability that a neuron is kept in the network. Dropout is a regularization technique to compromise the precision and the complexity of the neural network. We set the dropout to be 0.1 to 0.8. As shown in Figure 7, the optimal dropout ratio on Frappe is 0.3. The result shows that adding reasonable randomness to model can strengthen model's robustness.

The number of network layers h in the depth learning phase has a direct effect on the final estimate of the model. Therefore, this paper experimented with parameters to select the better combination of parameters. In the Frappe dataset, as presented in Figure 8, increasing number of hidden layers improves the performance of the models at the beginning. However, with the increasing of the number of hidden layers, the model performance is degraded. This phenomenon is because of overfitting. We can see from Figure 8, the highest AUC value is obtained when the number of hidden layers is 4 in Frappe dataset.

The number of iterations (iter) in the training phase have a direct effect on the final estimation of the model. In the KDD dataset, we used a set of sampled data training models

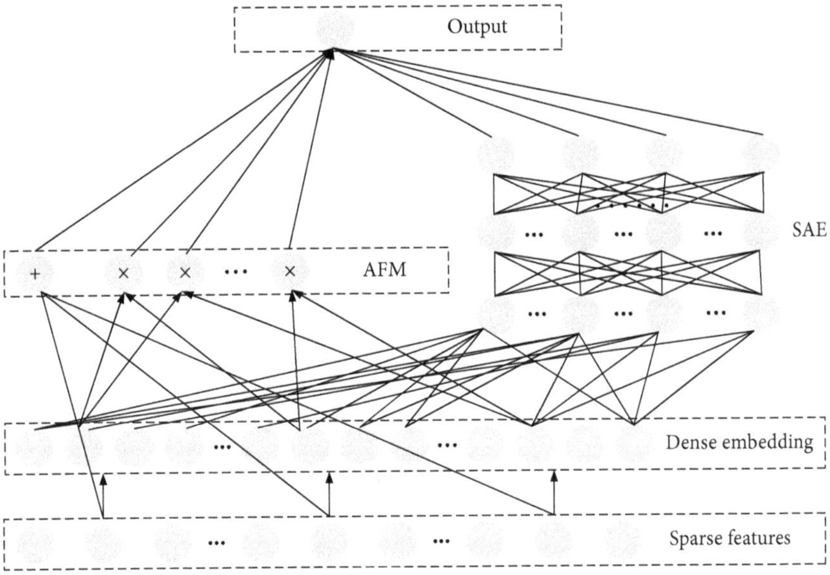

FIGURE 6: The architecture of ASAE.

TABLE 1: Statistics of the evaluation datasets.

Dataset	Instance	Feature	User	Item
Frappe	276,672	5,479	1,028	5,183
KDD	3,500,000	192,886	127,385	150,672

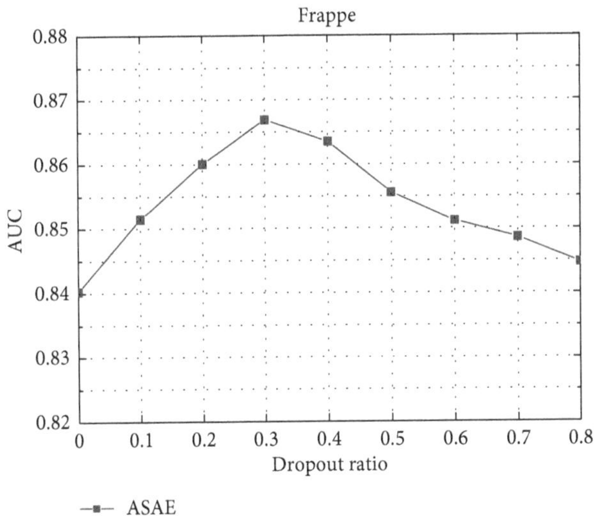

FIGURE 7: The influence of the dropout ratio.

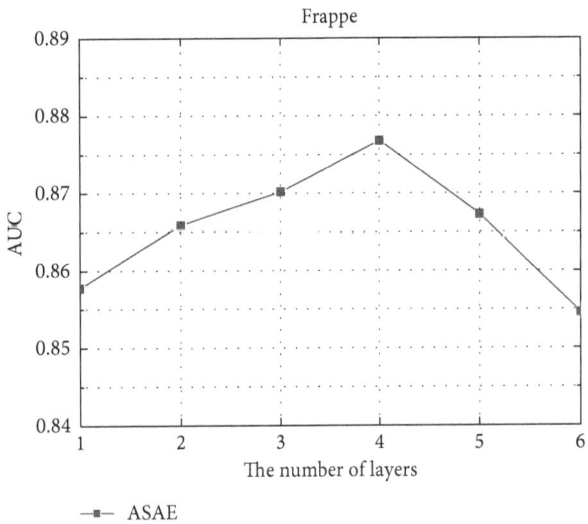

FIGURE 8: The influence of the number of hidden layers.

with a data size of 500,000 samples, tested on the test set, and we used it to select the best parameters. While fixing the network layer number ($h = 2, 3, 4, 5, 6$) of the model, we can analyze the effect of different iter on the model performance, and the results are shown in Table 2.

In accordance with Table 2, Figure 9 reflects the AUC change for different network hidden layers h and LR model iterations for iter. As seen in Figure 9, when the number of iterations is 90 to 120, the AUC values of several curves stabilized. Therefore, in the comparison experiment, 115 is

TABLE 2: The relationship between the number of network layers and iter.

iter	AUC				
	$h = 2$	$h = 3$	$h = 4$	$h = 5$	$h = 6$
10	0.6732	0.6717	0.6812	0.6771	0.6761
20	0.6834	0.6735	0.6968	0.6811	0.6814
30	0.6949	0.6812	0.7069	0.6869	0.6933
50	0.7132	0.7073	0.7231	0.7103	0.7058
70	0.7285	0.7191	0.7349	0.7215	0.7245
90	0.7372	0.7311	0.7482	0.7355	0.7401
100	0.7419	0.7429	0.7516	0.7481	0.7464
110	0.7432	0.7441	0.7543	0.7521	0.7508
130	0.7441	0.7452	0.7540	0.7529	0.7486
150	0.7456	0.7463	0.7524	0.7511	0.7502

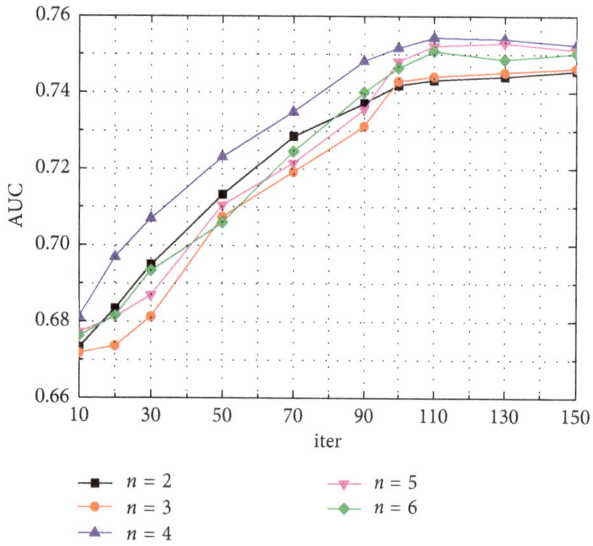

Figure 9: AUC comparison of different iterations and different network layers.

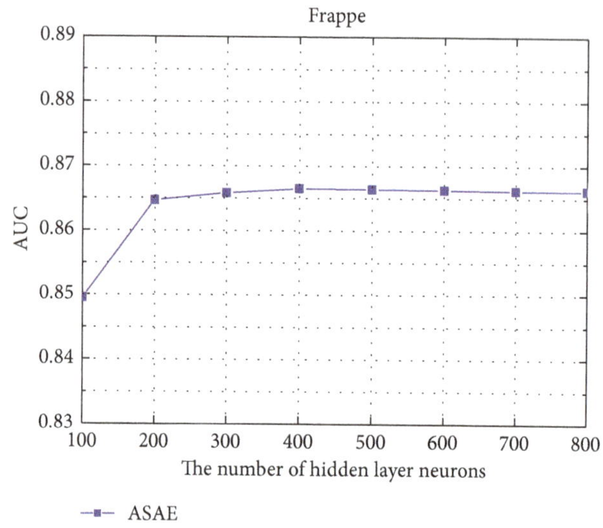

Figure 10: The influence of the number of neurons.

chosen as the number of iterations for training the prediction model. As shown in Figure 9, the curve fluctuates greatly with the change of iterations, and $h = 4$ is relatively stable, so we choose $h = 4$.

When other factors remain constant, the number of hidden layer units in the ASAE has a huge impact on network performance and the direct cause of the problem is extremely important. However, this figure does not have a general parameter adjustment method in theory. Therefore, in this part, we carry out experiments on the effect of the number of hidden layer neurons. As we can observe from Figure 10, increasing the number of neurons per layer does not always bring benefit. For example, when the number of neurons per layer is increased from 400 to 800, the ASAE performs stably. This is because the complicated model is easy to overfit. In our experiment, 200 to 400 neurons per layer is a good choice.

Table 3: The performance of AUC and Logloss in different models.

	Frappe					
	FM	FNN	CCPM	Deep cross	Wide and deep	ASAE
AUC	0.7935	0.8012	0.8178	0.8266	0.8293	0.8386
Logloss	0.03842	0.03473	0.03208	0.03127	0.03015	0.02813

Table 4: The AUC performance at different data sizes in KDD dataset.

Data size	AUC					
	FM	FNN	CCPM	Deep cross	Wide and deep	ASAE
150,000	0.6849	0.6941	0.6938	0.7135	0.7193	0.7212
200,000	0.6923	0.7012	0.7027	0.7263	0.7275	0.7355
300,000	0.7033	0.7186	0.7134	0.7397	0.7413	0.7482
500,000	0.7119	0.7235	0.7204	0.7434	0.7486	0.7547
600,000	0.7181	0.7398	0.7291	0.7503	0.7578	0.7672
750,000	0.7255	0.7490	0.7467	0.7549	0.7649	0.7793
1,000,000	0.7315	0.7517	0.7528	0.7652	0.7763	0.7981

4.4.2. Performance Comparison. We trained the models on the two datasets and evaluated the estimated results on the same test set. Tables 3–5 describe the estimated results for the different methods at different datasets.

Tables 3–5 show the overall performance. Compared with the other five methods, the ASAE model showed a better prediction effect. As the data size increased, the accuracy rose and the logloss declined.

FM: this model was successfully applied to the user response prediction task. It explores feature interaction, which is effective on sparse data. However, this model is limited in mining high-order latent patterns or learning quality feature representations. As shown in Tables 3–5, the performance of this model is worst in all comparison models.

FNN: FNN is a FM-initialized feedforward neural network. The FM pretraining strategy results in some limitations, such as the embedding parameters might be over affected by FM and the efficiency is reduced by the overhead introduced by the pretraining stage. From Tables 3–5, we can see that the performance of FNN ranked fifth.

CCPM: this model is based on convolution neural network for single and sequential advertising impression. However, this model highly relies on feature alignment and is a lack of interpretation. Thus, as shown in Tables 3–5, the performance of this model ranked fourth.

Deep cross: the deep cross is the deepest method that stacks 10 layers above the embedding layer in all compared methods. From Tables 3–5, we can see that the performance of this model ranked third due to the problems of overfitting.

TABLE 5: The Logloss performance at different data sizes in KDD dataset.

Data size	Logloss					
	FM	FNN	CCPM	Deep cross	Wide and deep	ASAE
150,000	0.02967	0.02925	0.02941	0.02826	0.02714	0.02733
200,000	0.02898	0.02893	0.02882	0.02793	0.02665	0.02704
300,000	0.02886	0.02872	0.02876	0.02768	0.02657	0.02658
500,000	0.02875	0.02864	0.02872	0.02759	0.02652	0.02639
600,000	0.02783	0.02715	0.02713	0.02674	0.02645	0.02630
750,000	0.02752	0.02739	0.02728	0.02631	0.02627	0.02543
1,000,000	0.02677	0.02642	0.02634	0.02628	0.02606	0.02501

Wide and deep: wide and deep combines a linear model and a deep model. It learns high- and low-order feature interactions. There is a need for expertise feature engineering on the input to the "wide" part. Thus, as shown in Tables 3–5, the performance of this model ranked second.

ASAE: the ASAE model performed best. The reasons are as follows. (1) The input of the model exploits dimension reduction based on decomposition and reduces the sparseness of data. (2) The model takes advantage of the attention mechanism in neural network modelling and improves FM to make feature interactions contribute differently to the prediction. (3) We use improved FM for low-order feature interactions, and stacked autoencoder is used for high-order feature interactions. The model more effectively mines the relationship between features, which can improve the CTR.

5. Conclusions

In this paper, based on the search advertising click data, we proposed a sparse feature learning method for advertising data from the perspective of feature learning (DLSAE). We used the reduced dimension method to cluster similar advertisements, queries, and users and established a three-dimensional tensor model for the trial after dimension reduction. Then the low-order approximate tensor was obtained using the Tucker decomposition. Aiming at the highly nonlinear relation between the features, we proposed a hybrid model (ASAE) for advertising CTR estimation based on the stacked autoencoder from the perspective of feature learning. The ASAE model trains a deep component and an AFM component jointly. Performance improved based on these advantages. First, this model does not need any pre-training. Second, it learns both high- and low-order feature interactions, introduces a sharing strategy of feature embedding, and more effectively mines the relationship between features. Last but not least, the proposed model distinguishes the importance of features and makes click-through rate predictions more accurate. More importantly, the importance of feature interactions is automatically learned from the data with any human-domain knowledge. We conducted extensive experiments in two datasets to compare the effectiveness of ASAE with other models. There are two interesting directions for future study. One is exploring a convolutional click prediction model based on CNN for single and sequential advertising impression. And another we are interested in exploring the pooling for recurrent neural networks (RNNs) for sequential data modelling.

Conflicts of Interest

The authors declare no conflicts of interest.

Acknowledgments

This work was supported by the National Natural Science Foundation of China (nos. 61572301 and 61772321), the Innovation Foundation of Science and Technology Development Center of Ministry of Education and New H3C Group (2017A15047), CERNET Innovation Project (NGII20170508), and the Open Research Fund from Shandong Provincial Key Laboratory of Computer Network (no. SDKLCN-2016-01).

References

[1] H. T. Cheng, L. Koc, J. Harmsen et al., "Wide & deep learning for recommender systems," in *Proceedings of Workshop on Deep Learning for Recommender Systems*, pp. 7–10, ACM, Boston, MA, USA, September 2016.

[2] O. Chapelle, R. Rosales, and R. Rosales, "Simple and scalable response prediction for display advertising," *ACM Transactions on Intelligent Systems and Technology*, vol. 5, no. 4, pp. 1–34, 2015.

[3] T. Graepel, T. Borchert, and R. Herbrich, "Web-scale Bayesian click-through rate prediction for sponsored search advertising in Microsoft's Bing search engine," in *Proceedings of International Conference on Machine Learning*, pp. 13–20, Omnipress, Haifa, Israel, June 2010.

[4] M. Richardson, E. Dominowska, and R. Ragno, "Predicting clicks: estimating the click-through rate for new ads," in *Proceedings of 16th International Conference on World Wide Web*, Alberta, Canada, May 2007.

[5] X. He, L. Liao, H. Zhang, L. Nie, X. Hu, and T.-S. Chua, "Neural collaborative filtering," in *Proceedings of 26th International Conference on World Wide Web*, Perth, Australia, April 2017.

[6] H. B. McMahan, D. Golovin, S. Chikkerur et al., "Ad click prediction: a view from the trenches," in *Proceedings of 19th ACM SIGKDD International Conference on Knowledge Discovery and Data Mining*, Chicago, IL, USA, August 2013.

[7] M. J. Effendi and S. A. Ali, "Click through rate prediction for contextual advertisement using linear regression," *International Journal of Computer Science and Information Security*, vol. 14, no. 11, 2017.

[8] O. Chapelle, "Modeling delayed feedback in display advertising," in *Proceedings of 20th ACM SIGKDD International Conference on Knowledge Discovery and Data Mining*, New York, NY, USA, August 2014.

[9] K. Dave and V. Varma, *Predicting the Click-through Rate for Rare/New Ads*, Center for Search and Information Extraction Lab International Institute of Information Technology, Hyderabad, India, 2010.

[10] X. He, S. Bowers, J. Q. Candela et al., "Practical lessons from predicting clicks on ads at Facebook," in *Proceedings of Eighth International Workshop on Data Mining for Online Advertising*, New York, NY, USA, August 2014.

[11] Y. Shan, T. R. Hoens, J. Jiao, H. Wang, D. Yu, and J. Mao, "Deep crossing: web-scale modeling without manually crafted combinatorial features," in *Proceedings of 22nd ACM SIGKDD International Conference on Knowledge Discovery and Data Mining*, pp. 255–262, San Francisco, CA, USA, June 2016.

[12] S. Rendle, "Factorization machines with LIBFM," *ACM Transactions on Intelligent Systems and Technology*, vol. 3, no. 3, p. 57, 2012.

[13] Y. Juan, Y. Zhuang, W.-S. Chin, and C.-J. Lin, "Field-aware factorization machines for CTR prediction," in *Proceedings of 10th ACM Conference on Recommender Systems*, Boston, MA, USA, September 2016.

[14] G. E. Hinton, S. Osindero, and Y.-W. Teh, "A fast learning algorithm for deep belief nets," *Neural Computation*, vol. 18, no. 7, pp. 1527–1554, 2006.

[15] Y. LeCun, Y. Bengio, and G. Hinton, "Deep learning," *Nature*, vol. 521, no. 7553, pp. 436–444, 2015.

[16] A. Graves, A. Mohamed, and G. Hinton, "Speech recognition with deep recurrent neural networks," in *Proceedings of IEEE International Conference on Acoustics, Speech and Signal Processing (ICASSP)*, Vancouver, Canada, March 2013.

[17] A. Krizhevsky, I. Sutskever, and G. E. Hinton, "Imagenet classification with deep convolutional neural networks," *Communications of the ACM*, vol. 60, no. 6, pp. 84–80, 2012.

[18] Y. Shen, X. He, J. Gao, L. Deng, and G. Mesnil, "A latent semantic model with convolutional-pooling structure for information retrieval," in *Proceedings of 23rd ACM International Conference on Conference on Information and Knowledge Management*, Shanghai, China, November 2014.

[19] Q. Liu, F. Yu, S. Wu, and L. Wang, "A convolutional click prediction model," in *Proceedings of 24th ACM International on Conference on Information and Knowledge Management*, Melbourne, Australia, October 2015.

[20] W. Zhang, T. Du, and J. Wang, "Deep learning over multi-field categorical data," in *Proceedings of European Conference on Information Retrieval*, Padua, Italy, March 2016.

[21] X. He and T.-S. Chua, "Neural factorization machines for sparse predictive analytics," in *Proceedings of 40th International ACM SIGIR Conference on Research and Development in Information Retrieval*, Tokyo, Japan, August 2017.

[22] M. Genzel and G. Kutyniok, "A mathematical framework for feature selection from real-world data with non-linear observations," http://arxiv.org/abs/1608.08852, 2016.

[23] J. Chen, H. Zhang, X. He, L. Nie, W. Liu, and T.-S. Chua, "Attentive collaborative filtering: multimedia recommendation with item-and component-level attention," in *Proceedings of 40th International ACM SIGIR Conference on Research and Development in Information Retrieval*, Tokyo, Japan, August 2017.

[24] L. Chen, H. Zhang, J. Xiao et al., "SCA-CNN: spatial and channel-wise attention in convolutional networks for image captioning," http://arxiv.org/abs/1611.05594, 2016.

[25] J. A. Hartigan and M. A. Wong, "Algorithm AS 136: A k-means clustering algorithm," *Journal of the Royal Statistical Society: Series C (Applied Statistics)*, vol. 28, no. 1, pp. 100–108, 1979.

[26] T. G. Kolda and J. Sun, "Scalable tensor decompositions for multi-aspect data mining," in *Proceedings of Eighth IEEE International Conference on Data Mining*, pp. 363–372, Pisa, Italy, December 2008.

[27] A. Szwabe, P. Misiorek, and M. Ciesielczyk, "Tensor-based modeling of temporal features for big data CTR estimation," in *Proceedings of International Conference: Beyond Databases, Architectures and Structures*, pp. 16–27, Ustron, Poland, April 2017.

[28] J. Chen, B. Sun, H. Li, H. Lu, and X.-S. Hua, "Deep CTR prediction in display advertising," in *Proceedings of 2016 ACM on Multimedia Conference-MM'16*, New York, NY, USA, October 2016.

[29] P. Vincent, H. Larochelle, I. Lajoie et al., "Stacked denoising autoencoders: learning useful representations in a deep network with a local denoising criterion," *Journal of Machine Learning Research*, vol. 11, pp. 3371–3408, 2010.

[30] M. Jahrer, A. T¨oscher, J.-Y. Lee, J. Deng, H. Zhang, and J. Spoelstra, "Ensemble of collaborative filtering and feature engineered model for click through rate prediction," in *Proceedings of KDD Cup 2012 Workshop*, Beijing, China, August 2012.

[31] W. S. Chin, Y. Zhuang, Y. C. Juan et al., "A learning-rate schedule for stochastic gradient methods to matrix factorization," in *Advances in Knowledge Discovery and Data Mining*, Springer International Publishing, Cham, Switzerland, pp. 442–455, 2015.

[32] L. Baltrunas, K. Church, A. Karatzoglou, and N. Oliver, "Frappe: understanding the usage and perception of mobile app recommendations in-the-wild," http://arxiv.org/abs/1505.03014, 2015.

[33] T. Akimova, M. H. Levine, U. H. Beier, and W. W. Hancock, "Standardization, evaluation, and area-under-curve analysis of human and murine treg suppressive function," in *Methods in Molecular Biology*, pp. 43–78, Springer, Berlin, Germany, 2016.

[34] S. Rendle, "Factorization machines," in *Proceedings of International Conference on Data Mining*, pp. 995–1000, Vancouver, Canada, December 2011.

[35] H. Guo, R. Tang, Y. Ye, Z. Li, and X. He, "DeepFM: a factorization-machine based neural network for CTR prediction," in *Proceedings of Twenty-Sixth International Joint Conference on Artificial Intelligence*, pp. 1725–1731, Melbourne, Australia, August 2017.

[36] N. Srivastava, G. Hinton, A. Krizhevsky, and R. Salakhutdinov, "Dropout: a simple way to prevent neural networks from overfitting," *Journal of Machine Learning Research*, vol. 15, no. 1, pp. 1929–1958, 2014.

18

The Causality Research between Syndrome Elements by Attribute Topology

Tao Zhang,[1,2] **Mengqi Liu**,[1,2] **and Wenyuan Liu**[1]

[1]*School of Information Science and Engineering, Yanshan University, No. 438, Hebei Avenue, 066000 Qinhuangdao, China*
[2]*Hebei Key Laboratory of Information Transmission and Signal Processing, No. 438, Hebei Avenue, 066000 Qinhuangdao, China*

Correspondence should be addressed to Tao Zhang; zhtao@ysu.edu.cn

Academic Editor: Jenny M. Wilkinson

Background. The traditional Chinese medicine (TCM) is an empirical medical system and has its own diagnosis and treatment method. The syndrome elements are atoms to modern TCM diagnosis proposed by Professor Zhu Wenfeng. Researching and analyzing the syndrome element system is one of the active issues for TCM research. At present, most related researches focus on the correlativity and hierarchical relationship of the diseases and symptoms, but the causality researches between syndrome elements themselves have not been reported so far. *Methods.* To explore the causality between syndrome elements, a method named causality by attribute topology (CAT) is proposed. Based on the subordinate relations in attribute topology, the inference method analyzes and reasons the dependency relationship between the sets of objects which contain attributes. Through the removal of attributes in the attribute topology, the formal context is updated constantly. Thus, the causal relationship among the attributes is deduced. In this method, 500 records are mathematically transferred to a binary context for syndrome element analysis. Through the analysis and verification of the potential causal relationship between the syndrome elements, knowledge discovery of the diagnostic data of traditional Chinese medicine based on attribute topology structure diagram is conducted. *Results.* This paper has verified the causal transformation between these syndrome elements. The experimental results between the female group data and the male group data show that different genders have different characteristics and relations of syndrome elements. The experimental results are basically consistent with the traditional Chinese medicine theory. *Conclusion.* The experiment shows that causality by attribute topology (CAT) is feasible to describe the causality between TCM syndrome elements. Further research on possible knowledge discovery in TCM diagnostic data should be conducted through the analysis of the potential causal relationship between TCM diagnostic data and each syndrome element.

1. Introduction

In the field of traditional Chinese medicine (TCM), treatment based on syndrome differentiation is the basis for preventing and treating diseases [1, 2]. As the premise of TCM treatment, the accuracy of syndrome differentiation will have a necessary influence on the effect of treatment [3, 4]. The fundamental inference methods of traditional Chinese diagnostics include the following: to infer inner changes from outer phenomenon, to deduce overall status from partial changes, and to identify syndromes in the standard of a healthy person [5]. These are classic methods for the discovery of syndrome elements in TCM, which have been fully studied by Zhu [6] and applied to clinical diagnosis by Hong [7, 8].

Zhu [9, 10] introduced the term "syndrome element" based on the research of syndrome differentiation and quantitative correlation relation between pairs of syndrome elements in TCM. "Syndrome element" was defined [11] as the basic element of syndrome differentiation; the identification of "syndrome" to determine disease location and disease-natures; the basic element of "syndrome name." Following the previous studies, he further studied the syndrome elements of disease-natures and disease location and proposed a novel system of syndrome differentiation based on syndrome elements [12]. This new system identifies syndrome elements from clinical symptoms and then determines the syndrome name according to the identified syndrome elements [13]. Therefore, the relationship between syndromes, syndrome elements, and the syndrome name has become the focus of the study on the syndrome element system [14].

Zhu [6, 15, 16] obtained the standard weights between syndromes and syndrome elements by double frequency weight scissors fork algorithm. Li Candong [17–22] discussed the correlation between five viscera identification and facial lesion distribution. Five viscera have a relative position in the face. The disease location of puberal acne is closely related to liver and kidney. Xiong Liping [23] analyzed a lot of cases and found that syndrome elements have an influence on syndromes. Dai [24] found and verified that there is a relationship between pale tongue and some syndrome elements. Hong analyzed the relationship between syndromes, syndrome elements, and syndrome names by the principle of attribute partial order and formed a syndrome analysis system which further standardized the syndrome differentiation system. According to the theory of traditional Chinese medicine, there is a certain causal relationship between syndrome elements, such as the syndrome element Yin Deficiency and the syndrome element Exterior which are both the causes of the syndrome element Fire-Heat. However, the mathematical analysis of the causal relationship between syndrome elements has not been reported yet.

As a branch of attribute partial order, attribute topology [25–28] is a tool focused on formal concept analysis [29, 30], cognitive computing, and relationship analysis [31–34]. In this paper, we propose a causal inference method by attribute topology under the representation framework of attribute partial order graphs. The method is applied to clinical data analysis, and the causal relationship between syndrome elements in clinical data is derived, which will be the basis for further knowledge discovery in syndrome elements system [35–38].

2. Materials and Methods

2.1. Attribute Topology. Attribute topology (AT) and attribute partial ordering graph belong to the framework of formal structure analysis, which is a graph description for formal context. Formal context, which acts as the research object and data representation, is an important basic aspect in FCA. Here are a few notions about formal context.

Definition 1. A formal context $K = (G, M, I)$ consists of two sets G and M and a relation I between G and M. The elements of G are called the objects and the elements of M are called the attributes of the context. In order to express that an object g is in a relation I with an attribute m, we write $g\text{I}m$ or $(g, m) \in I$ and read it as "the object g has the attribute m".

Definition 2. In a formal context $K = (G, M, I)$, for $A \subseteq G$ of objects, $f(A) = \{m \in M \mid \forall g \in A, (g, m) \in I\}$. Correspondingly, for $B \subseteq M$ of attributes, $g(B) = \{g \in G \mid \forall m \in B, (g, m) \in I\}$.

From the perspective of graph theory, attribute topology shows a weighted graph that depicts the relationships between attribute pairs. Thus the storage method of the graph can be borrowed. This section carries out a description of adjacency matrix of AT from the perspective of inclusive relationship of attribute pairs.

Definition 3. In context $K = (G, M, I)$, $AT := (V, Edge)$ is defined as adjacency matrix of AT in which $V = M$ is the set of vertex in AT and Edge represents the weight of edges in AT. Edge is expressed as follows:

$$Edge\left(m_i, m_j\right) = \begin{cases} \varnothing, & g\left(m_i\right) \cap g\left(m_j\right) = \varnothing \\ \varnothing, & g\left(m_i\right) \cap g\left(m_j\right) = g\left(m_i\right) \\ g\left(m_i\right) \cap g\left(m_j\right), & others \end{cases} \quad (1)$$

2.2. Attribute Topology and Causal Analysis. By the definition of attribute topology, the attribute topology itself emphasizes the correlation between attributes. At the same time, the relationship between superordinate attributes (SPA) and subordinate attributes (SBA) provide a way for causal analysis.

Definition 4. In context $K = (G, M, I)$, $m_i \in M$ and $m_j \in M$, $i \neq j$. m_i is the subordinate attribute of m_j and satisfied $g(m_i) \subset g(m_j)$.

From the definition of SBA, Property 5 is included obviously.

Property 5. In context $K = (G, M, I)$, $m_i \in M$, $m_j \in M$, and m_i is the SBA of m_j; then m_i is a necessary condition for m_j.

Definition 6. In context $K = (G, M, I)$, $m_i \in M$, $m_j \in M$, and m_i is a necessary condition for m_j; then m_j is part cause of m_i and m_i is the result of m_j, recorded as $m_j \rightarrow m_i$.

Definition 7. In context $K = (G, M, I)$, $m_m, m_n, m_p, m_q, m_i \in M$ and $m_m \rightarrow m_i, m_n \rightarrow m_i, m_p \rightarrow m_i, m_q \rightarrow m_i$. There is no m_k, which makes $m_k \rightarrow m_i$ and then set $m_c = \{m_m, m_n, m_p, m_q\}$ is cause of m_i, the subsets of m_c is the part cause of m_i, and m_i is the result of m_c, recorded as $(m_m, m_n, m_p, m_q) \rightarrow m_i$.

Property 8. In context $K = (G, M, I)$, $m_i \in M$, $m_j \in M$, and $m_j \rightarrow m_i$; then $g(m_i) - g(m_j) = \Phi$ and $g(m_j) - g(m_i) \neq \Phi$.

Proof. $\because m_j \rightarrow m_i, \therefore g(m_i) \subseteq g(m_j), \therefore g(m_i) - g(m_j) = \Phi$, and $g(m_j) - g(m_i) \neq \Phi$. □

Definition 9. In context $K = (G, M, I)$, the vertex whose outdegree is 0 and the nonzero indegree is the leaf node and its attribute is called leaf attribute.

Property 10. In context $K = (G, M, I)$, $m_i \in M$ and m_i is a leaf attribute, and its set of causes is the set of all adjacent vertices in the attribute topology.

Proof. According to Definition 9, in context $K = (G, M, I)$, $m_i \in M$ and $m_j \in M$. m_i is a leaf attribute and m_j is the adjacent vertex of m_i. From Definition 4 and Definition 6,

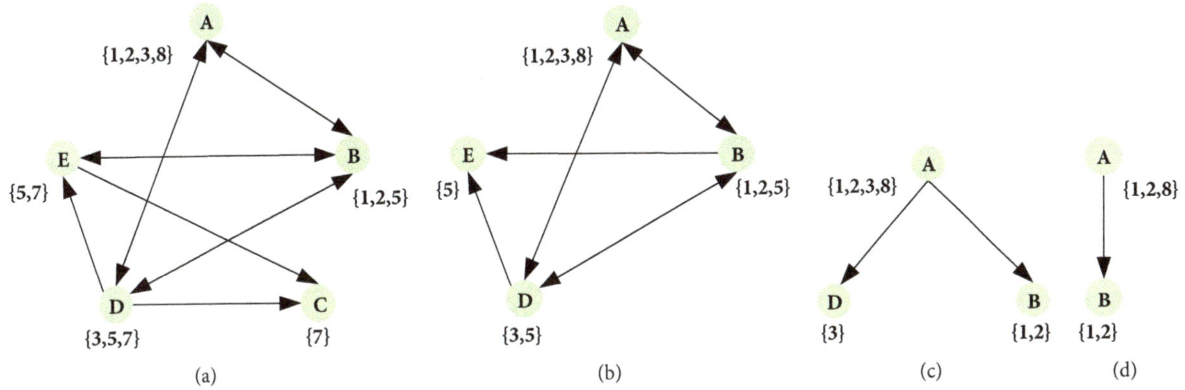

FIGURE 1: The attribute topology. (a) The AT of Table 1, (b) the AT of Table 2, (c) the AT of Table 3, and (d) the AT of Table 4.

TABLE 1: The original formal context.

	A	B	C	D	E
1	X	X			
2	X	X			
3	X			X	
4					
5		X		X	X
6					
7			X	X	X
8	X				

the conclusion $m_j \to m_i$ is obtained obviously. So the set of causes is m_i. □

2.3. The Algorithm. According to the theory of the previous section, the algorithm of causal analysis by AT is designed as follows.

Step 1. Getting $AT := (V, Edge)$ by a context, if there are leaf nodes, proceed to Step 2; otherwise, proceed to Step 4;

Step 2. If there is $m_j \to m_k$, the set of causes m_c is calculated for m_k. m_c is a set of attributes that are not null in the set of matrices $Edge(m_i, m_j)$. Then get causality $m_c \to m_i$.

Step 3. Update the context, $K = (G, M, I)$, $G = G - g(m_i)$, $M = M - f(m_i)$, and $I = G \times M$, and then go to Step 1, until there is no relationship $I = G \times M$.

Step 4. Finish.

Here is an example.

For a context as Table 1, its AT is Figure 1.

In Figure 1(a), the attribute C is a leaf attribute obviously, and the cause of C is $\{D, E\}$. As a result of that, there is $(D, E) \to C$. Update the AT by Step 3, $G = G - g(C) = \{1, 2, 3, 4, 5, 6, 7, 8\} - \{7\}$. Then the updated context is Table 2 and its AT is Figure 1(b).

TABLE 2: The first updated formal context.

	A	B	D	E
1	X	X		
2	X	X		
3	X		X	
5		X	X	X
8	X			

TABLE 3: The second updated formal context.

	A	B	D
1	X	X	
2	X	X	
3	X		X
8	X		

TABLE 4: The third updated context.

	A	B
1	X	X
2	X	X
8	X	

In Figure 1(b), the attribute E is a leaf node and can conclude $(B, D) \to E$ by Property 10. Update Table 2, the context shown in Table 3, and the AT shown in Figure 1(c).

In Figure 1(c), the attribute D is a leaf and the cause of D is $\{A, B\}$. Then get the following causality: $(A) \to D$. Update Table 3, and then get the context shown in Table 4 and the AT shown in Figure 1(d).

In Figure 1(d), the attribute B is a leaf and the cause of B is $\{A\}$. Then get the following causality: $(A) \to B$. And the causality between all attributes in the attribute topology can be inferred; there is $(D, E) \to C$, $(B, D) \to E$, $(A) \to D$, and $(A) \to B$.

Table 5: The 10 most used syndrome elements.

Gender	Half-exterior half-interior	Cold	Fire-heat	Yin deficiency	Yang deficiency	Yang floating	Dampness	Phlegm	External wind

Table 6: Data analysis.

No.	Gender	Exterior	Half-exterior Half-interior	Cold	Fire-Heat	Yin Deficiency	Yang Deficiency	Yang Hyperactivity	Yang Floating
17	Female	1345.77	5585.71	0.11	0.09	634.36	5647.99	345.91	0.05
53	Male	16.35	0.00	35.06	100.86	123.87	141.78	56.02	19.08
54	Female	10.76	15.28	30.56	42.02	94.69	75.03	32.92	0.00
98	Female	10.76	15.28	30.56	42.02	94.69	75.03	32.92	0.00
119	Female	45.76	0.00	36.21	0.00	0.00	0.00	44.91	0.00

2.4. Methods

2.4.1. Date Acquisition. The experimental data in this paper are from the clinical data, collected by the team of Hong [39]. In the system, for the clinical manifestations mapping to syndrome elements, 177 inquiry questions (related to symptoms or signs common in clinical diagnosis) are designed for males, while 194 inquiry questions are designed for females. 670 medical records were analyzed by Hong team. According to the method of Section 2.4.2 Data Processing, the data is filtered and 500 valid data are retained as the experimental data of this paper. A total of 500 cases are collected, of which are 300 males and 200 females and their age distribution is 40 to 70 years old. The syndrome element date we have measured in the experiment includes forty-five indicators: Yang Deficiency, Yang Hyperactivity, Exterior, Qi Deficiency, Qi Sinking, Qi Counterflow, Blood Stasis, Blood Cold, Stirring Wind, Retained Fluid, Fluid Depletion, External Wind, Summer Heat, Dryness, Food Accumulation, Gallbladder, Large Intestine, Stomach, Bladder, Small Intestine, Chest and Diaphragm, Sinew and Bone, and so on. The analytical methods involved in the experiment include analysis of eight principle syndrome differentiations [40], analysis of Qi-blood-fluid-humor syndrome differentiation [41], analysis of disease cause syndrome differentiation [42], and analysis of visceral syndrome differentiation [43] in the traditional Chinese medicine.

The data used in this experiment are collected according to the user's online examination in the diagnostic system. Due to some nonstandard behaviors in data storage, the data are incomplete, inconsistent, and noisy. In order to obtain high quality datasets to improve the accuracy of the algorithm, the first task is to preprocess the collected data.

Each row in the experimental data represents questionnaire results by a user. Each column is the measurements for all users under a certain syndrome element.

It is not convenient to display all the original data because the amount of data is large and involves a large number of syndrome elements. The 10 most used syndrome elements are shown in Table 5.

A total of 500 cases are collected and analyzed in four aspects. The collected data are expressed in the form of formal context. The causal relationships of syndrome elements can be obtained through the corresponding attribute topological graph. Figure 2 shows the overall analysis process.

2.4.2. Data Processing. The data processing method is as follows: (1) if all questions are answered identically by a user through online examination in the diagnostic system (namely all the values of the syndrome elements are the same), only one of these records is taken as valid data for the same answer in this paper (the same data is an invalid data for causal analysis), as shown in No.54 user and No.98 user in Table 6; (2) if some results of measurement are too extreme (i.e., values of all syndrome elements are either maximum or minimum), these results are excluded in data processing as certain patients have made extreme answers, as shown in No.17 user in Table 6; (3) if there are numerous 0 values in the diagnostic result of a user (possibly caused by mistakes in data collection), this record is invalid and should be excluded (it is less possible to obtain such a result through diagnostic system for syndrome elements are interrelated), as shown in No.113 user in Table 6.

As shown in No.53 in Table 6, the value of syndrome element Half-Exterior Half-Interior is 0. Considering that it is possible that the users do not have the relevant feature of this syndrome element, rather than the error caused by data extraction, the above situation has not been removed directly.

Since there is no normal value or normal range of the indicators, it is difficult to discover the above data only according to the size of the data rather than compare with the normal value. In order to reasonably indicate the normal values of the indicators, the method used in this paper is as follows: the average value of all data in each column is used as a critical value of the indicators, greater than or equal to this value has the index considered not normal, and less than this value is considered normal indicators. In order to analyze the impact of gender on the indicators more comprehensively and rationally in this paper, the above raw data is divided into two parts, one part male data and the other female data.

Because the data processed by causal inference method based on attribute topology and qualitative reasoning are formal context, the collected data is expressed in the form of the formal context. The index value corresponding to

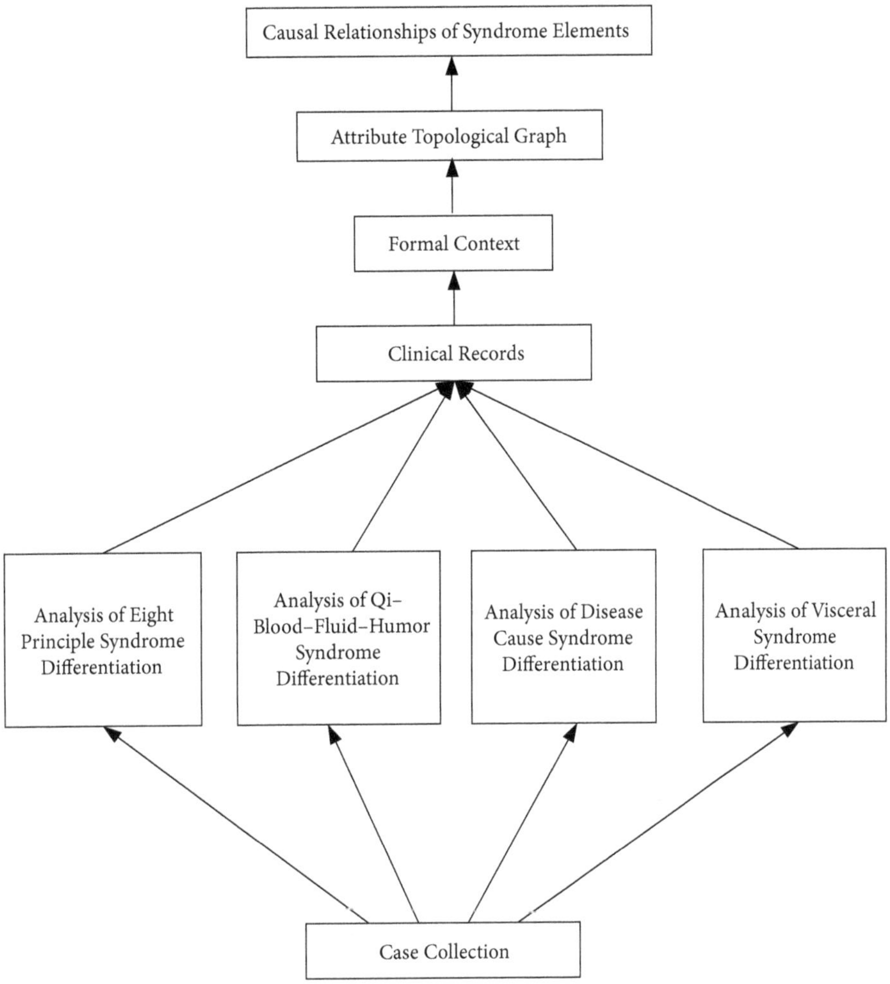

FIGURE 2: The overall analysis process.

TABLE 7: Part of the female-male data after purification.

No	Exterior	Half-exterior Half-interior	Cold	Fire-Heat	Yin Deficiency	Yang Deficiency	Yang Hyperactivity	Yang Floating
2	0	0	1	1	0	0	0	0
3	0	0	0	0	0	0	0	1
5	0	0	0	0	0	0	0	0
6	1	1	0	0	1	0	1	0
7	1	1	0	1	1	1	1	1
1	1	1	1	1	1	1	1	1
4	0	0	0	0	0	0	0	1
9	1	1	1	1	1	1	1	1
12	0	0	0	0	0	1	1	0
13	1	0	1	1	0	1	0	0

the attribute of all the date which is greater than or equal to the average value is assigned to 1, and the index value corresponding to the attribute which is smaller than the average value is assigned a value of 0. No.2, No.3, No.5, No.6, and No.7 in Table 7 show the data of female after processing. No.1, No.4, No.9, No.12, and No.13 in Table 7 show the data of male after processing. After the purification (remove duplicate data, remove the row with the same content, etc.) of objects and attributes by formal context, the number of male data changed to 300 sets and the female data changed to 200 sets.

2.4.3. The Cause Analysis. The number of syndromes in the experimental data is relatively large; consequently, the

TABLE 8: Original formal context.

No.	Exterior	Half-exterior Half-interior	Cold	Fire-Heat	Yin Deficiency	Yang Deficiency	Yang Hyperactivity	Yang Floating
1	1	0	0	1	1	0	0	0
2	0	0	0	0	0	0	0	1
3	0	0	0	0	0	0	0	0
4	1	1	1	0	0	1	0	1
5	1	1	1	0	1	1	1	1
6	0	1	0	0	0	0	1	0

TABLE 9: The formal context after purification.

No.	Exterior	Half-exterior Half-interior	Cold / Yang Deficiency	Fire-Heat	Yin Deficiency	Yang Hyperactivity	Yang Floating
1	1	0	0	1	1	0	0
2	0	0	0	0	0	0	1
3	1	1	1	0	0	0	1
4	1	1	1	0	1	1	1
5	0	1	0	0	0	1	0

attribute topological graph may be very complex if all the experimental data are converted into attribute topological graph. Taking it into account, this paper chooses the first eight syndrome elements of the first six objects in the inference process as a demonstration.

The original formal context after preconditioning is shown in Table 8.

It can be seen clearly from Table 8 that the values of each syndrome element in object 3 are 0, so object 3 is removed. Syndrome element Cold has the same content as syndrome element Yang Deficiency, so these two can be combined. The formal context after purification is shown in Table 9.

The formal context shown in Table 9 is transformed into the original attribute topological graph shown in Figure 3(a).

By analyzing the original attribute topological graph, the syndrome element Heat should be the starting syndrome element to infer its causal relationship with other syndrome elements. It can be concluded that the syndrome element Yin Deficiency and the syndrome element Exterior are the cause of the syndrome element Fire-Heat and the syndrome element Fire-Heat is the result.

After the first update, syndrome element Cold and syndrome element Exterior belong to the same set of objects $\{3, 4\}$, so this paper combines syndrome element Exterior and syndrome element Cold together. The resulting attribute topological graph is shown in Figure 3(b).

After the first update, the causal relationship is inferred: (Exterior, Yin Deficiency)→Fire-Heat. Remove the syndrome elements Fire-Heat, and update the object set of syndrome element Exterior and syndrome element Yin Deficiency. Secondly, the syndrome element Yin Deficiency was selected to judge the causal relationship between itself and other syndrome elements. The attribute topology is updated a second time, and the update results are shown in Figure 3(c). After updating twice, the causal relationship is inferred: (Exterior, Half-Exterior Half-Interior, Yang Hyperactivity, and Yang Floating) →Yin Deficiency. Execute the above update loop until all causal relationships are inferred.

3. Results and Discussion

Representation and determination of causality, the relationship between an event (the cause) and a second event (the effect), where the second event is understood as a consequence of the first, are a challenging problem [44]. From the experimental data study, the causality between the syndrome elements in female group is shown in Table 10. The causality between the syndrome elements in male group is shown in Table 11. For example, as shown in the third line of Table 11, Qi counterflow and insecurity of Qi are the causes of syndrome elements. Qi deficiency is the effect of syndrome elements. Generally speaking, Qi counterflow and insecurity of Qi are special characterization of Qi deficiency [45]. This association analysis of syndrome elements can show which syndrome elements appear more frequently and identify possible relationships between syndrome elements.

For ease of analysis, the female group data were expressed as the first set of data and the male data were represented as second set of data. It is found that the experimental results of the two groups of data were extremely different, not only in the relationship between the two syndrome elements, but also in the relationship between the syndrome elements of combination. Table 12 lists the comparison of the two sets of data.

Due to the two groups of data involved in the number of syndrome elements which is relatively large, Table 12 does not list all the syndrome elements involved. Table 12 lists relatively few syndrome elements that are not involved, as

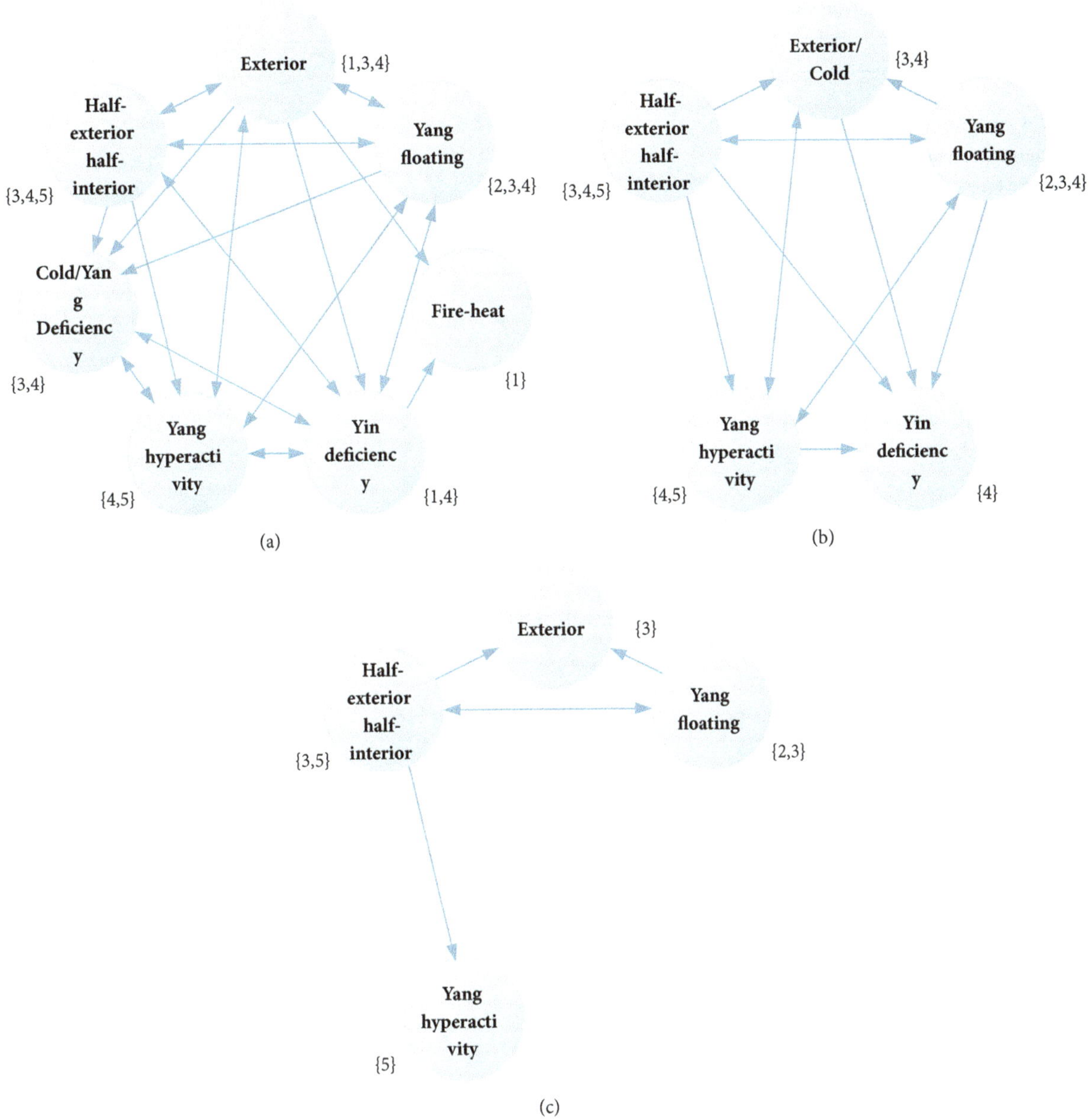

FIGURE 3: Attribute topological graph. (a) The original attribute topological graph by the conversion of Table 9, (b) the first updated attribute topological graph, and (c) the second updated attribute topological graph.

well as the ratio of its total syndrome elements. It turns out, a total of 12 syndrome elements are not involved in the first group, accounting for 26.67% of the total syndrome elements. There are 12 syndrome elements in the second group which are not involved, accounting for 8.89% of the total syndrome elements. Obviously, there are little syndrome elements which are not involved in the second group compared with the first group; the proportion of the total syndrome element is 8.89%. The syndrome elements which are not involved in the first group account for a high proportion of total syndrome elements, but only 26.67%.

In order to compare two groups of data more intuitively, the inference group number is taken as an index for comparative analysis, shown in Table 13.

The group of inference is the group of syndrome elements with causal relationship inferred from the causal relationship. Analysis shows that the two groups of data inference group are similar; the first group is 26 and the second group is 29. The proportion of syndrome elements involved in the two sets of data is also more than 70%.

These unequal results indicate that the male patient would be more likely to involve more syndrome elements. From these two groups, we can see that the contribution for

TABLE 10: The causality between the syndrome elements in female group.

Cause of syndrome elements	effect of syndrome elements
Heart spirit	Qi deficiency
Kidney, Fluid depletion	Heart
Blood deficiency, Uterus	Liver
Summer heat, Heart, Heart spirit, Retained fluid	Spleen
Lung, Fluid depletion, Food accumulation	Heart
Exterior	Yin deficiency
Fluid depletion, Large intestine	Fire-heat
Half-exterior half-interior	Essence deficiency
Yang floating	Qi sinking
Retained fluid, Small intestine, Bladder	Yang hyperactivity
Food accumulation, Blood cold, Heart spirit, Small intestine, Uterus	Stirring wind
Summer heat	Dryness
Skin	Fluid depletion
Water retention, Blood cold, Stirring blood, Uterus	Blood deficiency
Food accumulation	Small intestine
Summer heat	Half-exterior half-interior
Blood cold, Food accumulation, Retained fluid, Half-exterior half-interior	Lung
Blood heat, Stirring blood, Uterus	Blood cold
Water retention, Uterus	Stirring blood
Uterus	Summer heat
Blood heat, Stirring blood, Uterus	Blood cold
Water retention, Uterus	Stirring blood
Uterus	Food accumulation
Uterus	Water retention
External wind, Retained fluid, Lung, Skin	Exterior
Blood heat, Retained fluid, Uterus, Spleen	Lung
Yang floating, Blood heat, Uterus, Skin	External wind
Uterus	Skin
Uterus	Blood heat

syndrome elements research of male group was greater than that of female group. In the meantime, it can be inferred that the difference of data between male group and female group is the life habit, skin constitution, viscera function, hormone content, and so on. This kind of analysis would provide an objective basis for standardization of dialectic diagnosis.

Through the above analysis, we can see that the results of causal inference involve fairly comprehensive syndrome elements, which can be used as a data support for further causal analysis. Because of the more combinations of syndrome elements and the lack of theoretical basis for some combinations of syndrome elements, this paper mainly analyzes the relationship among various syndrome elements in the existing mature theory of traditional Chinese medicine: the relationship between the five internal organs. Five internal organs include heart, lung, spleen, liver, and kidney. The relationship between the five internal organs and their relationship in TCM is shown in Table 14.

Tables 15 and 16 give the inference results of two sets of data to analyze the relationship among five internal organs.

It can be seen from Tables 14–16 that although the experimental results are different from the traditional Chinese medicine theory, the whole is basically consistent.

The main cause for the difference is that all the syndrome elements are considered as a whole when proving the causality between each syndrome elements in this paper; however, TCM theory is only analyzed from the five internal organs. There is also a need to combine more clinical data as well as syndromes and syndrome elements for the causal relationship between other syndrome elements.

4. Conclusions

In this paper, a visual inference method of causal relation between TCM syndrome elements is proposed based on the theory of attribute topology. The main purpose of this

TABLE 11: The causality between the syndrome elements in male group.

Cause of syndrome elements	Effect of syndrome elements
Fire-heat, Dryness	Liver
Summer heat, Stirring blood	Blood deficiency
Qi counterflow, Insecurity of Qi	Qi deficiency
Half-exterior half-interior, Yang floating, Dampness, Summer heat, Qi sinking, Water retention, Heart spirit, Heart, Small intestine	Spleen
Food accumulation	Dampness
Phlegm, Dryness, Fluid depletion	External wind
Exterior, Lung	Liver
Summer heat, Blood heat	Yin deficiency
Summer heat, Qi counterflow	Fire-heat
Yang deficiency, Dryness, Essence deficiency, Qi sinking, Insecurity of Qi, Fluid depletion, Heart spirit, Small intestine, Large intestine, Bladder, Skin	Yang hyperactivity
Half-exterior half-interior, Uterus	Heart spirit
Yang deficiency, Yang floating, Dryness, Qi sinking, Insecurity of Qi, Retained fluid, Lung, Kidney, Bladder, Skin	Essence deficiency
Dryness, Fluid depletion, Bladder, Skin	Qi sinking
Yang floating, Gallbladder, Uterus	Yang deficiency
Skin	Fluid depletion
Qi counterflow, Insecurity of Qi, Lung	Dryness
Stirring wind, Insecurity of Qi, Lung	Kidney
Yang floating	Gallbladder
Bladder	Half-exterior half-interior
Food accumulation, Uterus	Small intestine
Cold, Blood cold, Uterus, Spleen	Lung
Meridian–collateral, Sinew and bone	Cold
Retained fluid	Qi counterflow
Insecurity of Qi, Bladder	Sinew and bone
Yang floating	Bladder
Stirring wind	Yang floating
Insecurity of Qi	Stirring wind
Blood cold, Meridian–collateral	Insecurity of Qi
Chest and diaphragm	Uterus

TABLE 12: The comparison of the two group of date sets.

Group number	Syndrome elements not involved	Ratio
First group	Cold, Dampness, Qi stagnation, Blood stasis, Insecurity of Qi, Stirring wind, Kidney, Gallbladder, Stomach, Sinew and bone, Chest and diaphragm, Meridian–collateral	26.67%
Second group	Qi stagnation, Blood stasis, Stirring blood, Stomach	8.89%

TABLE 13: The comparative analysis after adding inference group number.

Group name	Inference group number	Number of syndrome element	Ratio
First group	26	33	73.33%
Second group	29	41	91.11%

algorithm is to verify the causal transformation of TCM syndrome elements through clinical data collection. Through this experiment, we have preliminarily verified the causal transformation between these syndrome elements. This paper is discussed from the male-female perspective. As for future work, the analysis can further be expanded to account for other aspects of causality.

The next steps include the following: (1) to scale up clinical data collection and extend the mathematical expression of the relationship between TCM syndrome elements and promote the objectivity of TCM; (2) to discover the relationship between syndrome elements uninvolved in the classical literature from the clinical data for the development of the TCM syndrome element system.

TABLE 14: Five internal organs and their relationship.

The five internal organs	Relationship
Heart and Lung	The heart governs the blood and the lung governs Qi. Although the normal operation of blood is the heart leading, it must promote with the help of the Lung Qi.
Heart and Spleen	The heart governs the blood and the spleen governs the blood. The spleen can govern the blood if the function of the spleen is normal.
Heart and Kidney	The heart and kidney are interactive and inter-conditioned so as to maintain the relative balance of physiological functions
Liver and Spleen	The liver storing blood. The spleen controlling digestion as well as essence of water and grain to produce blood
Liver and Lung	The meridian and vessels of liver passes through the fat and is injected into the lung.
Liver and Kidney	The liver stores blood, and the kidney stores the essence. Liver blood needs to depend on the nourishment of kidney essence and the kidney essence needs the supplement of liver blood continuously, both are interdependent and mutual promotion.
Spleen and Kidney	Spleen Yang rely on warm nourishing of kidney yang to play the role of transportation and transformation.
Lung and Kidney	lung governing purification and descending and regulation of water passages so that water metabolism inferior to the kidney
Lung and Spleen	lung being reservoir of phlegm. Spleen being source of phlegm.

TABLE 15: The inference results of five internal organs in female group.

Cause of syndrome elements	Effect of syndrome elements
Kidney, Fluid depletion	Heart
Summer heat, Heart, Heart spirit, Retained fluid	Spleen
Lung, Fluid depletion, Food accumulation	Heart
Blood heat, Retained fluid, Uterus, Spleen	Lung

TABLE 16: The inference results of five internal organs in male group.

Cause of syndrome elements	Effect of syndrome elements
Half-exterior half-interior, Dampness, Summer heat, Qi sinking, Water retention, Heart spirit, Heart, Small intestine	Spleen
Exterior, Lung	Liver
Stirring wind, Insecurity of Qi, Lung	Kidney
Cold, Blood cold, Uterus, Spleen	Lung

Abbreviations

TCM: Traditional Chinese medicine
FCA: Formal concept analysis
AT: Attribute topology.

Conflicts of Interest

The authors declare no conflicts of interest.

Authors' Contributions

Tao Zhang proposed the thought and framework of this paper. Mengqi Liu implemented the thought with software, analyzed the data, and wrote this paper. Wenyuan Liu provided medical guidance of this paper. All authors read and approved the final manuscript.

Acknowledgments

This work is supported by the National Science Foundation of China (NSFC) under Grant no. 61603327 and the Science Foundation of Hebei Province under Grant nos. F2015203013 and F2016203443. It is also supported by the Youth Talent Support Program of Hebei. The authors gratefully acknowledge the supports. The authors are grateful to RenQiang for the helpful comments and suggestions.

References

[1] C. Li, "Discussion on clinical thinking of traditional Chinese medicine from, five differentiation," *Journal of Traditional Chinese Medicine*, vol. 14, no. 03, pp. 1–6, 2015.

[2] L. Jiang, B. Liu, Q. Xie et al., "Investigation into the Influence of Physician for Treatment Based on Syndrome Differentiation," *Evidence-Based Complementary and Alternative Medicine*, vol. 2013, Article ID 587234, 16 pages, 2013.

[3] G.-Z. Li, S. Sun, M. You, Y.-L. Wang, and G.-P. Liu, "Inquiry diagnosis of coronary heart disease in Chinese medicine based on symptom-syndrome interactions," *Chinese Medicine*, vol. 7, no. 1, article 9, 2012.

[4] X. Chen and C. Li, "Analysis of, syndrome , in traditional Chinese medicine," *Journal of Traditional Chinese Medicine*, vol. 15, no. 01, pp. 38–4145, 2016.

[5] W. Yi, "Discussion on teaching methods of diagnostics of traditional Chinese medicine," *Education of Chinese Medicine*, vol. 24, no. 3, pp. 35-36, 2005.

[6] W. Zhu, *Diagnostics of Traditional Chinese Medicine*, China Press of Traditional Chinese Medicine, 2004.

[7] "Cognition and practice on the factors of curative effect of moxibustion," *hinese Acupuncture and Moxibustion - Chin Acup Moxib*, vol. 20, no. 5, pp. 297-298, 2000.

[8] S. Li, "On the traditional Chinese medical four diagnostic in treatise on febrile diseases," *Traditional Chinese Medicine Journal*, vol. 10, no. 3, pp. 7–11, 2011.

[9] W. Zhu, "The establishment of syndrome differentiation based on syndrome elements," *Journal of Traditional Chinese Medicine University of Hunan*, vol. 24, no. 6, pp. 38-39, 2004.

[10] J. He, W. Zhu, B. Huang et al., "The practice of syndrome differentiation system and measurement diagnosis algorithm," *Traditional Chinese Medicinal Research*, vol. 21, no. 1, pp. 1–3, 2008.

[11] W. Zhu and H. Zhang, "The basic characteristics of "syndrome element"," *Chinese Journal of Basic Medicine in Traditional Chinese Medicine*, vol. 11, no. 1, pp. 17-18, 2001.

[12] W. Zhu and J. Yan, "The content and scientific significance of the new syndrome differentiation system," *Journal of Medicine and Philosophy*, vol. 26, no. 1, pp. 69-70, 2005.

[13] W. Zhu, "The significance of building a new system of, syndrome differentiation," *Journal of Zhejiang University of Traditional Chinese Medicine*, vol. 30, no. 2, pp. 135-136, 2006.

[14] B. Huang, W. Zhu, and J. Yan, "Using "double frequency weight scissors fork algorithm" to determine the contribution degree of syndrome differentiation," *World Chinese Medicine Journal*, vol. 3, no. 2, pp. 119-120, 2008.

[15] W. Zhu, B. Huang, and J. He, "New exploration on information processing method of traditional Chinese medicine syndrome differentiation-double frequency weight scissors fork algorithm," *Asia-Pacific Traditional Medicine*, vol. 3, no. 9, pp. 23–26, 2007.

[16] W. Zhu, J. Yan, J. He et al., "Double frequency weight scissors fork algorithm of traditional Chinese medicine syndrome differentiation," *Chinese Journal of Information on Traditional Chinese Medicine*, vol. 15, no. 1, pp. 108-109, 2008.

[17] C. Li, B. Gao, G. Cheng et al., "Correlation between adolescent acne vulgaris area and five internal organs," *Journal of Traditional Chinese Medicine*, vol. 48, no. 7, pp. 633-634, 2007.

[18] C. Li, L. Ji, Y. Lu, and H. Gan, "On the logical starting point of the theory of health cognition in traditional Chinese medicine," *China Journal of Traditional Chinese Medicine and Pharmacy*, vol. 26, no. 1, pp. 109–111, 2011.

[19] C. Li, *Syndrome Differentiation and Disease Discrimination*, 2010.

[20] C. Li, "On the clinical thinking of traditional Chinese medicine from, five discrimination," *Journal of Traditional Chinese Medicine*, vol. 14, no. 03, p. 1, 2015.

[21] C. Li, *Basic Characteristics of Disease*, 2006.

[22] L. Yang, C. Li, and D. Lin, "Study on the relationship between traditional Chinese medicine syndrome and renal function," *CHinese Archives of Traditional Chinese Medicine*, vol. 24, no. 8, pp. 1473-1474, 2006.

[23] L. Xiong, C. Li, W. Wu et al., "Study on the correlation between traditional Chinese medicine syndrome and carotid intima of metabolic syndrome," *Guangming traditional Chinese Medicine*, vol. 25, no. 8, pp. 1317–1319, 2010.

[24] Y. Tang, W. Jia, and F. Dai, "Study on the relationship between pale white tongue and syndrome factors of disease," *Lishizhen Medicine and Materia Medica Research*, vol. 2014, no. 2, pp. 505-506, 2014.

[25] T. Zhang and H. Ren, "Attribute topology of formal context," *Journal of Chinese Mini-Micro Computer Systems*, vol. 35, no. 3, pp. 590–593, 2014.

[26] T. Zhang and X. Wei, "Association rules detecting based on attribute topology," *Journal of Chinese Mini-Micro Computer Systems*, vol. 38, no. 3, pp. 548–552, 2017.

[27] T. Zhang, H. Li, and H. Ren, "Attribute topology analysis of blogger data," *Journal of Yanshan University*, vol. 39, no. 1, pp. 42–50, 2015.

[28] X. Wei, T. Zhang, and D. Bai, "Attribute topology granular analysis based on topology split," *Journal of Chinese Mini-Micro Computer Systems*, vol. 37, no. 8, pp. 1751–1754, 2016.

[29] P. Cimiano, A. Hotho, and S. Staab, "Learning concept hierarchies from text corpora using formal concept analysis," *Journal of Artificial Intelligence Research*, vol. 24, no. 1, pp. 305–339, 2008.

[30] Z. Pei, D. Ruan, D. Meng, and Z. Liu, "Formal concept analysis based on the topology for attributes of a formal context," *Information Sciences*, vol. 236, pp. 66–82, 2013.

[31] M. Van der Heijden, M. Velikova, and P. J. F. Lucas, "Learning Bayesian networks for clinical time series analysis," *Journal of Biomedical Informatics*, vol. 48, pp. 94–105, 2014.

[32] J. Pearl, "Probabilistic reasoning in intelligent systems: networks of plausible inference," *Computer Science Artificial Intelligence*, vol. 70, no. 2, pp. 1022–1027, 1988.

[33] B. Yet, A. Constantinou, N. Fenton, M. Neil, E. Luedeling, and K. Shepherd, "A Bayesian network framework for project cost, benefit and risk analysis with an agricultural development case study," *Expert Systems with Applications*, vol. 60, pp. 141–155, 2016.

[34] A. Formica, "Ontology-based concept similarity in formal concept analysis," *Information Sciences*, vol. 176, no. 18, pp. 2624–2641, 2006.

[35] Y. Iwasaki and H. A. Simon, "Theories of causal ordering. Reply to de Kleer and Brown," *Artificial Intelligence*, vol. 29, no. 1, pp. 63–72, 1986.

[36] Y. Liu, B. Yan, and Y. Zhou, "Urbanization, economic growth, and carbon dioxide emissions in China: A panel cointegration and causality analysis," *Journal of Geographical Sciences*, vol. 26, no. 2, pp. 131–152, 2016.

[37] L. Dagher and T. Yacoubian, "The causal relationship between energy consumption and economic growth in Lebanon," *Energy Policy*, vol. 50, pp. 795–801, 2012.

[38] L. Hsu, "Multipartite information causality," *Physical Review A: Atomic, Molecular and Optical Physics*, vol. 85, no. 3, 2012.

[39] E. Yan, J. Song, C. Liu, and W. Hong, "A research on syndrome element differentiation based on phenomenology and mathematical method," *Chinese Medicine*, vol. 12, no. 1, 2017.

[40] Y. Wu, C. Zhou, and Z. Zhang, "Nn knowledge based to traditional Chinese medicine tongue diagnosis eight principal syndronmes," *Application Research of Computers*, vol. 23, no. 6, pp. 188-189, 2006.

[41] J. Li, L. Ma, and Y. Liu, "Discussion on the scope and connotation of traditional Chinese medicine syndrome differentiation," *Journal of Guangzhou University of Traditional Chinese Medicine*, vol. 21, no. 2, pp. 85–87, 2004.

[42] Z. Zheng, "Dialectical research on etiology and qi-blood-body fluid," *Chinese Journal of Modern Drug Application*, vol. 06, no. 5, pp. 113-114, 2012.

[43] D. Zhou, "Treatment of psoriasis with viscera syndrome differentiation," *Journal of Emergency in Traditional Chinese Medicine*, vol. 14, no. 11, p. 1080, 2005.

[44] J. Pearl, *Causality: Models, Reasoning, and Inference*, Cambridge University Press, Cambridge, UK, 2000.

[45] B. Huang and Z. Zhu, "Relationship between syndrome element and its related concepts," *Traditional Chinese Medicinal Research*, vol. 18, no. 6, pp. 6-7, 2005.

19

Skin Disease Recognition Method Based on Image Color and Texture Features

Li-sheng Wei, Quan Gan, and Tao Ji

School of Electrical and Engineer, Anhui Polytechnic University, Wuhu 241000, China

Correspondence should be addressed to Li-sheng Wei; lshwei_11@163.com

Academic Editor: John Mitchell

Skin diseases have a serious impact on people's life and health. Current research proposes an efficient approach to identify singular type of skin diseases. It is necessary to develop automatic methods in order to increase the accuracy of diagnosis for multitype skin diseases. In this paper, three type skin diseases such as herpes, dermatitis, and psoriasis skin disease could be identified by a new recognition method. Initially, skin images were preprocessed to remove noise and irrelevant background by filtering and transformation. Then the method of grey-level co-occurrence matrix (GLCM) was introduced to segment images of skin disease. The texture and color features of different skin disease images could be obtained accurately. Finally, by using the support vector machine (SVM) classification method, three types of skin diseases were identified. The experimental results demonstrate the effectiveness and feasibility of the proposed method.

1. Introduction

Composed of epidermis, dermis, and subcutaneous tissues, skin is the largest organ of human body, containing blood vessels, lymphatic vessels, nerves, and muscles, which can perspire, perceive the external temperature, and protect the body. Covering the entire body, the skin can protect multiple tissues and organs in the body from external invasions including artificial skin damage, chemical damage, adventitious viruses, and individuals' immune system. Besides, skin can also avoid the loss of lipids together with water within epidermis and dermis so that skin barrier function can be stabilized [1]. In spite of defense and barrier function, skin is not indestructible in that skin tends to be constantly influenced by a variety of external and genetic factors. Currently, there are three main types of skin diseases appearing in human body, including viral skin diseases, fungal skin diseases, and allergic skin disease. Despite the fact that these types of skin diseases can be cured at present, these diseases indeed have brought trouble to patients' life. Nowadays, the majority of conclusions on the patients' existing symptoms are drawn mainly based on doctors' years of experience or their own subjective judgments, which may lead to misjudgments and consequently delay the treatment of these. Therefore, it is of great theoretical significance and practical value to study how to extract symptoms of diverse skin diseases on the basis of modern science and technology. Under this circumstance, effective and accurate identification of the types of skin diseases can be achieved to prescribe treatment according to patients' symptoms [2].

Over the past few years, the image processing technique has achieved rapid development in medicine. Some equipment based on digital image technology has also been widely applied to people's everyday life, for instance, computed tomography (CT), digital subtraction angiography (DSA), and magnetic resonance imaging (MRI). Deeper research on this direction has been carried out by scholars all over the world. For example, the skin disease varicella was detected by Oyola and Arroyo [3] through image processing technique's color transformation, equalization as well as edge detection, and the image of varicella was eventually collected and classified through Hough transform. The final empirical results demonstrated that a better diagnosis was received in terms of detection on varicella, and preliminary test was also conducted on varicella and herpes zoster on that basis. Chung and Sapiro [4] put forward a method to detect the image of skin lesions based on partial differential equation (PDE), with which a contour model of skin lesions was extracted on the basis of its morphological filtering through PDE. The final empirical results demonstrated that skin diseases could be accurately identified through this algorithm. Yu et al. [5] made a diagnosis on herpes simplex, varicella, and herpes zoster through reflectance confocal microscopy (RCM). The final empirical results demonstrated that specificity could be extracted from all the three different types of herpes. Zhong et al. [6] diagnosed psoriasis vulgaris through three-dimensional computed tomography (CT) imageological technique of skin. The final empirical results demonstrated that diagnosis on psoriasis vulgaris of Munro's microabscess was highly sensitive and specific. Sumithra et al. [7] proposed a novel approach for automatic segmentation and classification of skin lesions by using SVM and k-nearest neighbor (k-NN) classifier. Yasir et al. [8] proposed a detection system, which could be used in computers or mobile phones, based on computer vision techniques. Arivazhagan et al. [9] proposed an automated system based on texture analysis for recognizing human skin diseases by independent component analysis of skin color image. The minimum distance classifier was used to classify the type of human skin diseases. Niu et al. [10] carried out an experiment on the color image of skin erythema through HSV color space and fuzzy C-means clustering algorithm. The final empirical results demonstrated that the time and accuracy of final segmentation could be improved through this method of dimensionality reduction. Liu and Guo [11] diagnosed and identified skin tumors, vascular dermatosis, and psoriasis through computed tomography (CT) based on the principle of optical confocal, which has been widely applied. Luo et al. [12] conducted diagnosis and identification on vitiligo through an independently developed analytic system of skin digital images. Both scientific and objective, the quantitative evaluation method was worth studying. Lu et al. [13] classified smooth pixels by employing two-dimensional digital image segmentation and resizing combined with Markov random field (MRF), which demonstrated a reliable segmentation. The final results illustrated that the diagnosis and identification of psoriasis can be well solved. Jaleel et al. [14] researched the skin cancer diagnosing methodology by using back propagation neural (BPN) network classifier. Salimi et al. [15] presented the pattern recognition method to classify the skin disease. Ganeshkumar and Vasanthi [16] researched the melanoma disease detection technique by using preprocessing and edge detection. Kolkur et al. [17] proposed a new human skin detection algorithm to improve the recognition of skin pixels, such as RGB (red, green, and blue), HSV (hue, saturation, and value), and YCbCr (luminance and chrominance) color models. Gindhi et al. [18] and Kotian and Deepa [19] studied the problem of skin disease automated diagnosis system based on the techniques of image border identification and feature data mining by matlab software. Kumar and Singh [20] established the relationship of skin cancer images across different types of neural network. Then, medical images were collected into this skin cancer classification system for training and testing based on the matlab image processing toolbox.

The above literature has made significant achievements on the identification of skin diseases. However, the proposed methods mainly aim at the identification of one type of skin disease, which makes them difficult to apply to the precise identification of multitype skins. Therefore, in this paper, a method based on vertical image segmentation, GLCM, and SVM is proposed to identify three various types of skin diseases, namely, herpes, paederus dermatitis, and psoriasis. Firstly, the sample images of three skin diseases need to be preprocessed. Secondly, the vertical image is segmented and made corresponding geometric transformation. Based on this, three types of skin diseases' features are extracted, and their correlated parameters of feature texture and pixels of lesion areas are collected through image segmentation. Finally, the symptoms of herpes, paederus dermatitis, and psoriasis are identified by utilizing the support vector machine (SVM) method in order to improve identification accuracy.

The remaining structures of this paper are as follows. The Preliminaries section presents the identification method of skin diseases based on GLCM and SVM. The feature extraction and disease recognition are presented in detail. Experimental results, comparisons, and discussions are provided in the Experiments and Analysis section, and finally conclusions are given in the Conclusions section.

2. Preliminaries

The main steps of proposed methodology to skin disease recognition are shown in Figure 1. The flow chart contains three phases: (1) processing of the original image; (2) feature extraction; and (3) classification based on SVM. The first stage is image processing. Because image may contain some unwanted noise, it becomes necessary to filter the image to remove the noise. Then, by using image rotation and segmentation, the representation of an image into something that is more meaningful and easier to analyze. The specific areas of skin lesions are precisely divided, and the identification accuracy is improved through vertical image segmentation. The second stage is feature extraction, in which the image texture features and color features of the skin lesions are further extracted. In texture feature extraction, GLCM is used to find the mathematical parameters of like contrast, correlation, entropy, uniformity, and energy. The third stage is to identify the three various types of skin diseases according to obtained features based on SVM.

2.1. Image Prepocessing. Three common skin diseases are selected in this paper as the main research objects, which are herpes, paederus dermatitis, and psoriasis, respectively. Due to differences in ways of acquiring the source images of skin epidermis and in resolution and size, image preprocessing

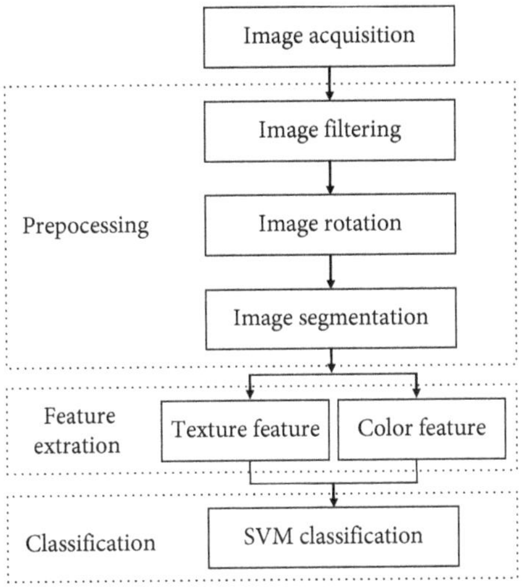

FIGURE 1: The identification process of skin diseases.

on source images are needed for the sake of subsequent vertical image segmentation. The specific processing is as follows.

Firstly, considering that noise constantly exerts a negative impact on acquired samples of skin epidermis's source images, it is necessary to denoise through median filtering for the reduction of the impact on skin segmentation and identification brought by irrelevant background in the images. The median filter is a popular way to remove "salt-and-pepper" noise from an image and at the same time preserve edges and keep useful information. In this paper, the median filter is adopted to preprocess and smoothen the source images. The used formula is as follows:

$$F'(x_0, y_0) = \left[\underset{(x,y)\in S}{\text{sort}}\, F(x, y)\right]_{(N+1)/2}, \quad N \geq 0, \tag{1}$$

where $F'(x_0, y_0)$ is the median of the gray value of the image; S is the neighborhood collection of pixel; (x, y) is the element of S; $F(x, y)$ is the gray value of (x, y); N is expressed as the number of the elements in the set of S; sort is expressed as sequencing; and $[g(\bullet)]_{(N+1)/2}$ is the median of the function $g(\bullet)$. The main idea of the median filter algorithm is to run through the signal gray value by gray value, replacing each gray value with the median of neighboring gray values. The pattern of neighbors is called the "window," which slides, gray value by gray value, over the entire signal. By using median filtering algorithm, the original images are denoised and their qualities are enhanced.

Secondly, denoise images are rotated in order to get the medial axis to segment the images. The denoised skin epidermis's source images are processed via neighborhood sampling with the intention of better obtaining the highlight line through the transformation of Euclidean distance. The denoise image I is rotated to a specific angle θ, in which the width and height of new image I_c as shown in Equation (2) via transformation of neighborhood sampling are acquired:

$$\begin{aligned} \text{width}_{\text{new}} &= 2 \times \left(\left| \left(\frac{\text{width}}{2 \times \cos\theta} \right) \right| + \left| \left(\frac{\text{height}}{2 \times \sin\theta} \right) \right| \right), \\ \text{height}_{\text{new}} &= 2 \times \left(\left| \left(\frac{\text{height}}{2 \times \cos\theta} \right) \right| + \left| \left(\frac{\text{width}}{2 \times \sin\theta} \right) \right| \right), \end{aligned} \tag{2}$$

where width and height each represent width and height of the original image, while $\text{width}_{\text{new}}$ and $\text{height}_{\text{new}}$ refer to width and height of the rotated image and θ is the angle needing rotating.

The corresponding coordinates of image I and image I_c is as follows:

$$\begin{aligned} (x_0 - x_{r1}) &= (x_1 - x_{r2}) \times \cos\theta - (y_1 - y_{r2}) \times \sin\theta, \\ (y_0 - y_{r1}) &= (x_1 - x_{r2}) \times \sin\theta - (y_1 - y_{r2}) \times \cos\theta, \end{aligned} \tag{3}$$

where x_0 and y_0 are the original coordinates, while x_1 and y_1 are transformed ones; x_{r1} and y_{r1} are the central coordinates of the original image, while x_{r2} and y_{r2} are the ones of transformed image. Then, the image of neighborhood sampling transformation is processed via transformation of Euclidean distance which is widely applied to the binary image and is particularly effective for the extraction of skeleton. Accordingly, in this paper, the sample new transformation image I_c is binarized first to obtain the brightest pixel line, and the transformation of Euclidean distance comes second. The specific process of the transformation of Euclidean distance is as follows: through defining a binary image S under a two-dimensional (2D) $M \times N$, a background point set $B(B \in S)$ is first assumed, and then a foreground point set $F(F \in S)$ meeting the premise $B \cup F = S$ is also assumed. Based on this, M is taken as a set in the medial axis transform meeting the premise $M \in F$. As a result, the shortest distance from a random point $p_{x,y} \in M$ to B can be calculated through Equation (4):

$$d\left(p_{x,y}, b_{m,n}\right) = \sqrt{(x - m)^2 + (y - n)^2}, \tag{4}$$

where $d(\bullet)$ represents the shortest Euclidean distance; $p_{x,y}$ is a random point in the medial axis transform set; x and y are the coordinate values of the current point; $b_{m,n}$ is a point belonging to the background point set; and m and n are the coordinate values of the current point. Binary mask image I_c^B is obtained from new image I_c of neighborhood sampling transformation, and I_c^E is eventually acquired via the transformation of Euclidean distance. The brightest pixel line can be obtained from the transformation of via the application of I_c^E Euclidean distance, which is medial axis. In Figure 2(b), the image I_c is rotated to the horizontal coordinate. The binary mask image I_c^B is shown in Figure 2(c). And, the Euclidean distance transform image I_c^E is presented in Figure 2(d). In order to obtain the vertical segment image, the medial axis is marked out in Figure 2(d).

Finally, medial axis of the segmented region is located and divided into ten vertical segments. It is necessary to create a bounding rectangle r_{box} on the image I_c in order to limit the image zone that is dotted with the white line. In Figure 3, it is shown that the medial axis is marked in light green after being extracted from the transformation Euclidean distance.

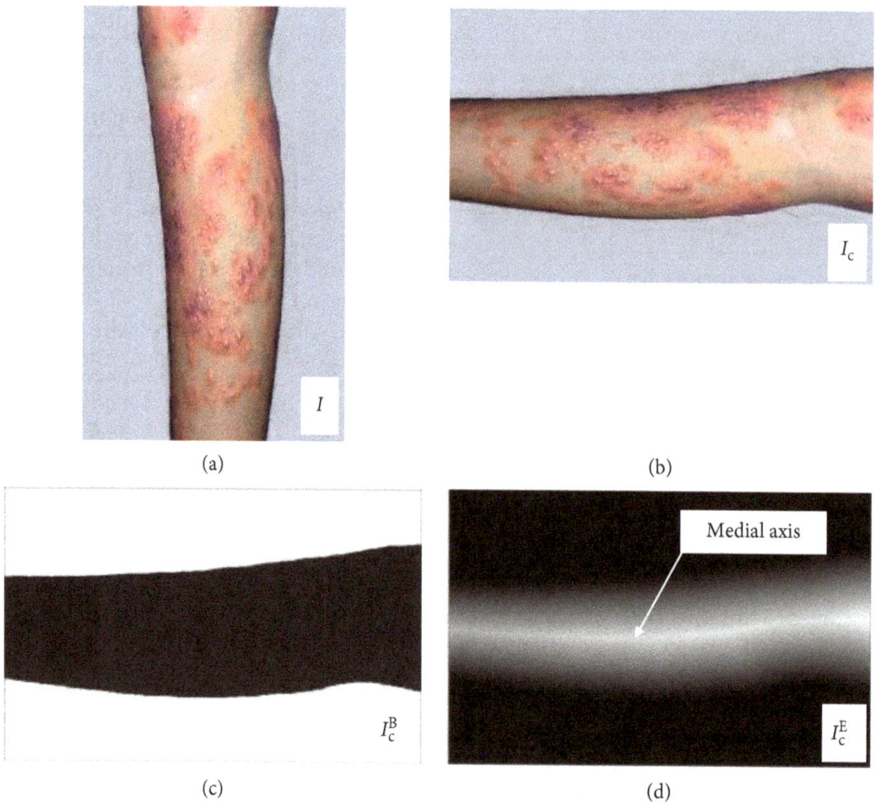

FIGURE 2: Image of the transformation of Euclidean distance. (a) Filtering image. (b) Rotated image. (c) Binary mask image. (d) European transformation image.

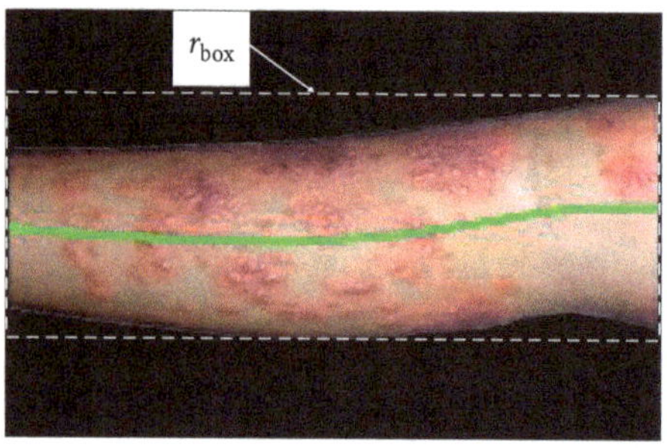

FIGURE 3: The boundary rectangle of the image and medial axis.

The perpendicular line for each point on the main axis can be determined. So, the epithelium can be divided into ten vertical image regions, as done in [21]. The correspondingly original medial axis is changed into ten short straight lines, which is shown in Figure 4. After the ten vertical segment images are determined, we will deal with these local images to recognize the images $L_1, L_2, \ldots, L_{10}$ of skin disease.

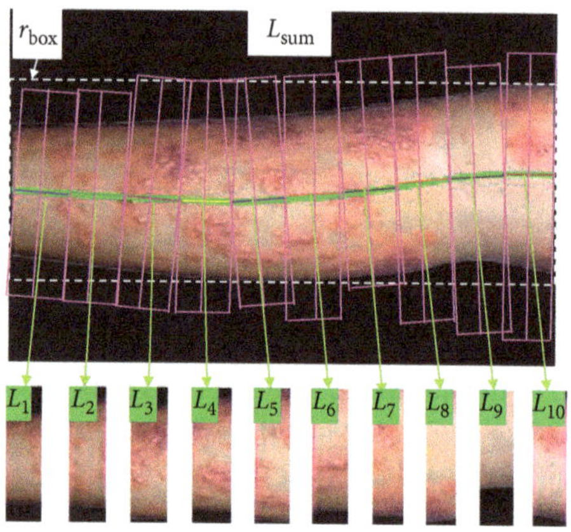

FIGURE 4: Ten vertical image regions ($L_1, L_2, \ldots, L_{10}$).

2.2. Texture Feature Extraction. Compared with the traditional way, GLCM is an effective tool to analyze the features of texture. The textures of different diseases in the skin epithelial image can be obtained, such as contrast, correlation, entropy, uniformity, and energy. In this paper, three common skin diseases are selected as the main research objects, which are herpes, paederus dermatitis, and psoriasis, respectively. All the pictures as shown in Figure 5 are extracted from photo gallery of sino-medicine website.

2.2.1. Contrast

$$A_1 = \sum_{i}^{L-1} \sum_{j}^{L-1} (i-j)^2 G(i,j), \tag{5}$$

where $|i-j|$ refers to the gray level difference between adjacent pixels; $G(i,j)$ is the distribution probability of gray-level difference between adjacent pixels; and A_1 refers to the contrast, mainly used to describe the degree of depth of the

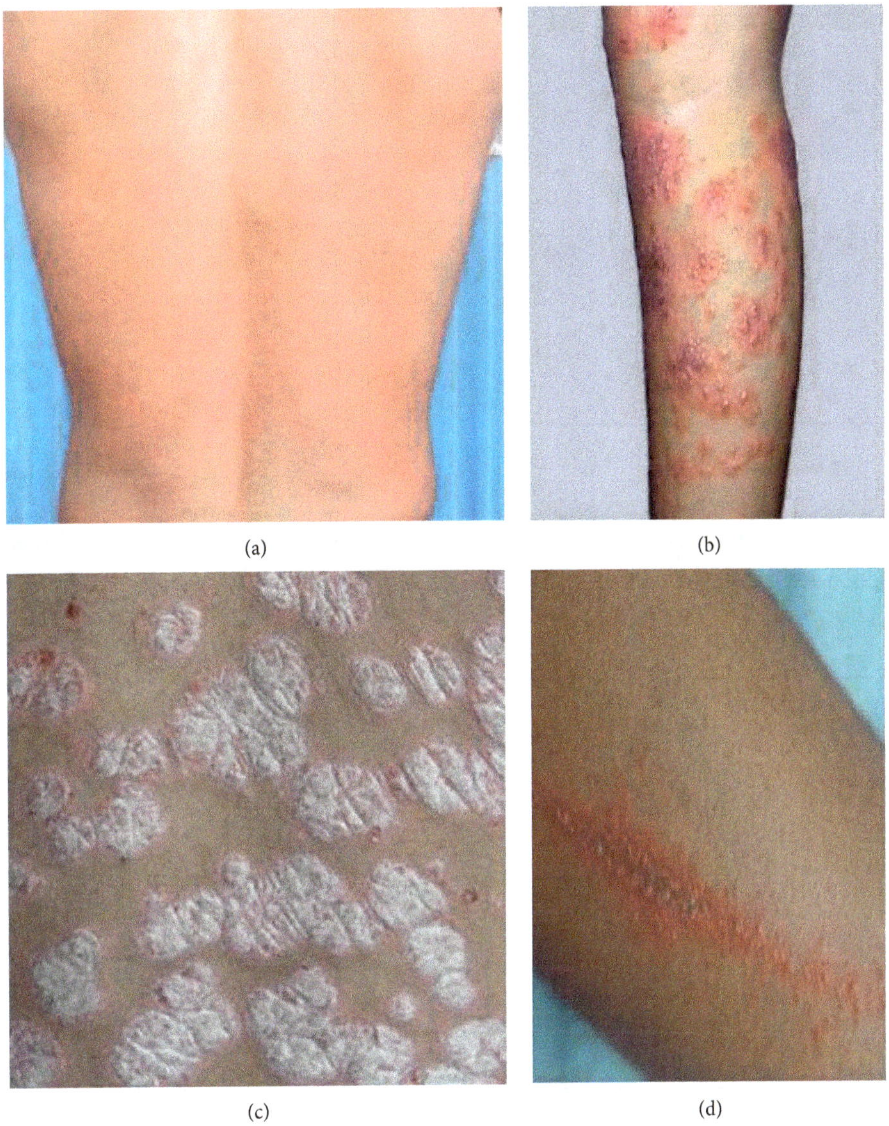

FIGURE 5: Types of skin diseases. (a) Normal skin. (b) Herpes. (c) Paederus dermatitis. (d) Psoriasis.

image textile grooves. The higher the contrast value goes, the deeper the grooves, and vice versa. Through experiment, we can get the texture parameter of contrast of normal skin, herpes, paederus, and psoriasis which is depicted in Figure 6.

2.2.2. Correlation

$$A_2 = \frac{\sum_i^{L-1}\sum_j^{L-1}(i-\overline{x})(j-\overline{y})G(i,\ j)}{\sigma_x\sigma_y}, \quad (6)$$

where A_2 refers to correlation, $\overline{x}$ is the mean value of the sum of the elements in each column in the square, y_1 is the mean value of the sum of the elements in each row, σ_x the standard deviation of the sum of the elements in each column, and σ_y the standard deviation of the sum of the elements in each row. Correlation is mainly used to describe the details of relevant elements in each row and column in the process of vertical image segmentation. Then, we have the texture parameter of contrast in Figure 7.

2.2.3. Entropy

$$A_3 = -\sum_i^{L-1}\sum_j^{L-1}[G(i,j)\cdot\log G(i,j)], \quad (7)$$

where A_3 refers to entropy, which measures the quantity of information within the image and is changed with different textures. As A_3 increases, the textures of speck would be arranged sparsely, and vice versa. When the entropy becomes zero, there is no texture (Figure 8).

2.2.4. Uniformity

$$A_4 = \sum_i^{L-1}\sum_j^{L-1}\frac{1}{(i-j)^2+1}G(i,j), \quad (8)$$

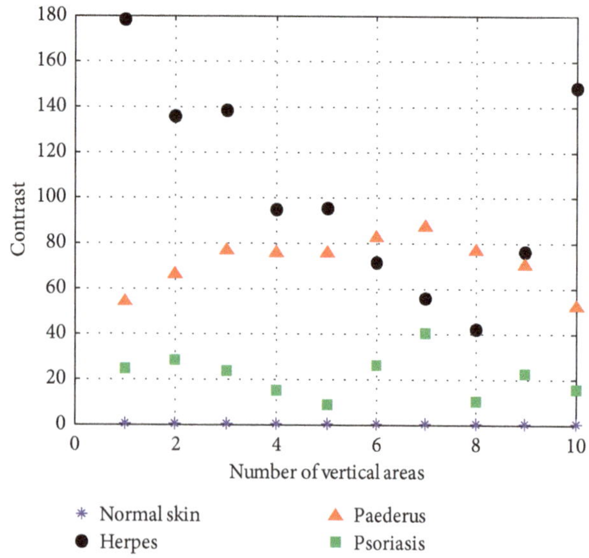

Figure 6: Contrast.

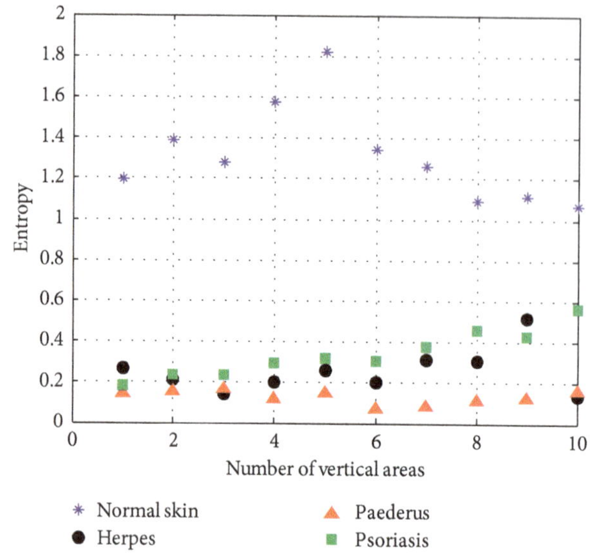

Figure 8: Entropy.

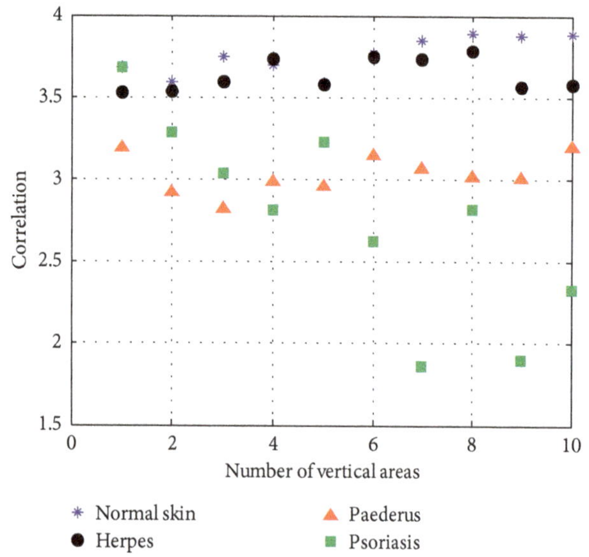

Figure 7: Correlation.

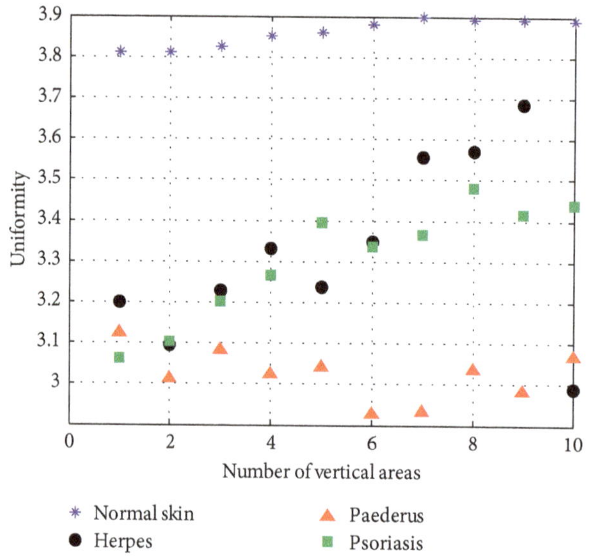

Figure 9: Uniformity.

where A_4 stands for the moment of inertia that is used for the description of the roughness of image texture. The larger the value of A_4, the rougher the texture of endemic area will be, and vice versa (Figure 9).

2.2.5. Energy

$$A_5 = \sum_{i}^{L-1} \sum_{j}^{L-1} G^2(i, j), \quad (9)$$

where A_5, the energy value mainly applied to describe the thickness of texture, is the quadratic sum of elements of gray-level co-occurrence matrix in the horizontal and vertical directions. We can get the texture parameter of energy as it is depicted, respectively, in Figure 10.

From the above five figures, it can be seen that there are discrepancies in the discrete spots presented by different lesions. Table 1 shows the texture features of different skin disease for the convenience of comparison.

2.3. Color Feature Extraction. In order to improve identification and treatment, the disease area of different skin diseases can be extracted on the basis of identifying the texture features so that the area of herpes, paederus, and psoriasis can be calculated, correspondingly. Given the fact that most skin diseases appear in the shape of a circle, the watershed algorithm can be applied for image segmentation. In this paper, in order to avoid oversegmentation, marker-controlled watershed, combined with clustering, was performed to segment the image. That is to say, to segment the

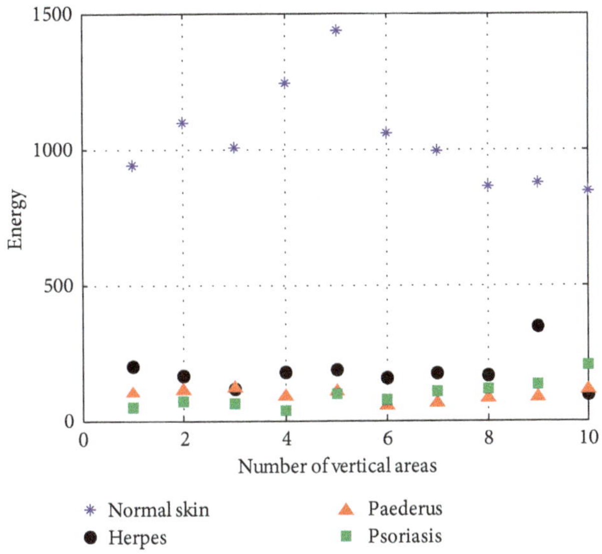

FIGURE 10: Energy.

TABLE 1: Texture features of different skin disease.

	Contrast		Correlation		Entropy		Uniformity		Energy	
	Minimum	Maximum	Minimum	Maximum	Minimum	Maximum	Minimum	Maximum	Minimum	Maximum
Normal skin	0.2053	0.3805	3.589	3.8917	1.0696	1.8224	3.8105	3.8974	843.7668	1437.6
Herpes	41.7036	178.2016	3.5308	3.7852	0.1355	0.5163	2.9878	3.6857	93.0058	345.2991
Paederus	51.676	87.3321	2.8167	3.2025	0.0741	0.1691	2.928	3.125	53.2181	121.5074
Psoriasis	8.713	40.3506	1.8612	3.6789	0.1849	0.5666	3.0617	3.4794	36.9206	205.6441

image by means of watershed algorithm that goes well with a proper threshold, cut each image into relatively small zones, as S_i, $i = 1, 2, \ldots, W$. In the meantime, areas that are mistakenly divided need to be highlighted. Then, it is required to initialize the feature data of these areas such as S_i, $i = 1, 2, \ldots, W$ into four clusters, namely, the normal skin zone, the skin disease zone, and the zone that is mistakenly cut. Lastly, take each one in the four clustering forms as a large unit, among which each contains a smaller number of units. Then, look for one of the best smaller unit in the four larger units and take it as cluster center; continue to update the cluster center until the last center is no longer changed. The results are shown in Figure 11.

Figure 11 shows the result of segmentation. Marker-controlled watershed segmentation that is fusion-based can divide the lesion of skin disease in an effective way. Considering there are still some black flecks which will have a certain impact on the accuracy of experiments, in this paper, morphological approach is utilized to process images, thus making the segmented image smoother so that it will be easier to count the black and white pixel in the certain area. In Figure 12, I_O and I_M of segmented area are picked up for inspecting after morphology processing. The variation of different skin disease can be recognized by means of calculating the pixel area, as it is shown in Figure 13.

In Figure 13, according to the statistics of these three kinds of skin diseases, it is shown that the pixel area of normal skin is nearly a straight line. Besides the pixel area of normal skin, the pixel area of psoriasis is of the greatest, herpes the second, and with that of paederus dermatitis being the lowest. These three diseases can be accurately identified through clear and distinct variation. However, if some herpes once gets severe, the area of injured skin could become larger. In this case, some certain errors may emerge if we solely adhere to the statistics of pixel area. Therefore, only by combining the features of texture with pixel area can the recognition rate of skin disease be improved.

2.4. Classification Based on SVM. SVM is the statistical method based on the statistical learning theory, which is suitable for the classification of small sample numbers. It can obtain the minimized training error and a confidence interval term by analyzing the given training set to predict the test set. In this paper, we use SVM to identify three different skin diseases. Firstly, the sample number and training number are selected from the extracted features (such as color feature, and texture feature), and then by using the rational kernel function of support vector machine, the classification model can be established.

In this paper, three common skin diseases herpes, dermatitis, and psoriasis are represented as Class I, Class II, and Class III, respectively. Texture feature classifier is the SVM1, lesion area feature classifier refers to SVM2, and

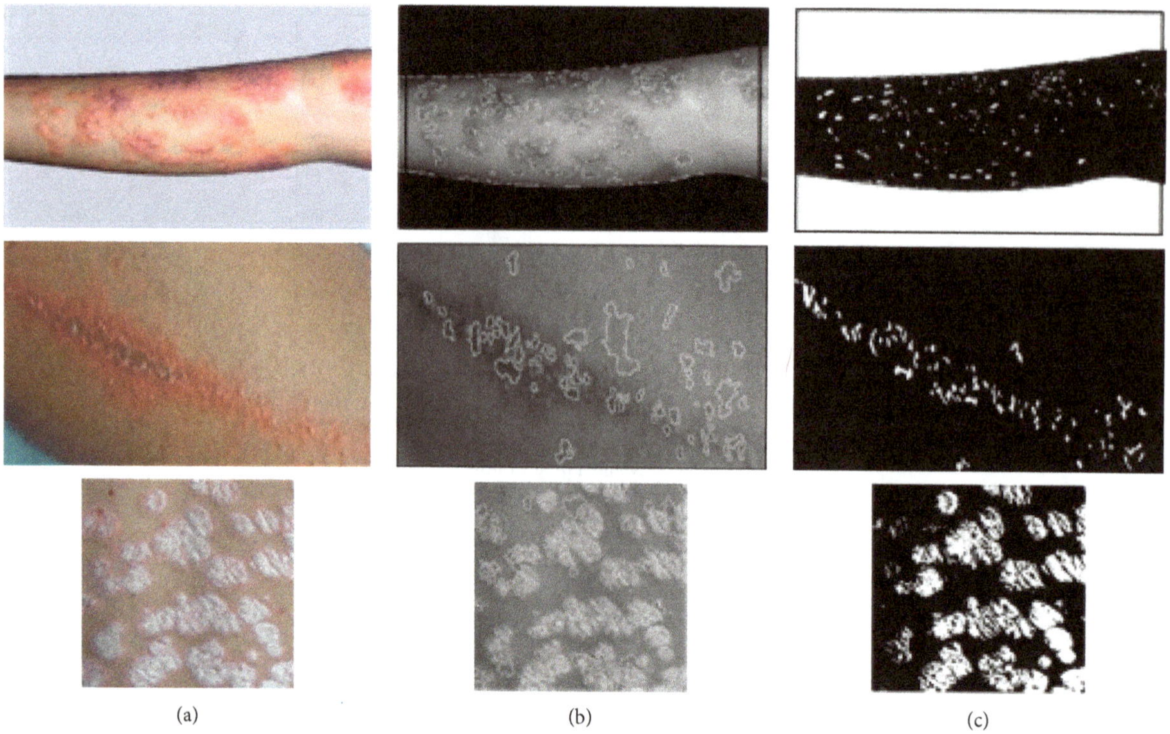

FIGURE 11: Image segmentation. (a) Original pictures. (b) Marker-controlled watershed segmentation. (c) Mark control + clustering segmentation.

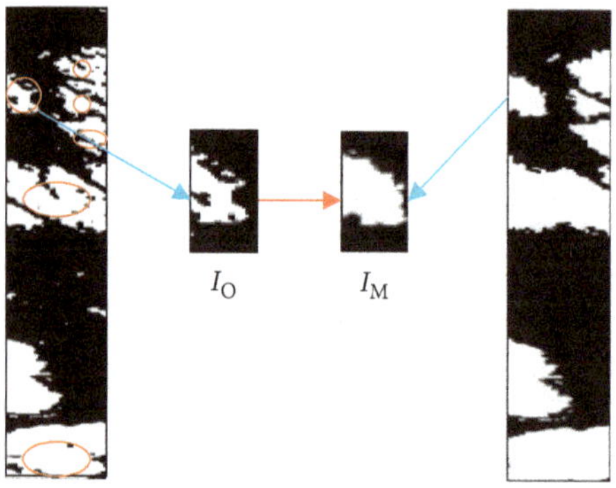

FIGURE 12: Morphological operation.

integrated classifier is represented as SVM. On this foundation, we change the different penalty factors to classify the maize diseases. As a result, the optimal recognition effect can be realized when the penalty factor $C = 50$. The optimal classification function and the kernel function are as follows:

$$f(x) = \operatorname{sgn}\left\{\left(\sum_{j=1}^{n} a_j y_j k(x_i, x_j)\right) + b^*\right\},$$
$$k(x_i, x_j) = \exp\left(-\frac{\|x_i - x_j\|^2}{2\sigma^2}\right), \tag{10}$$

where a_j is a Lagrange multiplier, b^* is the bias, and $k(x_i, x_j)$ is a kernel function; x_i refers to the eigenvector that is obtained from the characteristic model; y_j refers to the results; and σ is a parameter value in the radial basis function.

3. Experiments and Analysis

To identify the classification, in this experiment, common skin diseases (herpes, dermatitis, and psoriasis) are selected as the research objects. Ninety images are also selected to be identified accordingly, including herpes, dermatitis, and psoriasis, thirty cases in each, together with twenty test samples and ten standard samples. In this paper, the combination of the color feature and the texture feature is employed to conduct the experiment, the results of which are compared with those in [21, 22]. The comparison is presented in Table 2.

Table 1 presents the results of different recognition algorithms. It is used to identify images on twenty test samples and ten standard samples of each skin disease. In the study by Guo and Huo [22], the recognition accuracy is relatively low, with that of three diseases being at 75%, 80%, and 80%. De et al. [21] mainly applies the method of unidirectional texture feature recognition, which can cause the relatively large errors. While in the study by De et al. [21], the method of support vector machine and genetic algorithm, with a rather ideal efficiency, are utilized for recognition, with the recognition accuracy rate being 85%, 80%, and 85%. In this paper, the color feature and the texture feature can be used to make up for the weakness of the unidirectional recognition so that the recognition rate can reach to 90% and more, which greatly improves the accuracy. Ten local vertical images after segmentation can be obtained by the vertical

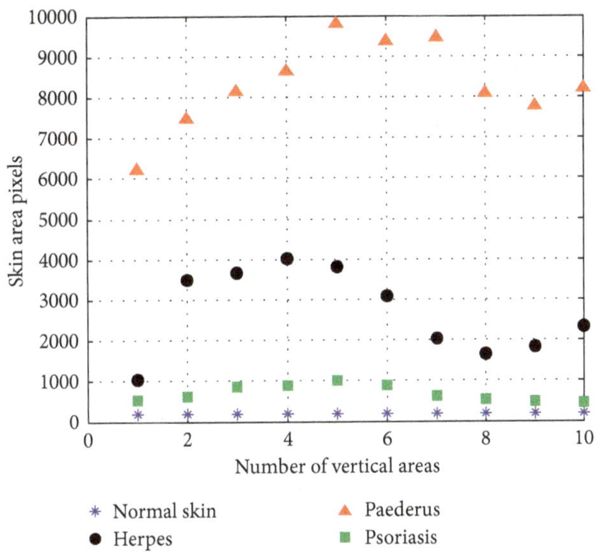

FIGURE 13: The calculation result of three skin diseases.

TABLE 2: The different methods to identify the results.

Method	Herpes			Dermatitis			Psoriasis		
	Number of test	Recognition numbers	Recognition rate (%)	Number of test	Recognition numbers	Recognition rate (%)	Number of test	Recognition numbers	Recognition rate (%)
Reference [22]	20	15	75	20	16	80	20	16	80
Reference [21]	20	17	85	20	16	80	20	17	85
The proposed method	20	17	85	20	18	90	20	19	95

image segmentation, therefore improving the accuracy of skin diseases identification.

4. Conclusions

In this paper, the analysis method of vertical image segmentation is employed to identify three common skin diseases. A number of irrelevant variables can be reduced through image filtering, image rotation, and Euclidean distance transformation applied in image preprocessing. Then, the perpendicular line for each point on the main axis can be determined. And, the epithelium can be divided into ten vertical image regions. Based on this, the grey-level co-occurrence matrix is adopted to extract the texture feature, and the area pixel method is applied to extract the characteristics of the lesion area. Finally, the support vector machine is utilized to classify the data of three different skin diseases according to the features of the texture and the lesion area, achieving a more ideal accuracy of recognition. Nevertheless, the paper concentrating on herpes, dermatitis, and psoriasis does not consider the different symptoms caused by the same kind of skin disease. For instance, eczema, herpes, and rubella all belong to the same series. Therefore, it will be the focus of next step to recognize different types of skin diseases of the same kind of series by using image processing technique.

Conflicts of Interest

The authors declare that they have no conflicts of interest.

Acknowledgments

This study was funded by the National Natural Science Foundation of China (Grant 61203033), the Natural Science Foundation of Anhui Province (Grant 1608085MF146), the Natural Science Research Programme of Colleges and Universities of Anhui Province (Grant KJ2016A062), the Visiting Study Foundation for Outstanding Young Talent of Anhui Educational Committee (Grant gxfxZD2016108), and the Foundation for Talented Young People of Anhui Polytechnic University (Grant 2016BJRC008).

References

[1] Z. Hu and C. S. Yu, "Functional research and development of skin barrier," *Chinese Journal of Clinicians*, vol. 7, no. 7, pp. 3101–3103, 2013.

[2] V. Vezhnevets, V. Sazonov, and A. Andreeva, "A survey on pixel-based skin color detection techniques," in *Proceedings of Graphicon 2003*, pp. 85–92, Moscow, Russia, September 2003.

[3] J. Oyola and V. Arroyo, "Detection of chickenpox vesicles in digital images of skin lesions," in *Proceedings of 17th Iber-*

oamerican Congress on Progress in Pattern Recognition, Image Analysis, Computer Vision, and Applications, pp. 583–590, Buenos Aires, Argentina, September 2012.

[4] D. H. Chung and G. Sapiro, "Segmenting skin lesions with partial differential-equations-based image processing algorithms," *IEEE Transactions on Medicine Image*, vol. 19, no. 7, pp. 763–767, 2000.

[5] S. R. Yu, X. H. Zhao, and X. M. Pu, "Image characteristics of dermopathic herpesvirus disease under reflectance confocal microscope," *China Journal of Leprosy and Skin Diseases*, vol. 31, no. 2, pp. 85–88, 2015.

[6] L. S. Zhong, X. Jin, and C. Quan, "Diagnostic applicability of confocal laser scanning microscopy in psoriasis vulgaris," *Chinese Journal of Dermatovenereology*, vol. 25, no. 8, pp. 607-608, 2011.

[7] R. Sumithra, M. Suhilb, and D. S. Guruc, "Segmentation and classification of skin lesions for disease diagnosis," *Procedia Computer Science*, vol. 45, pp. 76–85, 2015.

[8] R. Yasir, M. S. I. Nibir, and N. Ahmed, "A skin disease detection system for financially unstable people in developing countries," *Global Science and Technology Journal*, vol. 3, no. 1, pp. 77–93, 2015.

[9] S. Arivazhagan, R. N. Shebiah, K. Divya, and M. P. Subadevi, "Skin disease classification by extracting independent components," *Journal of Emerging Trends in Computing and Information Sciences*, vol. 3, no. 10, pp. 1379–1382, 2012.

[10] H. J. Niu, K. K. Shang, and Y. Liu, "Study of segmenting skin erythema images by reducing dimensions of color space," *Computer Engineering and Applications*, vol. 13, no. 3, pp. 219–221, 2006.

[11] F. Liu and H. X. Guo, "Research progress of CT in skin diseases," *Chinese Journal of Dermatovenerol Integrated Traditional Western Medicine*, vol. 313, no. 3, pp. 189–191, 2014.

[12] W. Luo, R. S. Meng, and R. K. Cai, "Application of CDIA system in measurement and research of vitiligo," *Chinese Medical Journal*, vol. 22, no. 12, pp. 1059-1060, 2011.

[13] J. Lu, E. Kazmierczak, and H. Jonathan, "Automatic segmentation of scaling in 2-D psoriasis skin images," *IEEE Transaction on Medical Imaging*, vol. 32, no. 4, pp. 719–730, 2013.

[14] J. A. Jaleel, S. Salim, and R. B. Aswin, "Artificial neural network based detection of skin cancer," *International Journal of Advanced Research in Electrical, Electronics and Instrumentation Engineering*, vol. 1, no. 3, pp. 200–205, 2012.

[15] S. Salimi, M. S. Nobarian, and S. Rajebi, "Skin disease images recognition based on classification methods," *International Journal on Technical and Physical Problems of Engineering*, vol. 22, no. 7, pp. 78–85, 2015.

[16] M. Ganeshkumar and J. J. B. Vasanthi, "Skin disease identification using image segmentation," *International Journal of Innovative Research in Computer and Communication Engineering*, vol. 5, no. 1, pp. 154–160, 2017.

[17] S. Kolkur, D. Kalbande, P. Shimpi, C. Bapat, and J. Jatakia, "Human skin detection using RGB, HSV and YCbCr Color models," *Advances in Intelligent Systems Research*, vol. 137, pp. 324–332, 2016.

[18] E. S. Gindhi, A. Nausheen, A. Zoya, and S. Ruhin, "An innovative approach for skin disease detection using image processing and data mining," *International Journal of Innovative Research in Computer and Communication Engineering*, vol. 5, no. 4, pp. 8135–8141, 2017.

[19] A. L. Kotian and K. Deepa, "Detection and classification of skin diseases by image analysis using MATLAB," *International Journal of Emerging Research in Management andTechnology*, vol. 6, no. 5, pp. 779–784, 2017.

[20] S. Kumar and A. Singh, "Image processing for recognition of skin diseases," *International Journal of Computer Applications*, vol. 149, no. 3, pp. 37–40, 2016.

[21] S. De, R. Joe Stanley, C. Lu et al., "A fusion-based approach for uterine cervical cancer histology image classification," *Computerized Medical Imaging and Graphics*, vol. 37, no. 8, pp. 475–487, 2013.

[22] H. Guo and H. T. Huo, "Research on the application of gray level co-occurrence matrix for skin texture detection," *Journal of Image and Graphic*, vol. 15, no. 7, pp. 1074–1078, 2010.

Permissions

All chapters in this book were first published in CMMM, by Hindawi Publishing Corporation; hereby published with permission under the Creative Commons Attribution License or equivalent. Every chapter published in this book has been scrutinized by our experts. Their significance has been extensively debated. The topics covered herein carry significant findings which will fuel the growth of the discipline. They may even be implemented as practical applications or may be referred to as a beginning point for another development.

The contributors of this book come from diverse backgrounds, making this book a truly international effort. This book will bring forth new frontiers with its revolutionizing research information and detailed analysis of the nascent developments around the world.

We would like to thank all the contributing authors for lending their expertise to make the book truly unique. They have played a crucial role in the development of this book. Without their invaluable contributions this book wouldn't have been possible. They have made vital efforts to compile up to date information on the varied aspects of this subject to make this book a valuable addition to the collection of many professionals and students.

This book was conceptualized with the vision of imparting up-to-date information and advanced data in this field. To ensure the same, a matchless editorial board was set up. Every individual on the board went through rigorous rounds of assessment to prove their worth. After which they invested a large part of their time researching and compiling the most relevant data for our readers.

The editorial board has been involved in producing this book since its inception. They have spent rigorous hours researching and exploring the diverse topics which have resulted in the successful publishing of this book. They have passed on their knowledge of decades through this book. To expedite this challenging task, the publisher supported the team at every step. A small team of assistant editors was also appointed to further simplify the editing procedure and attain best results for the readers.

Apart from the editorial board, the designing team has also invested a significant amount of their time in understanding the subject and creating the most relevant covers. They scrutinized every image to scout for the most suitable representation of the subject and create an appropriate cover for the book.

The publishing team has been an ardent support to the editorial, designing and production team. Their endless efforts to recruit the best for this project, has resulted in the accomplishment of this book. They are a veteran in the field of academics and their pool of knowledge is as vast as their experience in printing. Their expertise and guidance has proved useful at every step. Their uncompromising quality standards have made this book an exceptional effort. Their encouragement from time to time has been an inspiration for everyone.

The publisher and the editorial board hope that this book will prove to be a valuable piece of knowledge for researchers, students, practitioners and scholars across the globe.

List of Contributors

S. Fiocchi, E. Chiaramello, P. Ravazzani and M. Parazzini
CNR Consiglio Nazionale delle Ricerche, Istituto di Elettronica e di Ingegneria dell'Informazione e delle Telecomunicazioni IEIIT, Milan, Italy

Formijn van Hemert, Maarten Jebbink and Ben Berkhout
Laboratory of Experimental Virology, Medical Microbiology, Amsterdam UMC, University of Amsterdam, Amsterdam, Netherlands

Andries van der Ark
Research Institute of Child Development and Education, University of Amsterdam, Amsterdam, Netherlands

Frits Scholer
Medical Microbiology, Amsterdam UMC, University of Amsterdam, Meibergdreef 9, 1105 AZ Amsterdam, Netherlands

Linfei Ge and Jin Zhang
Department of Computer Science and Engineering, Southern University of Science and Technology, Shenzhen, Guangdong, China

Jing Wei
Department of Electrical and Electronic Engineering, Southern University of Science and Technology, Shenzhen, Guangdong, China

Idit Avrahami, Dafna Raz and Oranit Bash
Department of Mechanical Engineering and Mechatronics, Ariel University, Ariel, Israel

Liyuan Zhang, Jiashi Zhao, Huamin Yang, Zhengang Jiang and Qingliang Li
School of Computer Science and Technology, Changchun University of Science and Technology, No. 7089, Weixing Road, Changchun, China

Wei Zhou
College of Information Science and Engineering, Northeastern University, Shenyang, China

Hao Wu
Engineering Faculty, University of Sydney, NSW 2006, Austria

Chengdong Wu and Xiaosheng Yu
Faculty of Robot Science and Engineering, Northeastern University, Shenyang, China

Yugen Yi
School of Software, Jiangxi Normal University, Nanchang, China

Hanshu Cai
School of Information Science and Engineering, Lanzhou University, Lanzhou, China

Bingtao Zhang
School of Information Science and Engineering, Lanzhou University, Lanzhou, China
School of Electronic and Information Engineering, Lanzhou Jiaotong University, Lanzhou, China

Tao Lei
College of Electronical and Information Engineering, Shaanxi University of Science and Technology, Xi'an, China

Hong Liu
School of Information Science and Engineering, Shandong Normal University, Jinan, China

Shuiping Gou, Linlin Chen and Yu Gu
Key Laboratory of Intelligent Perception and Image Understanding of Ministry of Education of China, Xidian University, Xi'an 710071, China

Liyu Huang
Department of Biological Science and Engineering, Xidian University, Xi'an 710071, China

Meiping Huang
Catheterization Lab, Guangdong Cardiovascular Institute, Guangdong Provincial Key Laboratory of South China Structural Heart Disease, Guangdong General Hospital, Guangdong Academy of Medical Sciences, Guangzhou 510000, China

Jian Zhuang
Department of Cardiac Surgery, Guangdong Cardiovascular Institute, Guangdong Provincial Key Laboratory of South China Structural Heart Disease, Guangdong General Hospital, Guangdong Academy of Medical Sciences, Guangzhou 510000, China

Sanghun Choi
School of Mechanical Engineering, Kyungpook National University, Daegu, Republic of Korea

Shinjiro Miyawaki
Jacobs, 1100 NE Circle Blvd., Suite 300, Corvallis, Oregon 97330, USA

Department of Mechanical and Industrial Engineering, IIHR-Hydroscience and Engineering, University of Iowa, Iowa City, Iowa, USA

Ching-Long Lin
Department of Mechanical and Industrial Engineering, IIHR-Hydroscience and Engineering, University of Iowa, Iowa City, Iowa, USA

Jae Hun Jung, Anna Park and Il Hyo Jung
Department of Mathematics, Pusan National University, Busan 46241, Republic of Korea

Yang Li
School of Mathematics and Statistics, Northeast Normal University, Changchun, Jilin 130024, China
School of Computer Science and Engineering, Changchun University of Technology, Jilin 130012, China

Zhichuan Zhu
School of Mathematics and Statistics, Northeast Normal University, Changchun, Jilin 130024, China
Faculty of Statistics, Jilin University of Finance and Economics, Changchun, Jilin 130117, China

Alin Hou, Qingdong Zhao, Liwei Liu and Lijuan Zhang
School of Computer Science and Engineering, Changchun University of Technology, Jilin 130012, China

Math J. J. M. Candel
Department Methodology and Statistics, CAPHRI Care and Public Health Research Institute, Maastricht University, Maastricht, Netherlands

Kuryati Kipli, Mohammed Enamul Hoque, Siti Kudnie Sahari, Rohana Sapawi, Nordiana Rajaee and Annie Joseph
Department of Electrical and Electronics Engineering, University Malaysia Sarawak (UNIMAS), 94300 Kota Samarahan, Kuching, Malaysia

Lik Thai Lim
Department of Ophthalmology, Faculty of Medicine and Health Sciences (FMHS), University Malaysia Sarawak (UNIMAS), 94300 Kota Samarahan, Sarawak, Malaysia

Muhammad Hamdi Mahmood
Department of Para-Clinical Sciences, Faculty of Medicine and Health Sciences (FMHS), University Malaysia Sarawak (UNIMAS), 94300 Kota Samarahan, Sarawak, Malaysia

Yifu Xu, Bin Yan, Jian Chen, Lei Zeng and Linyuang Wang
National Digital Switching System Engineering and Technological R and D Centre, Zhengzhou 450002, China

Jingfang Zhang
153 Central Hospital of Henan Province, Zhengzhou 450002, China

E. L. Rex
Biomedical Engineering, University of Calgary, Calgary, Canada
McCaig Institute for Bone and Joint Health, University of Calgary, Calgary, Canada

J. Werle, B. C. Burkart, J. R. Mac Kenzie and K. D. Johnston
McCaig Institute for Bone and Joint Health, University of Calgary, Calgary, Canada
Section of Orthopaedic Surgery, University of Calgary, Calgary, Canada

C. Anglin
Biomedical Engineering, University of Calgary, Calgary, Canada
McCaig Institute for Bone and Joint Health, University of Calgary, Calgary, Canada
Dept. of Civil Engineering, University of Calgary, Calgary, Canada

Jinxiang Xi
Department of Biomedical Engineering, California Baptist University, Riverside, CA 92504, USA
Department of Aerospace, Industrial, and Mechanical Engineering, California Baptist University, Riverside, CA 92504, USA
School of Engineering and Technology, Central Michigan University, Mt. Pleasant, MI 48858, USA

Mohamed Talaat
Department of Aerospace, Industrial, and Mechanical Engineering, California Baptist University, Riverside, CA 92504, USA

Hesham Tanbour
School of Engineering and Technology, Central Michigan University, Mt. Pleasant, MI 48858, USA

Khaled Talaat
Department of Nuclear Engineering, +e University of New Mexico, Albuquerque, NM 87131, USA

Qianqian Wang, Fang'ai Liu and Shuning Xing
School of Information Science and Engineering, Shandong Normal University, Jinan, China

Xiaohui Zhao
School of Mathematical Science, Shandong Normal University, Jinan, China

Wenyuan Liu
School of Information Science and Engineering, Yanshan University, No. 438, Hebei Avenue, 066000 Qinhuangdao, China

Tao Zhang and Mengqi Liu
School of Information Science and Engineering, Yanshan University, No. 438, Hebei Avenue, 066000 Qinhuangdao, China
Hebei Key Laboratory of Information Transmission and Signal Processing, No. 438, Hebei Avenue, 066000 Qinhuangdao, China

Li-sheng Wei, Quan Gan and Tao Ji
School of Electrical and Engineer, Anhui Polytechnic University, Wuhu 241000, China

Index

www.ingramcontent.com/pod-product-compliance
Lightning Source LLC
LaVergne TN
LVHW060517210926
848485LV00005B/1092
9781632428882